Lecture Notes in Networks and Systems

Volume 130

The series "Lecture Notes in Networks and Systems" publishes the latest developments in Networks and Systems—quickly, informally and with high quality. Original research reported in proceedings and post-proceedings represents the core of LNNS.

Volumes published in LNNS embrace all aspects and subfields of, as well as new challenges in, Networks and Systems.

The series contains proceedings and edited volumes in systems and networks, spanning the areas of Cyber-Physical Systems, Autonomous Systems, Sensor Networks, Control Systems, Energy Systems, Automotive Systems, Biological Systems, Vehicular Networking and Connected Vehicles, Aerospace Systems, Automation, Manufacturing, Smart Grids, Nonlinear Systems, Power Systems, Robotics, Social Systems, Economic Systems and other. Of particular value to both the contributors and the readership are the short publication timeframe and the world-wide distribution and exposure which enable both a wide and rapid dissemination of research output.

The series covers the theory, applications, and perspectives on the state of the art and future developments relevant to systems and networks, decision making, control, complex processes and related areas, as embedded in the fields of interdisciplinary and applied sciences, engineering, computer science, physics, economics, social, and life sciences, as well as the paradigms and methodologies behind them.

**** Indexing: The books of this series are submitted to ISI Proceedings, SCOPUS, Google Scholar and Springerlink ****

More information about this series at http://www.springer.com/series/15179

D. Jude Hemanth · G. Vadivu ·
M. Sangeetha · Valentina Emilia Balas
Editors

Artificial Intelligence Techniques for Advanced Computing Applications

Proceedings of ICACT 2020

Springer

Editors
D. Jude Hemanth
Karunya Institute of Technology
and Sciences
Coimbatore, India

G. Vadivu
SRM Institute of Science
and Technology
Chennai, India

M. Sangeetha
SRM Institute of Science and Technology
Chennai, India

Valentina Emilia Balas
Aurel Vlaicu University of Arad
Arad, Romania

ISSN 2367-3370 ISSN 2367-3389 (electronic)
Lecture Notes in Networks and Systems
ISBN 978-981-15-5328-8 ISBN 978-981-15-5329-5 (eBook)
https://doi.org/10.1007/978-981-15-5329-5

This Springer imprint is published by the registered company Springer Nature Singapore Pte Ltd.
The registered company address is: 152 Beach Road, #21-01/04 Gateway East, Singapore 189721, Singapore

Preface

Artificial intelligence (AI) refers to the creation of intelligent machines that simulate human behavior through the acquisition and application of knowledge. With the recent development and advancement of technologies such as big data, networks and machine learning, the field of artificial intelligence is also advancing. With new developments and technological advancements, AI has attracted more attention for introducing it to several areas.

AI, big data and machine learning are three of the emerging technologies that are used in network sector extensively, helping network service providers (NSPs) manage, optimize and maintain not only their infrastructure but their customer support operations as well.

The purpose of this book is to inform the broad audience about the major concerns of adopting AI, big data and machine learning in networking systems. It begins with an introductory tutorial followed by a collection of surveys describing traditional networking approaches such as reactive, centrally managed, one-size-fits-all approaches and provides major insights and results in conventional data analysis tools. Special emphasis is placed on novel approaches such as proactive, self-aware, self-adaptive and predictive networking.

This book focuses on four major areas. The first one is artificial intelligence. We look at the role of AI in making network intelligent systems in terms of self-aware networks, self-adaptive networks, and reactive and proactive networks.

The second area is big data. We examine the development and access to large amounts of data, specifically from network service providers (NSPs). Examination and analysis of the big data greatly help in making intelligent and smart networks and optimizing such networks. We also look at advanced data analytics for data-driven next-generation networks. We also discuss the data sources and strong drivers for the adoption of data analytics.

The third area covered in this book is the networks. We look at the evolution of next-generation networks to complex systems and how the diversified service requirements, heterogeneity in applications, devices and networks contributed to the complexity. We also examine how network service providers should make use of the available resources such as infrastructure and spectrum for best network optimization.

The last specialized area is machine learning. We examine how machine learning plays a critical role in data analytics for next-generation networks. We take a look at how ML provides insights on root causes, network performance, traffic reports, etc. Contributions to this topic include descriptive, diagnostic, predictive and prescriptive analytics.

This book not only describes the role of AI and big data in next-generation networks, but also highlights goals and areas that need further research. This book will serve as an introduction to the field. The primary audience will be researchers and graduate students in big data who wish to start new research in artificial intelligence. The secondary audience includes network engineers and data scientists.

Coimbatore, India — Dr. D. Jude Hemanth
Chennai, India — Dr. G. Vadivu
Chennai, India — Ms. M. Sangeetha
Arad, Romania — Valentina Emilia Balas

Contents

Editors and Contributors

About the Editors

Dr. D. Jude Hemanth received his B.E. degree in ECE from Bharathiar University in 2002, M.E. degree in Communication Systems from Anna University in 2006 and Ph.D. from Karunya University in 2013. His research areas include computational intelligence and image processing. He has published more than 100 research papers in respected SCIE-indexed international journals and Scopus-indexed international conferences, as well as 27 edited books with leading publishers such as Elsevier, Springer and IET. He is currently an Associate Professor at the Department of ECE, Karunya University, India.

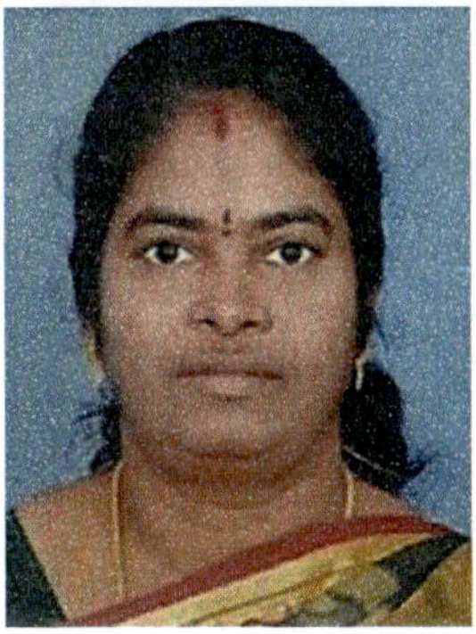

Dr. G. Vadivu received her B.E. degree in CSE from IRTT, affiliated to Bharathiar University, in 1993; M.E. degree in CSE from SRM Institute of Science and Technology (formerly known as SRM University) in 2007; and her Ph.D. from SRM Institute of Science and Technology in 2013. Her research areas include big data analytics, machine learning, streaming analytics and Semantic Web. She has also been a member of the organizing committees of several international conferences. She is a member the IET, ISC, ACM-W, CSI and IEEE. She has also received funding from the Scheme for Promotion of Academic and Research Collaboration (SPARC), MHRD, to investigate the use of deep learning to identify diabetic from retina images.

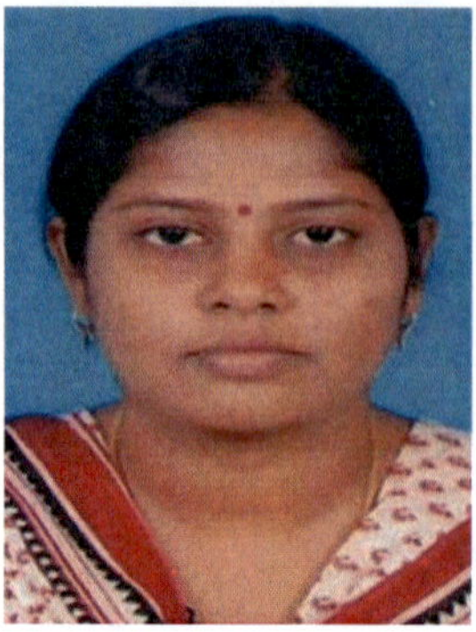

M. Sangeetha received her B.E. degree in CSE from Sakthi Mariamman Engineering College, affiliated to Anna University, in 2006 and her M.E. degree from Anna University, Coimbatore, in 2010. Her research areas include machine learning, big data analytics and deep learning. She holds professionally qualifications in Big Data Visualization Tool (TABLEAU), machine learning and Python language. She has also been a member of the organizing committees of several national and international conferences. She is a member of IAENG, CSI, IEEE, ISTE, the Indian Science Congress Association and ISTE.

Valentina Emilia Balas is currently a Full Professor at the Department of Automatics and Applied Software at the Faculty of Engineering, "Aurel Vlaicu" University of Arad, Romania. She holds a Ph.D. in Applied Electronics and Telecommunications from the Polytechnic University of Timisoara. Her research interests include intelligent systems, fuzzy control, soft computing, smart sensors, information fusion, modeling and simulation, and she has published over 300 research papers in refereed journals and international conferences. She is a member of EUSFLAT, ACM, a senior member of IEEE, and a member of TC–Fuzzy Systems (IEEE CIS), TC–Emergent Technologies (IEEE CIS), and TC–Soft Computing (IEEE SMCS).

Contributors

A. P. Aakash Department of Information Technology, SRM Institute of Science and Technology, Chengalpattu, Tamil Nadu, India

A. Aarthi Department of Information Technology, SRM Institute of Science and Technology, Chennai, India

M. Aasritha SRM Institute of Science and Technology, Chennai, India

S. Amutha Computer Science and Engineering, Chennai Institute of Technology, Chennai, Tamil Nadu, India

M. Anand SRM Institute of Science and Technology, Kattankulathur, Kancheepuram, Tamil Nadu, India

N. Arivazhagan Department of Information Technology, SRM Institute of Science and Technology, Chengalpattu, India

S. Aswini Department of Computer Science and Engineering, Rajalakshmi Institute of Technology, Anna University, Chennai, Tamil Nadu, India

P. Baskaran Information Technology, Sri Venkateswara College of Engineering, Sriperumbudur, Chennai, India

Kaaviya Baskaran Information Technology, Sri Venkateswara College of Engineering, Sriperumbudur, Chennai, India

N. Bharathi Department of Computer Science and Engineering, SRMIST, Chennai, India

K. Boopathi National Institute of Wind Energy, Chennai, India

Arathi Boyanapalli SRM Institute of Science and Technology, Chennai, India

Srinidhi Bulusu Department of Electronics and Communication Engineering, Amrita School of Engineering, Amrita Vishwa Vidyapeetham, Bengaluru, India

Samir Kumar Das Department of CSE, UIU, Dhaka, Bangladesh

R. Dayana SRM Institute of Science and Technology, Kattankulathur, Tamil Nadu, India

S. Deepanjali Department of Information Technology, SRM Institute of Science and Technology, Kattankulathur, Chennai, India

K. Dhinakaran Department of Computer Science and Engineering, Rajalakshmi Institute of Technology, Anna University, Chennai, Tamil Nadu, India

S. Dhivya Department of Information Technology, SRM Institute of Science and Technology, Kattankulathur, Chennai, India

V. Sathya Durga Department of CSE, Hindustan Institute of Technology and Science, Padur, Chennai, Tamil Nadu, India

S. Durgadevi Department of Computer Science and Engineering, Rajalakshmi Institute of Technology, Anna University, Chennai, Tamil Nadu, India

C. Fancy SRM Institute of Science and Technology, Kattankulathur, Kancheepuram, Tamil Nadu, India

G. Geetha Sathyabama Institute of Science and Technology, Chennai, India

R. Gnanavel Department of Computer Science and Engineering, Rajalakshmi Institute of Technology, Anna University, Chennai, Tamil Nadu, India

J. Godwin Ponsam Department of Information Technology, SRM Institute of Science and Technology, Kattankulathur, Chennai, Tamil Nadu, India

Rahul Gupta Department of Computer Science and Engineering, SRM Institute of Science and Technology, Chennai, India

C. Harini SRM Institute of Science and Technology, Kattankulathur, Chennai, India

K. Hemanth Reddy Department of Information Technology, SRM Institute of Science and Technology, Chengalpattu, India

D. Hemavathi SRM Institute of Science and Technology, Chennai, India

Najia Hossain Department of CSE, UIU, Dhaka, Bangladesh

Chaitanya Lakshmi Indukuri SRM Institute of Science and Technology, Chennai, India

Md. Adnanul Islam Department of CSE, UIU, Dhaka, Bangladesh

K. Janaki Raman Department of Information Technology, SRM Institute of Science and Technology, Chengalpattu, India

Kayalvizhi Jayavel Department of Information Technology, School of Computing, SRMIST, Chennai, India

P. Jenifer Department of Information Technology, SRM Institute of Science and Technology, Kattankulathur, Chennai, Tamil Nadu, India

Thangakumar Jeyaprakash Department of Computer Science and Engineering, Hindustan Institute of Technology and Science, Padur, Chennai, Tamil Nadu, India

V. Joseph Raymond Department of Information Security and Cyber Forensics, SRM Institute of Science and Technology, Chennai, India

Haripriya Kaduvu SRM Institute of Science and Technology, Kattankulathur, Chennai, India

G. M. Karthik Department of Information Technology, SRM Institute of Science and Technology, Chennai, India

Aditya Katari Department of Information Technology, SRM Institute of Science and Technology, Chengalpattu, India

Nidhi Kawale SRM Institute of Science and Technology, Chennai, India

Tanmay Khokle Department of Computer Science, SRM Institute of Science and Technology, Kattankulathur, India

T. Kirthiga Devi Department of Information Technology, SRM Institute of Science and Technology, Kattankulathur, Chennai, Tamil Nadu, India

Kottilingam Kottursamy SRM Institute of Science and Technology, Chennai, India

G. Krishna Chaitanya Department of Information Technology, SRM Institute of Science and Technology, Kattankulathur, India

Abhishek Kumar Department of Computer Science, SRM Institute of Science and Technology, Kattankulathur, India

M. Lakshmi Saveetha School of Engineering, Chennai, India

M. C. Lavanya National Institute of Wind Energy, Chennai, India

R. Lavanya Department of Computer Science, SRM Institute of Science and Technology, Kattankulathur, India

V. Lavanya SRM Institute of Science and Technology, Chennai, India

Utkarsh Maheria SRM Institute of Science and Technology, Kattankulathur, Kancheepuram, Tamil Nadu, India

Abhideep Maity Department of Computer Science, SRM Institute of Science and Technology, Kattankulathur, India

P. Malarvezhi SRM Institute of Science and Technology, Kattankulathur, Tamil Nadu, India

G. Malvika Department of Information Technology, SRM Institute of Science and Technology, Chengalpattu, India

G. Maragatham Department of Information Technology, SRM Institute of Science and Technology, Kattankulathur, India

Musara Maronge Information Security and Cyber Forensics, SRM Institute of Science and Technology, Kattankulathur, Tamil Nadu, India

K. Meenakshi Department of Information Technology, SRM Institute of Science and Technology, Chengalpattu, India

Joyita Das Mela Department of CSE, UIU, Dhaka, Bangladesh

K. Mohana Prasad Sathyabama Institute of Science and Technology, Chennai, India

S. Monica Catherine Department of Information Technology, SRM Institute of Science and Technology, Kattankulathur, Chennai, India

S. Murugaanandam Department of Computer Science and Technology, SRMIST, Chennai, India

K. Navin Department of Computer Science and Technology, SRMIST, Chennai, India

R. Nedunchelian Department of Computer Science and Engineering, Karpaga Vinayaga College of Engineering and Technology, Chengalpattu, Tamil Nadu, India

N. Neelima Department of Electronics and Communication Engineering, Amrita School of Engineering, Amrita Vishwa Vidyapeetham, Bengaluru, India

S. Nithiya Department of Computer Science and Technology, SRMIST, Chennai, India

S. Nivetha SRM Institute of Science and Technology, Chennai, India

A. Pandian Department of Computer Science Engineering, SRM Institute of Science and Technology, Kattankulathur, Kancheepuram, India

Rutvi Patel Department of Information Technology, SRM Institute of Science and Technology, Chengalpattu, India

J. Prathipa Department of Computer Science and Engineering, SRM Institute of Science and Technology, Chennai, India

A. Pravin Department of Computer Science and Engineering, Sathyabama Institute of Science and Technology, Chennai, India

Jeberson Retna Raj Sathyabama Institute of Science and Technology, Chennai, India

V. Rajaram Information Technology, Sri Venkateswara College of Engineering, Sriperumbudur, Chennai, India

Surya Rajendran Department of Information Technology, School of Computing, SRMIST, Chennai, India

S. Rajes Kannan Computer Science and Engineering, Chennai Institute of Technology, Chennai, Tamil Nadu, India

Dewang Rajkakati Department of Computer Science and Engineering, SRM Institute of Science and Technology, Kattankulathur, Chennai, India

S. Ravishankar Department of Computer Science, SRM Institute of Science and Technology, Kattankulathur, India

R. Jeberson Retna Raj Faculty of Computer Science & Engineering, Sathyabama Institute of Science and Technology, Chennai, India

M. S. Roobini Sathyabama Institute of Science and Technology, Chennai, India

Neelima Roy Department of Information Technology, SRM Institute of Science and Technology, Chengalpattu, India

Saumya Roy Department of Computer Science, SRM Institute of Science and Technology, Kattankulathur, India

Sumaya Saadat Department of Information Security and Cyber Forensics, SRM Institute of Science and Technology, Chennai, India

M. Safa Department of Information Technology, SRM Institute of Science and Technology, Kattankulathur, Kancheepuram, India

Raghavarapu Sai Surya Siva Prasad Department of Electronics and Communication Engineering, Amrita School of Engineering, Amrita Vishwa Vidyapeetham, Bengaluru, India

B. V. R. Sai Teja Department of Information Technology, SRM Institute of Science and Technology, Chengalpattu, India

Shadikun Nahar Sakiba Department of CSE, UIU, Dhaka, Bangladesh

Anushka Saksena Department of Computer Science, SRM Institute of Science and Technology, Kattankulathur, Chennai, India

G. Saranya Department of Computer Science and Engineering, Sathyabama Institute of Science and Technology, Chennai, India

R. Saranya National Institute of Wind Energy, Chennai, India

M. Saravanan SRM Institute of Science and Technology, Chennai, India

Binoy Sasmal Department of Computer Science and Engineering, SRM Institute of Science and Technology, Chennai, India

Aditya Sattiraju Department of Computer Science, SRM Institute of Science and Technology, Kattankulathur, India

D. Saveetha Information Technology, SRM Institute of Science and Technology, Kattankulathur, Tamil Nadu, India

M. Sayeekumar Department of Information Technology, SRM Institute of Science and Technology, Chennai, India

P. Selvaraj Department of Information Technology, SRM Institute of Science and Technology, Chengalpattu, Tamil Nadu, India

Anuradha Sengupta SRM Institute of Science and Technology, Kattankulathur, Chennai, India

A. Shanthini Department of Information Technology, SRM Institute of Science and Technology, Chengalpattu, India

Chinmay Shelukar Department of Computer Science and Engineering, SRM Institute of Science and Technology, Kattankulathur, Chennai, India

A. Shobana Devi Department of Information Technology, SRMIST, Chennai, India

Md. Mahatab Uddin Shuvo Department of CSE, UIU, Dhaka, Bangladesh

C. Sindhu Department of Computer Science and Engineering, SRM Institute of Science and Technology, Chennai, India

S. Sindhu Department of Information Technology, SRM Institute of Science and Technology, Chengalpattu, India

Aparnika Singh Department of Computer Science, SRM Institute of Science and Technology, Kattankulathur, Chennai, India

S. Sivashankari Department of Computer Science and Technology, SRM Institute of Science and Technology, Chennai, India

K. Sornalakshmi Department of Information Technology, SRM Institute of Science and Technology, Chengalpattu, India

Ravi Sridharan Department of Computer Science and Engineering, Hindustan Institute of Technology and Science, Padur, Chennai, Tamil Nadu, India

L. N. B. Srinivas SRM Institute of Science and Technology, Chennai, India

T. V. S. Subhash Chandra SRM Institute of Science and Technology, Kattankulathur, Tamil Nadu, India

G. Sujatha SRM Institute of Science and Technology, Chennai, India

V. Tejaswini SRM Institute of Science and Technology, Kattankulathur, Chennai, India

Pavan Telluri Department of Electronics and Communication Engineering, Amrita School of Engineering, Amrita Vishwa Vidyapeetham, Bengaluru, India

S. Theebaajan Department of Information Technology, SRM Institute of Science and Technology, Kattankulathur, Chennai, Tamil Nadu, India

K. Venkatesh SRM Institute of Science and Technology, Chennai, India

D. Viji Department of Computer Science, SRM Institute of Science and Technology, Kattankulathur, India

A Novel Approach for Analysing and Detection of Obfuscated Malware Payloads in Android Platform Using DexMonitor

Anuradha Sengupta and S. Sivasankari

Abstract The developers of Android applications and malware use complex obfuscation and encryption tools and techniques to hinder the mobile applications that they develop from being repackaged and analysed. These tools and techniques obfuscate and encrypt the strings and classes, API calls and control flows in the Dalvik bytecode. The obfuscated and encrypted Android applications which are obtained from sources such as Koodous, Virus Total, etc. need to be analysed using DexMonitor and the countermeasures for packed or obfuscated malware have to be found. A decision support system is then constructed and the guidelines and countermeasures for packed malware are provided.

Keywords Encrypted malware · DexMonitor · Byte code · Decision support system

1 Introduction

The usage of mobile and handheld devices is rapidly increasing. The manufacturers of these device operating systems allow third-party application vendors to develop applications for these mobile devices. Google Play for Android is one of the most popular repositories from where users can download applications. Till 2017, it has been estimated that there are about 3.5 million Android applications in Google Play. These Android applications are available as both free and paid applications.

The Android applications are written in Java programming language and use Java core libraries. The Java source code is converted to bytecode and then compiled to .dex format which runs on the Dalvik virtual machine. The dex files containing the bytecode can be converted to human-readable form (smali code) using a disassembler baksmali. The unauthorised or malicious developers perform repackaging attacks on

A. Sengupta (✉) · S. Sivasankari
SRM Institute of Science and Technology, Kattankulathur, Chennai, India
e-mail: anuradhasengupta.as@gmail.com

S. Sivasankari
e-mail: sivasans2@srmist.edu.in

D. J. Hemanth et al. (eds.), *Artificial Intelligence Techniques for Advanced Computing Applications*, Lecture Notes in Networks and Systems 130,
https://doi.org/10.1007/978-981-15-5329-5_1

Android applications. The repackaging attacks take place when a free or paid application is reverse-engineered and the code is modified and is again redistributed for the users to download. Code obfuscation is the process of modifying the source code so that it becomes difficult to reverse engineer and analysis of the code is difficult. The common methods of Android obfuscation or encryption techniques are string encryption, class encryption, control flow obfuscation, instruction pattern transformation, etc. There are two analysis techniques for analysis of Android applications name static and dynamic analysis. The static analysis involves manually observing and analysing the code and is difficult and a tiresome and can by code encryption techniques. Therefore, various other methods have to be devised for the effective and accurate analysis of the encrypted Android applications.

In this paper, the encrypted malware datasets are obtained from data repositories such as Koodous, Virus Total, Symantec, etc. and is analysed. A Dex Monitor is designed for analysis of the Android applications and a decision support system is then constructed. The guidelines and countermeasures for the encrypted or obfuscated malware are provided.

2 Review of Literature

There are various research works that have already been carried out in the area of Android application security. Each of the research work provides an insight into Android application security, various frameworks and detection, classification and analysis of malware.

Enck et al. [1] studies the ded decompiler which extracts the source code directly from the installation image. The authors have also used the ded decompiler for the analysis of the Android applications.

Rastogi et al. [2] has developed a framework called DroidChameleon which studies the various transformation techniques. The paper also proposes solutions for effective malware detection and analysis.

Jung et al. [3] studies the repackaging attacks taking place on Android Banking applications without obtaining the sender's personal information and their countermeasures. The paper also studies the causes of the attacks and some countermeasures have also been suggested.

Bichsel et al. [4] have proposed a new method for statistical analysis of deobfuscated malware based on probabilistic learning of large codebases. The tool called DEGUARD has been implemented and used to study the various features.

Garcia et al. [5] has designed and implemented a framework called RealDroid based on machine-learning for analysis and malware detection without the need to perform complex operations or extraction of features.

Lim et al. [6] has studied and analysed two code hiding tools in Android namely Bangcle and DexProtector. The study also involves the extraction of the original hidden code.

Fei et al. [7] develops a hybrid approach for static and dynamic analysis of Android malware. The sample malicious and benign applications are collected using net link technology to study the pattern of system calls.

Cho et al. [8] proposes and studies the various levels of API hiding by the static analysis of a Dex file. The dynamic extraction of the API call code in Dalvik is done and the classification is done.

Bhattacharya et al. [9] has proposed a method in which the permissions are extracted from the manifest file and classifies the Android applications using WEKA tool. A total of 170 applications has been used and the values have been calculated.

Kang et al. [10] has proposed an effective and accurate framework for the analysis and detection of malware. The n-opcode analysis approach uses machine-learning algorithms to classify and detect the type of malware.

Fing et al. [11] has designed and proposed a framework called EnDroid for dynamic analysis. The framework is used to extract multiple dynamic behaviour features.

Souri et al. [12] has presented a survey of the various malware detection schemes using data mining methods and are based on two types signature-based and behaviour-based. The results have then been compared with other crucial factors.

Zhang et al. [13] has proposed to construct a Dalvik opcode graph and study the global topology properties. The approach can be used to classify the malware based on graph theory and information theory.

Kumar et al. [14] uses the naïve Bayes classifier for extracting the different features of Android malware. The proposed scheme uses the blockchain based on permissions to store the accurate information of the extracted features.

Zhang et al. [15] has proposed a new malware detection model based on method-level correlation relation of the API calls. The various machine-learning techniques are used to identify the type of malware.

3 Existing System

In the existing system, the Android developers use sophisticated encryption tools and techniques, packers, etc. which makes it difficult for the malware analyst to analyse the codes. The reverse-engineering of Android applications becomes difficult. The obfuscation can be done in the following ways:

- String Encryption: The strings are encrypted in the code.
- Class Encryption: The classes are encrypted and remove it from the classes .dex.
- Renaming of Identifiers: The identifiers of the application are changed and it makes the process of reverse-engineering difficult.
- API Hiding: The API is encrypted and the API requests are hidden.

4 Proposed System

4.1 Basic Idea

The Dex Monitoring system takes the encrypted malware as an input and the output of the dex monitor is written to a file and log-based vulnerabilities are performed.

4.2 Architecture

The encrypted malware datasets are taken from resources such as Koodous, Virus Total, Symantec, etc. The encrypted malware datasets are then given as an input to the dex file loader DexMonitor and the dex analysis is performed. The Dex-Monitor is implemented of the Dex Monitoring system is written to a file and the log-based vulnerabilities are performed. A decision support system is constructed and the guidelines and countermeasures for packed malware are provided (Fig. 1).

4.3 Dataset

The encrypted malware dataset is obtained from resources such as Koodous, Symantec, etc. The encrypted malware dataset is downloaded as a zip folder and is then uploaded in VirusTotal to check the details of the encrypted malware such as the number of engines which have detected the particular malware, the behaviour of the malware, etc. The Santoku OS installed in VMW is then used for analysing and studying the particular obtained encrypted dataset. The zipped file is first converted to a .apk file and is then unzipped by using command-line instructions in the in-built LXTerminal. A separate directory folder is created and then the directories and files present in the file can be viewed and navigated such as AndroidManifest.xml, META-INF, etc. In the META-INF directory, the various certificates can be viewed such as CERT.RSA, CERT.SF, etc. The classes .dex file can be analysed by using commands such as 'dexdump' and 'hexdump' with a variety of options. The bytecode can be viewed and analysed by using these two commands from the LXTerminal. Similarly, the classes .dex file can also be viewed using the 010 Editor which is a professional text/hex editor in Santoku OS can be installed and the bytecode can be analysed. The hex bytecode can be further analysed by using the DEX.bt template. The template displays the fields such as name, value, starting, size, colour and comment.

During the analysis of the bytecode of classes .dex, it can be viewed that there are encoded fields, encoded strings, encoded methods, encoded parameters, etc. which suggests that it has been modified and encrypted by the Android application and malware developer. The strings, fields and methods have been encrypted by the developer so that it cannot be reverse-engineered easily (Figs. 2, 3, 4, 5 and 6).

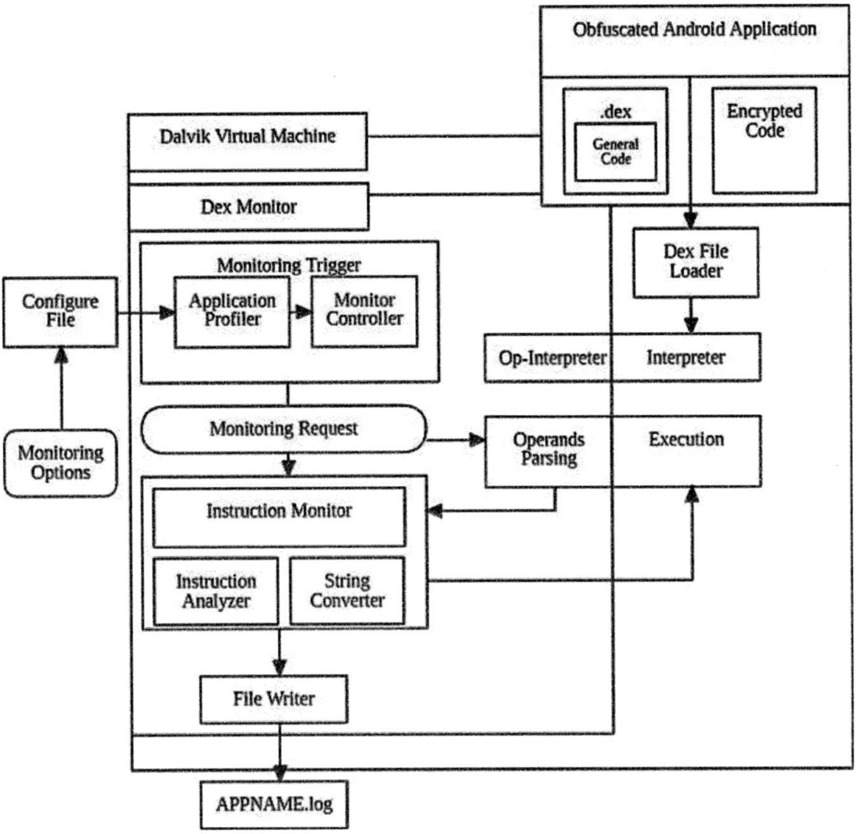

Fig. 1 System architecture of dex monitoring system

5 Conclusion

In this paper, the encrypted dataset has been obtained from resources such as Koodous, Symantec, etc. and has been viewed and analysed in Santoku OS in the LXTerminal. For further analysis, the classes .dex has been loaded and analysed in the 010 Editor and the presence of encoded fields, encoded parameters, etc. has been found which suggests that the developers of the Android applications have encoded the bytecode so that the reverse-engineering cannot be done easily.

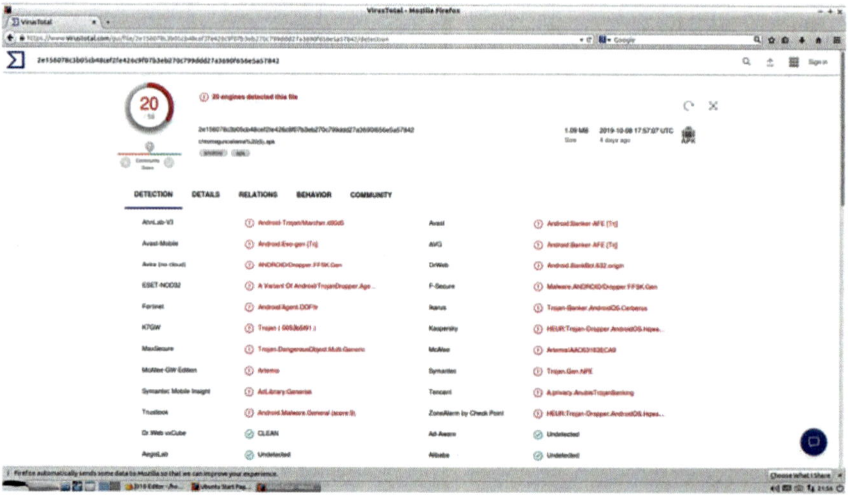

Fig. 2 Uploading of the encrypted malware in virus total

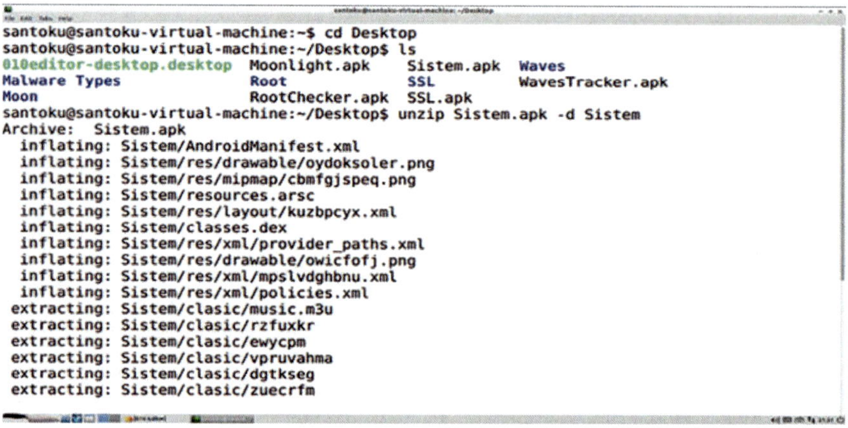

Fig. 3 Unzipping of the .apk file

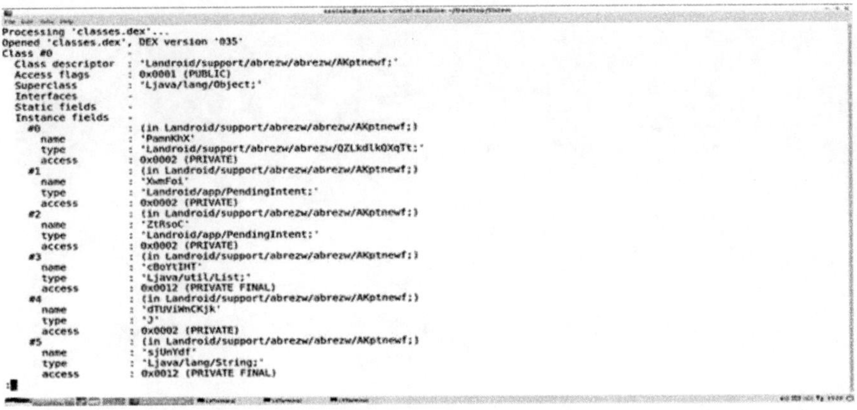

Fig. 4 Dexdump analysis of classes .dex in LXTerminal

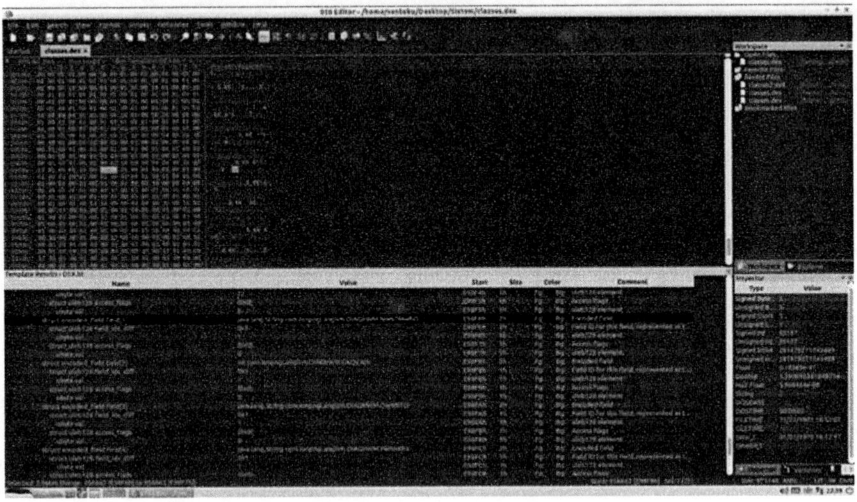

Fig. 5 Encoded fields present in bytecode while analysing in 010 Editor

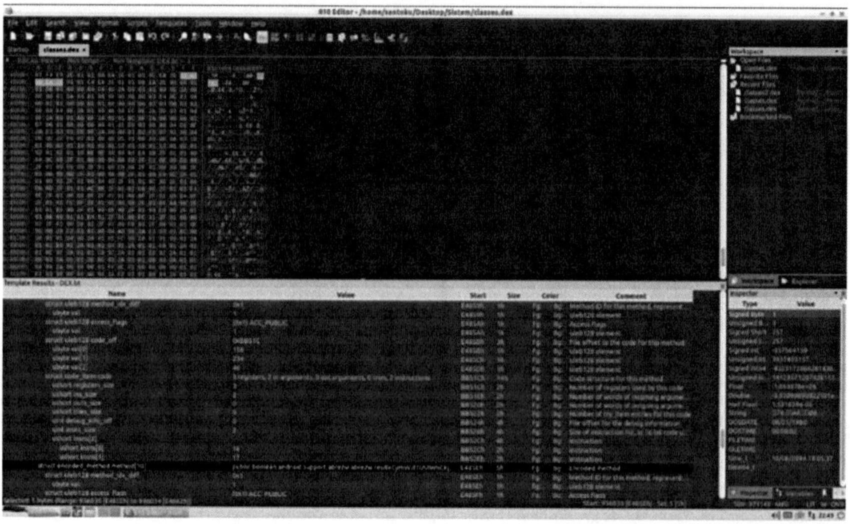

Fig. 6 Encoded methods present in bytecode while analysing in 010 Editor

References

1. Enck W, Enck DW, Octeau D, McDaniel P, Chaudhuri S (2011) A study of Android application security. Proc USENIX Secur Symp 2:2
2. Rastogi V, Chen Y, Jiang X (2013) DroidChameleon: evaluating android anti-malware against transformation attacks. In: ASIA CCS 2013 – Proceedings of the 8th ACM SIGSAC Symposium of Information, Computer and Communications Security, pp 329–334. https://doi.org/ 10.1145/ 2484313.2484355
3. Jung J-H, Kim JY, Lee HC, Yi JH (2013) Repackaging attack on android banking applications and its countermeasures. Wireless Pers Commun 73. https://doi.org/10.1007/s11277-013-1258-x
4. Bichsel B, Raychev V, Tsankov P, Vechev M (2016) Statistical deobfuscation of android applications. Proc ACM SIGSAM Conf Comput Commun Secur: 343–355
5. Garcia J, Hammad M, Malek S (2016) Lightweight, obfuscation-resilient detection and family identification of android malware. ACM Trans Softw Eng Methodol 9, 4, 39: 27
6. Lim J, Yi JH (2016) Structural analysis of packing schemes for extracting hidden codes in mobile malware. EURASIP J Wireless Commun Netw: 221. https://doi.org/10.1186/s13638-016-0720-3
7. Fei T, Yan Z (2017) A hybrid approach of mobile malware detection in android. J Parallel Distrib Comput 103:22–31
8. Cho T, Kim H, Yi JH (2017) Security assessment of code obfuscation based on dynamic monitoring in android things in special section on security and privacy in applications and services for future internet of things, vol 5
9. Bhattacharya A, Goswami RT (2017) DMDAM: data mining based detection of android malware. In: Mandal JK, Satapathy SC, Sanyal MK, Bhateja V (eds) Proceedings of the first international conference on intelligent computing and communication springer Singapore. Singapore, pp 187–194
10. Kang B, Yerima SY, Mclaughlin K, Sezer S (2017) N-opcode analysis for android malware classification and categorization. Proc Int Conf Cyber Secur Prot Digit Serv Cyber Secur: 1–7

11. Fing P, Ma J, Sun C, Xu X, Ma Y (2018) A novel dynamic android malware detection system with ensemble learning. IEEE Access 6
12. Souri A, Hosseini R (2018) Hum Cent Comput Inf Sci 8:3. https://doi.org/10.1186/s13673-018-0125-x
13. Zhang J, Qin Z, Zhang K, Yin H, Zou J (2018) Dalvik opcode graph based android malware variants detection using global topology features. IEEE Access 6:51964–51974
14. Kumar R, Zhang X, Wang W, Khan RU, Kumar J, Sharif A (2019) A multimodal malware detection technique for android IOT devices using various features. IEEE Access 7
15. Zhang H, Luo S, Zhang Y, Pan L (2019) An efficient android malware detection system based on method-level behavioral semantic analysis. IEEE Access 7

A Study on the Prevention Mechanisms for Kernel Attacks

C. Harini and C. Fancy

Abstract The kernel of the operating system is the lifeline of a system. It controls all the components of the system and is needed for a problem-free execution of all the processes in the system. In the past decade, attackers are launching attacks that modify the kernel and remain undetected due to less security protection to the kernel. Kernel rootkits were used to manipulate the code to gain root-level access. Now, with anti-viruses and firewalls blocking malicious codes in the system, attackers have started targeting the kernel memory to subvert the running behavior of the system without any code injection. This is known as a kernel data attack. This paper discusses the different types of attacks that have been launched by the attackers to manipulate the kernel in the operating system and classifies the different attacks based on their types. It also discusses the currently existing techniques to prevent kernel attacks and their characteristics.

Keywords Kernel · Linux · Rootkits · Privilege escalation · Memory corruption

1 Introduction

When the operating systems were first developed, security was not the major concern for the developers. There was nearly no security measure established to protect the operating system from attacks. As technology became widespread, attackers exploited this to spread malware that were able to change the control flow of the operating system. They were able to obtain privilege escalation by executing at ring 0 level. This made significant changes to the security issues in the operating system with more researches undertaken to provide better kernel security. This led to ideas like Address Space Layout Randomization (ASLR), Executable Space Protection (ESP), Control Flow Integrity (CFI), etc. ASLR randomizes the kernel address space of a process every time it is executed. ESP prohibits the execution of machine code in the region which has been marked non-executable. CFI prevents the subverting of

C. Harini (✉) · C. Fancy
SRM Institute of Science and Technology, Kattankulathur, Chennai, India
e-mail: harinic192@gmail.com

D. J. Hemanth et al. (eds.), *Artificial Intelligence Techniques for Advanced Computing Applications*, Lecture Notes in Networks and Systems 130,
https://doi.org/10.1007/978-981-15-5329-5_2

the control flow of the kernel. These techniques help in detecting any changes to the code [1], [2].

These techniques have caused the attackers to find other mechanisms to attack the kernel. Attackers have also found ways to manipulate kernel behavior without injecting any new code by altering the existing kernel data. This type of attack is known as kernel data attack. This attack is not easily detectable as the existing kernel rootkits. The kernel codes have limited change during their lifetime but most kernel data are changeable making them difficult to track [3].

This paper discusses the various types of attacks used by the attackers to exploit the kernel. It also classifies these attacks into different categories based on their functionality. Further, the existing techniques to prevent these kinds of attacks are also discussed.

2 Background

Operating systems have two modes of operation: Kernel mode and user mode. User mode is lower privileged mode which has restrictions on the access to I/O devices and memory, etc. The kernel mode is the higher privileged mode which has access to all the underlying hardware and can reference any memory address and can execute any instruction. Any malfunctioning in the kernel mode has the potential to crash the entire system. If a user mode process tries to perform something out of its purview, instead of crashing the system only that process is thrown an exception. These two modes secure the operating system from illegal access.

Additionally, there are protection rings that protect the data and functionality from malicious behavior. These protection rings are two or more layers of privilege based on the computer architecture. Typically, in the x86 architecture, there are four rings ranging from ring0 of the highest privilege (kernel mode) to ring3 of the lowest privilege (user mode) (Figs. 1 and 2).

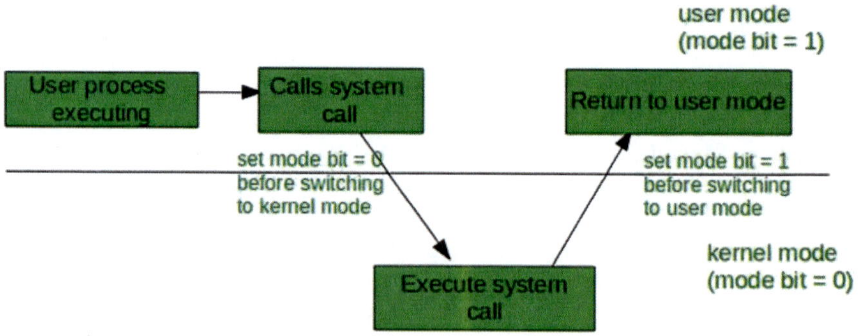

Fig. 1 Dual-mode operation in the OS

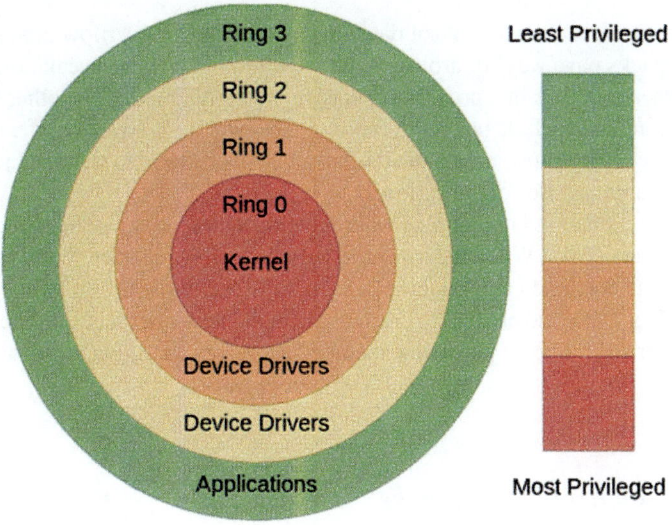

Fig. 2 Protection ring levels

This paper discusses the privilege escalation attacks in the kernel. Attackers exploit a user mode process to perform kernel level tasks by hoodwinking the mode bits, protection rings, etc. They try different techniques to perform this without any detection. Rootkits are used by the attackers to spread malware which affects the kernel without any detection [4], [5]. They are able to run in the ring0, i.e., with root privileges. They inject code into the kernel which subvert the flow of the process.

3 Categorizations of Attacks

This paper categorizes the exploits into two different types based on their mode of operation as control data attacks and non-control data attacks. Control data attacks are those in which the flow of the process is changed by modifying the control data (function pointers, return addresses, etc.) to execute the injected code. Non-control data attacks are those attacks that do not alter the control flow of the process.

3.1 Control Data Attacks

In control data attacks, the attacker tries to take control of the system by injecting code into it. Through this, the attacker can have access to the entire system (privilege escalation) or they can crash the system (denial of service).

The most widely known control data attack is the buffer overflow attack. Buffer overflow attacks pave way for arbitrary code execution. Privilege escalation is made possible when it occurs in a process running in the kernel mode. The other forms of control data attacks include integer overflow attack which is a type of buffer overflow attack. In integer overflow attack, the buffer can be overflowed to return to a memory address containing code for execution.

Double free attack is another type where freeing a memory more than once makes way for memory leaks. When the same memory is freed twice, the memory management data gets corrupted and could let a malicious user write arbitrary code into it. This can either cause the process to crash or alters the flow of execution. The attacker can also alter the value of particular memory addresses or registers to gain higher privileges.

3.2 Non-control Data Attacks

Non-control data attacks are lesser in existence when compared to the control data attacks. This could be due to the reason that the non-control data attacks are more complex to be implemented. The increase in the number of protection mechanisms for control data attacks has turned the attention of the attackers to the non-control data attacks. Non-control data are more sophisticated in nature. They are classified into different types based on their availability, functionality, and access type.

Based on availability, the data can be either global data or local data. Though both global and local data are available in the system, global data can be exploited feasibly as they are easily identified. The global data is stored by Linux in a proc system file/proc/kallsyms which contains the symbol to virtual memory mapping [6]. Local data is usually stored in kernel heap making it difficult to identify.

The function pointers and variables are the categories of non-control data based on their functionality. The function pointers are usually used in control data attacks to change the control flow.

The data is categorized into two types as read-only and read-write data based on the access type. The read-only data can only be read during the run time and no process can write to it while the read-write data can be modified during the run time. These read-write data are more challenging to be handled as they are huge in number and difficult to identify.

Vulnerabilities in the kernel codes can cause non-control data attacks. Recently, a vulnerability CVE-2019-13272 was found in the Linux kernel which allowed local users to gain root-level access by manipulating the parent-child relation of the process to make the parent drop the privileges and call the execve function which can be used by an attacker. Using this vulnerability, the attacker can gain full control over the system [7].

Use-after-free attacks are also being used by the attackers to perform non-control data attacks. The vulnerability CVE-2019-11815 was discovered in rds_tcp_kill_sock in net/rds/tcp.c in Linux which can cause a use-after-free attack [8].

4 Protection Mechanisms

There have been a lot of suggested mechanisms to protect against kernel attacks. Address Space Layout Randomization is a technique in which the process executes at a different random address space every time it executes. This is done to avoid the attacker from gaining access to the memory from which the process is being executed [9]. Supervisor Mode Execution Protection disables the execution of code in a user mode page when the system is running in a higher privileged mode. This prohibits the privilege escalation of injected codes in the memory. Another technique used to protect against control data attacks is Control Flow Integrity. Control Flow Integrity prevents malware from redirecting the flow of the program. Control Flow Integrity has been implemented in many different ways. Canary values are also used to check if there are any unwanted changes in the memory [10].

There are fewer techniques available to protect against non-control data attacks. Researchers have proposed different techniques to provide solutions for non-control data attacks.

A research suggests combining the static binary analysis and virtualization technology to detect kernel heap overflow attacks [11]. Srivastava et al. propose a mechanism in which the kernel memory is partitioned to provide memory isolation. This separates the kernel mode and user mode pages thus protecting against dynamic kernel data attacks [12]. Prakash et al. showcased that most semantic fields of the kernel can be freely mutated by testing 41 fields of Ubuntu using fuzz technique [13]. In another proposal, researchers were able to detect 23 rootkits using data structure invariants. Virtual machine monitor policies can also be used to monitor the kernel memory access [14]. The system calls used in the user mode to kernel mode escalation can be hooked to detect the changes made to the kernel data [1]. Also, most of these techniques have been implemented in the older versions of the Linux kernel and they may not be successful against later versions. Yet, they all have some disadvantages like large overhead or protection against only particular kinds of attacks.

5 Discussion

This paper explores the different kinds of attacks that can target the kernel and their defense mechanisms present within the kernel. It is observed that with more defense techniques being found, the attackers are trying to be one step ahead with another kind of attack. This emphasizes a need for a protection system that can prevent the kernel attacks at the root level. Security of an operating system is mostly concentrated on the application level with the usage of anti-virus software, firewalls, etc., but all these cannot protect the inner workings. With the kernel data attacks being split into two types, it is observed that the control data attacks are more familiar than the non-control data attacks. Non-control data attacks are becoming the preferred attack mode for the attackers due to them being unfamiliar. Thus, more research is required in the field of kernel security to prevent kernel attacks. Kernel being the lifeline of an operating system when guarded properly can secure the whole system from compromise.

6 Conclusion

Based on the inputs from previous researches, it is seen that there is a huge scope for exploration in the field of kernel attacks. It is also seen that the non-control data attacks are more unknown compared to the control data attacks. They are still unexplored with their impact potential still undetermined. It is also seen that the existing protection mechanisms against the attacks can be improved. Kernel being the lifeline of the operating system can have a large impact if failed and thus there is a need to keep updating the kernel protection mechanisms frequently.

7 Future Work

Non-control data attacks are increasing and their protection mechanisms aren't enough. Thus, there is a need for newer techniques to protect the kernel against privilege escalation attacks. We plan to work on protecting the kernel against privilege escalation attacks in the future.

References

1. Qiang W, Yang J, Jin H, Shi X (2018) PrivGuard: protecting sensitive kernel data from privilege escalation attacks. IEEE Access 6:46584–46594
2. Zhai G, Li Y (2008)Analysis and study of security mechanisms inside Linux Kernel. In: International conference on security technology
3. Feng X, Yang Q, Shi L, Wang Q (2018) BehaviorKI: behavior pattern based runtime integrity checking for operating system kernel. In: IEEE international conference on software quality, reliability and security
4. Behrozinia S, Azmi R (2014) KLrtD: kernel level rootkit detection. In: The 22nd Iranian conference on electrical engineering
5. Rhee J, Riley R, Xu D, Jiang X (2009) Defeating dynamic data kernel rootkit attacks via VMM-based guest-transparent monitoring. In: International conference on availability, reliability and security

6. Xiao J, Huang H, Wang H (2015) Kernel data attack is a realistic security threat. In: Thuraisingham B, Wang X, Yegneswaran V (eds) Security and privacy in communication networks. SecureComm 2015. Lecture Notes of the institute for computer sciences, social informatics and telecommunications engineering, vol 164. Springer, Cham
7. https://cve.mitre.org/cgi-bin/cvename.cgi?name=CVE-2019-11815
8. https://nvd.nist.gov/vuln/detail/CVE-2019-13272
9. Hund R, Willems C, Holz T (2013) Practical timing side channel attacks against kernel space ASLR. IEEE symposium on security and privacy
10. Wang L, Wu D, Liu P (2016) iCruiser: protecting kernel link-based data structures with secure canary. In: IEEE international conference on software quality, reliability and security companion
11. Tian DX et al (2016) A practical online approach to protecting kernel heap buffers in kernel modules. China Commun 13(11):143–152
12. Srivastava A, Erete I, Giffin J (2009) Kernel data integrity protection via memory access control. Georgia Institute of Technology
13. Prakash A, Venkataramani E, Yin H, Lin Z (2013) Manipulating semantic values in kernel data structures: Attack assessments and implications. In: Proceedings of the 2013 43rd annual IEEE/IFIP international conference on dependable systems and networks (DSN), pp 1–12, 24–27 June 2013
14. Baliga A, Ganapathy V, Iftode L (2011) Detecting kernel-level rootkits using data structure invariants. IEEE Trans Dependable Secure Comput 8(5):670–684

Enhancing Data Security Using DNA Algorithm in Cloud Storage

P. Jenifer and T. Kirthiga Devi

Abstract With exponential growth of data, it is hard to store the data locally. To overcome this issue, the individuals and organizations move forward to store their data in cloud. In some cloud storage systems, like electronic health record system, the cloud contains several sensitive information that should not be exposed to unauthorized users. To ensure this, we propose a strong encryption technique using an efficient DNA encryption algorithm. To secure the data, even if an attack occurred it is not possible for an attacker to leak the sensitive information of the user, because of the strongly proposed DNA algorithm. In this scheme, a private key generator is used for an individual's authentication process to generate a unique key to log in. For sanitizing the sensitive information, the sanitizer verifies the corresponding data and signature of the EHR to be a valid one to store them in the cloud. Third-party auditor is used to verify and audit the integrity of the signature stored in cloud on behalf of the organization. To ensure the encrypted data is safe, we propose a decrypting tool, where if the attacker tries to attack or decrypt the sensitive information, the data could not be decrypted and will not be in a user-readable format.

Keywords EHR · TPA · PKG · Sanitizer · DNA

1 Introduction

This cloud storage platform has been used by all the enterprise and also individuals to access to store and access the data remotely. In healthcare industry, they have large amount of patient's sensitive information been stored in cloud where any unauthorized users can access their data. To address this issue, we propose a DNA encryption technique to securely store the data in cloud, where the third party cannot access it.

P. Jenifer (✉) · T. Kirthiga Devi
Department of Information Technology, SRM Institute of Science and Technology, Kattankulathur, Chennai, Tamil Nadu 602103, India
e-mail: jeniferpaulraj11@gmail.com

D. J. Hemanth et al. (eds.), *Artificial Intelligence Techniques for Advanced Computing Applications*, Lecture Notes in Networks and Systems 130,
https://doi.org/10.1007/978-981-15-5329-5_3

In this, the doctor has to register them to store the data in cloud, and once they have registered, they will get a key to log in from the private key generator, where all the sensitive information of a particular user can be stored. The doctor will upload the electronic health record where the sensitive information will be encrypted by DNA encryption algorithm when it reaches the sanitizer [1]. The sanitizer will receive the encrypted signature and data that have been generated from doctor, where the sanitizer cannot view the sensitive information of the user received from the doctor. So, here the sanitizer will sanitize the data and stores it in cloud. If the user wants to view their record, they can register and get a login key from the private key generator. We can keep track of the data by checking the system logs by whom the data is modified and from where the data gets modified in order to achieve data security.

2 Contribution

This project aims to secure the information that has been held in cloud. This cloud storage platform has been utilized by all the enterprise and additional people to access to store and access the information remotely. In healthcare trade, they need great amount of patient's sensitive information been hold on in cloud where ever any unauthorized user's will access their data. To deal with this issue, we have a tendency to propose DNA encryption algorithm to powerfully secure the information in cloud whenever the third party cannot access it. During this, the doctor must register them to store the information in cloud, and once they have registered, they will get a key to log in from the private key generator, where all the sensitive information of a specific user are often hold on. The doctor can transfer the EHR whenever the sensitive information is going to be encrypted once it reaches the sanitizer. The sanitizer will receive the encrypted data with the file signature that has been generated from doctor where the sanitizer cannot read the sensitive user information. Thus, sanitizer will store the EHR and the data in cloud after sanitized. If the user needs to read their health record data, the user has to get registered and obtain a login key from the private key generator. Once the user logs in, they will only see the encrypted information. If the user wants to download the data, they will request the sanitizer and the sanitizer will verify the signatures, and once the signature gets matched, the user will get the access to download the data. To verify and check the integrity of the signatures in cloud, third-party auditor (TPA) is used. They make a challenge audit to the cloud to perform an auditing proof. Once the auditing proof is verified by the TPA, finally, the TPA will send a report to the organization after auditing and verifying the integrity check. We can keep track of the data by checking the system logs by whom the data is modified and from where the data gets modified in order to achieve data security.

3 Scope

The main scope of the project is to protect the user's sensitive data in cloud. If the attacker tries to attack the encrypted data, they could not view the sensitive information with user's readable format. Hence, sensitive data leakage can be prevented using an efficient encryption algorithm.

4 Existing System

The doctor uses his organization identity to enter the portal, and when he logs in using his identity, the login credentials reach the private key generator (PKG) which provides access to the doctor by generating a private key for the authentication process [2, 3]. PKG generates a new and random key whenever the doctor needs to enter the organization portal. Now, the doctor uploads the electronic health record (EHR) data to sanitizer which includes the patient's sensitive information like name, ID, and also organization's sensitive information. When the doctor uploads the data with its signature to the sanitizer, then the sensitive information inside it is encrypted using an efficient algorithm and reaches the sanitizer. The sanitizer sanitizes the information as well as the signature as a valid one to store signature in the EHR record and uploads the sanitized information and signature to cloud. The signatures stored are to verify the integrity of the file during the auditing process. Here comes the third-party auditor (TPA) for auditing and checking the integrity of the signatures [4]. TPA is implemented here to audit and verify the integrity check, where organizations cannot audit the signatures all the time stored in the cloud. TPA can be scheduled or set upon the priority to verify and audit the file. The TPA sends an auditing challenge request to cloud and it responds with the proof of auditing that it has been successfully finished. When TPA receives the auditing proof, it checks the signature whether there is any unwanted modification of data or the data is as the same from the time of upload and verifies that the integrity is not lost. After this process, the TPA sends the verification and the auditing report to the organization in the scheduled manner or priority-based customization of the organization. In the download process, if the user or the doctor wants to download a file from the cloud, the individuals need to request the PKG to log in. Once they have given access to the portal, they will be able to view their files in encrypted format. To download those files, individual needs to request the sanitizer for downloading the sensitive information from cloud. Once the sanitizer receives the request, the sanitizer will compare the signature received from the individual with the signature stored in cloud. If the signature matches with the electronic health record table which is managed by the sanitizer and the sanitized signature stored in the cloud, the sanitizer will provide the access to pull the data from the cloud by the individual's request to view their information (Fig. 1).

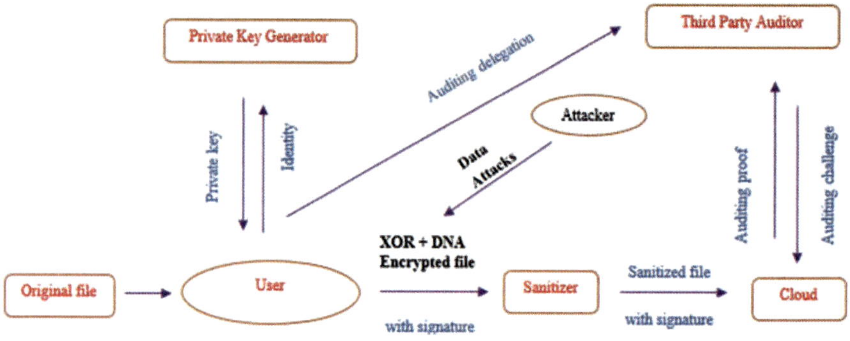

Fig. 1 Architecture of existing system

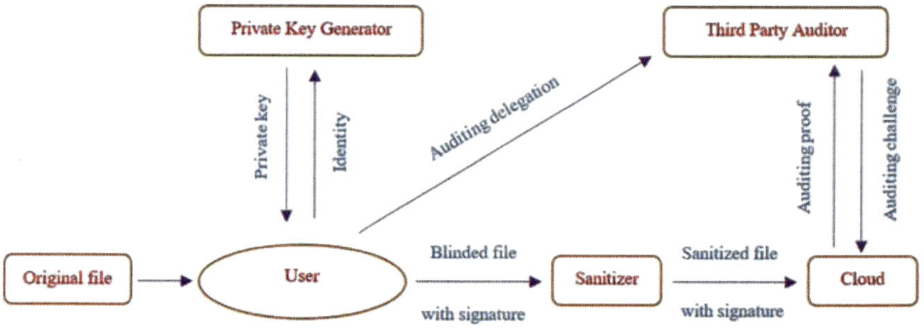

Fig. 2 Architecture of proposed system

5 Proposed System

The main idea of the project is to implement an efficient DNA algorithm which can encrypt the sensitive data, and even if the attacker tries to decrypt the data, it will not be in a readable format [5]. Hence, we are proposing a decrypting tool to show how the data cannot get decrypted when the data is encrypted using DNA encryption algorithm [6, 7]. In this, we can audit it by compromising the system logs to check the modified date or time or by which user the sensitive data gets modified to achieve data security (Fig. 2).

6 Decomposition of the System

The main scope of the project is to protect the user's sensitive data in cloud by using an efficient DNA encryption algorithm.

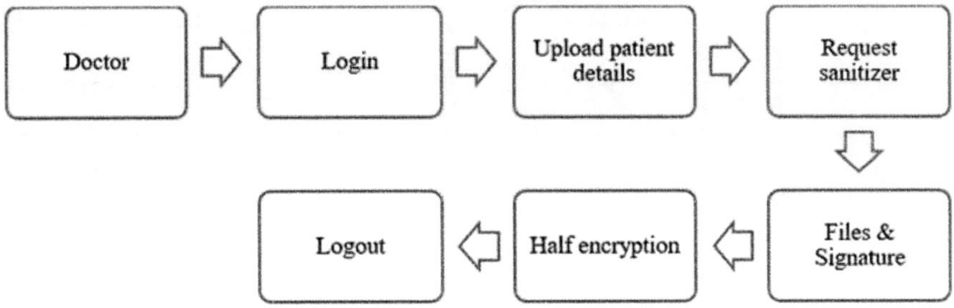

Fig. 3 Uploading data to sanitizer

6.1 *Organization Login*

The private key generator (PKG) is trusted by all other entities. They are used to generate random private keys for users, considering their identity ID for the authentication process to enter the organization portal.

6.2 *Uploading Data to Sanitizer*

When the doctor uploads the data to sanitizer, the data is been encrypted by DNA encryption algorithm technique and sent to sanitizer with signature of the uploaded file (Fig. 3).

6.3 *Uploading Data to Cloud*

The sanitizer will receive the encrypted data with the file signature that has been generated from doctor. The sanitizer cannot view the sensitive user information from the doctor's side. So, here the sanitizer will upload the encrypted electronic health record and signature to be stored in cloud (Fig. 4).

6.4 *TPA Auditing Files in Cloud*

If the user or the doctor wants to audit their information in cloud, they will have a third-party auditing user for verifying the integrity of the sanitized file stored in cloud. It sends a request to the cloud which in turn responds with the respective auditing proof [8]. Finally, the verification for the integrity of the sanitized file is done by checking the correctness of the auditing proof.

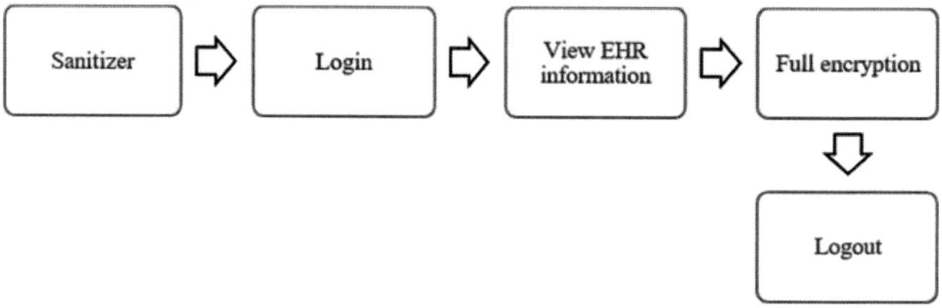

Fig. 4 Uploading data to cloud

6.5 Downloading the Data from Cloud

When the doctor needs electronic health record (EHR), a request is generated to the sanitizer for EHR information system from which they can download the encrypted EHR file. After the file is downloaded into the EHR data system, it can be send to the doctor.

The doctor can easily decrypt the original EHR by using DNA decryption technique.

7 Future Implementation

The healthcare industry is a leading industry where many sensitive information of patients gets stored in cloud. Here, we propose an efficient DNA encryption technique where it encrypts all the sensitive information before the data gets stored on to the cloud. There are possibilities that the data can get attacked or decrypted by the attacker and misuse them wrongly, or the attacker might use the data by targeting the hospital and steal the information and expose the data online and they can even demand for the ransom for decrypting the data, so to avoid this, we are using the DNA encryption where the data can be encrypted strongly, and even if the attacker tries to attack the data or break the encryption algorithm, they cannot get the sensitive information, it will be scrambled and the attacker cannot view the sensitive information in user's readable format [9]. So here, we propose a decrypting tool to show how the attack cannot be possible when the attacker tries to decrypt the data to achieve data security by using efficient and strong encryption algorithm techniques. The auditor helps to audit the data by verifying the signatures to secure the data. If we need to audit the files, we can check the logs of the encrypted data, and if the data or sensitive information is modified, we can verify it by taking system logs to achieve data security to verify data authentication and authorization of the system. So, in this project, we focus on achieving the sensitive data gets encrypted by DNA encryption technique before the data gets stored on the cloud. To achieve data security, the data must be strongly encrypted before it gets stored on to the cloud.

8 Expected Results

In this paper, we proposed a strong DNA encryption algorithm where the attacker cannot decrypt the sensitive information. If the attacker tries to attack the data using different attacking techniques or by using decrypting tool, the sensitive data cannot be decrypted or it cannot be useful for the attacker, the sanitizer will get notified when they access the cloud for the security and to safeguard the data privacy of the individuals. We deploy a cloud to store the data in cloud, the sensitive information now gets encrypted by the DNA encryption algorithm technique before it reaches the sanitizer, and the sanitizer sanitizes all the sensitive data before they store it in cloud.

We propose a decrypting tool to show how the sensitive information cannot be decrypted even when the attacker tries to attack the data or uses different decrypting tools for fetching the sensitive data. The main purpose of this is to achieve data security by implementing efficient encryption algorithm and show how the decryption is not possible by an attacker who tries to attack or tamper the sensitive data.

9 Conclusion

To achieve data security, when the data gets stored in the cloud, it must be strongly encrypted with efficient DNA encryption so that the attacker or any unauthorized user cannot access the sensitive data which is encrypted to secure the data before it gets stored on cloud. By applying this DNA patch-up algorithm to the existing algorithm, we can secure our sensitive data more securely.

References

1. Rathi M, Bhaskare S, Kale T, Shah N (2016) Data security using DNA cryptography. IJCSMC 5(10)
2. Raut SD, Bogiri N (2018) A survey on identity-based integrity auditing and data sharing with sensitive information hiding for secure cloud storage. Int J Adv Sci Res Eng Trends 3(9)
3. Shen W, Qin J, Yu J, Hao R, Hu J (2019) Enabling identity-based integrity auditing and data sharing with sensitive information hiding for secure cloud storage. IEEE Trans Inf Forensics Secur 14(2)

4. Vijayalakshmi Giri T, Naveen Kumar N (2019) Enabling identity-based integrity auditing and data sharing with sensitive information hiding for secure cloud storage. Int J Manage Technol Eng 9(2)
5. Tornea O, Borda ME Cryptographic algorithms. Communication Department, Technical University
6. Terec R, Vaida M-F, Alboaie L, Chiorean L DNA Security using symmetric and asymmetric cryptography. Department of Communications
7. Sinhgad SJ, Sinhgad RK Secure data communication and cryptography based on DNA based message encoding. Department of Computer Engineering, Lonavala Pune, University of Pune
8. Yu Y, Au MH, Ateniese G, Huang X, Susilo W, Dai Y, Min G (2017) Identity-based remote data integrity checking with perfect data privacy preserving for cloud storage. IEEE Trans Inf Forensics Secur 12(4)
9. Chouhan P, Singh R Security attacks on cloud computing with possible solution

Stable Multi-agent Clustering Method to Improve the Security in VANET

S. Theebaajan and J. Godwin Ponsam

Abstract Ad hoc vehicle network is method of the mobile communication network which transfers the information between vehicle groups to give vehicle owner safety, traffic information, performance, data splitting, etc. Clustering is an effective method for dealing with the regular shift in vehicular ad hoc network topology by regional coordination. We present stable clustering based on multi-agent to give stable, which in convert combination the life of the cluster. This approach consists of all static and also mobile agents to establish information between the vehicles and road side unit using multi-agent interaction method. Road side unit agent responsible for deciding the size of the cluster and choosing the correct group occurs based on vehicle data transmission and neighboring node joining. In this adjust the size of the cluster according to the speed of the vehicle, which in turn improves the life of the cluster and reduces the losses of the routing. Group maintenance is handed over to cluster head vehicles after cluster maintenance has been established. The proposed approach is generated in Network Simulator-2 by using some of the performance evaluation such as vehicle rate, vehicle size, Group selection time, contact distance, etc.

Keywords VANET · MANET · Radius cluster · Rate · Density of nodes

1 Introduction

The VANET is the sub-class of all mobile specially designated network, correspondence among vehicles. In vehicular technology correspondence development the vehicle is not just a moving object or an entity, vehicular ad hoc network is joining learning into the vehicle system. The critical motivation behind acquainting vehicular ad hoc network is with upgrade driver wellbeing and evades crash by sending threat zone and mishap data. It likewise encourages amusement applications, traffic the board, Internet providers, climate data and so on. It comprises of RSU

S. Theebaajan (✉) · J. Godwin Ponsam
Department of Information Technology, SRM Institute of Science and Technology, Kattankulathur, Chennai, Tamil Nadu 602103, India
e-mail: tsn210568@gmail.com

D. J. Hemanth et al. (eds.), *Artificial Intelligence Techniques for Advanced Computing Applications*, Lecture Notes in Networks and Systems 130,
https://doi.org/10.1007/978-981-15-5329-5_4

what's more, vehicles with Web as emotionally supportive network. In vehicular ad hoc network building data dispersal happens between vehicles (Vehicle to Vehicle), among Road Side Unit and vehicle and hybrid correspondence through IEEE 802.11p standard. Since in vehicular ad hoc network, the vehicles are moving with different speed there is perpetual contrast in framework topology which extends the orchestrate unpredictability. Pack course of action is huge part in vehicular ad hoc network for gathering a data and also for collecting the data. This method is gathering system subject to vehicle movement for capacity of correspondence in vehicular ad hoc network. We go through programming administrators to fabricate affiliation and message spread among vehicles and Road Side Unit. The range of gathering changed in perspective on typical vehicle speed to improve gathering head life time and structure stable gathering.

The investigated works are based on Paper [1] depicts the steady gathering can be molded using multi-pros, these administrators structure the gathering reliant on relative speed and Cluster Head decision is considering accessibility of vehicle. The audit of vehicular ad hoc network is discussed in paper [2] that is need of vehicular ad hoc network, different sorts of coordinating in vehicular ad hoc network, security issues, for instance, data change, disparaging, tunneling, etc. The pack head assurance time increases with increase in center point thickness are found in paper [3]. Paper [4] considers channel clouding to maintain a strategic distance from impact of impedance of channel using Reyleigh obscuring model.

The major inspiration driving proposed method is to edge stable gathering in perspective on vehicle movement and neighbor index. Here in this use adaptable what are progressively, static programming authorities to working dispensed data. Street Side Unit authorities make the pack and select all the gathered head by then leave the gathering to make it work by itself. Pack the board are achieved by bundle main node and it in like manner manages gathering to move along the all other part nodes. At last bit of paper is dealt with as seeks after Sect. 2 presents Multi-administrator based stable packing in vehicular ad hoc network. Section 3 delineates estimation. Territory IV discusses the diversion and output. Region V wraps up the proposed method.

1.1 *Stable Clustering Multi-agent in Vehicular Ad Hoc Network*

This part describes the model of the topology, the mathematical evaluation, the scheme developed and the algorithm.

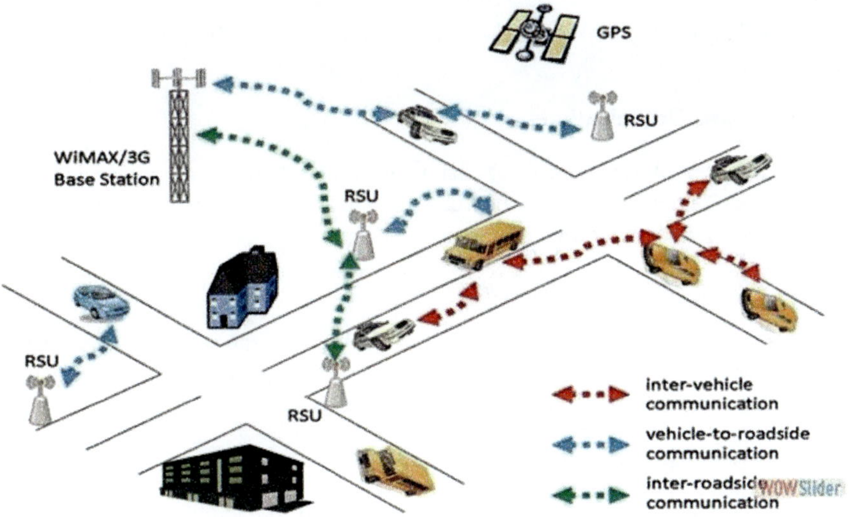

Fig. 1 Network environment. https://www.researchgate.net/figure/VANET-Communication_fig1_305243551

1.2 The State of the Network

Network model includes support for infrastructure and deployment of vehicles. The vehicles deployed push in the same direction as shown in Fig. 1. It is believed that all cars are fitted with GPS and sensors to track the location of the car.

1.3 Maintenance of Clusters

For the following cases, maintenance of the group maintenance group is required. In the new example, if first node is inserted into the current cluster contact array, then group head must join this first node to the list of (group member). In the another case, if the current cluster member is out of Group head contact range, then Group head would exclude the cluster member from the list of cluster member. CH fails in the third case, then one of CM with the lowest 'WT' will be selected as CH. When two CHs are similar to each other, then cluster merging occurs in the forth case.

1.4 Agency

Operators are independent programming method; they need the condition and follow up through condition to accomplish given objectives. The use multi specialists, are vehicle method office and the RSU board operator to frame the steady groups are depicted underneath.

1. **Node Board Organization**

Node board organization situated for each node comprises of locally available manager agent, locally available black board, locally available registration agent, bunch maintenance agent are recorded as pursues.

a. *Locally available Managing Agent*

The specialist is static, makes on black board and controls the activity of clustering support operator. In the event that vehicle chose group head, at that point locally available manager agent conjures bunch head operator.

b. *Locally available black board*

This is also a static operator, and It stores the area of itself and the neighbor vehicle data, for example, vehicle label, speed, course, neighbor list and it is occasionally refreshed by other specialists.

c. *Locally available Registration Agent*

It continually monitors for accessibility of Road side unit or Group member of other group also, register itself by communicating with locally available registration agent or locally available black board of group member and this information is put away in chalkboard specialist of road side unit or nodes.

2 Related Works

Creators in [3, 4] used the genuine maps imported from Google maps, and vehicle portability examples are obliged or limited to a genuine street topology from maps, yet the source hub and goal hubs are arbitrary produced. In useful or genuine circumstances, the city's traffic stream will be adjusted as for time, i.e., during office hours, during odd timings and exceptional occasions and so on and it isn't totally produced by arbitrary. A genuine traffic generator has been utilized in [5], yet it doesn't give the traffic lights, street greatest speed and different subtleties. Creators in [6] incorporate VISSIM with NS2 and accomplish an understanding under genuine traffic conditions; however VISSIM isn't open source, so this technique isn't broadly utilized.

3 Methodology

3.1 Agency of the Board of RSU

a. *Vehicle Transmitting Agent*: This is portable specialist. Vehicle Transmitting Agent consistently track for accessibility of vehicle, on the off chance that vehicle goes into Road Side Unit go it registers that vehicle and stores data into Road side unit Black Board.
b. *Road Side Unit Manager Agent*: Road Side Unit Manager Agent is static operator put in Road side unit. It structures Road side unit Black Board and synchronizes activity among Road side unit and vehicle. The significant capacity of Road side unit Cluster Formation Agent is to select the Group head dependent on vehicle speed and neighboring vehicles.
c. *Road side unit Cluster Formation Agent*: It collaborates with Road side unit black board and registers the normal speed of vehicles in the range at that point chooses the size of bunch at that point stores and updates the Road side unit Black Board
d. *Road side unit Black Board*: Road side unit black board is the static operator which stores data and also gives data identified with all vehicles in the range. It is routinely refreshed by different operators of Road side unit. It stores data like label, position also, data of vehicles in the range, for example, vehicle label, position, speed.

  - When new vehicle goes into group Cluster Formation Agent includes that vehicle in Group member rundown and updates the Road side unit Cluster Formation Agent
  - When Group member leaves correspondence scope of group Cluster Formation Agent erases that Group member from Group member rundown.
  - If there should arise an occurrence of disappointment of group head, Cluster Formation Agent of group head vehicle chooses the other vehicle in bunch which fulfills group head condition. At that point data of rancher Road Side Unit Manager Agent is moved to new group member.

3.2 Security Attacks in Vanet

1. *Wormhole Assault*

Two gatecrasher hub is make a way misfortune in a system is called worm opening assault. Means two interloper hubs that are more separation and they are joined by a gatecrasher giving an impersonation that they are neighbors.

2. *Open Wormhole Assault*

In this assault vindictive hub keep inspect the remote medium to process the finding Route Request bundles, within the sight of pernicious hub in the net- work other hub on the system guess that malevolent hub are available on way and they are their immediate neighbors.

3. *Performance Parameters*

Quality Parameters are calculated using certain of the quality methods defined below

- Clustering creation time: The time given by Road side unit to shape the cluster.
- Time for selecting the cluster head: it is time for Road side unit to pick the group head based on certain criteria.
- Lifetime of the Group: it took the vehicle time to remain as the head of the cluster.

The proposed model is evaluated across performance evaluation such as vehicle movement, number of nodes, cluster size, node time, group head selection time, simulation time of the cluster, contact distance (Fig. 2).

Throughput is given as number of data transmitted from source to destination in a particular simulation (m/sec) (Fig. 3).

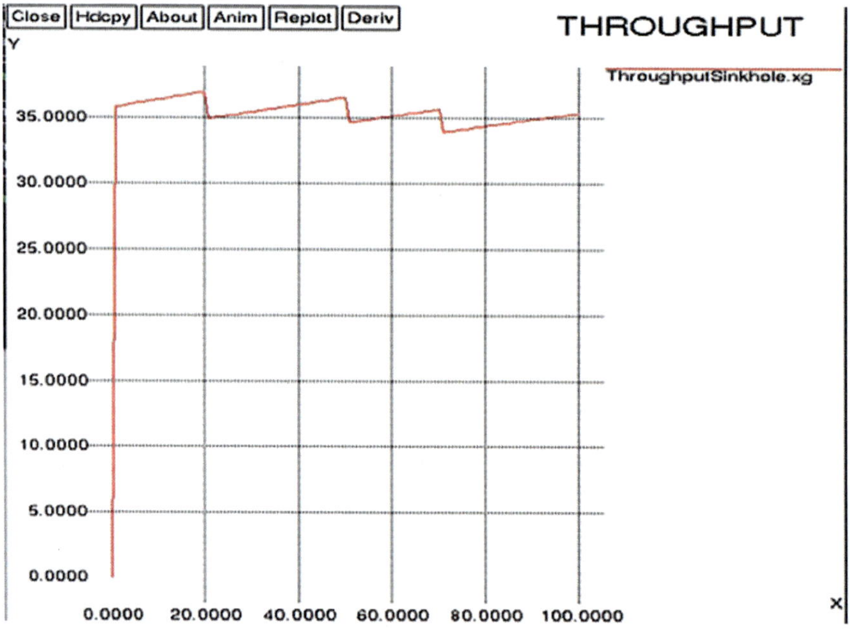

Fig. 2 Throughput

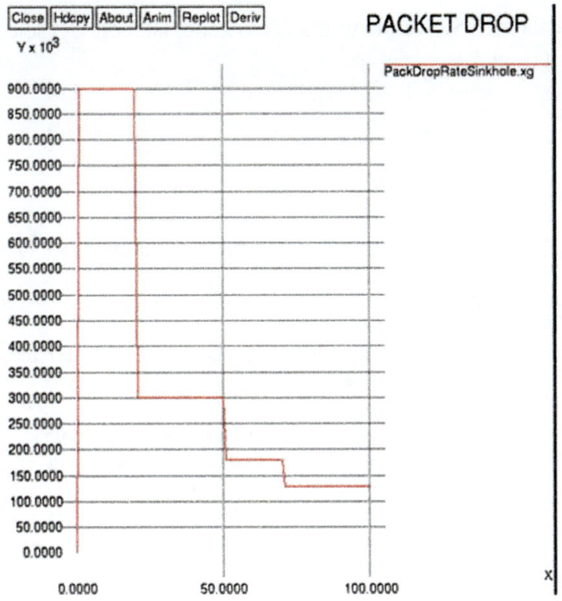

Fig. 3 Packet drop

The Packet drop of the expect method is low bit increased than the existing approach.

Packet drop (kbps) = (Receive bits/(stop simulation time − start simulation) $*$ 1/60.

4 Conclusion

We introduced robust clustering techniques based on multi-agent in vehicular ad hoc network. In this system gives methods from Road side Unit and agents from vehicles. The size of the cluster varies depending on the relative mobility of the Road side unit agent, which is calculated by the average movement of all nodes in range. The Road side unit agent selects and correct clustering head based on the weight method after clustering has been formed, vehicle with the highest weight factor chosen as the group head. During creation of cluster and cluster head choice, the exchange of control messages between vehicles introduces the routing overhead into the network. Because we are using Road side unit to pick the group and group neck, which helps to lower the network overhead routing. Selecting different cluster sizes based on low vehicle speed would help build long-living clusters. The presented approach performs simulation in NS2 well in terms of cluster existence, CH selection time and cluster formation time.

References

1. Kakkasageri MS, Manvi SS (2014) Agent based multicast routing protocol. Int J Future Comput Commun 3(1):188–192
2. Bhoi SK, Khilar PM (2014) Vehicular communication, a survey. Inst Eng Technol IET Netw 3:204–217. ISSN 2047-4954, Oct 2014
3. Ramakrishnan B, Rajesh RS, Shaji RS (2011) A cluster based VANET for simple highway communication. Int J Adv Netw Appl 2(04):755–761
4. Jahanbakhsh SK, Hajhosseini M (2008) Improving performance of CBRP using cross-layer design. Int J Comput Res Repository CoRR. http://www.abs.com/0802.0543
5. Luo Y, Zhang W, Hu Y (2010) A new Cluster based routing protocol for VANET. In: Second international conference on networks security, wireless communications and trusted computing, Wuhan, Hubei, China, pp 176–180, Apr 2010. https://doi.org/10.1109/nswctc.2010.48
6. Poonia RC, Bhargava D, Suresh Kumar B (2015) Cluster based dynamic routing as a development of AODV in VANETs. In: International conference on signal processing and communication engineering systems (SPACES), Guntur, pp 397–401, Jan 2015. https://doi.org/10.1109/spaces.2015.7058293

Anonymous Block Chain, Electronic Voting Using Linkable Ring Signatures

D. Saveetha and Musara Maronge

Abstract Electronic voting is the use of electronic means to cast and count votes. The system must be able to offer fairness, trust, transparency, privacy, integrity, and availability of votes. The application of block chain in electronic voting seems to match the above qualities of an electronic voting system. However, recent researches have shown that block chain technology lacks anonymity. Transactions are pseudonymous and not anonymous. For this cause, a proposed method to use linkable ring signatures in block chain electronic voting system provides anonymity. With this system, a voter can verify votes with the highest level of anonymity from other voters. Voters can also tally the votes without the use of centralized third party. The proposed system will be implemented using Ethereum protocol. It uses elliptic curve cryptography to provide authentication and non-repudiation, immutability of vote records through hashing algorithms, transparency, and accessibility due to its decentralized nature as well as voter anonymity.

Keywords Block chain · Anonymity · Electronic voting · Immutability · Linkable ring signatures

1 Introduction

The rapid growth of digital environment and wide use of the Internet seems to be replacing paper-based systems within organizations. Electronic voting systems have not been spared in this digital wave, aiming to minimize issues to do with redundancies and inconsistencies caused by heavy dependencies on paper-based systems

D. Saveetha
Information Technology, SRM Institute of Science and Technology, Kattankulathur, Tamil Nadu, India
e-mail: saveethd@srminst.edu.in

M. Maronge (✉)
Information Security and Cyber Forensics, SRM Institute of Science and Technology, Kattankulathur, Tamil Nadu, India
e-mail: marongemusara@gmail.com

D. J. Hemanth et al. (eds.), *Artificial Intelligence Techniques for Advanced Computing Applications*, Lecture Notes in Networks and Systems 130,
https://doi.org/10.1007/978-981-15-5329-5_5

used in traditional voting. Its implementation has also been hindered by security loopholes resulting in vote fraud and manipulation and lack of transparency in the voting process and in handling and announcing the votes by third parties involved in the election system [1].

The people are no longer feeling more interested in casting their vote because of one reason or the other. First among the reasons is that the digital world seems to be transforming everything and sticking to traditional voting mechanisms of going to a physical voting station may not bring comfort and easy for voters. The major reason is also that the voters have lacked trust on those handling elections since the voting system cannot be verified and traced. This lack of transparency has resulted in people who are eligible to vote to think that their will through their vote will be a waste and claim it will be legitimizing process of elections [2, 3].

Researches are being done that are trying to improve the whole voting process in terms of improving transparency, legitimacy of the vote outcome, vote verification, decentralization of the voting system, tamper-proof election results, and a coerced voting free environment. [4].

Block chain technology with more researches can provide all the critical requirements for a trustable electronic voting system. This is because by its nature it provides immutability of votes, vote verification, decentralization of the voting process by using cryptographic process to provide security, transparency, and flexibility [5]. Block chain simply is a decentralized, distributed set of digital records in network nodes. The records are stored as a series of blocks where each block is hashed, and except the first block, subsequent blocks are connected by the hash value of the previous block making it difficult to change data stored in a particular block as a recalculation of the older block hashes is needed and takes more computing power [6].

2 Literature Review

A lot of variations on electronic voting systems have been proposed, and some have been implemented. Most of them are still facing challenges in meeting the requirement for a free, fair, and credible electoral process during parliamentarian elections.

2.1 The Estonia E-Voting System (i-Voting)

Estonia became the first country to use Internet voting called i-Voting for nationwide or parliamentary elections since 2005. The system allowed its voting citizens to cast their unique vote online from any place. It is believed that the system saved a lot of time because of its simplicity, security, and elegance. Voters would log on to the Web-based system using the Internet during the election period using their national identity card or mobile identity to uniquely identify each voter. A voter would insert his or her national identity card into a card reader and then authenticate by entering a personal identity number. Voters would also use the mobile identity for authentication if their computer did not have the card reader. If successfully authenticated, the voter will be allowed to vote; else, he or she will be rejected to cast the ballot. The system also allowed voters to change their votes until four days before the day of elections. The voter's identifying information will be removed off the ballot, to ensure voter anonymity before reaching the Commission for National Electoral [7].

This Estonian voting model used three servers Vote Forwarding Server (VFS), Vote Storage Server (VSS), and Vote Counting Server (VCS). When a voter cast his or her ballot, the vote passes through VFS and VSS that are publicly accessible. The VSS stores the encrypted votes until the elections are done. All the votes in the VSS will be cleared from the identifying information and then moved to the VCS using a digital versatile device (DVD) for decryption and vote counting and later providing results [8]. However, a lot of security risks have been unearthed due to the centralization of election data. It may allow third parties to manipulate the votes and denial of services attacks on the servers leaving the whole system inaccessible affecting availability. The system has not provided mechanism for vote verification by voters, poor procedure controls and lack of transparency may also a breeding point for election outcome that is disputed [8].

2.2 *New South Wales i-Vote System*

This system is an improvement from the Estonian i-Voting system by allowing a voter to choose a personal identity number (PIN) of 6 digits. The voter would then log into the system using the identity number (ID) and PIN. Upon successful authentication, each voter receives a number of 12 digits for voting. After voting, if the voter wants to perform verification of the ballot, the voter has to submit the ID, PIN, and receipt number [8].

2.3 *West Virginia Voatz App*

West Virginia is planning to use the Voatz block chain mobile application for 2020 national elections. It was developed to allow its citizens who are staying overseas to vote. A pilot program was done during the 2018 midterms using this private block chain technology mobile application and seemed to have gained favorable response. However, criticisms have emerged from cryptographers, politician, and other researchers about the technical feasibility citing that protecting the device against hacking is very difficult. Anonymity in this case also is not guaranteed [9].

2.4 Helios Voting

Helios Voting is a Web-based electronic voting system and is a free, open source system used in elections. It was developed by Adida and tested during the University of Louvain elections. Helios implements the homomorphic additive properties and properties of distributed decryption used by ElGamal together with Sako-Killian's protocol for mix nets. It uses the protocol called Chaum-Pedersen for proof of decryption [10]. For a voter to vote, he or she has to fill out an online ballot and then click a button that encrypts the vote afterward, hiding the contents of a vote. After clicking the button, a tracking number will be sent to a voter to identify a vote and finally submit the vote to the server. If a voter wants to check whether his vote has been counted, he or she has to go to the election's ballot-tracking Web site, where voters' names are matched with specific tracking numbers. Anyone can also access the election data and verify using open auditing facility [11].

However Panizo et al. [12] recommended that the use of Helios Voting is not advisable for a public elections that have to be legally binding, considering lack of end-to-end verifiability and coercion resistance. Election authorities can illegally perform ballot stuffing, and this affects the integrity of the vote. The Adida developers said that they are not trying to solve the coercion problem, and for those using the system to ensure privacy, more trustees have to be recruited and this has attracted a lot attacks on the ballot privacy of the Helios Voting.

3 Aim and Objectives

3.1 Problem Definition

The above literature has exposed that each of the electronic voting system in place has some challenges ranging from lack to transparency, vote manipulation, and centralization of data. Votes are also very difficult to verify in some of the systems in place. Block chain technology like Voatz, a private block chain of West Virginia, whether private or public lacks anonymity without compromising transparency of the general voting process. Transactions for Ethereum block chain are pseudo anonymous every, node in the block chain network can see the public key to which a vote may have been generated. It is possible that while real identity of the voter is secure, peers in the network can still be able to see all the transactions performed by a particular public address [13]. Authorities that register users for elections might possess records that link public keys to voters. This linking affects anonymity of voters because the authorities will try can match voters and votes when results are decrypted [6, 14–16].

3.2 Requirement Analysis

Every election system must be capable of proving the following features [6]:

Availability

- An electronic voting system must be readily available for use during and after election period voters should be able to vote using diverse of devices.

Authentication

- A voter must be authenticated to allow only eligible voters to vote, and a voter should vote only once.

Integrity

- An electronic system for voting should provide vote integrity; no vote manipulation must be allowed.

Anonymity

- There has to be a complete disassociation of the vote and the voter.

Correctness

- The tallying and publication of results must be done without errors.

Robustness

- The voting system must withstand some attacks.

Transparency

- The system should provide a mechanism for voters to verify that his or her vote was counted as well as verification that the published results are correct. The system must be subjected to some audits to meet requirements.

Coercion Freeness

- The electronic voting system must have security mechanisms that prevent voters from being forced or threatened to vote for a particular candidate.

3.3 Proposed System

This project seeks for designing and implementing a secure electronic voting system based on block chain technology and cryptographic linkable ring signatures to provide voter confidentiality, voter anonymity through asymmetric key cryptography and ring signatures, vote tamper proof through hashing algorithms linking each block, authentication through elliptic digital signatures algorithms, transparency and availability of voting system and results due to decentralization of the system, and reduction in the involvement of electoral commissions in every task from voter registration to vote counting. Thus, the use of block chain technology with linkable ring signatures provides anonymity, privacy, integrity, authentication, availability, and verifiability which are necessary requirements for a proper voting system. The proposed system will be tested using Ethereum protocol [17, 18].

3.4 Proposed System Objectives

Following the design requirements of the electronic voting systems, the objectives of the system will be:

- To develop a block chain electronic voting system that will enhance privacy anonymity of voters
- To develop a block chain electronic voting system that produces unforgeable vote outcome
- To develop a block chain electronic voting system that ensures transparency, verifiability and that is auditable
- To develop a block chain electronic voting system that is widely accessible.

System architecture
See Fig. 1.

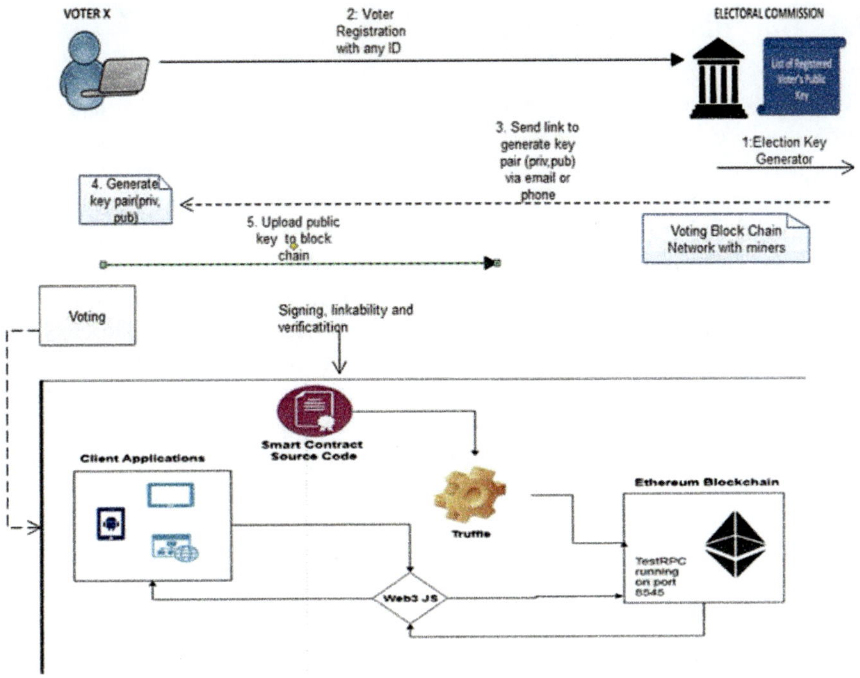

Fig. 1 Proposed system architecture diagram

3.5 Block Chain Overview

Block chain is a distributed database of digital information stored in a chain of blocks. Decentralization means that the nodes act as peer to peer network, and same data is stored across all nodes of a network. Each block has a body and a header, and the body stores transactions, through smart contracts that are written to the system. A smart contract, sometimes called the chain code is a set of code that encapsulates the business rules into logic and run when invoked. The header consists of data such as hash value if it is the genesis block and previous block hash value, timestamp, nonce values, and the level of difficulty of the block [19].

Block chain is grouped into three types, namely private, public, and consortium block chain [19] Public block chain is public. It does not have any restrictions on who should join and leave the network. Examples include Ethereum and Bitcoin. Private block chain network has restrictions on who should join the network and joining is regulated by a private company offering the block chain services. Example includes Horizen. Consortium is a semi-private block technology that can work across different organizations but has a set of user-controlled groups [20].

3.6 How Block Chain Works

Every block chain starts with a transaction which represents a task that has to be stored in a block. A transaction is done on a node which is just a computer on a block chain network. Each time a transaction takes place, it generates a cryptographic hash. For a transaction to be done, a user needs to have a wallet consisting of a public key called an address and a private key generated by asymmetric cryptographic key generators. The private key has to be kept private.

The transaction is performed using currency called ether for Ethereum protocol or coin for Bitcoin protocol. When a user performs a transaction, he or she has to sign the transaction message using his or her private key for authentication by creating a digital signature. A block of transaction is verified and approved by the nodes before being added to the block chain network by nodes called miners through the proof of work concept. A proof of work creates block in a way that there has to be a proof that a significant effort has to resolve a mathematical problem as introduced during a block creation in block chain [19].

A transaction block will contain a timestamp, a reference hash value to the previous block, the transactions, and the computational problem called nonce that had to be solved before the block adds to the block chain. Once the transaction is approved, it is placed on the block chain network by computers called miners and updated to every node and that is why it is called a distributed database. The records of the transactions will be immutable making forgery close to impossible [21].

Overview diagram of block chain

See Fig. 2.

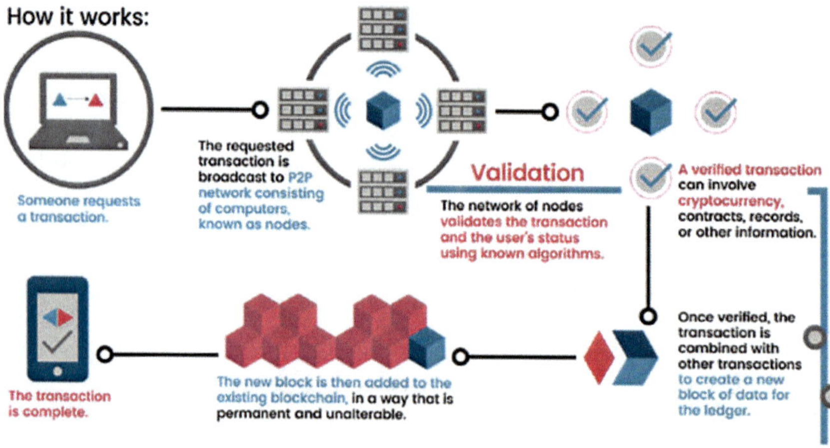

Fig. 2 Block chain concepts (*Source* the future of democracy: blockchain voting [p. 7]) [22]

3.7 *Linkable Ring Signatures*

This involves an algorithm that allows signing of messages by voters on behalf of a voting group without revealing the identity of the voter who signed the vote. A vote is given a tag that links signature and uniquely identifies votes from a particular voter. Identical signatures means a particular voter has voted more than once. It is because of this tag that third parties can identify that the signatures came from the same signer or voter without knowing who the signer is. This will prevent double voting and ensure anonymous during voting [18, 19].

3.8 *Ethereum*

It is a platform that allows programmers to develop and deploy applications that run on an Ethereum network called decentralized applications. The decentralized applications are sometimes called smart contracts. Smart contracts are pieces of programming code that perform a particular task. The codes are written using any language, but we use python, go and solidity. Once the code is written, it is compiled into byte code that will be read and executed by Ethereum Virtual Machine (EVM) and executed the nodes. These smart contracts are deployed onto the block chain network using Truffle and geth. The amount of computational time that a contract can use on the network is imposed by the gas limit [23, 24].

4 System Design

The system shall be developed using Ethereum platform as it allows developers to create customized smart contracts. The smart contracts will be written in solidity language for Ethereum block chain. Java scripts will be used to write some backend functionalities that will integrate with Hypertext Markup Language (html) pages. To develop some interfaces for the voting, html will be used. The html will interface with Ethereum block chain via Web3 framework. Other tools to be used are Metamask plugin and Truffle to run nodes on local machines. Python will also be used to develop some modules to encrypt and digitally sign the votes.

4.1 *Constraints*

- For the purpose of illustration and testing the system, the system shall assume that the registration and verification phase will have to be conducted manually or by some other means by the electoral body.

- The system will be tested using fewer nodes during implementation which does not represent the actual traffic during a national election. More sophisticated hardware will be required.
- Generation of private and public key pairs will be done using the elliptic curve algorithm.

4.2 Voting Protocols

4.2.1 Voting Registration

A voter has to register first with the election body to prove his or her identity and eligibility. The voter will use the parameters sent by election administrators to generate a key pair (**vsk*i*, vpk*i***) which are voter's private key and public key, respectively. The voter then uploads the public key to the smart contract and keeps the private key somewhere. Smart contract then puts the voter's public key onto the block chain network to finish voter registration. The election authorities must also produce a master key pair, namely master public key and master private keys (**mPub, mPriv**). The master public key will be used to encrypt the votes and private key for decrypting votes.

Algorithm

$$\text{KeyGen(parameters)} \rightarrow (\mathbf{vsk}\boldsymbol{i}, \mathbf{vpk}\boldsymbol{i}) \tag{1}$$

4.2.2 Message Creation

The voter will create a ballot **m** with timestamp, a nonce value, and public address.

4.2.3 Message Encryption

After creating a ballot, voter will encrypt it with the master public key.

Algorithm

$$\mathbf{M} = \text{Enc(m, mPub)} \tag{2}$$

4.2.4 Signing

The function has to generate a signature with a tag by using all voters' public keys Vpks, encrypted transaction ballot to be signed (M) and voter's private key **vsk*i***. The signing algorithm produces a ring signature on a particular message. To achieve this, the Koblitz curve secp256k1 that uses elliptic curve digital signature algorithm to produce signatures on transactions will be used. The curve does have structure in particular, which allows super-fast performance and increased efficiency when point addition and multiplication by a scalar is implemented in elliptic curve cryptography [25].

Algorithm

$$\sigma \leftarrow \text{Sign}(\text{Vpks}, \mathbf{vsk}\boldsymbol{i}, M) \tag{3}$$

4.2.5 Verification

The function takes encrypted message M, a signature σ, and public keys of voters Vpks, as parameters to test if the vote is valid. The block chain will then accept if the signature is legitimate; otherwise, it will be rejected.

Algorithm

$$\text{Accept/Reject} \leftarrow \text{Verify}(\sigma, \text{Vpks}, M) \tag{4}$$

4.2.6 Linkability

During voting, this function examines the signatures to determine if that another vote has the same tag. If the tag is the same, it means that the voter has already voted and the system must return linked and therefore the vote has to be rejected; else, the vote is posted on the block chain.

Algorithm

$$\text{Link}(\sigma 1, \sigma 2) \rightarrow \text{linked or unlinked} \tag{5}$$

5 Conclusion

The concept of block chain technology and its application is gaining a lot of interests for researchers. The anonymous block chain, electronic voting system is an improvement in the use of block chain technology in electronic voting systems that aim to provide free, fair, and credible elections. The proposed system aims to satisfy tamper-proof votes due to the immutability of block chain. Among other properties, the system will provide authenticity through the use of digital signatures to sign votes, verifiability, anonymity through the use of linkable ring signatures algorithm, availability as it is a distributed system. Outside other requirements, the system itself will provide a more democratic space and transparency in conducting elections, thus respecting the will and rights of the people.

References

1. Weaver N (2016) Secure the vote today. www.lawfareblog.com/secure-vote-today
2. Why people don't vote. https://www.raconteur.net/
3. Youth and elections: I refuse to vote, but I still have a right to complain. https://www.dailymaverick.co.za/opinionista/
4. Hjálmarsson FÞ (2018) Blockchain-based e-voting system. In: IEEE 11th international conference on cloud computing, p 983
5. Alam A, Rashid SMZU, Salam MA, Islam A (2018) Towards blockchain-based e-voting system. In: IEEE 2nd international conference on innovations in science, engineering and technology (ICISET), Chittagong, Bangladesh, pp 351–354, 27–28 Oct 2018
6. Adiputra CK, Hjort R, Sato H (2018) A proposal of blockchain-based electronic voting system. In: IEEE second world conference on smart trends in systems, security and sustainability (WorldS4), pp 22–27 (2018)
7. i-Voting. https://e-estonia.com/solutions/e-governance/i-voting/
8. Kumar D, Chandini DV, ReddyBD Bhattacharyya D, Kim T (2018) Secure electronic voting system using blockchain. Int J Adv Sci Technol 118:14
9. West Virginia will allow "blockchain voting" in the 2020 election. That's a risky idea. https://www.technologyreview.com/f/613358/
10. Yeregui, D, del Blanco M, Gascó M (2019) IEEE a protocolized, comparative study of helios voting and Scytl/iVote, pp 32–38
11. Secret ballots, verifiable votes. https://harvardmagazine.com/
12. Panizo L, Gascó M, David Y, del Blanco M, Hermida JA, Barrat J, Aláiz H (2019) E-voting system evaluation based on the Council of Europe recommendations: helios voting. In: IEEE transactions on emerging topics in computing, special issue on e-government development and applications (Siegda), p 10
13. What is blockchain technology? A step-by-step guide for beginners. https://blockgeeks.com/guides/what-is-blockchain-technology/
14. Block Chain: transactions, blocks and adding data. http://asecuritysite.com
15. Alam A, Rashid SMZU, Salam MA, Islam A (2018) Towards blockchain-based e-voting system. In: 2nd international conference on innovations in science, engineering and technology (ICISET), Chittagong, Bangladesh, 27–28 Oct 2018
16. Salman T, Zolanvari M, Erbad A, Jain R, Samaka M (2018) Security services using blockchains: a state of the art survey, p 19
17. Lu X, Au1 MH, Raptor ZZ (2018) A practical lattice-based (linkable) ring signature
18. Zhang R, Xue R, Liu L (2019) Security and privacy on blockchain. ACM Comput Surv 1(1)
19. Shahzad B, Crowcroft J (2019) Trustworthy electronic voting using adjusted blockchain technology 7: 24481
20. What are consortium blockchains, and what purpose do they serve? https://openledger.info/insights/consortium-blockchains/

21. How blockchain technology works. Guide for beginners. https://cointelegraph.com/bitcoin-for-beginners/
22. Osgood R (2016) The future of democracy: blockchain voting 7
23. How Ethereum works. https://www.coindesk.com/information/
24. What is Ethereum? The most comprehensive step-by-step-guide! https://blockgeeks.com/guides/ethereum/
25. A closer look at Ethereum signature. https://hackernoon.com

A Comprehensive Study on Ransomware Attacks in Online Pharmacy Community

V. Joseph Raymond and R. Jeberson Retna Raj

Abstract In recent trends, there are many cyber-attacks done across various organizations around the globe. The largest pharmaceutical companies found to be comprised of a ransomware Petya malicious virus. Online pharmacy communities are doing their best for delivering care and treatment with Internet support, but the attackers exploit vulnerabilities in operating systems to identify software flaws. The attackers access patient summary records and threaten pharmacy owners to pay a substantial sum to prevent leaking confidential information and henceforth locking the system by sending phishing e-mails across network and seed themselves. In this paper, we create an awareness of the potential threat and defend against these types of phishing attacks.

Keywords Phishing attacks · Eternal blue exploit · Entropy · Malware payload · Patch

1 Introduction

Ransomware will demand a random sum of money from the victim user preventing him to access device or data which is a type of malware [1]. They encrypt confidential files and lock screens as well as display a group of messages that are threatening. Locking screen is used to extort the victim implemented using various techniques. Hijacking activity detects top activity and restarts itself; to overwrite the previous activity using adds Flags. Home and Menu buttons can be made unusable using addView (), certain keys also can be disabled using OnKeyUp () and onKeyDown

V. Joseph Raymond (✉)
Sathyabama Institute of Science and Technology, Chennai, India
e-mail: josephrv@srmist.edu.in

R. J. Retna Raj
Faculty of Computer Science & Engineering, Sathyabama Institute of Science and Technology, Chennai, India
e-mail: jebersonretnaraj.it@sathyabama.ac.in

D. J. Hemanth et al. (eds.), *Artificial Intelligence Techniques for Advanced Computing Applications*, Lecture Notes in Networks and Systems 130,
https://doi.org/10.1007/978-981-15-5329-5_6

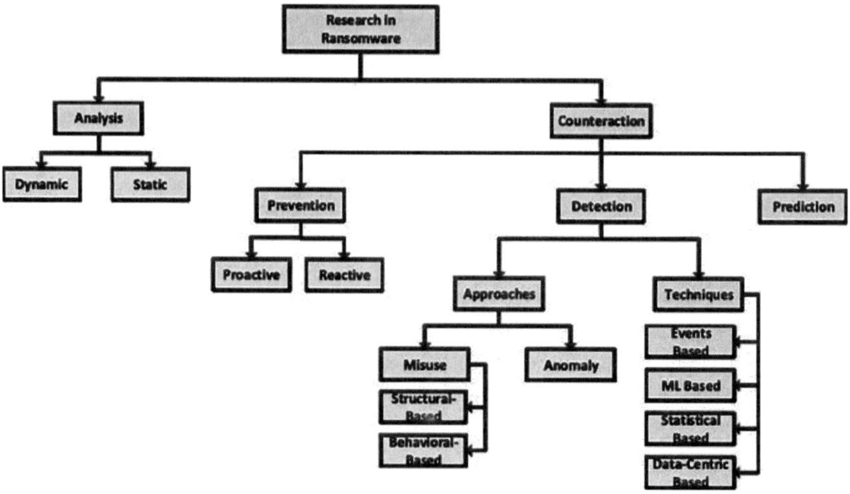

Fig. 1 Research on ransomware

() method and returns true. Standard cryptosystems or even customized can be used [1].

Contribution. The main contribution aims in improving novelty by giving a comprehensive analysis of important security of Ransomware giving us an overall analysis of how cyber-attack impact society. Mostly these attacks are happening in developed countries. Much more efficient work can produce better and accurate results that can be used for future research. Based on the survey paper and results further research will be carried on an Android-based platform. The figure is given below how research can be carried under Ransomware (Fig. 1).

The encrypted file which is entropy is more than that of non-encrypted files, the frequency of bytes shown in the graph (Fig. 2).

2 Phishing Attacks

The hackers use phishing by making them credible; create personalized e-mail and intimate communications from source to target. These e-mails look convincing for the victim as they include information like personal details that hackers try to obtain from fingerprinting attacks. We have analyzed the online community pharmacy site and found for the local domain using open source Vega subgraph tool in Debian X.8 64 bit Linux (Fig. 3).

The subgraph Vega detected contents on a server that includes java script where the code is fetched from legitimate user load the data into DOM and protection offered by the origin policy will not be considered. The server is compromised shown in Fig. 4 hence forth script program trusted by the Web administrator. Many employees

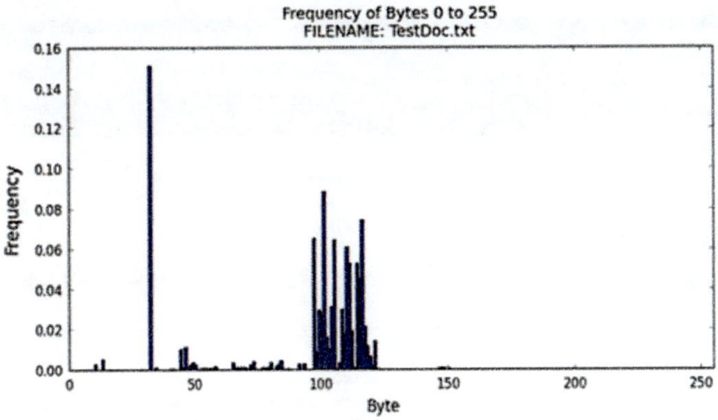

Fig. 2 Entropy of encrypted file

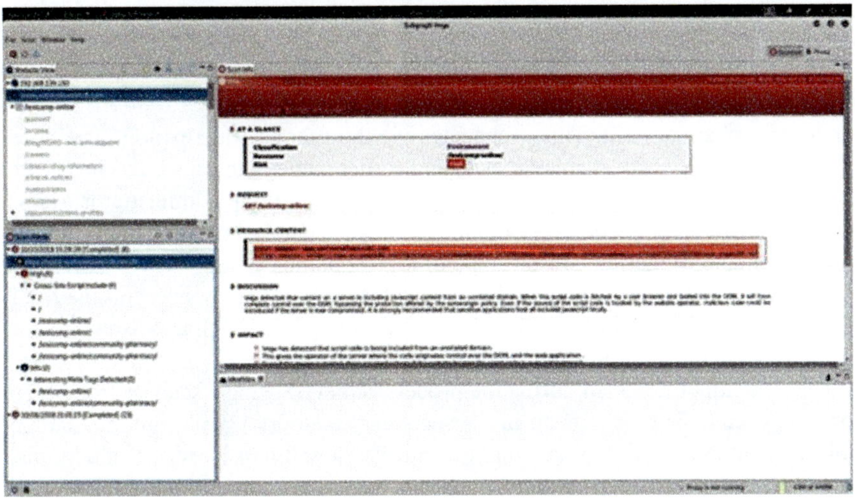

Fig. 3 Subgraph Vega Lexi-comp online—community Pharmacy

in the organization affected by phishing attacks where the percentage is 55% and many fall as a victim.

Local domain: www.wolterskluwercdi.com.

3 Variants of Petya

The e-mail attachments which have payload propagate in Petya. In recent days, a new Petya was used for a cyber-attack around the globe, targeting Ukraine's online

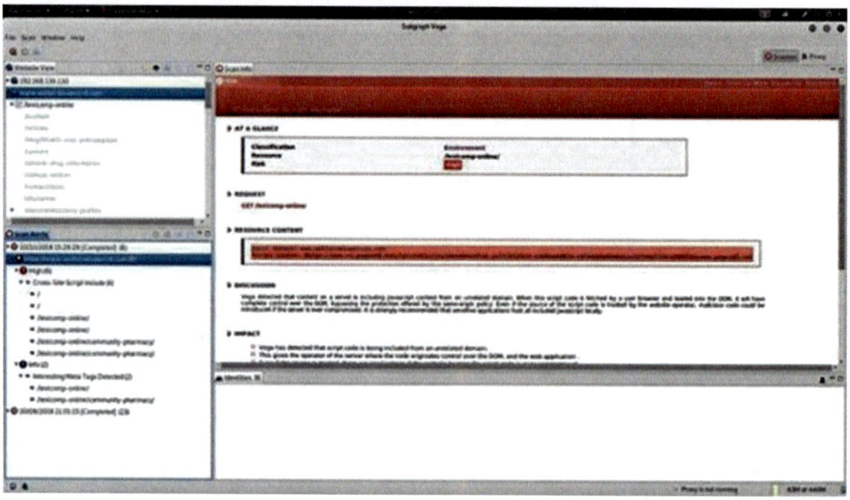

Fig. 4 Subgraph Vega high-risk cross side script

community pharmacy [2]. This propagates via the Eternal Blue exploit developed by the National Security Agency (NSA) and earlier named ransomware. The crypto wall is a secondary payload if the malware could not get the administrator access level. Petya infects the boot record of an operating system, overwriting the Windows boot loader, and triggering a restart type of logical bomb.

On the next startup of the computer, the payload comes to play, encrypts the MFT of the network file system, and then displays ransom pay through a message via the technology bitcoin as shown in Fig. 5.

The text output expected during the process achieved using a scanner from Windows suggesting there is a repair in the secondary memory [3]. The alternate malicious content known as Mischa bundled with Petya will request for grant administrative privileges considering the original payload. This conventional ransomware payload is an executable file pretend to as a PDF file attached via e-mail and found by Malware Initial Finding Report (MFIR).

The "NotPetya" variant released in the 2017 attack uses EternalBlue, taking advantage over a vulnerability in protocol SMB comes from the server of Windows. This malware spread across various computers may be called worm as do harvesting of passwords. The encryption routine is modified and made tougher for the victim to analyze the payload as shown in Fig. 6.

Cyberwarfare influenced by state intelligence helps in improving performance. There is an expectancy that Russian computers also affected.

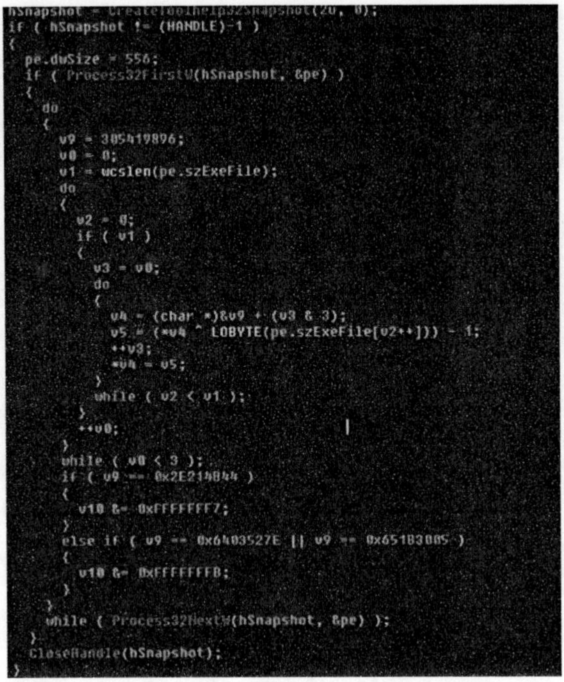

Fig. 5 Master file table encryption

4 Petya Microsoft Patch

This hole was patched in recent days named as MS17-010 for serving the ecosystem an important vulnerability patched in the series CVE-2017-0144 and CVE-2017-0145.

5 Methodology

The ransomware drops a credential dumping tool that comes in two different variants of the bits from the Operating System [4]. The credentials are stolen from multiple machines having admin privileges. After this, the local network gets scanned by getting connections on the port tcp/53 as shown Fig. 7.

The malware can be executed remotely using either PSEXEC or WMIC tools. The ransomware tries to drop the legitimate psexec.exe using the CredEnumerateW function to get all other user credentials [5]. The name starts with "TERMSRV/" and generic typeset as 1 uses that credential to pass through the network as shown in Fig. 8.

The ransomware encrypts all files extensions in all folders in fixed drives except for Windows C Drive shown in Fig. 9.

```
wsprintfW(&Name, L"\\\\%s\\admin$", a1);
NetResource.dwScope = 0;
memset(&NetResource.dwType, 0, 0x1Cu);
NetResource.lpRemoteName = &Name;
NetResource.dwType = 1;
get_current_module_name_convert_tows(&v23);
wsprintfW(&FileName, L"\\\\%ws\\admin$\\%ws", a1, &v23);
while ( 1 )
{
  pszPath = 0;
  v11 = v4;
  v18 = WNetAddConnection2W(&NetResource, lpPassword, lpUserName, 0);
  wsprintfW(&pszPath, L"\\\\%ws\\admin$\\%ws", a1, &v23);
  v5 = PathFindExtensionW(&pszPath);
  if ( v5 )
  {
    *v5 = 0;
    if ( PathFileExistsW(&pszPath) )
    {
      v13 = 1;
      goto exit;
    }
    dwErrCode = GetLastError();
  }
  v6 = 0;
  if ( write_file(dword_1001F11C, &FileName, (LPCVOID)dword_1001F0FC, 1u) )
    break;
  v7 = GetLastError();
  dwErrCode = v7;
  if ( v7 == 80 || v7 == 53 || v7 == 67 || v18 != 1219 )
    goto exit;
  if ( v11 )
    goto LABEL_61;
  v4 = 1;
  WNetCancelConnection2W(&Name, 0, 1);
}
```

Fig. 6 Ransomware code for accessing different machines

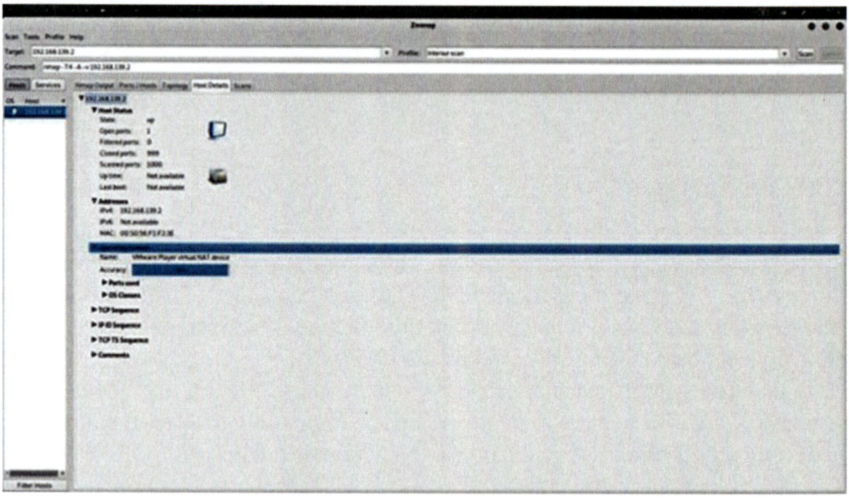

Fig. 7 ZenMap tool showing OpenPort

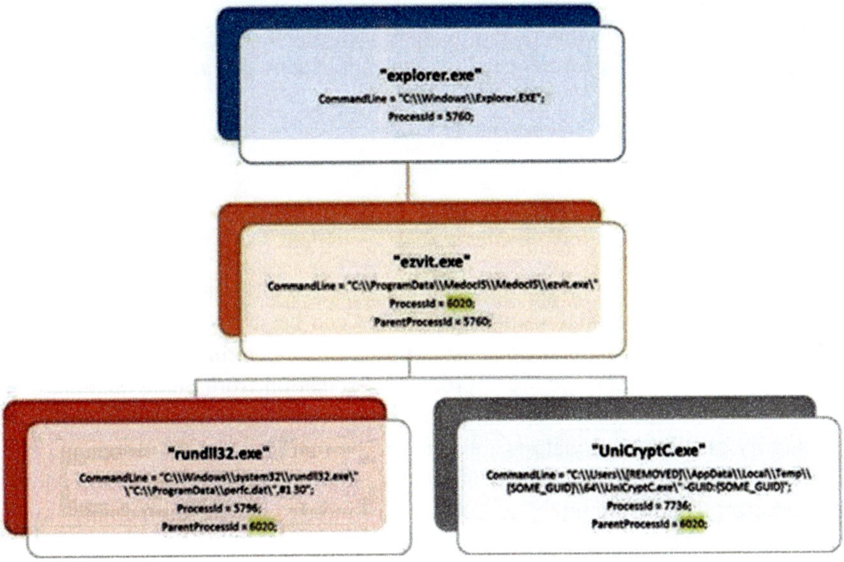

Fig. 8 Working principle of .exe system calls

This ransomware attempts to encrypt all files with the following file name extensions in all folders in all fixed drives, except for *C:\Windows*:

.3ds	.7z	.accdb	.ai
.asp	.aspx	.avhd	.back
.bak	.c	.cfg	.conf
.cpp	.cs	.ctl	.dbf
.disk	.djvu	.doc	.docx
.dwg	.eml	.fdb	.gz
.h	.hdd	.kdbx	.mail
.mdb	.msg	.nrg	.ora
.ost	.ova	.ovf	.pdf
.php	.pmf	.ppt	.pptx
.pst	.pvi	.py	.pyc
.rar	.rtf	.sln	.sql
.tar	.vbox	.vbs	.vcb
.vdi	.vfd	.vmc	.vmdk
.vmsd	.vmx	.vsdx	.vsv
.work	.xls	.xlsx	.xvd
.zip			

Fig. 9 Infected file extensions

6 Result and Discussion

Our goal is making a comprehensive study on recent ransomware attacks done in various platform mostly vendor-supported operating systems. This makes research directions today's performing static and dynamic analysis on malicious as stated in the introductory chapter. This understanding gives us a way to understand the structural and statistical analysis for further carrying the path of activities and contribute to the cybersecurity system.

7 Conclusion and Future Works

The cleverness of ransomware can be handled by three different approaches. First, the delivery malware through e-mail attachment can be stopped by using a better operating system doing regular updates and patching. Operating system protection plays a vital role [6]. The second approach is protection from being written over or deleted by malware should be immediately noticed. The third and important is the integrity of file system data is subjected to be more toward development. The majority of the online community pharmacy server's including mobile apps don't have these protections inbuilt. The ransomware industries will flourish and infect many patients then demand ransom payments.

References

1. Chen J, Wang C, Zhao Z, Chen K, Du R, Ahn G-J (2018) Uncovering the face of android ransomware: characterization and real-time detection. IEEE Trans Inform Forens Secur 13(5)
2. Types of ransom ware accessible. http://www.vinransomware.com/types
3. Ransomware with social engineering accessible. www.continuum.cisco.com
4. Orman H, Streak P (2016) Evil offspring—ransomware and crypto technology. Published by the IEEE Computer Society, Sept/Oct 2016
5. Types of Ransomware. http://mobile.esucurityplanet.com/malware/types
6. Detecting Ransomware activity at www.netfort.com/blog/methods

A Comprehensive Survey on RF Module at 433 MHz for Different Applications

P. Malarvezhi, R. Dayana, and T. V. S. Subhash Chandra

Abstract There have been a large number of experiments being done on RF (433 MHz). This paper focuses on the survey of various RF applications, uses and comparison of those applications. This paper has also a glimpse of various components in RF. It also talks about the significance of various modules of RF. The RF that is discussed in this paper works on 433 MHz frequency, and various applications using the same are mentioned. RF technology is widely used in many situations to ease the difficulty. The RF has a transmitter section and a receiver section. The data is analysed differently based on the application.

Keywords RF(433 MHz) · Transmitter · Receiver · Applications · Comparisons

1 Introduction

The wireless technology has been around in the use over a century now. The groundwork for the same has been laid by Alexander Popor and Sir Oliver Lodge [1]. Since so many advancements being done, the wireless technology is becoming more and more prominent to encounter the daily needs of the consumers as well as the commercial purposes. A large section of the world is seeing wireless communication as more handy, useful and easy technology as well. One of the eminent wireless technologies is being done in RF(433 MHz). RF stands for radio frequencies and as the name suggests RF refers to the rate of oscillation of electromagnetic radio-frequency waves ranging from 3 kHz to 300 GHz. Since RF primarily focuses on short-range access, many experiments are done, handy projects are being materialized, and even more are encouraged and funded. This paper focuses on the survey on various explicit applications of RF (433 MHz). There are four standard modules of RF [2].

- Transmitter module
- Receiver module

P. Malarvezhi (✉) · R. Dayana · T. V. S. Subhash Chandra
SRM Institute of Science and Technology, Kattankulathur, Tamil Nadu, India
e-mail: malarvip@srmist.edu.in

D. J. Hemanth et al. (eds.), *Artificial Intelligence Techniques for Advanced Computing Applications*, Lecture Notes in Networks and Systems 130,
https://doi.org/10.1007/978-981-15-5329-5_7

- Transceiver module
- System on a chip module.

As the title suggests, this RF system transmits on 433 MHz frequency. The basic system consists of a transmitter module and receiver module. Both of the modules have antennas connected to them which help to transmit the data. RF can be paired with an encoder/decoder that helps to encode and decode the data. Detailed working will be discussed in the later part of this paper. One should consider output noise power, phase noise and unwanted signals while using RF.

The various sections listed in this paper are as follows

- RF modules
- Role of encoders and decoders
- Applications of RF
- Comparison of applications
- Pros and cons
- Future scope of RF
- Conclusion.

2 RF Modules

2.1 Transmitter

The transmitter is an RF module that is noticeably small in size as shown in Fig. 1. This transmitter works on 433 MHz frequency. The basic transmitter can transmit data for shorter distances, not more than 100 m. Factors like exposed environment, antenna design play significant role in effective distance that the transmitter is capable

Fig. 1 Transmitter

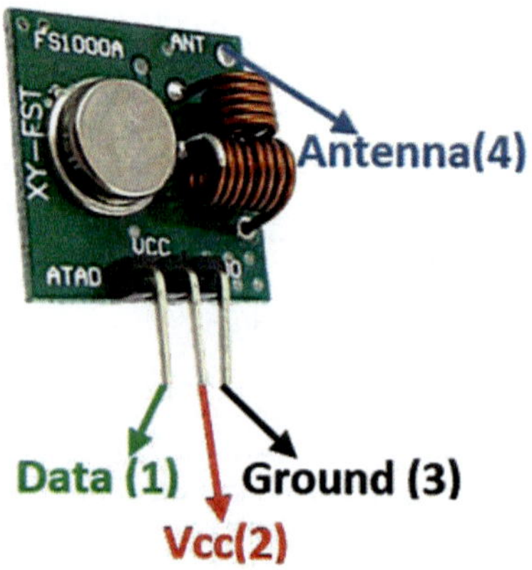

Table 1 Pin description of TF transmitter

S. no.	Pin	Description
1	Data	The data that should be transferred is given to this pin
2	Vcc	Power supply
3	Ground	Grounded in circuit
4	Antenna	Data is sent wirelessly through antenna

of transmission. The transmitter can be paired with an encoder which is discussed in detail in the later part of the paper.

From Table 1, we can see that there are 4 pins in the transmitter. Each pin has its distinct functionality. Data we provide is given to the data pin, Vcc is the power supply, ground pin is grounded in the circuit board, and antenna is the 4th pin through which data is transferred wirelessly [3–5].

2.2 *Receiver*

The receiver is similar in size to transmitter as shown in Fig. 2. The receiver retrieves the data transmitted through antenna. Receiver can be paired with a decoder also. Receiver analyses the data received.

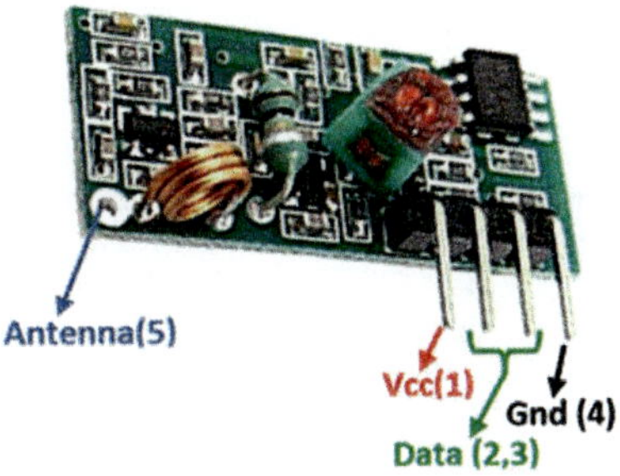

Fig. 2 Receiver

Table 2 Pin description

S. no.	Pin	Description
1	Vcc	Power supply
2	Ground	Grounded in circuit
3	Data	Received data is obtained in this pin
4	Data	This is also data pin
5	Antenna	Data is received through antenna

Table 2 explains the same as that of the previous table except for the antenna here receives the data, and the second data pin is also used for the data reception but both cannot be used at a time [4–6].

3 Role of Encoders and Decoders in Rf

The RF modules can function without encoders and decoders. Just powering the RF modules itself will be able to establish a connection between transmitter and receiver. But this is restricted to one button each on transmitter and receiver sides.

Now, when encoders and decoders like HT12E and HT12D come into play, the scale of the inputs increases [7]. These HT12E and HT12D are 4-bit data encoders and decoders from which we can get 16 various combinations. Having just 1 input to 16 inputs encoder/decoder made this possible to scale up the RF projects. One another such significant pair is HT640/HT648 being used in the market currently.

Figure 3 describes the working of RF transmitter/receiver when paired with encoder/decoder. As shown, first the input data otherwise called parallel data is passed to the encoder. The data is processed and converted to serial data in the encoder and is forwarded to the transmitter. The transmitter transmits the data and is received by the receiver. This serial data received is converted into parallel data by the decoder in the receiver part. Encoder used- HT12Eand Decoder used- HT12D.

There is another factor that affects the transmission is the antenna. The structure of antenna is important, and many developments are being made for efficient transmission.

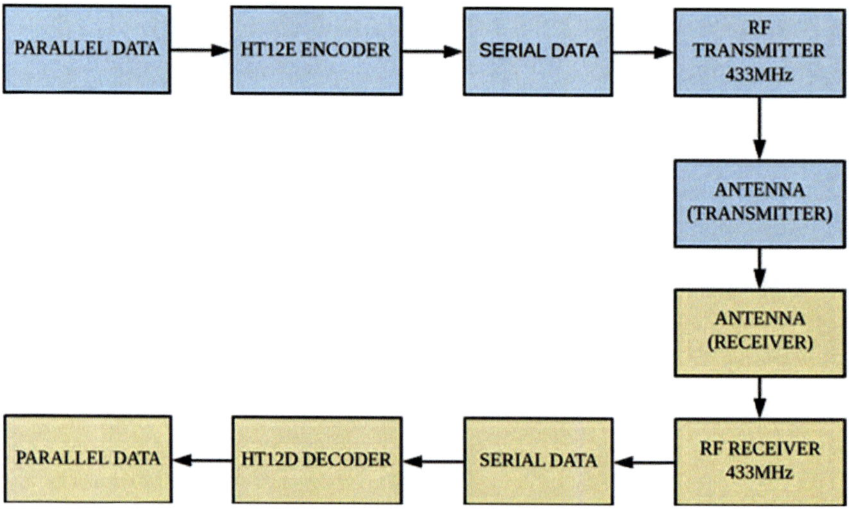

Fig. 3 Block diagram of working of RF with encoder and decoder

4 Applications

There are many applications of RF(433) out of which some are listed here which are used currently. Though it is less known, it is used in industrial, medical areas as well as apart from household applications. The following are some applications [5, 8–24]

- Portable indicator for underground mine
- Temperature, humidity and air pressure measurement system
- Underwater communication systems
- In vehicle, electronic infotainment applications using RF
- In vehicle, wireless sensor network using RF
- Home automation
- Wireless fire protecting system using RF
- Remote-controlled toys
- Telemetry
- Contactless smart cards using RF
- Automatic control of electrical load using RF
- Wireless communication for gas detection using RF
- Gesture-controlled robot.

The above mentioned are the some applications. There are plenty of projects done on RF, and many more experiments are being done.

These applications can be in any field and in any medium. Some of them are made to communicate on land and some underwater.

5 Comparision of Applications

There are not too many factors to compare among the applications as all of them use same Tx and Rx modules. The difference lies in the usage of these modules for different purposes. In this section, we compare some applications mentioned in the previous section based on different factors.

Table 3 [1, 13, 16, 25, 26] explains the comparison among some of the applications based on the sensors used in the respective applications. From the above table, we can see that sensors like temperature sensor, pressure sensor, accelerometer, gas sensor, humidity sensor, flex sensors, Arduino sensor are used. There are various applications that use the above-mentioned sensors, the above table illustrates various sensors being used using some applications.

Table 4 [12] describes the efficient range of the transmission of signals in different mediums. The above-mentioned range is for good transmission and reception even with unsuitable conditions for RF. The range can be increased by a good environment. But the water-to-water transmission remains the same.

We can transmit up to 100 m in very likely conditions.

Table 3 Comparison table based on sensors used

Application	Sensors used
Gesture controlled robot	Flex sensor, accelerometer
Portable indicator for underground mines	Pressure sensor, temperature/humidity sensor
Wireless communication for gas detection	MQ-2 gas sensor
Temperature, humidity, air pressure monitor system	Temperature sensor, air pressure sensor, humidity sensor
Wireless remote-controlled generator system	Arduino sensor

Table 4 Comparison based on medium of transmission

Transmission medium	Range (m)
Air to air	0–50
Air to water	0–30
Water to water	0–1.85

Table 5 Comparison of various technologies

Technology	Operating frequency	Range
Rf	433 MHz	Up to 100 m
Lora	433 MHz, 868 MHz, 915 MHz, 923 MHz	>10 km
NBIOT	700 MHz, 800 MHz, 900 MHz	<10 km
Zigbee	868 MHz, 915 MHz, 2.4 GHz	10–100 m
LoRWAN	865–867 MHz (India) 867–869 MHz (Europe) 920–925 MHz (Japan, Korea) 902–928 MHz (North America) 470–510 MHz (China)	>10 km

Table 5 [27–29] compares the RF with various latest technologies[30, 31]. Parameters like operating frequencies and ranges are compared. The above-mentioned frequencies are operational depending upon the geographical locations. Also, the ranges can be varied depending on the environment and surroundings.

6 Pros and Cons

There are many disadvantages as many advantages we have using RF(433 MHz). The RF is influenced by external factors. This can cause some disadvantages. Also, most of the population are moving towards wireless technology and RF comes into play which has many advantages too.

Pros:

- It is portable. Because of its portability, it can have various uses in day-to-day life.
- It is easy to install. Because of its noncomplex nature, it is easy to install wherever needed even by someone who is not professional.
- Suitable for indoors and outdoors. It is suitable for indoors as well as outdoors but for a limited range.
- More penetration levels. RF has more penetration levels as compared to IR. It can penetrate through obstructions easily without any fuss.
- Simple circuit. It has a simple circuit which makes it easy for many young aspirants to focus on the development of RF.
- Easy reinstallation. Because it is simple, it is easy to reinstall the circuit or it is easy to replace the defective part.

Cons:

- Constant power supply is needed. If there is no constant power supply, it cannot function properly.
- The quality of data transmission and reception can be greatly affected by the external factors as said earlier.
- The external noise and radiation can deeply affect the transmission. The area of transmission should be free of radiation primarily for effective use of RF.
- Though the range is good but not useful for long range. The range is limited to a certain distance.
- It is suitable primarily for small- and medium-scale projects. Since it has limitations, it is not suitable for large-scale projects.

7 Future Scope

Since the people prefer the wireless technology over the others which makes the life easier, RF has a great scope. One such application is home automation. Though the home automation is currently in use, there are many others that can be automated in household that has a vast use of RF. Even in outdoors, RF can greatly affect the traffic control. There are some projects being done on traffic control but most of them are not used in the daily life. They also use RF technology for efficient transmission of the data. One other area is in parking, where RF can be used. Most of the crowded areas have parking issues in which the advancements are in making and RF can come in handy for such projects. One more application is it can be used for in house GPS system. It is not easy in multi-storeyed buildings to track things. Projects are being done in this area which has a greater use in the future, in which RF can be used. The RF can be used in IOT projects too that has significant uses. Also, many more upcoming projects need RF for transmission.

8 Conclusion

RF is a wireless technology which the world is gradually moving towards. Advancements are being made in RF so that it is easily accessible for large masses. We can easily use RF for a range of 100 m max. There are many applications that are currently in use, and many more are being materialized. Though there are disadvantages, there are many advantages of RF. Because of its salient features, it is being recognized as one of the useful wireless technologies worldwide.

References

1. Onengiye G, Chukwunazo E (2016) Design and implementation of RF based wireless remote control generator system. Int J Eng Adv Technol (IJEAT) 5(4). Accessed 2 Nov 2016
2. Agarwal T (2019) RF wireless radio frequency working and applications. Edgefx Kits Official Blog. [Online]. Available: https://www.edgefxkits.com/blog/rf-wireless-radio-frequency-working-application/
3. 433 MHz RF transmitter module pinout, specifications, equivalent & datasheet. Components101.com. [Online]. Available: https://components101.com/433-mhz-rf-transmitter-module. Accessed: 09 Nov 2019
4. Ahmed F, Md S, Alim A, Shafiqul Islam M, Kawshik K, Islam S (2016) 433 MHz (wireless RF) communication between two Arduino UNO. Am J Eng Res (AJER) 5(10):358–362
5. How to interface with RF transmitter and receiver. Instructables, 2019. [Online]. Available: https://www.instructables.com/id/How-to-Interface-With-RF-Transmitter-and-Receiver/
6. 433 MHz RF receiver module pinout, specifications, equivalent & datasheet. Components101.com. [Online]. Available: https://components101.com/433-mhz-rf-receiver-module. Accessed: 09 Nov 2019
7. Wireless Transmitter and Receiver using ASK RF Module. *electroSome*, 2019. [Online]. Available: https://electrosome.com/wireless-transmitter-and-receiver-using-ask-rf-module/
8. Weiss MD, Smith J, Bach J (2009) RF coupling in a 433-MHz biotelemetry system for an artificial hip. IEEE Antennas Wireless Propag Lett 8:916–919
9. Rusia J, Naugarhiya A, Majumder S, Majumdar S, Acharya B, Verma S (2016) RF based wireless data transmission between two FPGAs. In: 2016 international conference on ICT in business industry & government (ICTBIG)
10. Balachander D, Rao T, Tiwari N (2013) In-vehicle RF propagation measurements for wireless sensor networks at 433/868/915/2400 MHz. In: 2013 international conference on communication and signal processing
11. Yonemoto N, Yamamoto K, Yamada K, Hirata T (2007) RF emission measurement of 433 MHZ RFID tags for EMI evaluation to onboard instruments of aircrafts. In: 2007 7th international symposium on electromagnetic compatibility and electromagnetic ecology
12. Maher S, Ali Z, Mahmoud H, Abdellatif S, Abdellatif M (2019) Performance of RF underwater communications operating at 433 MHz and 2.4 GHz. In: 2019 international conference on innovative trends in computer engineering (ITCE)
13. Indra S, Barik S, Pati U (2018) Design of portable indicator for underground mines using 433 MHz wireless communication. In: 2018 2nd International Conference on Electronics, Materials Engineering & Nano-Technology (IEMENTech)
14. RF module, En.wikipedia.org, 2019. [Online]. Available: https://en.wikipedia.org/wiki/RF_module#Module_physical_connection. Accessed: 09 Nov 2019
15. Sharma T, Saini G (2016) A survey on RF energy harvesting from mobile towers. Int J Electr Electron Eng 3(1)

16. Kiran D, Singh H, Saxeriya K (2019) Gesture control robot using Arduino. Int J Trend Sci Res Develop 3(3):1467–1469
17. Balachander D, Rao T (2013) In-vehicle RF propagation measurements for electronic infotainment applications at 433/868/915/2400 MHz. In: 2013 International conference on advances in computing, communications and informatics (ICACCI)
18. Kim S et al (2013) Evaluation of a 433 MHz band body sensor network for biomedical applications. Sensors 13(1):898–917
19. Wireless RF module. RF Transmitter and receiver. Latest applications. ElProCus—Electronic Projects for Engineering Students, 2019. [Online]
20. Alam M, Jamil I, Mahmud K, Islam N (2014) Design and implementation of a RF controlled robotic environmental survey assistant system. In: 16th international conference on computer and information technology
21. Verma G, Sharma V (2016) A survey on hardware design issues in RF energy harvesting for wireless sensor networks (WSN). In: 2016 5th international conference on wireless networks and embedded systems (WECON)
22. Narayan M, Kumar M, Kumar A, Raja K (2017) Radio frequency based automatic control of electrical loads. Int J Sci Res Comput Sci Eng Inform Technol 2(1)
23. Zagade A, Jamkhedkar V, Dhakane S, Patankar V, Kasture P, Gaike P (2018) A study on gesture control Ardiuno robot. Int J Sci Develop Res (IJSDR) 3(5)
24. Peiris V et al (2005) A 1v 433/868 MHz 25 kb/s-FSK 2 kb/s-OOK RF transceiver SoC in standard digital 0.18 μm CMOS. In: IEEE International Digest of Technical Papers. Solid-State Circuits Conference (ISSCC 2005)
25. Taha Z (2017) Wireless communication for gas detection using 433RF modules and Arduino processor. Int J Comput Appl 165(4):18–20
26. Setiyono B, Sumardi, Harisuryo R (2015) Measurement system of temperature, humidity and air pressure over 433 MHz radio frequency: an application on quadrotor. In: 2015 2nd international conference on information technology, computer, and electrical engineering (ICITACEE)
27. What is ZigBee technology. Architecture and its applications? ElProCus—Electronic Projects for Engineering Students, 2020. [Online]. Available: https://www.elprocus.com/what-is-zigbee-technology-architecture-and-its-applications/
28. LoRa (2020) En.wikipedia.org. [Online]. Available: https://en.wikipedia.org/wiki/LoRa
29. What is LoRa? (2020) SemtechLoRa Technology. Semtech. Semtech.com. [Online]. Available: https://www.semtech.com/lora/what-is-lora
30. Randy B, Hariharan M, Kumar R (2014) Secured wireless power transmission using radio frequency signal. Int J Inform Sci Tech 4(3):115–122
31. Kaddoum G (2016) Wireless chaos-based communication systems: a comprehensive survey. IEEE Access 4:2621–2648

A Memory-Efficient Tool for Bengali Parts of Speech Tagging

Shadikun Nahar Sakiba, Md. Mahatab Uddin Shuvo, Najia Hossain, Samir Kumar Das, Joyita Das Mela, and Md. Adnanul Islam

Abstract Detecting parts of speech of words in a sentence efficiently is an inseparable task in natural language processing (NLP). The goal of this study is to detect the parts of speech of words in a Bengali sentence with higher accuracy and memory optimization. In this study, we develop such a parts of speech tagging tool specifically for those people who want to learn or research on Bengali grammar. Besides, it has an impact on the quality of education as our tool aims to provide knowledge about Bengali grammar. Although there are different existing studies on Bengali parts of speech tagging such as hybrid model, global linear model, maximum entropy model, hidden Markov model, unsupervised model, Bangla parsing algorithm, rule-based classifier, decision tree algorithm, Baum–Welch algorithm, unsupervised word segmentation algorithm, etc., none of these studies consider memory optimization technique. Therefore, in this study, we propose and endeavor to implement our own rule-based parts of speech tagging mechanism with an intent to achieve higher accuracy by optimizing memory overhead.

Keywords NLP · Levenshtein distance

S. N. Sakiba (✉) · Md. M. U. Shuvo · N. Hossain · S. K. Das · J. D. Mela · Md. A. Islam
Department of CSE, UIU, Dhaka, Bangladesh
e-mail: ssakiba153082@bscse.uiu.ac.bd; sakiba256@gmail.com

Md. M. U. Shuvo
e-mail: mshuvo153093@bscse.uiu.ac.bd

N. Hossain
e-mail: najiahossain133@gmail.com

S. K. Das
e-mail: bipondas516@gmail.com

J. D. Mela
e-mail: jmela153038@bscse.uiu.ac.bd

Md. A. Islam
e-mail: adnanul@cse.uiu.ac.bd; islamadnan2265@gmail.com

D. J. Hemanth et al. (eds.), *Artificial Intelligence Techniques for Advanced Computing Applications*, Lecture Notes in Networks and Systems 130,
https://doi.org/10.1007/978-981-15-5329-5_8

1 Introduction

Bengali is one of the top ten most widely spoken languages in the world [1, 2]. However, Bengali lacks in some crucial areas of research in NLP. More specifically, there is no efficient tool available yet for Bengali parts of speech (PoS) tagging in spite of realizing its importance specifically for the large group of Bengali speakers worldwide [3, 4]. Besides, it will be immensely beneficial for those who want to learn and study on Bengali language.

Undeniably, there are many uses of NLP in our life such as sentimental analysis, translation, voice testing, spell checker, [5–7] etc. Nowadays, many E-Commerce sites use some software to get reviews and information, which encompasses one of the most important applications of NLP. PoS tagging is a very basic and amusing research topic in NLP, which also enhances the opportunities in other important fields of NLP such as translation, text categorization, etc. In this study, our objective is to create an open-source PoS tagging tool for Bengali texts by improving the performance over the existing ones along with memory optimization. Since we plan to make this tool absolutely free to access and use for common purpose, we expect a large group of people to be directly benefitted from it, especially the new learners and researchers. Since there are hardly any such tool available for Bengali, our objective is to make a memory-efficient Bengali PoS tagging tool attainable.

2 Motivation

In spite of Bengali being native language, even most of the native Bengali speakers have minor knowledge about Bengali grammar (specifically PoS). Besides, Bengali PoS tagging can be considered to be pre-requisite for translating Bengali sentences (rule-based translation), grammar checking, text categorization, etc. Apart from this, unfortunately, there exists no efficient tool for recognizing Bengali grammars including parts of speech of words in a Bengali sentence. These realistic aspects lead to our investigation and implementation of a novel rule-based PoS tagging approach. Besides, we take the memory overhead of our proposed PoS tagging approach into consideration with a view to optimizing it. Make the tool available to aid the people to gather knowledge about Bengali grammar so easily. Our target is not only to create such a tool for Bengali PoS tagging but also to avail it online for common use.

3 Related Work

There is a plethora of real-life applications of NLP such as translation, speech recognition, voice testing, etc. Although significant studies have been performed for English parts of speech tagging, same has not happened for Bengali language. Notable studies in this regard include several models and algorithms such as hybrid model [5, 8], global linear model [3, 9, 7], maximum entropy model [1], hidden Markov model [10], unsupervised model [11, 12], Bangla parsing algorithm [13, 6], Rule-based classifier [14, 2], decision tree algorithm [15, 16], Baum–Welch algorithm [17, 18], unsupervised word segmentation algorithm [19, 4], etc.

Mukherjee and Mandal [3] used global linear model to find the best tags for prefix or suffix and root and compare accuracy with other models. It acquired 91.2% accuracy.

Alam et al. [14] proposed to develop tokenizer, splitter, and classifier of a sentence from text input and normalize it using rule-based classifier and decision tree with 93% accuracy. Moreover, Alam et al. [17] checked whether an input sentence was grammatically correct or not using n-gram based analysis and statistical approach. Apart from this, Kabir et al. [20] created their own dictionary and a PoS tagger using datasets and deep belief network. Then, they compared the dictionary with PoS tagging using corpus to find the best tags with good accuracy.

None of the existing studies considered memory constraint while determining PoS of words in a sentence. Besides, existing studies ignored some basic grammatical rules related to Bengali PoS tagging. Thus, we propose our own rule-based PoS tagging approach addressing the memory constraint.

4 Proposed Mechanism

Initially, we analyze various existing related tools for understanding how they work. Afterward, we start analyzing Bengali grammar properly as we have to generate rules for Bengali grammar. More specifically, first, we take an input Bengali sentence or paragraph. If paragraph, we first identify the sentences using delimiters (। or ;). Next, we use Unicode to read Bengali text efficiently. Next, we tokenize the sentences and determine the positions (subject, object and verb) of words (tokens) using basic grammatical rules. Then, we analyze parts of speech using Bengali grammar and apply rules to generate the output mentioning the words with corresponding parts of speech tag. Here, we use Levenshtein distance algorithm to find out the closest word to avoid storing similar types of words (serving similar purpose) such as different forms of Bengali verbs, in our system's memory. Additionally, we add some novel rules for Bengali PoS tagging, and finally, evaluate the accuracy of our system. Figure 1 illustrates our proposed mechanism for Bengali PoS tagging.

We explain each step of our methodology in the next subsections.

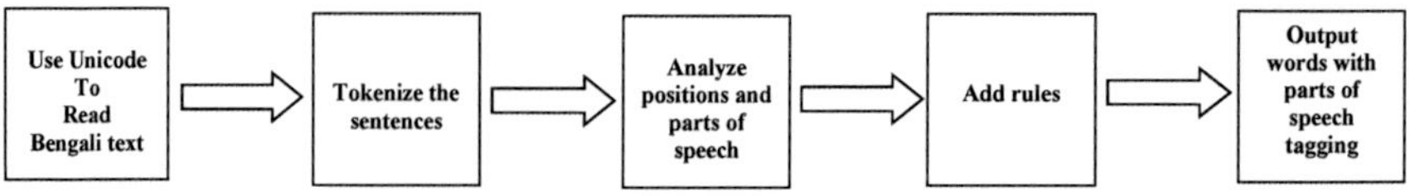

Fig. 1 Proposed PoS tagging methodology

4.1 Detect Bengali Sentences

To find out PoS of words in a Bengali sentence we have to first ensure that we are getting only Bengali characters as input. Otherwise, our tool will not work correctly as it deals with Bengali sentences consisting of only valid Bengali characters. Therefore, to detect sentences with only Bengali characters, we check input sentences with Bengali valid characters' set to detect whether each character in a sentence is Bengali character. Thus, we remove any invalid (unmatched) character to clean the input sentences. We show some examples regarding this in Table 1.

4.2 Predefined List

Next, we develop some predefined lists of pronouns (PR), verbs (VB), adjectives (ADJ), prepositions (PRE), adverbs (RB), principal verbs (PV), conjunctions (CON), cardinal digits (CD), punctuations (PUNC), etc., to primarily detect the PoS of each of the words in an input sentence. We show some of our predefined lists as follows.

Table 1 Examples of detecting Bengali sentences through cleaning

User input	Output Bengali sentences
আমার English বইটি তোমাকে দিয়ে দিব	আমার বইটি তোমাকে দিয়ে দিব
আমার football অনেক পছন্দ	আমার অনেক পছন্দ
আমি Dairy milk চকলেট খাবো	আমি চকলেট খাবো

PR = ['আমি', 'তুমি', 'সে', 'তারা', 'তার'] etc.

VB = ['যাবো', 'যাবে', 'খাবো', 'খাই', 'দৌড়ায়'] etc.

ADJ = ['খুব', 'সুন্দর', 'ভাল', 'দ্রুত', 'সত্য', 'ধীরে'] etc.

RB = ['অবশ্যই', 'নিশ্চয়ই', 'ইতিপূর্বে', 'শীঘ্রই'] etc.

PRE = ['উপরে', 'এ', 'মদ্ধে', 'থেকে', 'সাথে'] etc.

CON = ['কিন্তু', 'তবু', 'বিনা', 'অতএব', 'অথবা'] etc.

CD = ['এক', 'দুই', 'তিন', 'চার', 'পাঁচ', 'ছয়'] etc.

PUNC = ['+', '-', ',', ':', '"', '_', '!', '/'] etc

4.3 *Modify by Applying Rules*

After we detect PoS of words in a sentence using our predefined lists, we apply different grammatical rules related to Bengali PoS tagging so that we can get more accurate PoS tagging by modifying the initial PoS tagging (using predefined lists). Next, we discuss some of our most impactful rules for Bengali PoS tagging.

Consecutive Adjectives. When we have consecutive adjectives in a Bengali sentence, all the adjectives except the last one need to be tagged as adverbs. Here, we tag an adverb as 'RB'. We show an example of this rule in Table 2.

Adjectives before Verb (Adverbs). When we have an adjective before any verb, which is actually modifying the verb; then the adjective becomes an adverb (RB) as shown in Table 3.

Consecutive Adjectives before Verb. When we have consecutive adjectives before any verb, which are actually modifying the verb; then all the adjectives before the verb become adverbs (RB) as shown in Table 4.

Table 2 Consecutive adjectives

User Input	PoS without Rule	PoS with Rule
সে খুব ভাল ছেলে	সে / PR খুব/ ADJ (Wrong) ভাল/ ADJ ছেলে/ NN	সে/ PR খুব/ RB (Correct) ভাল/ADJ ছেলে /NN

Table 3 Adjective before a verb

User Input	PoS without Rule	PoS with Rule
সে দ্রুত দৌড়ায়	সে/ PR দ্রুত/ ADJ (Wrong) দৌড়ায়/ VB	সে/ PR দ্রুত/ RB (Correct) দৌড়ায়/ VB

Table 4 Consecutive adjectives before a verb

User Input	POS without Rule	POS with Rule
সে খুব দ্রুত দৌড়ায়	সে/ PR খুব/ ADJ (Wrong) দ্রুত/ ADJ (Wrong) দৌড়ায় /VB	সে/ PR খুব/ RB (Correct) দ্রুত/ RB (Correct) দৌড়ায়/ VB

Table 5 Principal verb with rule

User Input	POS without Rule	POS with Rule
আমি খেতে থাকব	আমি/ PR খেতে / VB (Wrong) থাকব / VB (Wrong)	আমি/ PR খেতে থাকব/ PV (Correct)

Principal Verb. When we have a verb in future continuous (খেতে থাকব)tense and future perfect (গিয়ে থাকব) tense, then that we need to carefully identify the principal verb since the principal verb generally constitutes of more than one word (in a specific pattern though) in these two tenses. We also cover this rule as shown in Table 5.

String Distance Measurement. This rule is perhaps the most significant contribution by our proposed PoS tagging approach. Here, we find out the closest word by calculating the minimum distance value. Specifically, we apply this algorithm for verbs only because Bengali verbs can take different forms based on number, person, and tense. To do so, we utilize a popular string distance measurement algorithm, known as Levenshtein distance. The motive behind using this algorithm is to avoid storing all the different forms of the same verb (root verb). Instead, we store only the root verbs in the predefined list of verbs so that the storage (memory overhead) does not become so large, and then, we detect these root verbs for different forms of verbs (not found in the verbs' list) using this algorithm. Thus, to optimize memory consumption, we use Levenshtein distance algorithm to find the closest word (verb). To determine the root verb more accurately, we remove the suffix from a verb before finding its closest root verb. We show some examples regarding this in Table 6.

4.4 Accuracy Measurement

After getting PoS of Bengali sentences we find out the accuracy of our system. First, we take PoS tag for each word in an input sentence as reference. Then, we match

Table 6 Finding root verbs from verbs having different forms

Input verb	Root Verb
খেলতেছি	খেলা
খেলছিলাম	খেলা
খাচ্ছিলাম	খাই
খাইতেছি	খাই

our output PoS tags with the reference POS tags. Next, we count the matched PoS tags, and find out accuracy for a sentence using the following equation:

$$\text{Accuracy} = (\text{Number of Matched PoS tags}/\text{Total words}) * 100\%$$

Finally, we can find the accuracy for the full dataset of Bengali sentences, which we actually the average accuracy using following equation

$$\text{Average Accuracy} = \text{Total Accuracy}/\text{Total sentences}$$

5 Experimental Evaluation

Next, we discuss the tools and settings that we use to carry out our experimentation. Finally, we present our experimental results in details.

5.1 Tools and Settings

To find out Bengali PoS, we use Python language and Google Colab. Besides, we need NLP Toolkit to understand how the PoS tagging tool works. We use UTF-8 encoding to read Bengali text as input. Nonetheless, it appears a bit difficult to deal with Bengali characters as it requires extra time to check (match) Bengali characters.

5.2 Dataset

It is quite tough to collect a proper dataset for Bengali since Bengali is a low resource language. We collect Bengali sentences from different websites, online news portals, and movie subtitles. Our dataset consists of 2000 Bengali sentences currently. We show a snippet of our dataset in Fig. 2. Here, our dataset contains non-Bengali characters initially, which we need to clean next.

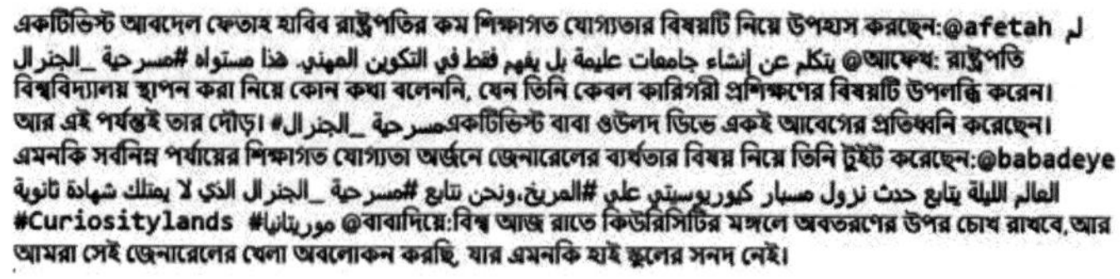

Fig. 2 Snippet of dataset

5.3 Results and Findings

Next, we present the experimental results reflecting our different outputs step by step. First, we detect Bengali characters in an input sentence. After that, we tokenize the sentences, and perform the PoS tagging by applying rules. We present the results corresponding to our rules discussed earlier in this study. Besides, we present how the application of Levenshtein distance algorithm reduces the memory overhead in our system. Finally, we calculate and show the accuracy of our system.

Detection of Bengali Characters. As discussed earlier, we clear all different invalid characters (from other languages) to ensure that whatever user gives as input, our tool works with Bengali language only as shown in Fig. 3.

Predefined List. Next, we show a snippet of our predefined lists in Fig. 4.

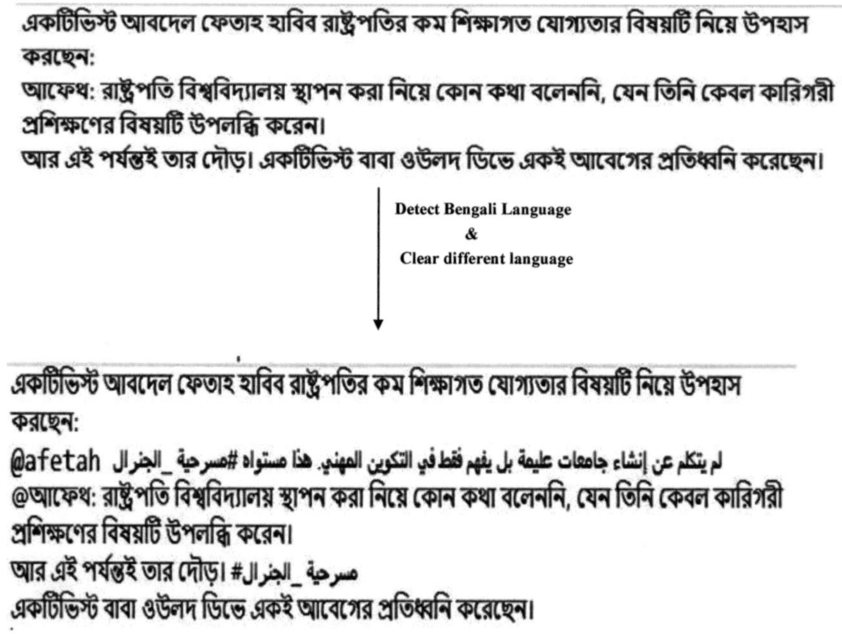

Fig. 3 Bengali language detection

```
PR=[ 'আমি','তুমি','সে','তারা','আমার','তিনি','তোমার','আমাকে','তার']
VB=[ 'যাবো','যাবে', 'খাবো','খাই', 'দৌড়ায়','পড়ে','খেতে','যেতে']
ADJ=[ 'খুব','সুন্দর','ভাল','দ্রুত','সত্য','ধীরে','অসুস্থ','একটি লাল']
RB=[ 'চরম','একেবারেই','ফলন্ত','বেশ','স্পষ্টভাবে','ন্যায়ভাবে','একা']
PRE=[ 'উপরে','এ','মন্ধে','থেকে','সাথে','এর','পরে','নিচে','আগে']
CD=[ 'এক','দুই','তিন', 'চার','পাঁচ', 'ছয়', 'সাত','আট','নয়','দশ']
```

Fig. 4 Predefined lists

```
['সে', 'খুব', 'ভাল', 'ছেলে']
Generating Output:  ['PR']
Generating Output:  ['PR', 'ADJ']
Generating Output:  ['PR', 'ADJ', 'ADJ']
Generating Output:  ['PR', 'ADJ', 'ADJ', 'NN']
Final Output ['PR', 'RB', 'ADJ', 'NN']
```

Fig. 5 Consecutive adjectives

```
['সে', 'দ্রুত', 'দৌড়ায়']
Generating Output:  ['PR']
Generating Output:  ['PR', 'ADJ']
Generating Output:  ['PR', 'ADJ', 'VB']
Final Output ['PR', 'RB', 'VB']
```

Fig. 6 Adjective before verb

Adding Rules. We have discussed our four implemented rules and a string distance measurement algorithm in the previous section. Next, we present their corresponding results one by one.

Consecutive Adjectives. Figure 5 shows that consecutive adjectives become adverbs except the last one in our implementation.

Adjective before Verb. Figure 6 shows that adjective before verb becomes adverb in our implementation.

Consecutive Adjectives before Verb. Figure 7 shows consecutive adjectives before verb becomes adverbs.

Principal Verb. Next, we address principal verbs in future continuous and future perfect tense in Fig. 8.

String Distance Measurement. We use Levenshtein distance to find the closest word (Fig. 9).

```
['সে', 'খুব', 'দ্রুত', 'দৌড়ায়']
Generating Output:  ['PR']
Generating Output:  ['PR', 'ADJ']
Generating Output:  ['PR', 'ADJ', 'ADJ']
Generating Output:  ['PR', 'ADJ', 'ADJ', 'VB']
Final Output ['PR', 'RB', 'RB', 'VB']
```

Fig. 7 Consecutive adjectives before verb

```
['আমি', 'গিয়ে', 'থাকব']
Generating Output:  ['PR']
Generating Output:  ['PR', 'VB1']
Generating Output:  ['PR', 'VB1', 'VB2']
Final Output ['PR', 'PV']
```

Fig. 8 Principal verb

```
Input: খেলছিলাম
Root word:  খেল
Distance:  2
Distance:  7
Distance:  3
Distance:  4
Distance:  3
Distance:  3
Distance:  3
Distance:  3
Distance:  4
Distance:  5
Distance:  3
Distance:  5
Distance:  3
Distance:  1
Distance:  5
lowest distance:  1
Closest word with lowest distance value : খেলা
```

Fig. 9 Taking word with minimum distance value with Levenshtein distance

Accuracy Measurement. We get 75, 100, 66, 40% accuracy here by as shown in Figs. 10, 11, 12, 13 respectively.

Fig. 10 75% accuracy

```
['আমি', ',', 'তুমি', 'ও', 'রাহিম', 'খেলতে', 'পছন্দ', 'করি']
Final Output ['PR', 'PUNC', 'PR', 'CON', 'NN', 'VB', 'VB', 'VB']
Reference ['PR', 'PUNC', 'PR', 'CON', 'NN', 'VB', 'VBI']
Accuracy----- 75.0 %
Time taken:  77.00472593307495
```

Fig. 11 100% accuracy

```
['সে', 'খুব', 'ভাল', 'ছেলে']
Final Output ['PR', 'RB', 'ADJ', 'NN']
Reference ['PR', 'RB', 'ADJ', 'NN']
Accuracy----- 100.0 %
Time taken:  77.0161247253418
```

Fig. 12 66% accuracy

```
['সে', 'দ্রুত', 'দৌড়ায়']
Final Output ['PR', 'RB', 'VB']
Reference ['PR', 'RB', 'ADJ']
Accuracy----- 66.66666666666666 %
Time taken:  77.01230311393738
```

Fig. 13 40% accuracy

```
['আমার', 'বইটি', 'তোমাকে', 'দিয়ে', 'দিব']
Final Output ['PR', 'NN', 'NN', 'NN', 'VB']
Reference ['PR', 'NN', 'PR', 'VBI']
Accuracy----- 40.0 %
Time taken:  77.00689625740051
```

6 Future Works

Nowadays, any kind of translator tool plays a vital role in this region. Developing a translator tool using NLP is one kind of necessity. Our aim is to develop a translator which is translating a sentence into Bengali to English and also corresponding English to Bengali. Initially we develop a tool that is Bangla parts of speech tagger which is detecting words with their corresponding parts of speech. It is a very important thing for those who are weak in Bengali.

7 Conclusion

There are many real-life applications of natural language processing (NLP) and parts of speech plays an important role in NLP. Nowadays many companies concentrating on NLP like Microsoft and Google. We are trying to make such a tool that has a huge market value and provide the best tool with higher accuracy. Generating each rules are difficult. Any rule is going wrong then the tool will not working properly. It is risky but still we are trying to give our best to make this tool accurately. We also use an algorithm which is Levenshtein distance algorithm. We have developed this algorithm. We will give the best tags with higher accuracy which will help the people.

References

1. Sarkar K, Gayen V (2012) A practical part-of-speech tagger for Bengali. In: 2012 third international conference on emerging applications of information technology. IEEE, pp 36–40
2. Hoque MN, Seddiqui MH (2015) Bangla parts-of-speech tagging using Bangla stemmer and rule based analyzer. In: 2015 18th international conference on computer and information technology (ICCIT). IEEE, pp 440–444
3. Mukherjee S, Mandal SKD (2013) Bengali parts-of-speech tagging using global linear model. In: 2013 Annual IEEE India Conference (INDICON). IEEE, pp 1–4
4. Mridha MF, Saha AK, Das JK (2014) New approach of solving semantic ambiguity problem of Bangla root words using universal networking language (UNL). In: 2014 international conference on informatics, electronics & vision (ICIEV). IEEE, pp 1–6
5. Nakagawa T, Uchimoto K (2007) A hybrid approach to word segmentation and pos tagging. In: Proceedings of the 45th annual meeting of the Association for Computational Linguistics Companion Volume Proceedings of the Demo and Poster Sessions, pp 217–220
6. Schmid H (1994) Part-of-speech tagging with neural networks. In: Proceedings of the 15th conference on computational linguistics, vol 1. Association for Computational Linguistics, pp 172–176
7. Saharia N, Das D, Sharma U, Kalita J (2009) Part of speech tagger for Assamese text. In: Proceedings of the ACL-IJCNLP 2009 conference short papers. Association for Computational Linguistics, pp 33–36

8. Chowdhury MSA, Uddin NM, Imran M, Hassan MM, Haque ME (2004) Parts of speech tagging of Bangla sentence. In: Proceeding of the 7th International Conference on Computer and Information Technology (ICCIT)
9. Chakrabarti D (2011) Layered parts of speech tagging for Bangla. Language in India, www.languageinindia.com, Special Volume: Problems of Parsing in Indian Languages
10. Dandapat S, Sarkar S (2006) Part of speech tagging for Bengali with hidden Markov model. In: Proceeding of the NLPAI machine learning competition
11. Roy MK, Paull PK, Noori SRH, Mahmud SH (2019) Suffix based automated parts of speech tagging for Bangla language. In: 2019 international conference on electrical, computer and communication engineering (ECCE). IEEE, pp 1–5
12. Ismail S, Rahman MS (2014) Bangla word clustering based on n-gram language model. In: 2014 international conference on electrical engineering and information & communication technology. IEEE, pp 1–5
13. Hasan K, Mondal A, Saha A et al (2012) Recognizing Bangla grammar using predictive parser. arXiv preprint arXiv: 1201.2010, 2012
14. Alam F, Habib S, Khan M (2008) Text normalization system for Bangla. BRAC University, Technical Report
15. Anwar MM, Anwar MZ, Bhuiyan MA-A (2009) Syntax analysis and machine translation of Bangla sentences. Int J Comput Sci Network Secur 9(8):317–326
16. Rahman MS, Mridha MF, Poddar SR, Huda MN (2010) Open morphological machine translation: Bangla to English. In: 2010 international conference on computer information systems and industrial management applications (CISIM). IEEE, pp 460–465
17. Alam M, UzZaman N, Khan M et al (2007) N-gram based statistical grammar checker for Bangla and English
18. Ekbal A, Haque R, Bandyopadhyay S (2007) Bengali part of speech tagging using conditional random field. In: Proceedings of seventh international symposium on natural language processing (SNLP2007), pp 131–136
19. Ali H (2010) An unsupervised parts-of-speech tagger for the Bangla language. Department of Computer Science, University of British Columbia, vol 20, pp 1–8
20. Kabir MF, Abdullah-Al-Mamun K, Huda MN (2016) Deep learning based parts of speech tagger for Bengali. In: 2016 5th international conference on informatics, electronics and vision (ICIEV). IEEE, pp 26–29

Long-Term Wind Speed Forecasting—A Review

A. Shobana Devi, G. Maragatham, K. Boopathi, M. C. Lavanya, and R. Saranya

Abstract Wind speed plays a predominant role in the wind energy system. Forecasting of long-term wind speed in certain diverse areas, for example, the optimal design of wind farms, energy management, and restructured electricity markets has always been a popular spot for research. A reliable wind speed forecast can mitigate the errors in scheduling and, in effect, improve the stability of the electrical grid power and reduce the ancillary service costs of the energy market. Short-term updates are being considered as less important to few sites than reliability and reasonable start-up time, while successful long-term forecasting wants not only a trigger of present events but similarly in-depth knowledge of historical patterns of seasonal winds and site-specific weather parameters. This paper describes insight into the leading forecasting models, related to wind speed and wind power, and depends on numerical weather prediction (NWP) models, statistical methods, artificial neural network (ANN) models, and hybrid models over various time horizons. This chapter also discussed an overview of the comparative study of several available forecasting models. However, this paper discusses the key challenges and problems related to the forecasting of wind speed. This survey addresses the existing functionality measures for various approaches by splitting them into different techniques: time-series models, artificial intelligence models, and hybrid models. All these methods include specific models that are explored with different parameters, benefits, and disadvantages.

A. Shobana Devi (✉) · G. Maragatham
Department of Information Technology, SRMIST, Chennai, India
e-mail: Shobanak07@gmail.com

G. Maragatham
e-mail: maragatg@srmist.edu.in

K. Boopathi · M. C. Lavanya · R. Saranya
National Institute of Wind Energy, Chennai, India
e-mail: boopathi.niwe@nic.in

M. C. Lavanya
e-mail: mclavanya.niwe@nic.in

R. Saranya
e-mail: saranyaraj1701@gmail.com

D. J. Hemanth et al. (eds.), *Artificial Intelligence Techniques for Advanced Computing Applications*, Lecture Notes in Networks and Systems 130,
https://doi.org/10.1007/978-981-15-5329-5_9

Keywords Artificial neural network · Autoregressive moving average · Fuzzy logic · Numeric weather prediction · Persistence method · Statistical methods · Wind speed forecasting · Wind power forecasting

1 Introduction

India being the world's third largest energy producing and consuming country faces an energy crisis due to some sort of looming in sustainable energy supply. The country's endurance with high quality of energy sources extremely depends on renewable sources and the grid integration policies. Wind power is of high standard in industrial output relative to solar, biomass, and some other sources of renewable energy, low cost on generating power, and strong physical and social impact on the environment and plays a major role in supporting the energy needs of the country. Wind speed commonly shows nonlinear, non-stationary, and unpredictable manner. Based on the study of existing historical data of wind speed, it has been found that Indian wind speed is seasonal by means of one cycle every year. Seasonality is still among the most frequently encountered sensations in many areas, such as power load and wind speed. Hence, employing the most suitable site-specific forecasting model lays the path towards the accurate long-term prediction followed by strategic planning and decision-making activities.

1.1 Wind Speed Forecasting

Several methods for forecasting wind speed have been proposed. However, they are commonly separated into two essential gatherings: the forecasting horizon and the used forecasting model. For the forecasting horizon, the following classification is proposed based on a literature survey: the techniques of wind speed forecasting can be clustered into very short-term, short-term, medium-term, and long-term models such as (Fig. 1)

- **Very short-Term forecasting:** This technique is being used to predict a few seconds to minutes ahead values of wind speed/power. The key use of this method is clearing and regulatory decisions on the electricity sector.
- **Short-Term forecasting**: The main persistence of forecasting short-term period of wind speed is to direct wind turbine power output in a limited time to meet customer needs. The horizon of time varies from 30 min to hours ahead.
- **Medium-Term forecasting**: A relatively medium-term method of timescale is based on the decision on/off wind generator, operational safety, and the purposes of the electrical market. The forecast period's duration ranges from 6 h to one day ahead period.
- **Long-Term forecasting**: This wind forecast strategy is being used mostly for decisions of unit commitment; turnaround time of maintenance planning and

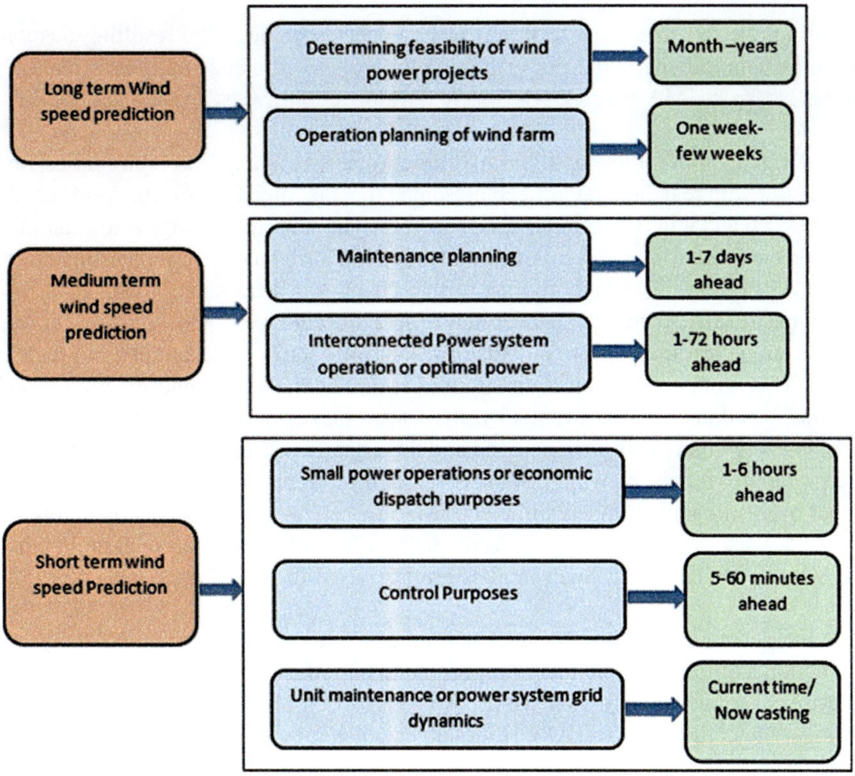

Fig. 1 Classification of wind energy forecasting based on time horizon

scheduling, long-term forecasting duration varies from one day to one week or one year ahead period.

Many works of literature have recently been published on wind speed forecasting. However, most studies focused on predicting wind speed, while a few studies focused on predicting wind power generation. The methodologies of these investigations could be organized into three classes: time series models, artificial intelligence models, and time series-based artificial intelligence (hybrid) models shown in Fig. 2.

2 Time Series Models

Many updated studies focus primarily on wind speed data, and its related errors of estimation are beyond the standards of the industry. Both wind speed and wind directions were evaluated in this paper to establish a predictive technique based on a statistical model. Yatiyana et al. [1] have proposed an autoregressive integrated moving average model for constructing the wind prediction system calculated in the region of Western Australia to produce the expected values. The resulting design could be able to boost wind power production system efficiency and performance. In this work, ARIMA-based models were developed to forecast speed and direction of wind.

The explanation for this option is its shorter time to respond. Research studies are carried out with real data. It has been presented that the forecast error for wind speed is below 5% and wind direction is less than 16%. The subsequent stage is to examine the collected data for a one-year cycle and create a hybrid model for forecasting built on the ARIMA system to incorporate further changes in season and trend [1].

Ambach et al. [2] provides an overview of specific wind speed and wind power prediction strategies. In addition, models of recent time series are examined in greater detail. This work focuses on accurate predictions of short term and medium term of wind speed and power. Eventually, the latest wind speed data and those out-of-sample energy outcomes are discussed and this chapter addresses the issue of asymmetric failure. Precisely, wind power forecasts need to be evaluated differently, for either overestimated or underestimated. Two time series-based models are therefore checked, a seasonal univariate ARFIMA-APARCH model and a seasonal multivariate VAR-TARCH model. Such models are also used to provide forecasts of wind power.

A measure of asymmetric accuracy is adopted to identify the best method of power prediction. A brief example that covers forecasts up to 24 h is provided to cover the effect of asymmetric losses on wind power and speed predictions. Ultimately, the multivariate periodic VAR-TARCH model, calculated by the iteratively reweighted lasso process, performs better than the other two models and thus helps to reduce the asymmetric loss of inaccurate predictions of wind power [2].

Radziukynas et al. addresses short-term wind speed prediction for the wind farm Laukžemė (Lithuania) utilizing the time series method. The model ARIMA has chosen, and the best design for it has been calculated with historical data of (4 months) wind speed and adjusting the model's training interval (3–5 days) as well as the average time of factual data (1–6 h). Predictive accuracy was measured with respect to RMSE values and values of absolute error. The forecast outcomes have presented and discussed for 39 successive time intervals using 6–48 h hourly forecasts. The structured statistical ARIMA model (3, 1, 4) shown the best fit for the Laukzeme wind farm's wind speed forecasting.

The 6-h and 12-h predictions are quite appropriate and close to the physical predictive model's accuracy. About 50% of the 6-h forecast errors, respectively, absolute error and RMSE are equivalent to 1 m/s and 35% of the portion is 12-h absolute errors. We think it is a pretty positive outcome. It seemed that the 24-h predictions were too conservative estimates. The additional practices, therefore, should be applied to the ARIMA method to enhance the predictions for a whole day cycle [3].

Verma et al. [4] have proposed the Markov chain method for forecasting wind power for day-ahead utilizing wind speeds in selected wind farms. Due to their robustness in sequential data modelling, models from Markov are a choice that is more appropriate. These are dependent on some of the non-restrictive assumptions that can be measured and applied to actual data. Therefore, accurate short-term forecasting of wind energy (30 min to 6 h ahead) is critical for the wind farm's optimum scheduling. At Jodhpur station in Rajasthan state of India, the geographical region for study is taken. The wind speed information can be divided into diverse seasons with proper analysis so that it can be trained and analysed as efficiently as possible. Just one parameter is utilized for the model training—wind speed. The model is verified with actual 1-year (2015) measured data for the area of Jodhpur, Rajasthan, India. Based on performance measures, namely MAPE, MAE, and RMSE, the outcomes have been evaluated by measuring the error difference among the real and the expected data [4].

3 Artificial Intelligence Models

Palomares et al. [5] have established three ARIMA models that suit the wind speed time series short-term behaviour. Compared to the neuronal networks, the ARIMA models have provided some pretty similar findings but the computation times are very low. Results indicate that for a short period (10 min, 1, 2, and 4 h) ARIMA is a better approach than NNT to predict. Data was collected from a weak orography system in Southern Andalusia (Peñaflor, Seville) (10 min' time stamp measurements). This approach makes the ARIMA model and the NNT model output are very similar, making it possible to use a simple predictive model to control the sources of energy. The study presents the validation process of the model together with an evaluation of regression based on real-life results. The best performance shown in the study is ARIMA(2,0,0). The validation was carried out using three specific performance indexes, based on procedures for correlation. [5].

Mezaache et al. [6] proposed a new artificial intelligence method, in this work for forecasting time series data of wind speed, it consists of two blocks: In which the block one constitutes of a variant of auto-encoder deep neural network whose function is to decrease dimensionality of the time series, this new dimension of the wind speed time series provided as a result of auto-encoder is supplied as input for the block two which provides the expected wind speed as output. Two categories of neural network architecture, Elman recurrent neural network (ENN) and extreme learning machine (ELM), are selected for this entire second block. The following error metrics were used to assess our approach: determination coefficient (R2), root mean square error (RMSE), mean absolute percentage error (MAPE) and mean absolute bias error (MABE). The results obtained on a specific data set indicate that our method using the Elman recurrent neural network (ENN) more effectively forecasts wind speed compared to which used the extreme learning machine (ELM) network. It applies to the Elman network's temporary dynamics. It also provides robust performance in predicting wind speed, even when adjusting the test site [6].

Guo Z et al. [7] suggested a customized EMD-FNN network (empirical mode decomposition (EMD) driven feed-forward neural network (FNN)) based learning a framework for forecasting wind speed. The initial wind speed time series has both nonlinear and non-stationary and is being decomposed into something like a predictable and frequently limited range of intrinsic mode functions (IMFs) and a residual sequence with EMD technique for a keen intuition into data structures. In spite of empirical pieces of evidence, we noticed that for both types of wind speed time series such as, monthly mean wind speed and daily mean wind speed—model named M-EMDFNN achieves the best performance according to various parameters (MAPE, MAE, and MSE). In all of those test cases, showing that the forecasting model for M-EMDFNN can be used in together for long-term and short-term time horizons as a very positive approach for forecasting wind speed [7].

Li and Shi [8] have proposed a detailed comparison analysis on the application within 1-hour-ahead wind speed forecasting of various artificial neural networks. Three forms of traditional neural networks (i.e. BP, RBF, and ADALINE) are studied, i.e. back propagation, radial basis function, and adaptive linear component. The wind data utilized is the mean hourly data of wind speed obtained at two North Dakota sites of observation. Depending on three measures, mean absolute percentage error, mean absolute error, and root-mean-square error, the performance is examined. It was realized that the predictability was directly influenced by various inputs and learning levels along with model architectures. Indeed, even with a similar data set of wind collection from a particular site, the decision of perhaps the top performing model is not equivalent with various assessment criteria. Using various data sets of wind, the decision turns out to be increasingly conflicting, even in terms of similar statistics. Along these lines, a powerful technique for consolidating models from various ANN methods is expected to defeat the irregularity issue in model determination. This shows the need for creating a solitary reliable, strong, and robust strategy by applying a post-processing technique [8].

Azad et al. [9] have suggested in this chapter a wind speed forecast solution for the long term. Two completely different methods have designed, for this reason, the neural network-based and statistical methods, to predict the hourly wind speed of the upcoming year. The innovation of this analysis is by constructing a data merging technique through many neural networks so as to predict the overall trend analysis of the upcoming year. A collection of most recent wind speed data has been utilized towards training and testing the data from two weather stations located in the Malaysia, respectively Kuala Terengganu and Mersing. The results show that the suggested system has enriched additional existing predictive techniques for the long-term forecast of wind speed. The suggested model shows about 0.8 m/s value for MAE, for instance, in context of MAE. By making comparisons with the actual and forecasted WSD, the hybrid technique has seen to be closely following the actual data series. Furthermore, as it adopts the overall trend recognized in the forecast method, it has used efficiently as a suitable alternative framework for forecasting wind speed of long-term activities [9].

Menezes et al. [10] have used the NARX neural system which can utilize its production criticism loop effectively to help model performance of prediction in difficult time series forecast tasks. NARX-based network was considered as a time delay feed-forward neural network (TDNN) while applied to the prediction of time series, i.e. excluding the criticism loop of deferred inputs, fundamentally diminishing its predictive effectiveness. In this chapter, we demonstrate that the NARX network's unique design can be effectively and proficiently extended out to univariate time series long-term forecasts (multistep-ahead). All the findings suggest that perhaps the proposed method constantly outperforms traditional neural network-based predictors like the architectures of TDNN and Elman [10].

Mohandes et al. [11] presented wind speed prediction, the latest neural network algorithm, and support vector machines (SVM) and contrasted their outcomes using the neural networks of multilayer perceptron (MLP). Average daily data of wind speed from Madina Town, Saudi Arabia region, is utilized for both model development and testing. Findings suggest that SVM compares the real and the expected data favourably with the MLP model based on the root-mean-square errors. For a framework with order 1–11, these results are confirmed. Depending on the efficiency of a data for cross-validation, the parameters are tailored for both algorithms. The minimum MSE for MLP testing data is 0.0090, while the SVM with order 11 data is 0.0078. SVM practically performs better than MLP for all the orders from 1 to 11 systems 11].

Surussavadee et al. [12] described a model of wind forecasting to be used in north-eastern Thailand for wind power management. The approach of the neural network has used. Neural networks were equipped and tested with observations data from the region's 17 wind stations. Two forecast periods are taken into consideration, i.e. 3 and 6 h ahead of time and 2 elevations, for example, 65 and 90 m over the ground. Neural network inputs comprise measured wind speeds data at the predicted wind station of 24 sequential hours before the date of the forecast. Neural network training data includes more than 174,000 samples in 2011 and 2012. In 2013, the expected accuracies will be measured using more than 83,000 specimens. For each task, ten diverse neural systems are prepared, and among them, best neural system is picked. Observations are well associated with predictions. The predictions for three hours seem to be more robust than the predictions for six hours. Wind speed forecasts beneath 4 m/s were biased positively, and some forecasts beyond 4 m/s were skewed negatively. For both periods of forecasts, forecasts of wind at 65 m and 90 m over the ground had extraordinary benefits for data of wind speeds over 2 m/s [12].

In the study of Osama [13], over the past decade, there have been several wind algorithms for forecasting speeds which have been designed to improve the accuracy of forecasting. Support vector regression (SVR) constraints such as penalty factor (C) and kernel parameter (K) have a major consequence on forecasting algorithm's accuracy and complexity. This chapter suggested a hybrid method dependent on the SVR and whale optimization algorithm (WOA), called WOASVR, to address issues that traditional methods are not able to handle efficiently and have exhibited superior performance in various aspects. The reliability of the proposed method (WOA-SVR) was analysed with various aspects and the space weather monitoring centre (SWMC) daily average wind speed data in Egypt as a research study. The outcomes of the proposed method were compared for validation with the particle swarm optimization (PSO) algorithm and the existing SVR excluding optimization of parameters. The findings revealed that the suggested WOA-SVR model is proficient in estimating optimum SVR parameter values, avoiding local optima problems, and is efficient in wind speed forecasting [13].

Barbounis et al. [14] managed a real-time application, for long-term forecasting of wind power and wind speed for the wind farm utilizing locally intermittent multilayered networks as methods of forecasting. To adapt to the unpredictability of the procedure and to enhance the exhibition of the methods, a set of ideal online networks learning frameworks is utilized for preparing the locally intermittent systems dependent on algorithm of recursive prediction error (RPE). A global RPE method is contrived and furthermore three neighbourhood learning algorithms have recommended by dividing the GRPE algorithm into lot of subissues on the level of neuron to diminish computational intricacy and capacity prerequisites. The LF-MLN model trained using the proposed algorithms is utilized to the issue of forecasting long-term wind speed and wind power in the wind farms. The examination results show that the forecasting models using recurrent networks give enhanced multistep ahead values of forecasts when contrasted with the time series, the atmospheric models, and persistence models [14].

Barbounis et al. [15] managed the issue of long-term forecasting of wind speed and wind power dependent on meteorological data. Forecasts of hourly values equal to 72-h ahead are created for the wind farm on Crete's Greek island. Three sorts of local recurrent neural systems are utilized as models for forecasting, to be specific, the diagonal recurrent neural network (RNN), the local activation feedback multilayer network (LAF-MLN), and the infinite impulse response multilayer perceptron (IIR-MLP). Two epic algorithms of learning are presented for the forecasting models training using recurrent network, the GRPE algorithm, and DRPE algorithm, which have extensively lesser computational and storage prerequisites. Broad experimentation is done where three systems of recurrent networks are moreover contrasted with two static-based models, a finite-impulse response NN (FIR-NN), and an ordinary static MLP network models. Results of simulation show that the models of a recurrent, network trained using the recommended techniques achieve better results than static models, while the process of persistence is critically improved [15].

In the study of Malik and Savita [16], the principle goal of this study is to forecast the short-term wind speed based on authentic time arrangement meteorological information. A nonlinear information driven system is utilized here for wind speed forecasting as it serves as a versatile and adaptive tool for forecasting modelling purpose. Five diverse ANN models are designed. The least mean square error (MSE) for those designed models is estimated based on the performance of the network. The designed model with minimum error could be utilized for wind speed forecasting and power generation forecasting from the wind turbines with error limit of $\pm 30\%$ (according to CERC). Various arrangements of ANNs were developed and contrasted over error metrics for assuring the chosen model's accuracy. The ANN 19 network system of 19 hidden 4 input and one output layers remained the best for wind speed forecasting of short term. The outcomes were contrasted with actual data and indicated generally excellent accuracy. As noticed, since the used data is totally random, processing more data is probably going to decrease the further error level and model refinement [16].

4 Time Series-Based Artificial Intelligence (Hybrid) Models

Guo et al. [17] recommends a novel hybrid model built on the initial definite series of season index technique and autoregressive moving average (ARMA) or generalized autoregressive conditional heteroscedasticity (GARCH) models of forecasting for long-term forecasting of wind speed. The errors of forecasting were examined and compared with those recorded using the models ARMA, support vector machine (SVM), and GARCH. This chapter provides separate seasonal factors for both approaches to forecast Zhangye area wind speed; simulation results specify the proposed models' adequacy utilized in the forecasting of wind speed. This procedure can improve the predictive capacity of the well-known models GARCH and ARMA for volatility. In addition, MAE and RMSE are two loss functions used as the parameters for the predictive quality measurement. From the values of loss functions, it indicates that whichever ES-ARMA or ES-GARCH technique could provide improved performance through the new forecasting procedure. Subsequently, this case study is also able to address the effective approaches for accomplishing the problems of long-term prediction [17].

Cadenas et al. [18] have developed hybrid models made up of artificial neural network (ANN) and the autoregressive integrated moving average (ARIMA) methods for wind speed forecasting. The ARIMA model was used first for wind speed prediction, and after that, ANN model was developed of the time series with the errors obtained and also considering the nonlinear tendencies that the ARIMA method was unable to define, thus minimizing the final errors. After the development of the hybrid models, the wind speed forecast has based on 48 sample data for each site, the findings are contrasted with ARIMA model, and the strategies of the ANN that operate separately. To compare the three methods, statistical error measurements named mean absolute error (MAE), mean square error (MSE), and mean error (ME) were calculated. The results revealed that the three techniques predict the time series behaviour of various sites in a reasonable manner, but it was evident with the analysis of the statistical error measure reveal hybrid methods forecast the wind speeds with greater accuracy compared to ARIMA model and ANN model simulations that are developed individually in the examined three sites. It can also decide that hybrid model proposed might be a superior alternative for forecasting wind speeds where nonlinear and linear patterns found [18].

Nair et al. [19] have done work about forecasting wind speeds for dissimilar time horizons at three various sites in the state of Tamil Nadu, in India (Nollur, Dharapuram, Kayathar) utilizing three diverse models, namely the autoregressive integrated moving average (ARIMA), artificial neural network (ANN), and the hybrid methods, in which ARIMA and ANN combined with their outcomes. The data of wind speed has taken as inputs for the considered model with a time interval of 1 h for three consecutive years. It is clear from the results of the ANN, hybrid, and ARIMA models, the hybrid technique indicates a much fewer error. The result showed that the hybrid model has an estimated error less than the ANN and the ARIMA models developed separately [19].

Shukur et al. [20] used the ARIMA model's inefficient forecasting which reflects modelling system instability. The objective of this study is to increase wind speed prediction accuracy by suggesting a more suitable approach. For handling issues of nonlinearity and ambiguity, Kalman filter (KF) and artificial neural network (ANN) have been applied. A hybrid method of KF-ANN built on the method of ARIMA would boost wind speed prediction accuracy. First, ARIMA's efficiency has supported in defining the construction of inputs for ANN, KF, and the hybrid model. A summary of the event has conducted using data from Iraq and Malaysia on the daily wind speed. Outcomes showed that the proposed model was effective for ARIMA, ANN, hybrid ARIMA-KF, and hybrid KF-ANN. The MAPE results showed, however, that the KF-ANN hybrid model was the operative tool to improve the wind speed forecasting accuracy. The hybrid model's benefits were the consequence of hybridization of the ARIMA model's nonlinear and linear parts and again combining it through a KF to cope with stochastic volatility and ANN for addressing nonlinearity in the hybrid KF-ANN system [20].

Soman et al. [21] delivers insight into the leading wind speed and wind power related forecasting techniques focused on numerical weather prediction (NWP), statistical techniques artificial neural network (ANN), and hybrid models over various time horizons. Also discussed is a summary of the benchmark analysis of different existing forecasting techniques. Furthermore, this paper highlights the key challenges and issues associated with the prediction of wind power. This paper provided an analysis of the wind speed prediction and power production, considering multiple time horizons. Many forecasting techniques have been deliberated, and many research works have been done on the methods with features of their own. The main attention was on demonstrating the variety of various available forecasting models and offering comparison of existing techniques to assess the best models among them [21].

Lei [22] provided a bibliographic survey on the overall background of forecasting wind speed and power research works and developments. The path for further research and development has proposed depending on the evaluation of existing wind power forecasting models. Different forecasting techniques were developed, and much research has been done on the forecasting models. Every one of the models has its own functionality. For short-term forecasting, some of the models are good, whereas some models perform better for long-term forecasting; most of them are quite basic and frequently used, while others are more accurate in complex models. Recently, a variety of new methods have been proposed by designing artificial intelligence and computational techniques. Many of the models are better than conventional models and have good prospects for development. Ultimately, the possible trend has recommended depending on the wind speed and generated power prediction development history [22].

Bhaskar et al. [23] have given for the past 15 years a comprehensive survey of wind speed prediction techniques which has been published. An endeavour has made to recognize distinctive wind forecasting suppliers accessible in the market. An itemized investigation of the forecasting wind speed methods has introduced in this paper for as far back as 15 years. There is additionally an attempt to distinguish various suppliers of wind forecasting available. This study will be of extraordinary advantage to the crisp researchers involved in this data analysis domain. This review would also encourage owners of every wind farm in order to know the current model functionality of wind forecast and offer them an indication of which approach will be appropriate for forecasting wind speed on their respective wind farms. The survey shows that popular models of wind prediction built with the integration using biologically inspired methods have very strong potential as those models bring lots of improvement in wind forecasting error [23].

Bali et al. [24] given a review on deep learning, which would be the subdiscipline of machine learning and can significantly increase the rate of accuracy on large amounts of data and predictions with deep learning using LSTM. Combining deep learning and LSTM could increase the forecasting rate because of the pattern memory characteristics of LSTM over a longer period. This study addresses the current usability metrics of various approaches by splitting them into different approaches: very short-term interval, short-term interval, and long-term interval models. All of these methods include some models that have explored with different parameters, benefits, and disadvantages. The aim of this detailed review is to provide a more efficient and better assessment of different approaches to assist the researcher in selecting the best approach from all current models. The procedure for the deep learning techniques using LSTM (long-term memory) could lead to better wind speed forecasting for energy production; later, the comprehensive examination of several researchers works to solve such problems that have not solved completely by the different predictions models. Therefore, new wind speed forecasting techniques need to develop with a lower error margin as well as a better and more effective performance. Such forecasts can sometimes be of significant use in different predictions of atmospheric activity [24].

Kavasseri et al. [25] explored the computational use of fractional ARIMA methods and forecast wind speeds data of the time horizons of day-ahead of (24 h) and the two-day-ahead of (48 h). The forecasting models have been implemented to measure wind speed gathered from North Dakota's four possible wind locations. The findings indicate that the proposed solution could increase forecast accuracy by an aggregate of 42% relative to the method of persistence. In combination with those of the power curve of an operating wind turbine generation, the projected wind speeds have been used to produce corresponding output power forecasts. Although prediction error values of wind speeds just under the cut-in wind speed are inconsistent with power forecasts, it has recognized that the proposed system is effective in variable wind speed regimes [25].

Colak et al. [26] have given an article which is the second part of a complete study focusing on multitime series and multitimescale modelling in forecasting wind speed and power. We discuss the medium- and long-term forecasting accuracy of ARIMA, ARMA WMA, and MA models in forecasting wind power and wind speed in this specific section of the entire report. Specifically, 3-h and 6-h time series predicting models have been built to perform 9-h and 24-h forecasting ahead, respectively. With respect to medium- and long-term forecasting time horizons, some useful comparisons are made for the statistical models used. Eventually, given a detailed comparison list of the whole research, some important accomplishments have been mentioned. Because of Part II, the ARIMA system has shown to improve predictive accuracy appropriately in 9-h ahead forecasting of wind speed and methods of ARMA in 9-h ahead forecasting of wind power and the 24-h ahead forecasting of wind speed and wind power. Furthermore, the forecasting accuracy has been seriously reduced by models of MA in 9-h ahead forecasting of wind speed and wind power and systems of ARIMA in the 24-ho ahead projection of wind speed and wind power. Furthermore, to these accomplishments, the durability method in the benchmark experiment was outperformed by all models used in medium-term and long-term forecasting of wind speed and wind power [26].

5 Discussions and Prospects

Wind speed prediction models discussed earlier had their own physical characteristics, and in different situations, they can behave much better. The models of NWP are being effective at forecasting wind speed on a large area and can accomplish better long-term forecasting performance. These have often been used as inputs to time series-based models like ANN, ARMA, and others; they have helped to achieve better outcomes. The persistence models are defined as the easiest models of the time series. In some very short-term forecasting, they will outperform many of the other models. Considering the unpredictable nature of forecasting, they are frequently used in study. Over the past 30 years, researchers have performed most work on time series-based models.

This sort of forecasting methods including Box–Jenkins techniques (for instance, ARIMA, AR, VAR, ARMA, and so on), neural network-based, and fuzzy logic-based models have used a vast amount of historical data to model inputs and can produce accurate outcomes from short-term forecasting. There are models like neural networks, and fuzzy logic is being the new model that is developed based on artificial intelligence.

For actual data input, neural networks behave well and have good learning skills and training skills. While dealing with reasoning problems, fuzzy logic systems outperform most, though understanding and modifying skills are mediocre.

New approaches merged the neural networks and fuzzy logic and produced outstanding performance. It is also quite harder to compare all models accurately because these models focus on different conditions and the collection of data is a daunting challenge. There has been some correlation and the preliminary findings, which showed that in short-term prediction the artificial intelligence-based models outperformed others.

We could see the phenomenon approximately from advancing wind speed and energy forecasting. The scenarios are as follows:

- More analysis of methods of artificial intelligence and strengthen their learning algorithm with a vision to producing more accurate results.

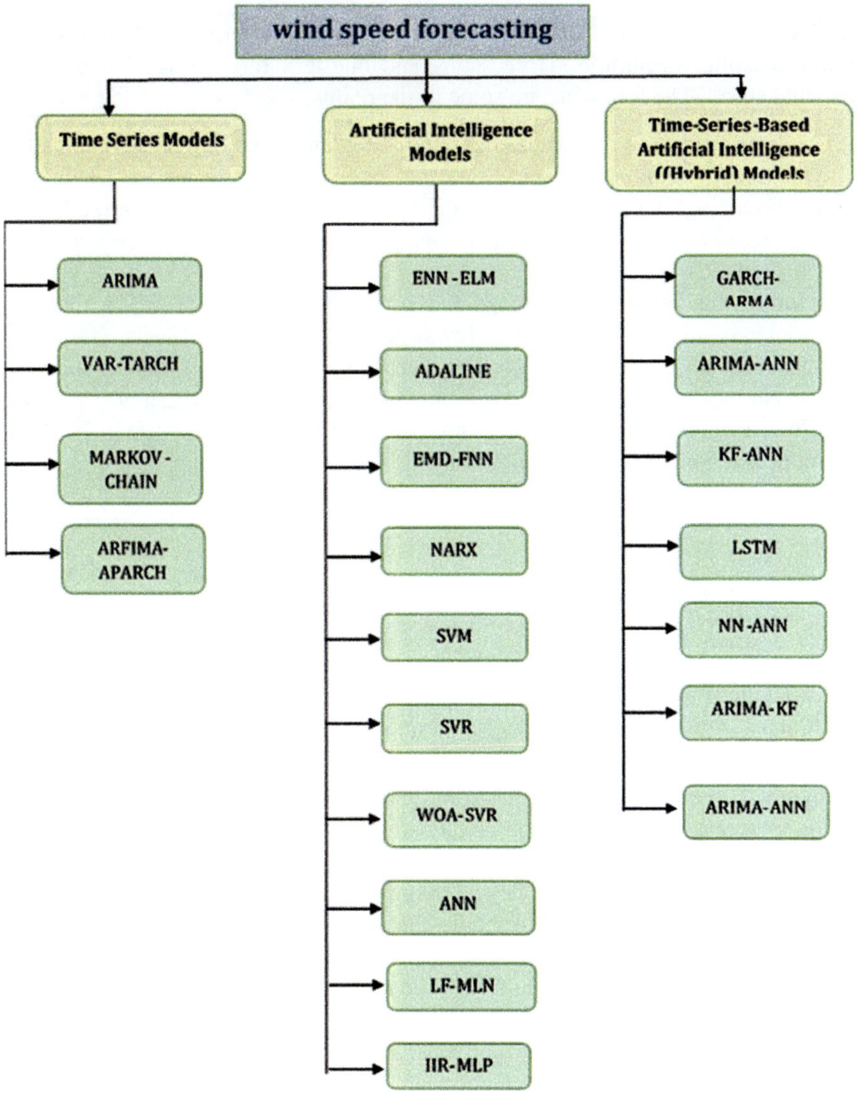

Fig. 2 Wind speed forecasting model classification

- To achieve better results in both long-term and short-term forecasting, combine different statistical and physical models.
- Additional research works on the practical implementation of designs, not just theoretically.
- Advance new methods in mathematical models, eventually, should be proposed.

Because of this comprehensive survey, the following comparative points allow researchers to make a variety of quantitative prediction quality assessments of

ARMA, WMA, MA, and ARIMA methods for very short-term forecasting, short-term forecasting, medium-term forecasting, and long-term forecasting of wind power and wind speed. The following are some of the points:

- The error level rates for forecasting wind speed and wind power rise in parallel to the time horizon.
- The percentage gain declines in wind speed prediction if the time period is shortened. Nevertheless, in wind power prediction, it demonstrates nonlinear characteristics.
- In forecasting of both wind speed and power, MA, as well as WMA models, cannot maximize prediction reliability as best prediction models.
- Most of the ARMA models have risen to prominence by having smaller forecast errors in together wind speed forecasting also wind power forecasting.
- In forecasting of both wind power and speeds, MA, along with ARIMA models, produces the poorest prediction results.

6 Conclusion

The detailed literature review deliberates the optimization of wind speed prediction by using hybrid models along with the statistical models. Existing data processing techniques are not appropriate for dealing with the immense amounts of data generated. The persistent forecasting models used either statistical or machine learning techniques to forecast wind speed, and some authors tried hybrid models such as combining optimization/decomposition methods with the algorithms. The authors also recommended hybrid models are giving better performance than individual models. The long-term data cannot be processed only with statistical analysis as it has to be interpreted with the physical site-specific characteristics. The intricacy evolving in long-range data structure makes these existing combinations of hybrid algorithms to fail in predicting the accurate long-term wind speed for random data sets.

This chapter provided a comprehensive review on forecasting of wind speed, taking into account varying time horizons. Many forecasting models have been addressed, and most of the research has been done on the models that have its own physical characteristics. The main emphasis was on demonstrating the variety of different available forecasting models and offering a comparison study of current techniques to assess the best among available methods. Through comparing the real and forecasted WSD, the hybrid model can be seen to be closely following the actual sequence. Therefore, as it fits the overall trend found in the forecast process, it can be efficiently utilized as a proper alternative template for the long-term wind speed prediction tasks.

Acknowledgements The authors would like to thank **National Institute of Institute of Wind Energy (NIWE),** Chennai, for the opportunities provided to conduct technical assistance for this study. The authors would also wish to extend their whole-hearted gratitude to **Dr. K. Balaraman,**

Director General, NIWE, and other colleagues for providing data and guidance to carry out this research work successfully in the institute.

Appendix

S.No	Author	NARX	ARIMA	ELM	ENN	SVM	VAR—TARCH	EMD with FNN	ANN	Hybrid KF—ANN	NN
1	Yatiyana et al. [1]		✓								
2	Ambach and Vetter [2]						✓				
3	Radziukynas and Klementavicius [3]		✓								
4	Verma et al. [4]										
5	Palomares et al. [5]		✓								
6	Mezaache and Bouzgou [6]			✓	✓						
7	Zhao et l. [7]							✓			
8	Li and Shi [8]								✓		
9	Azad et al. [9]								✓		
10	Menezes and Barreto [10]	✓									
11	Mohandes [11]					✓					
12	Surussavadee and Wu [12]										✓
13	Osama [13]										
14	Barbounis and Theocharis [14]										✓
15	Barbounis et al. [15]										
16	Malik and Savita [16]								✓		

(continued)

(continued)

S.No	Author	NARX	ARIMA	ELM	ENN	SVM	VAR—TARCH	EMD with FNN	ANN	Hybrid KF—ANN	NN
17	Dong et al. [17]										
18	Cadenas and Rivera [18]		✓						✓		
19	Nair et al. [19]		✓						✓		
20	Shukur and Lee [20]									✓	
21	Soman et al. [21]								✓		
22	Lei [22]										
23	Bhaskar et al. [23]								✓		
24	Bali et al. [24]										
25	Kavasseri and Seetharaman [25]		✓								
26	Colak et al. [26]		✓								

S.No	Author	Markov chain model	GARCH	WOA—SVR	SVM	ARMA	IIR-MLP	LF-MLN	RNN	MV	OF-MLN	LSTM
1	Yatiyana et al. [1]											
2	Ambach and Vetter [2]											
3	Radziukynas and Klementavicius [3]											

(continued)

(continued)

S.No	Author	Markov chain model	GARCH	WOA—SVR	SVM	ARMA	IIR-MLP	LF-MLN	RNN	MV	OF-MLN	LSTM
4	Verma et al. [4]	✓										
5	Palomares et al. [5]											
6	Mezaache and Bouzgou [6]											
7	Zhao et l. [7]											
8	Li and Shi [8]											
9	Azad et al. [9]											
10	Menezes and Barreto [10]											
11	Mohandes [11]											
12	Surussavadee and Wu [12]											
13	Osama [13]			✓								
14	Barbounis and Theocharis [14]						✓	✓			✓	
15	Barbounis et al. [15]						✓		✓		✓	
16	Malik and Savita [16]											
17	Dong et al. [17]		✓			✓						

(continued)

(continued)

S.No	Author	Markov chain model	GARCH	WOA—SVR	SVM	ARMA	IIR-MLP	LF-MLN	RNN	MV	OF-MLN	LSTM
18	Cadenas and Rivera [18]											
19	Nair et al. [19]											
20	Shukur and Lee [20]											
21	Soman et al. [21]					✓						
22	Lei [22]					✓						
23	Bhaskar et al. [23]				✓							
24	Bali et al. [24]								✓			✓
25	Kavasseri and Seetharaman [25]											
26	Colak et al. [26]					✓				✓		

References

1. Yatiyana E, Rajakaruna S, Ghosh A (2018) Wind speed and direction forecasting for wind power generation using ARIMA model. In: 2017 Australasian Universities power engineering conference (AUPEC 2017), pp 1–6, Nov 2017. https://doi.org/10.1109/aupec.2017.8282494
2. Ambach D, Vetter P (2016) Wind speed and power forecasting—a review and incorporating asymmetric loss. In: Proceedings of 2nd international symposium on stochastic models in reliability engineering, life science, and operations management (SMRLO 2016), pp 115–123. https://doi.org/10.1109/SMRLO.2016.29
3. Radziukynas V, Klementavicius A (2014) Short-term wind speed forecasting with ARIMA model. In: 2014 55th international scientific conference on power and electrical engineering of Riga Technical University (RTUCON). IEEE, pp 145–149
4. Verma SM, Reddy V, Verma K, Kumar R (2018) Markov models based short term forecasting of wind speed for estimating day-ahead wind power. In: 2018 international conference on power, energy, control and transmission systems (ICPECTS). IEEE, pp 31–35
5. Palomares-Salas JC, De La Rosa JJG, Ramiro JG, Melgar J, Aguera A, Moreno A (2009) ARIMA vs. neural networks for wind speed forecasting. In: 2009 IEEE international conference on computational intelligence for measurement systems and applications. IEEE, pp 129–133
6. Mezaache H, Bouzgou H (2019) Auto-encoder with neural networks for wind speed forecasting. In: Proceedings of international conference on communications and electrical engineering (ICCEE 2018). IEEE, pp 1–5. https://doi.org/10.1109/ccee.2018.8634551
7. Guo Z et al (2012) Multi-step forecasting for wind speed using a modified EMD-based artificial neural network model. Renew Energy 37(1):241–249. https://doi.org/10.1016/j.renene.2011.06.023
8. Li G, Shi J (2010) On comparing three artificial neural networks for wind speed forecasting. Appl Energy 87(7):2313–2320. https://doi.org/10.1016/j.apenergy.2009.12.013
9. Azad HB, Mekhilef S, Ganapathy VG (2014) Long-term wind speed forecasting and general. IEEE Trans Sustain Energy 5(2):546–553
10. Menezes JMP, Barreto GA (2008) Long-term time series prediction with the NARX network: an empirical evaluation. Neurocomputing 71(16–18):3335–3343. https://doi.org/10.1016/j.neucom.2008.01.030
11. Mohandes MA et al (2004) Support vector machines for wind speed prediction. Renew Energy 29(6):939–947. https://doi.org/10.1016/j.renene.2003.11.009
12. Surussavadee C, Wu W (2015) A neural network-based wind forecasting model for wind power management in northeastern Thailand. In: 2015 IEEE international geoscience and remote sensing symposium (IGARSS). IEEE, pp 3957–3960
13. Osama S et al (2018) Long-term wind speed prediction based on optimized support vector regression. In: 2017 IEEE 8th international conference on intelligent computing and information systems (ICICIS 2017), pp 191–196, Jan 2018. https://doi.org/10.1109/intelcis.2017.8260035
14. Barbounis TG, Theocharis JB (2006) Locally recurrent neural networks for long-term wind speed and power prediction. Neurocomputing 69(4–6):466–496
15. Barbounis TG, Theocharis JB, Alexiadis MC, Dokopoulos PS (2006) Long-term wind speed and power forecasting using local recurrent neural network models. IEEE Trans Energy Convers 21(1):273–284
16. Malik H, Savita (2016) Application of artificial neural network for long term wind speed prediction. In: Conference on advances in signal processing (CASP 2016), pp 217–222. https://doi.org/10.1109/CASP.2016.7746168
17. Guo Z, Dong Y, Wang J, Lu H (2010) The forecasting procedure for long-term wind speed in the Zhangye area. In: Mathematical problems in engineering
18. Cadenas E, Rivera W (2010) Wind speed forecasting in three different regions of Mexico, using a hybrid ARIMA-ANN model. Renew Energy 35(12):2732–2738. https://doi.org/10.1016/j.renene.2010.04.022

19. Nair KR, Vanitha V, Jisma M (2018) Forecasting of wind speed using ANN, ARIMA and hybrid models. In: 2017 international conference on intelligent computing, instrumentation and control technologies (ICICICT 2017), pp 170–175, Jan 2018. https://doi.org/10.1109/icicict1.2017.8342555
20. Shukur OB, Lee MH (2015) 'Daily wind speed forecasting through hybrid KF-ANN model based on ARIMA. Renew Energy 76:637–647. https://doi.org/10.1016/j.renene.2014.11.084
21. Soman SS et al (2010) A review of wind power and wind speed forecasting methods with different time horizons. In: North American power symposium 2010 (NAPS 2010), Oct 2010. https://doi.org/10.1109/naps.2010.5619586
22. Lei M et al (2009) A review on the forecasting of wind speed and generated power. Renew Sustain Energy Rev 13(4):915–920. https://doi.org/10.1016/j.rser.2008.02.002
23. Bhaskar M, Jain A, Srinath NV (2010) Wind speed forecasting: present status. In: 2010 international conference on power system technology: technological innovations making power grid smarter (POWERCON2010), pp 1–6. https://doi.org/10.1109/powercon.2010.5666623
24. Bali V, Kumar A, Gangwar S (2019) Deep learning based wind speed forecasting—a review. In: Proceedings of the 9th international conference on cloud computing, data science and engineering, confluence 2019. IEEE, pp 426–431. https://doi.org/10.1109/confluence.2019.8776923
25. Kavasseri RG, Seetharaman K (2009) Day-ahead wind speed forecasting using f-ARIMA models. Renew Energy 34(5):1388–1393
26. Colak I, Sagiroglu S, Yesilbudak M, Kabalci E, Bulbul HI (2015) Multi-time series and-time scale modeling for wind speed and wind power forecasting part II: Medium-term and long-term applications. In: 2015 International conference on renewable energy research and applications (ICRERA). IEEE, pp 215–220

Methods for Epileptic Seizure Prediction Using EEG Signals: A Survey

Srinidhi Bulusu, Raghavarapu Sai Surya Siva Prasad, Pavan Telluri, and N. Neelima

Abstract Epilepsy is one among the most common brain disorders where research is still going on to find the best solution towards its treatment. The phenomenon of occurrence of recurrent seizures is known as epilepsy. This condition is unpredictable, and the patient can suffer at any moment of time. This may lead to permanent nervous system breakdown or death. Several researchers have focused on predicting seizure activity from electrocardiogram (ECG) signals and electroencephalogram (EEG). This paper focuses on the different methods and models used to predict seizure from EEG signals to lessen the burden on patients because of the unpredictable nature of seizures. Furthermore, it also gives insights about deep learning approaches used to predict epilepsy using the EEG signals.

Keywords Epilepsy · Seizure · EEG · Signal processing · Deep learning

1 Introduction

A seizure can be described as an unmanageable, abrupt electrical disorder. Some of the common effects of seizures are drastic changes in emotional, mobility and behavioural patterns. Seizures are commonplace and manifest in various types. Periodic seizures, or greater than two seizures, are a case called epilepsy, a condition that affects about 50 million people on the earth today. Depending on the amount of time they last, severity of seizures can be classified. Unnatural electrical activity in a

S. Bulusu · R. Sai Surya Siva Prasad · P. Telluri (✉) · N. Neelima
Department of Electronics and Communication Engineering, Amrita School of Engineering, Amrita Vishwa Vidyapeetham, Bengaluru, India
e-mail: ptelluri@gmail.com

S. Bulusu
e-mail: bulususrinidhi@gmail.com

R. Sai Surya Siva Prasad
e-mail: suryaraghavarapu@gmail.com

N. Neelima
e-mail: n_neelima@blr.amrita.edu

D. J. Hemanth et al. (eds.), *Artificial Intelligence Techniques for Advanced Computing Applications*, Lecture Notes in Networks and Systems 130,
https://doi.org/10.1007/978-981-15-5329-5_10

region of the brain gives rise to focal seizures. Whether or not they occur with a loss of consciousness is the parameter to classify focal seizures. Broadly, the temporal dynamics depending on the seizure's occurrence are classified into four types: Ictal which is the state during seizure, preictal is the state observed right before an impending seizure, post-ictal is noted right after the seizure, and interictal state is between the seizures and normal state. Electroencephalogram (EEG) is a measurement mechanism of the electrical activity in brain which is non-stationary and nonlinear. EEG signals are useful in studying a lot of scenarios ranging from basic sleep cycle to crucial aspects of medical diagnosis. Based on the collection mechanism, the EEG waveforms are classified into intracranial electroencephalogram (iEEG) and scalp electroencephalogram (sEEG) where iEEG waveforms are considered to offer wider frequency range better signal-to-noise ratio [1]. The different states of the seizure occurrence can be identified and by distinguishing between interictal and preictal, from the recorded EEG signals. This paper presents the recent advances in seizure prediction using both iEEG and sEEG signals.

2 Methods and Models

2.1 Wavelets

Wavelets are the powerful mathematical tools that provide the detailed analysis of signals in the effective manner. With this advantage, Bhati et al. [2] have compared the performance of feature extraction from bispectral phase entropies and energies of the wavelet sub-bands of EEG signals to classify the EEG signals into non-seizure EEG signals (Interictal) and seizure EEG signals (preictal). Not just owing to the highly nonlinearity and non-stationary nature of the EEG signals, the authors have chosen wavelet transform which is the appropriate choice to note the fast-changing EEG impulses both in time and frequency domain but also because the wavelets can be tuned for better resolution in both the time and frequency domains simultaneously. Features are extracted from the wavelet sub-bands which are decomposed from the signal which again can be done from either bi-orthogonal filter banks or orthogonal filter banks. The authors have discussed the advantage of two-band bi-orthogonal filter banks over the two-band orthogonal filter banks as the higher degrees of freedom and proceeded in extracting the features with it.

The setup which provides an insight on the perfect reconstruction filter bank (PRFB) is shown in Fig. 1. This is used to perform wavelet decomposition.

The bispectral phase entropy is calculated using the following Eq. (1).

$$\beta(v1, v2) = E[F(v1)F(v2)F*(v1+v2)] \tag{1}$$

where $F(v)$ is the Fourier transform and the norm bispectral entropies are represented by H_{en1} and H_{en2} given by Eq. 2.

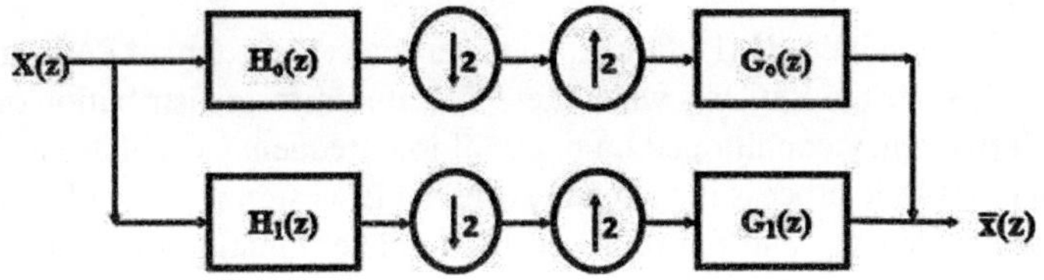

Fig. 1 Perfect reconstruction filter

$H_{\text{en1}} = -\sum_k (p_k \log(p_k))$ and $H_{\text{en2}} = -\sum_i (q_i \log(q_i))$ where p_k and q_i are calculated using Eq. 3

$$p_k = \frac{|\beta(v1, v2)|}{\Sigma_\Omega |\beta(v1, v2)|} \quad \text{and} \quad q_i = \frac{|\beta(v1, v2)|^2}{\Sigma_\Omega |\beta(v1, v2)|^2} \tag{2}$$

After the feature extraction, Levenberg–Marquardt backpropagation algorithm has been used as the neural network training algorithm with one hidden layer and ten hidden neurons in order to classify the EEG signals into seizure and non-seizure categories. These evaluations are performed on the data provided by Bonn University, Germany, which contains four sets of non-seizure data comprising of 400 recordings combined and a single set of 100 seizure signals. It is concluded that the performance of the bispectral phase entropy measure increases with the decrease of the regularity of the signal and the performance of the energy measure increases with the increase of the regularity of the signal. Moreover, it is noted that the highest classification accuracies achieved through energy measures and bispectral phase entropies are 98.2% and 96.4%, respectively, concluding energy measure performed better, but the overall complexity is high.

2.2 *Phase Amplitude Coupling*

In this paper [3], author gives the insights about the linear prediction error energy method [4] in time domain, and power spectral density (PSD) measure [5] in frequency domain is given in [3]. These methods lack in the consideration of the nonlinear features of the EEG signals and proceed with the proposal to implement nonlinear methodologies like the usage of entropy [6]. The proposed method in this study is based on the [7] which talks about CFC (cross frequency coupling) being a correlated pattern through distinct frequency ranges extracted from the same signal. Although there are numerous patterns of CFCs available like phase–phase coupling (PPC), phase amplitude coupling (PAC) and amplitude–amplitude coupling (AAC), the author particularly emphasizes on PAC citing. Schack et al. [8] concluded the fact that the phase of θ band modulates the amplitude of γ band during the times when the brain is crunching the tasks of memory mapping and perception. The author then brings in modulation index (MI) stating that it can detect the extent of PAC variations in EEG signals since the PAC lies when there is a non-uniform distribution of amplitude of high frequency conditioned on phase of low frequency oscillations. The MI calculation is a simple process in which the original EEG signal $y(t)$ is filtered at two different frequencies (f_a and f_p) followed by the determination of the time-series phases of $y_{fp(t)}$. The datasets used in this study are two open source datasets, and they are CHB-MIT and Bonn dataset. The extracted features by PAC are then classified by support vector machine (SVM), and the results are observed to be varying from 97.5 to 100% for the samples in Bonn dataset and an average of 97.5%. Furthermore, it is concluded from CHB-MIT database that there is an increase in the value of MI at the beginning point of the seizure.

2.3 Discrete Stationary Wavelet Transform

Anand et al. [9] presented a method to improve the detection of the epileptic seizures with better accuracy. The fundamental objectives of the proposed model are, noise removal on the EEG signal at the preprocessing step, feature extraction, feature classification and finally to detect the disease. Many other classification techniques are also exploited in the literature like backpropagation [10]. The preprocessed step is done using the discrete stationary wavelet-based Stockwell transform (DSWT), and its primary function is denoising. The development of DSWT is done by the combination of the stationary wavelet and Fourier transform. This step is followed by shift invariant in wavelet and discrete Fourier transform. Then, the ε-circular shift is being done for up sampling. The feature extraction method is done by considering the spectral feature, temporal feature and amplitude distribution estimation. By utilizing the above features, the spectral power is calculated based on periodogram. The feature selection process is done based on the best value of the provided signal classes. Here, the PSO model is developed which includes filter bank-based fitness function and swarm optimization to reduce the dimensionality of the data. Finally, classification is done with the help of the hybrid K-nearest-based SVM. Furthermore, classifications are stratified into four different classes to attain the best possible accuracy, sensitivity, precision, recall and other desired parameters.

2.4 Mean Phase Coherence

This work presented in [11] is solely based on the study that Mormann et al. [12] have conducted and concluded that the mean phase coherence (MPC) values before the ictal period drop significantly below the non-ictal EEG state rendering phase synchronization as a reliable parameter for seizure prediction. Mormann et al. [12] also have studied both linear and nonlinear techniques and concluded that variations in the dynamics were not entirely contributed by nonlinear features. However, Zheng

[13], owing to non-stationary feature of EEG signals, used Bi-variate empirical mode decomposition or BEMD in order to decompose pairs of samples into intrinsic mode functions (IMF) in which he observes that there is a fluctuation if not decrease in the phase synchronization. This study examines three types of samples which include normal EEG signals, generalized seizure samples along with focal epilepsy, where seizures are localized to the occipital lobe. The data used in this [11] study is collected from Neurology and Sleep Centre, Hauz Khas, Delhi, from two normal individuals and ten epileptic patients. The data was preprocessed by filter through a band-pass filter of the range from 0.5 to 40 Hz followed by a 50 Hz notch filter to eliminate the interference caused by the 50 Hz power supply. The data was sampled at 200 Hz before segmenting into distinct windows of size 5 s each. The phase synchronization is calculated using Eq. 3.

$$\phi_{n,n} = n\phi_a(t) - m\phi_b(t) = \text{constant} \tag{3}$$

Instantaneous phase of the signal can be calculated using

$$\phi(t) = \tan^{-1}\left(\frac{\tilde{s}(t)}{s(t)}\right);$$

where $\tilde{s}(t)$ is Hilbert transform of signal given by Eq. 4.

$$\tilde{s}(t) = -iFT^{-1}(FT(S(t)) * \text{sign}(w)) \tag{4}$$

To measure the relative phase between these two signals a and b and to confine the phase between $(-\frac{\pi}{2}, \frac{\pi}{2})$ the Eq. 5 is used with $m = n = 1$.

$$\phi_{1,1} = \phi_a(t) - \phi_b(t) = \arctan\frac{\tilde{s}_a(t)s_b(t) - s_a(t)\tilde{s}_b(t)}{s_a(t)s_b(t) - \tilde{s}_a(t)\tilde{s}_b(t)} \tag{5}$$

The mean phase coherence is calculated using Eq. 6.

$$R = \frac{\sum_{i=0}^{N-1} \mathrm{e}^{i\phi_{1,1}(j\Delta T)}}{N} \tag{6}$$

The threshold value of the running sum between the areas of the MPC and their mean is set to be −0.4 for the ten-window interval which means, for the past ten windows, if the threshold value is triggered, a seizure is imminent. The results have been categorized into three cases as per the samples collected, and they are pertaining to patients with focal epilepsy, patients with generalized epilepsy and normal individuals. From the results, it can be interpreted that the algorithm and features of the author's choice distinguished between the focal epilepsy and normal EEG signals. However, in the case of patients with general epilepsy, in a specific case, there is no significant drop in the curve below the mean for the detection of seizure. This might be interpreted as no imminent seizure.

2.5 Time–Frequency Analysis

Yang Li in the paper [14] discussed about the time frequency analysis of the EEG signals using the MRBF-MPSO technique which turned out to be the best method compared to other feature extraction techniques. Electroencephalography signals play a crucial in the automatic detection and classification of the epileptic seizures. Given the randomness of the seizure process, it is very challenging to predict the seizures accurately. Time–frequency analysis of the EEG signals is one of the many effective methods that has a predominant role in identifying the seizures. Many time domain approaches have been proposed for finding the periodic discharges in EEG signals during the seizures. Additionally, many techniques in frequency domain techniques like Fourier transform, Laplace transforms have also been proposed, but given the non-periodic, uncertain and non-stationary nature of the EEG signals, it is very challenging to identify the seizure onset. Hence, owing to the limitations of the individual techniques given above, an established method called time frequency analysis (TFA) has been used for extracting the desired features. The state of art techniques like multiscale radial basis functions (MRBF) and a modified particle swarm optimization (MPSO), logistic regressions are being used to ameliorate the precision of the epileptic EEG signals in both time and frequency spectrum. Care must be taken in reducing the dimensionality of the extracted features before they are fed into the classifiers. After the removal of the redundant features by the PCA, the most sensitive features are then fed into classifiers for classification. An SVM classifier based on RBF kernel is one of the extensively used classifiers in determining the signals. It outperformed many of the classifiers like KNN, RNN, LDA, LR and NB in terms of accuracy, minimum false positives and some other features [15, 16]. In SVM, a hyperplane separates the classes that are to be classified and has proven to be one of the best discriminative supervised learning algorithms in signals and systems.

3 Scalp Signals

3.1 Partial Directed Coherence

Partial directed coherence (PDC), used as a feature extraction mechanism for EEG recordings of the scalp, can help observe the brain activity before and after the start of a seizure, in the seizure prediction experiment [17]. The flow of information in regions of the brain was considered in this study. The original definition of PDC is as follows: $P_{ij}(f) = A_{ij}(f)/\sqrt{(a_j * (f)a_j(f)}$ where $P_{ij}(f)$ gives the intensity of the flow of information from channel j to i at frequency f along with its direction, $A(t)$ denotes MVAR model coefficients, A(f) denotes the Fourier transform of $A(t)$, and an element of $A(f)$ is given by $A_{ij}(f)$, '*' is the transpose of a complex conjugate operation. The range of PDC values is [0,1] after the procedure of normalization. PDC shows only the direct flow between channels, especially focussing on the sinks and not the sources. Paper [17] implemented a new approach using PDC for detecting seizure intervals in patients suffering from epilepsy. Ten patients suffering from intractable partial epilepsy were taken into study. The patients underwent pre-surgical evaluations that are non-invasive and video monitoring of scalp EEG. Scalp EEG recordings were taken at a sampling rate of 200 Hz. Forty-four seizures spanning 120.6 h, with the mean duration of ictal state of 98.7 s, was obtained, with the number of seizures ranging between 2 and 9 in different patients. The artifacts in each of the patients' data were eliminated and segmented based on window size for analysis, which is 2 s in this case. For every window, multivariate autoregressive (MVAR) model was carried out. PDC algorithm was implemented to extract the information flow and the direction. Feature dimensionality reduction was achieved by adding up all the outflow information of an EEG channel. The input to an SVM classifier was taken as the outflow information of a channel, and the interictal and ictal periods were classified. A five-fold cross-validation was utilized to evaluate the performance of the algorithm. The total ictal period duration was 72.38 min, the accuracy was reported as 98.3%, and the mean selectivity is reported as 67.88% along with the average positive rate of 95.39%. Paper 20 also carried out two more methods auto regression model and SVM classifier (AR-SVM) and approximate entropy and SVM (ApEn-SVM) and a combination method of PDC and artificial neural network-error backpropagation (ANN-BP). Among all of the tested methods, the proposed algorithm proved to be the most efficient. Yet, the drawback lies in the fact that the algorithm could not accurately identify every seizure interval due to their short-lasting time and low amplitudes for some seizures.

3.2 Spectral Power and Power Spectral Density

Prediction of the seizures is difficult as the EEG patterns changes from the patient to patient. Zhang and Parhi [18] in the paper presented a novel approach using the scalp EEG recording with hardware of minimal complexity. Here, spectral features along with their ratios are extracted. It is found that power spectral densities are varied before and after the seizures [16, 19]. The patient-specific algorithm is developed to predict the seizures using two electrodes. The main contribution of the paper is its highest sensitivity and lowest false positive rate in comparison with the papers using spectral powers as features. All the features are extracted by performing the FFT. Linear classifier is being used to separate the preictal from the interracial features because of its low power consumption compared to the nonlinear. The major key consideration taken in this paper is electrode selection, selecting the useful features and the choice of the classifier. The paper takes in account of the relative spectral powers and ratio of them. Scatter matrix is being exploited to select the feature basis. Linearly separable features are fed into a linear SVM. Several techniques like window-based signal processing, spectral power and spectral power ratio are used.

In the post-processing step, the Kalman filter is used to smooth the desired fluctuation. The different attributes that concerned the efficient selection of features are the feature basis selection, electrode selection and feature selection by BAB (Branch and Bound). Same feature is considered for the patients with same type of the features and classified using the receiver operating characteristics. For the different kind of the seizures at different parts of the brain, multi-dimensional feature is selected. Overall, the system utilizing SVM classifier is efficient energy consumption wise when contrasted with the any other classifier.

3.3 Empirical Mode Decomposition

Suitable for processing non-stationary and nonlinear series, empirical mode decomposition (EMD) is a time-space analysis method that is adaptive. It does not need a fixed basis in prior. Comparable to Fourier transforms and wavelet decomposition, EMD partitions a series into 'modes', intrinsic mode decompositions (IMFs). IMFs do not lose out on the properties of the original signal. Two principles are to be satisfied by each intrinsic mode function:

i. The sum of the total count of zero crossings and number of extrema should be less than or equal to one, at the most in the complete dataset.
ii. The mean value of the envelope defined by the local minima and the local maxima is zero.

The behaviour of IMFs changes with the changes or abnormalities in brain activity. Paper [11] and Paper [20] use the method of empirical mode decomposition for feature extraction in their approaches to seizure prediction. For Paper [11], the dataset is developed by University of Bonn, Germany. The dataset has five sets containing 100 single-channel EEG signals, each of 23.6 s duration. The five sets are Z, O, N, F, S. Z stands for the eyes open state, O for the eyes closed state, N and F seizure free data, S for seizure data. The data samples here are of intracranial EEG recordings, taken using intracranial recordings. The data sampling rate is 173.61 Hz, and there are 4096 samples in each segment. A low pass filter of 40 Hz was used to remove artifacts. Three classifiers, namely K-nearest neighbour classifier, linear discriminant analysis and Naive Bayesian classifier, were used. The performance was measured in sensitivity, specificity and accuracy. Using Renyi entropy for feature extraction and coherence measures for classification, the highest accuracy value, sensitivity value and specificity value were all found to be 96.97%. Paper [20] explains the EMD algorithm step-by-step: For a signal $m(t)$,

i. The maxima and minima of $m(t)$ are to be extracted
ii. To determine maximum and minimum envelopes, interpolation is to be done between them
iii. Using envelopes from step 2, the local mean, lm(t) is calculated by $\mathrm{lm}(t) = n_{\min}(t) + n_{\max}(t)/2$, where $n_{\min}$ and $n_{\max}$ are the minimum and maximum envelopes

iv. The detail $b(t) = m(t) - \text{lm}(t)$ is extracted
v. The procedure is iterated to step 1 if $b(t)$ does not match IMF criteria, with $b(t)$ as the new input. In this case, further steps are skipped.
vi. If $b(t)$ matches the IMF criteria, it is stored as an IMF, $v_i(t) = b(t)$ and $c(t) = m(t) - v_i(t)$, where i is the ith IMF.
vii. With $c(t)$, the procedure is again begun from step 1 and $v_i(t)$ stored as an IMF.

The above algorithm outputs an IMF series and a final residual signal. The algorithm can be stopped based on two criteria:

i. Based on IMF definition
ii. Number of produced IMFs.

The proposed method is cost-efficient, non-time consuming and efficient on the whole. The advantages of EMD are the possibility of reduction of the span of a signal, feasibility to change amplitude and frequency along the axis of time and no necessity of fixed basis functions in prior. Yet, the disadvantage lies in the fact that there is a change of mode-mixing, due to which some required data could be lost too.

3.4 *CSP and LDA for Scalp EEG Signals*

Epileptic seizure prediction system specified for a patient based on the common spatial pattern (CSP) for scalp EEG signals is proposed by Bou Assi [21]. CSP is extensively used in electromyography (EMG) [22]. CSP, being a statistical method, is used in distinguishing preictal and interictal activities by constructing a projection matrix. In this work, the long-term sEEG recordings of 24 patients are analysed for 987.85 h with 170 seizures. The proposed model comprises two main components, and they are feature extraction and classification. As discussed earlier, the training and testing features are extracted using the CSP. A linear discriminate analysis (LDA) trained with interictal and preictal vectors classifies the preictal stage as "1" and zero otherwise. Positive and negative alarms are used to predict the seizures. Alarm is positive if it is within the prediction horizon and negative otherwise. Three different prediction horizons used: 60, 90,120 min which are used by other authors. Parameters like sensitivity, false prediction rate and prediction time are used to measure the effectiveness of the model in this work. Periodic predictor raises alarm at a fixed time period T, whereas Poisson predictor gives an alarm according to the exponential distributed random time. The average sensitivity obtained was 0.89, average specificity was 0.39, and average prediction time was 68.71 min. Other means of successful evaluation are that when a patient gets a false positive rate of less than 0.2, wherein this paper has achieved an FPR of 0 for 10 out of 19 patients.

3.5 Hybrid Features

Paper 14 proposes integration of Hilbert-Huang transform (HHT) and Kraskov Entropy. To improve performance of classification, using hybrid features was observed to give better results. Paper [14] uses three datasets A, B, C. Data A developed by Bonn University has five subsets namely O, Z, F, S, N. Each subset has 100 segmented single-channel EEG signals. Subsets Z and O are for eyes opened and closed states, respectively. The sets F and N are from epileptic patients before attack. S set is from patients under the attack of a seizure. All the five sets were used for the purpose of this dataset A. Data B from CHB-MIT is a publicly available dataset at PhysioNet containing scalp EEG signals. The dataset has 23 child patients with seizures that are intractable, from Boston's Children's Hospital. With each signal lasting about a few hours and segmented into lengths of 23.6 s, the sampling rate of each signal was 256 Hz. Data C, a combined EEG of Data A and Data B, was used too. Signals identified as 'Normal' belong to five healthy people, i.e. from Z and O of Data A. 'Interictal' signals belong to epileptic patients of F and N of Dataset A and before seizure data from Data B. 'Ictal' signals are from Data A and B from patients during seizures. Signals identified as 'Non-Ictal' include 'Interictal' and 'Normal'. The classification was divided into three groups:

Group1—'Normal' and 'ictal' classification
Group2—'Ictal' and 'Interictal' classification
Group—'Ictal' and 'Non-ictal' classification

The approach in [23] mentions and explains three feature extraction methods. First, the proposed method is being, Kraskov entropy based on HHT. Second, Kraskov entropy based on tunable-Q wavelet transform. Third, instantaneous area of IMFs using HHT. In the first feature extraction method, before feature extraction, Hilbert-Huang transform is used to analyse time–frequency distribution of signals. Two steps are involved in the implementation of this transform. Firstly, using EMD to decompose EEG signals into IMFs, and next, applying Hilbert transform on these IMFs. The Hilbert transform for an IMF is as follows:

$$H[IMF_i(t)] = \frac{1}{\pi} \int^{+\infty} \frac{IMF_i(\tau)}{t-\tau} d\tau \tag{7}$$

In the second-mentioned method of feature extraction, Kraskov entropy decomposed by TQWT is used which is preferred transform for the analysis of non-stationary signals. In the third feature extraction method, the IMFs are obtained from HHT and their instantaneous areas calculated and used. The computation of instantaneous area for each IMF is given by

$$S_{IMF_m}(n) = \pi \sum_{i=n}^{L-1+n} r(i)^2 \tag{8}$$

All the above three were then combined to form a hybrid feature and input to the least squares version of support vector machine (LS-SVM) classifier with the radial basis function kernel. The results of the experiment show that hybrid features are a better choice for detection. For 'Ictal' and 'Normal' classification, accuracy was 77.72%. In the 'Interictal' and 'Ictal' EEG classification, the parameters were 82.8%, 74.93% and 88.56%. For classifying into 'Ictal' and 'non-Ictal', the parameters measured 85.30%, 59.93% and 94.55%, respectively, for accuracy, sensitivity and specificity. Paper [20] presents a detailed report of the method of detection of seizures using hybrid features and compares it with two other methods, namely wavelet and Fourier transform techniques. It gives the accuracy, sensitivity and specificity values for various comparisons and concludes that the hybrid feature method is the most efficient. The advantage of hybrid feature method is that it can be applied to complex databases with various components without compromising on efficiency of classification. But it is not fast enough for real-time applications and that still lies as a disadvantage.

4 Classification—Deep Learning Approach

Daoud et al. [24] give insights about the implementation of an automatic seizure detection system using FPGA for high accuracy. For dimensionality reduction of feature vectors, they proposed to apply mean power frequency (MPF) and calculating Hilbert's transform for extracting the features, and then, the multi-layer perceptron ANN classifies the signal into seizure and non-seizure categories. The dataset used in this work is developed by the University of Bonn on which Hilbert transform is applied followed by the usage of FIR filter as an analytical filter. MPF helps in selecting the most sensitive features. After the extraction of features, they were again subjected to a statistic analysis to verify the discrimination ability. The analysis used in this piece of work is Mann Whitney test which relies on relative ranks of continuous random variable in order to prove that both the tested features are equal and can be used for non-parametric distribution. Then, the classification is carried out by MLP with two hidden layers which is trained using backpropagation and was optimized by Adam optimization algorithm, and to ensure the least over-fitting, dropout regularization technique is used. The hardware is implemented on FPGA allowing a provision of reconfiguration. The FPGA used is M2GL025-VF256 from IGLOO2 by Microsemi. The Hilbert transform is represented by a FIR filter. The proposed hardware model is verified to have an accuracy of 100% and is evaluated to have 159.7 mW of total power.

4.1 Neural Network on SoC for Seizure Classification

Tianchan Guan, in the paper [25], gives an insight about seizure prediction using neural networks. Neural networks have become the state of art classifiers outperforming most of the existing models [26]. In this paper, the raw EEG signals are taken directly without explicitly removing the features in the pre-analysis step. In the training set, often the conundrum arises in the selection of the data. Usually, the raw data obtained is of non-seizure waveforms and only with minutes of the seizure states. This paper has modified the data by performing down sampling and obtained the desired set of the signals as an input to the NN classifier. The architecture is based on the multi-layer perceptron. The output is based on the consecutive input windows containing the seizures, and then, the signals are classified. 'The leave one out' validation scheme is used to train the neural networks. The performance on CHB-MIT dataset [25] showed a high average detection sensitivity and specificity of 82% and 95.5%, respectively. The model is comparable to the other existing models using energy hungry extraction features.

4.2 Feature Extractor Using FPGA

In this [27] work, an EEG feature extractor and classifier is designed and developed on Virtex 5 FPGA board, and the Verilog code was implemented on Xilinx ISE platform. The paper describes two categories of seizure detection, namely onset detection and event detection. Onset detection is the one where the classifier detects the seizure with the shortest time interval from the start of the seizure. Consequently, event detection is an implementation which detects whether or not if a seizure has occurred. The piece of work implemented classifier for event detection. First, in order to smoothen the EEG signals, FIR filter of order 64 is used without feedback after which the features are extracted using discrete wavelet transform (DWT) [28] which transforms a time domain signal to both time and frequency domains. After the feature extraction, the features are fed into linear classifier which classifies the EEG sample into two categories.

4.3 Classifying Epileptic Seizures Using Entropy Variants

Waqar Hussain in his work [29] throws light on the classification of the epileptic seizures using entropy variants. Withstanding with the randomness and non-stationary fashion of the EEG signals, it is very difficult to extract the desired features required to classify the seizures. In this paper, a novel approach to determine the hidden features of EEG signals is demonstrated. It is based on the entropy values, which resulted in the better performance compared to the existing state of art entropy techniques. A new index called PFuzzy is used for classification of EEG signal from the recorded data. In frequency attenuation, notch filter is used to remove the unwanted attenuation. Four entropies namely approximate entropy (ApEn), sample entropy (SampEn), permutation entropy (PE) and permutation fuzzy entropy (PFuzzy) are considered to perform the analysis. Out of these, the PFuzzy stood out in comparison with the other entropies. Anti-noise is the key feature in the PFuzzy entropy analysis and is insensitive to the embedded noise. After this step, the extracted features are sent to the SVM filter to classify both binary and multiclass problems. The PFuzzy entropy has acquired an overall accuracy of 95.1% in comparison with the adjacent entropies. There is a room for increasing the performance by adding a greater number of preictal samples.

4.4 Multichannel Feature Extraction

The proposed hardware [30] model consists of an SPI, FIFO, feature extractor, feature serializer, classifier and multichannel vote. The feature extractor yields the features in parallel to feature serializer, which buffers them and outputs them out serially to classifier. Of the four classifiers mentioned above, one is elected and others not put to action to optimize hardware resource usage. The multichannel vote block employs a simple voting method. If the aggregate of channels that group as a seizure surpasses a threshold, then it is categorized as a seizure. To counterbalance the performance and area was the superlative focus, and various architectures were exploited for the best performance of each classifier. The experimentation with each classifier led to deductions that pipelined architecture for SVM, parallel distance calculation for KNN and serial implementations for LR and Naive Bayes are the optimal choices. The best trade-off between performance and area could be spotted by simulating on MATLAB and comparing the real patient data and investigating accuracy. Results demonstrate that LR is the best classifier in terms of exercise of resources. NB classifier stands second. Though KNN has superior performance than SVM, it utilizes more resources comparatively. An accuracy of over 80% on data from ten patients and an F1 measure of 91% using LR was obtained.

5 Conclusion

The unexpected occurrence of seizures is the most frightening for the patients. This article presents significant information on various methods and models used to predict seizure activity from EEG signals. Numerous methods for the classification of seizure EEG signal and the non-seizure EEG data have been compared, and a few models of hardware implementation of the seizure prediction system also have been discussed. Highest accuracy is noticed when the features are extracted using the wavelet transforms, and KNN is used for training and classification. Furthermore, the performance of the hardware models is also compared, and the appropriate model can be chosen depending up on the features extracted. The objective of this article is to enhance the better understanding of existing models for researchers to encourage them towards further research to find optimum solution.

References

1. Edakawa K, Yanagisawa T, Kishima H, Fukuma R, Oshino S, Khoo HM, Kobayashi M, Tanaka M, Yoshimine T (2016) Detection of epileptic seizures using phase–amplitude coupling in intracranial electroencephalography. Sci Rep 6
2. Bhati D, Pachori RB, Sharma M, Gadre V (2019). Automated detection of seizure and nonseizure EEG signals using two band biorthogonal wavelet filter banks
3. Liu Y, Wang J, Cai L, Chen Y, Qin Y (2017) Epileptic seizure detection from EEG signals with phase–amplitude cross-frequency coupling and support vector machine. Int J Mod Phys 32(08)
4. Zhang C, Bin Altaf MA, Yoo J (2016) Design and implementation of an on-chip patient-specific closed-loop seizure onset and termination detection system. IEEE J Biomed Health Inform 20(4):996–1007
5. Faust O, Acharya UR, Allen A, Lin CM (2008) Analysis of EEG signals during epileptic and alcoholic states using AR modeling techniques, pp 44–52
6. Richman J, Moorman J (2000) Physiological time-series analysis using approximate entropy and sample entropy. Am J Physiol Heart Circul Physiol 278
7. Yamamoto Jun, Suh Junghyup, Takeuchi Daigo, Tonegawa Susumu (2014) Successful execution of working memory linked to synchronized high-frequency gamma oscillations. Cell 157(4):845–857
8. Schack B, Vath N, Petsche H, Geissler H-G, Möller E (2002) Phase-coupling of theta–gamma EEG rhythms during short-term memory processing. Int J Psychophysiol 44(2):143–163. ISSN 0167-8760
9. Anand Satyajit, Jaiswal Sandeep, Ghosh Pradip Kumar (2019) Epileptic seizure detection in EEG signal using discrete stationary wavelet-based stockwell transform. Majlesi J Electr Eng 13:55–63
10. Kumar TS, Kanhangad V, Pachori RB (2015) Classification of seizure and seizure-free eeg signals using local binary patterns. Biomed Signal Process Control 15:33–40
11. Aggarval G, Ghandi TK (2017) Prediction of epileptic seizures based on mean phase. BioArXiv. https://doi.org/10.1101/212563
12. Mormann F, Kreuz T, Andrzejak RG, David P, Lehnertz K, Elger CE (2003) Epileptic seizures are preceded by a decrease in synchronization. Epilepsy Res 53(3):173–185
13. Zheng Y, Wang G, Li K, Bao G, Wang J (2014) Epileptic seizure prediction using phase synchronization based on bivariate empirical mode decomposition. Clin Neurophysiol 125(6):1104–1111
14. Li Y, Wei HL, Billings SA, Liao XF (2012) Time-varying linear and nonlinear parametric model for granger causality analysis. Phys Rev E 85(4)
15. Meyer D, Leisch F, Hornik K (2003) The support vector machine under test. Neurocomputing 55(1/2):169–186
16. Park Y, Luo L, Parhi KK, Netoff T (2011) Seizure prediction with spectral power of EEG using cost-sensitive support vector machines. Epilepsia 52(10):1761–1770

17. Wang G, Sun Z, Tao R, Li K, Bao G, Yan X (2017) Epileptic seizure detection based on partial directed coherence analysis. IEEE J Biomed Health Inform 20(3):873–879
18. Zhang Z, Parhi KK (2016) Low-complexity seizure prediction from iEEG/sEEG using spectral power and ratios of spectral power. IEEE Trans Biomed Circuits Syst 10(3):693–706
19. Bandarabadi M, Teixeira CA, Rasekhi J, Dourado A (2014) Epileptic seizure prediction using relative spectral power features. Clin Neurophysiol
20. Parvez MZ, Paul M (2017) Seizure prediction using undulated global and local features. IEEE Trans Biomed Eng 64(1):208–217
21. Bou Assi E, Nguyen DK, Rihana S, Sawan M (2017) Towards accurate prediction of epileptic seizures: a review. Biomedical Signal Processing and Control, vol 34, pp 144–157, 8 Dec 2017
22. Hahne JM, Graimann B, Muller K-R (2012) Spatial filtering for robust myoelectric control. IEEE Trans Biomed Eng 59(5):1436–1443
23. Lu Y, Ma Y, Chen C, Wang Y (2018) Classification of single-channel eeg signals for epileptic seizures detection based on hybrid features. In: Technology and Health Care, no. Preprint, pp 1–10
24. Daoud HG, Abdel Hameed AM, Bayoumi M (2018) FPGA implementation of high accuracy automatic epileptic seizure detection system. In: 2018 IEEE 61st international midwest symposium on circuits and systems (MWSCAS), pp 407–410
25. Guan T, Zeng X, Huang L, GuanT, Seok M (2016) Neural network based seizure detection system using raw EEG data. 2016 International SoC design conference (ISOCC), pp 211–212
26. Madan K, Bhanu Anusha K, Neelima N (2019) Research on different classifiers for early detection of lung nodules. Int J Recent Technol Eng 8:2S3
27. Jacob R, Menon KP (2017) Implementation of EEG feature extractor and classifier for seizure detection on FPGA. In: International conference on intelligent computing and control systems (ICICCS), pp 307–310
28. Bhavana V, Krishnappa HK (2015) Multi-modality medical image fusion using Discrete Wavelet Transform. In: 4th international conference on eco-friendly computing and communication system (ICECCS 2015), Procedia Computer Science, pp 625–631
29. Waqar H et al (2019) Towards classifying epileptic seizures using entropy variants. In: 2019 IEEE fifth international conference on big data computing service and applications (Big Data Service), pp 296–300
30. Page A, Sagedy C, Smith E, Attaran N, Oates T, Mohsenin T (2015) A flexible multichannel EEG feature extractor and classifier for seizure detection. IEEE Trans Circ Syst II 62(2):109–113

Railway Wagon Health Monitoring System Using E-BMA Protocol in Wireless Sensor Networks

S. Rajes Kannan and S. Amutha

Abstract In railway wagon health monitoring system, sensors are worn to monitor the failures in the railway wagons and pass on the data to the source station. But sensors have restricted energy resources and their practicality continues till their energy is drained. Therefore, energy for detector networks ought to be managed and to attain energy potency within the detector network [1]. To achieve the energy potency in sensor networks, this study proposes a replacement cluster-based MAC scheme which is called Energy-Efficient Bit Map Assisted protocol. The main objective in planning the E-BMA MAC procedure is to avoid wasting the energy for every sensor node and to scale back the energy waste because of inactive listening and collisions whereas maintaining a decent low-latency achievement [8]. It additionally conserves energy once anode doesn't involve in the method of sending or receiving packets. In order to boost the energy potency more, DED clustering algorithm is implemented. Using this formula a node with higher residual energy, higher degree, and nearer to the base station is more possible elected as a cluster head which saves energy compared to exiting LEACH protocol. Proposed approach uses an E-BMA with DED clustering algorithm which is called Energy Cognizant Bit Map Assisted (EC-BMA) protocol.

Keywords Sensors · Energy · WSN · E-BMA

1 Introduction

WSNs usually comprises of base stations and an assortment of wireless sensors. Every sensor might be a unit with wireless networking capacity which will gather and process information automatically. Sensors are mostly used to watch the behavior

S. Rajes Kannan (✉) · S. Amutha
Computer Science and Engineering, Chennai Institute of Technology, Chennai, Tamil Nadu, India
e-mail: rajeskannans@citchennai.net

S. Amutha
e-mail: amutharajeswari95@gmail.com

D. J. Hemanth et al. (eds.), *Artificial Intelligence Techniques for Advanced Computing Applications*, Lecture Notes in Networks and Systems 130,
https://doi.org/10.1007/978-981-15-5329-5_11

of objects in an exact field and convey the information to a base station. The energy resources of these nodes are battery limited. So sensor nodes make power usage efficiently and may extend the lifetime of the utilization of the sensor nodes in sensor networks. MAC is one in each of the key areas wherever energy potency is achieved by coming up with such MAC protocol. MAC tries to crash by preventing two or fussier meddling nodes from getting the medium at a comparable minute, which is important to the thriving operation of shared-medium networks [2]. Disadvantages of the existing schemes such as EA-TDMA and TDMA are nodes or cluster heads with no data packets but keep their radio turned on throughout their regular slot and therefore energy is wasted in sensor network [10]. The Bit Map Assisted Protocol allows the problem of the source node doesn't reserve the spot following the information gets accessible. This proposed system overcomes these drawbacks by using E-BMA techniques, the supply nodes use piggybacking to create the stipulation of the equivalent information slot instead of distributing a control message throughout its allotted rivalry slot. The proposed system is to increase the energy efficiency, DED clustering algorithm is implemented. Clustering rule elects the sensing element having high degree, nearer to destination, and high energy-sensing element as a cluster head (CH). Proposed approach uses an E-BMA with DED clustering algorithm called Energy Cognizant Bit Map Assisted (EC-BMA) protocol. Cracks in rails are known to be the most common reason behind derailments and it is detected using LED/LDR Sensors.

2 Problem Statement

A device network may be deployed in an exceedingly hostile region to perform vital missions. The energy resources of these nodes are very restricted. In the majority of the cases, recharging or replacing these batteries isn't realizable. The most causes of energy waste in waterproof layer are collision, overhearing, inactive listening, and management packets overhead [9]. Collision happens once transmissions of two or more nodes overlap in time which ends up in failure of the communication and needs retransmission. Overhearing is the case that during which a node receives packets that don't seem to be destined for it. In inactive listening, a node keeps its receiver on with the hope of receiving one thing whereas the channel has nothing for it. Control packet transparency is created by all those packets communicated for network and link management functions. Ultrasonic will solely examine the core of materials; that's, the strategy cannot ensure for surface and near-surface cracking wherever several of the faults are placed.

To achieve energy efficiency within the wireless sensor networks and decrease the energy consumption due to inactive listening and collisions whereas maintaining a decent low latency performance. To boost the measurability, lifelong of the node within the network and minimize the amount of knowledge transmitted in the network.

2.1 Drawbacks of Existing System

In Time Divison Multiple Access, if supply node with no information packets, however, keeps their radio on throughout their regular slot. So dissipate a number of their energy within the network. Waterproof protocol wherever the transmission is split into many time slots, and every node is appointed a interval. Each node awakens and transmits information exclusively in its assigned interval and remnants in rest mode among the residual time slots. Anyway, this protocol exclusively utilizes the node energy expeditiously once the passage load is high. Nodes with unfilled buffers stay their radio turned on throughout their regular slot and thus, waste a number of their outstanding energy.

The Energy-Efficient Time Division Multiple Access (E-TDMA) decreases the energy consumption thanks to inactive listening. In BMA, a supply node doesn't create the reservation within the rivalry slot as before long because the information packet gets obtainable. During the disputation period of the BMA protocol, the transceiver of the supply node is turned on when it has no available data in the network.

2.2 Proposed System

The main proposal of the E-BMA algorithm is to ensure better performance in the core of the sensor network. The proposed E-BMA algorithm is separated into number of sessions. The period of every session is fixed. Each session comprises of a conflict period, an information transmission period and an inactive period. During contention period, each node is allotted a particular slot and transmits a 1-bit control message to a cluster head because it is allotted slot. After the rivalry period is completed, the cluster head is aware of all the nodes that have information to transmit. The proposed E-BMA algorithm is to increase the energy efficiency, DED clustering algorithm is implemented. During the info transmission period, every supply node activates its radio and sends its information to the cluster head over its allotted data slot-time, and keeps its radio off in the slightest degree alternative times [3]. If there are successive data packets, sensor nodes are set 1-bit field in each and every knowledge packet header to point whether or not supply node encompasses a serial. If a supply node has succeeding knowledge packets to send out a variety of uninterrupted frames, the booking is framed once for the underlying knowledge packet in its apportioned conflict opening slot, what's more, in this way the sequential affirmations will be made through piggybacking. Using GSM module SMS will be sent to the central system which transmits the message to PC which in results alarming the trains on the specific rail [4]. Automated model to monitor the tracks helps in prevention or alarm of such cracks on rails. The proposed IR sensor for detecting the cracks. Once a crack is detected the robotic trolley will stop automatically and alarm the officials.

3 Algorithms and Methodologies

3.1 DED Clustering Algorithm

LEACH is that the most well-liked formula. It chooses accidentally the cluster heads for just once consistent with the policy known as "Round Robin". However one downside that occur within the LEACH protocol like the election procedure (random) will lead cluster heads to own a weak energy reserve, which may have an effect on the information transmission and may direct to a reconfiguration of the engineered structure [5]. This difficulty has been conquering by the following techniques that used the DED clustering formula. The DED clustering algorithm elects the device having high outstanding energy, higher degree and nearer to the bottom station is a lot of possible elective as a cluster head. The members of every cluster converse directly with their ClusterHeads (CHs) and every ClusterHead aggregates the traditional messages and transmit them on to the bottom station [6].

3.2 LED/LDR Sensors

When the sunshine level is small the confrontation of the LDR is high. This keeps present from streaming to the help of the transistors [7]. Thus the LED doesn't have light weight. In any case, when light weight onto the LDR its resistance course and current streams into the assistance of the basic transistor and a while later the resulting transistor. The Light Emitting Diode lights, the preset electrical device will be turned up or option to down to augmentation or decreasing check, all through this philosophy it will collect the circuit some portion of or less sensitive. A Light Emitting Diode is utilized as a photodiode in light location. This capacity could likewise be utilized in a variety of uses adjacent to close light weight recognition and duplex correspondences. As a photodiode, a LED is keen to wavelengths equivalent to or shorter than the dominating wavelength it emanates. For instance, a green LED is agreeable to blue light and to some green light, anyway to not yellow or red light. This execution of LEDs could likewise be superimposed to styles with exclusively minor alterations in hardware. A LED will be multiplexed in such a circuit, indicated it will be utilized for both light discharge and detecting on totally various occasions. LED LDR is the sensor which detects the crack in daytime and night time, it detects the track and sends information to the transmitter.

4 Experimental Results

4.1 Hardware Image

This is a hardware image showing the transmitter and receiver unit and LCD display. Here an external hardware task will be shown in LCD display by a text or characters.

4.2 *Zigbee Transmission*

Network model represents 9 V battery for power supply through the serial cable. It is gotten in the Zigbee collector.

4.3 Control Message Transmission

Information is transmitted through Zigbee transmitter to the base station. Then crack detection is monitored.

4.4 Crack Detection

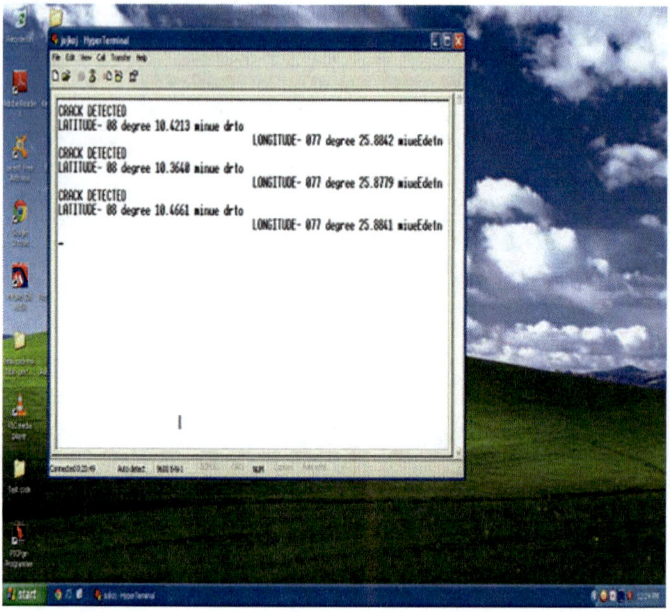

Uncovering of signal incline with 38.5 kHz sampling rate and interrupts. Then crack identification is detected and displayed in the monitor to base station.

5 Conclusion

Performance of an energy-efficient EC-BMA MAC strategy has been researched to achieve the energy efficiency in WSNs and to decrease the energy waste because of idle listening and collisions while keeping a superior low-latency performance. It is presented to detect the cracks in the tracks effectively with the LED/LDR sensors and Zigbee the cracks in the railway track are detected and by using wireless modules the information is approved to the control section. The control section is incessantly monitored by authority, and when there occurs an error a sound is produced and the authority can ask for the location of the crack. Therefore, it achieves low energy consumed so far implementing the crack detection using LED/LDR sensor.

References

1. Shafiullah GM, Thompson A, Wolfs P, Ali S (2010) Predicting vertical acceleration of railway wagons using regression algorithms. IEEE Trans Intell Transp Syst 11(2):290–299
2. Shafiullah GM, Thompson A, Wolfs P, Ali S (2008) Energy-efficient TDMA MAC protocol for wireless sensor networks applications. In: Proceedings of 5th ICECE, Dhaka, Bangladesh, 24–27, pp 85–90
3. Bleakley SS (2006) Time frequency analysis of railway wagon body accelerations for a low-power autonomous device. M.S. thesis, Faculty of Engineering and Physical Systems, Central Queensland University, Rockhampton, Australia
4. Wolfs PJ, Bleakley S, Senini ST, Thomas P (2006) An autonomous, low cost, distributed method for observing vehicle track interactions. In: Proceedings of the Joint Rail conference, Atlanta, GA, pp 279–286
5. Kottursamy K, Raja G, Padmanabhan J, Srinivasan V (2017) An improved database synchronization mechanism for mobile data using software-defined networking control. Comput Electr Eng 57:93–103
6. Bougard B, Catthoor F, Daly D, Chandrakasan A, Dehaene W (2005) Energy efficiency of the IEEE802.15.4 Standard in dense wireless microsensor networks: modeling and improvement perspectives. In: Proceedings of design, automation, and test in Europe conference and exhibition, pp 196–201
7. Tsang CW, Ho TK (2008) Optimal track access rights allocation for agent negotiation in an open railway market. IEEE Trans Intell Transp Syst 9(1):68–82
8. Heidemann W, Ye J, Estrin D (2002) An energy-efficient MAC protocol for wireless sensor networks. In: Proceedings of IEEEINFOCOM, New York, pp 1567–1576
9. Smith J, Russel S, Looi M (2003) Security as a safety issue in rail communications. In: Proceedings of the 8th Australian workshop on SCS, Canberra, Australia, pp 79–88
10. IEEE Standard 15.4 (2003) Wireless Medium Access Control (MAC) and Physical Layer (PHY) Specifications for Low-Rate Wireless Personal Area Networks (LR-WPAN). IEEE Computer Society, New York. IEEE Technical Report

An Interactive Virtual E-Learning Framework Using Crowdsourced Analytics

K. Dhinakaran, R. Nedunchelian, R. Gnanavel, S. Durgadevi, and S. Aswini

Abstract These days clients are keen on distance learning as there is a quick development in digital data because of everyday improvement in data just as computer technology. Presently, YouTube is the world-wide method for video sharing. It is having sure constraints such as, having idleness in Web-based learning. In online investigation learners anticipating some additional rules from given assets. This work first investigation dependent on dynamic learning and video-based learning ways to deal with stem training, universal learning and afterward exhibit the blend of micro lecture and Web learning to propose a novel method of micro-learning. Details are exhibited of a micro lecture Web learning framework that can support multi stages. The framework consolidates intelligent push, video explanation, Lucene full-text hunt, clustering analysis and different advancements. We propose a gathering commitment score that considers both individual movement and comparability of participation, consequently enables restorative moves to be made when not engaged learners or gatherings are distinguished. The stage enables learners to get to micro lecture videos and other excellent micro lecture assets wherever and whenever they like, in whatever time intervals they have accessible. Educators can get statistical analysis results of the micro lecture in MWLS (Micro Lecture Web Learning system) to give teaching/learning feedback and a powerful correspondence stage. MWLS advances the improvement of micro lecture.

Keywords Virtual learning · Crowd sourcing · Data analytics

K. Dhinakaran (✉) · R. Gnanavel · S. Durgadevi · S. Aswini
Department of Computer Science and Engineering, Rajalakshmi Institute of Technology, Anna University, Chennai, Tamil Nadu, India
e-mail: maildhina.k@gmail.com

R. Nedunchelian
Department of Computer Science and Engineering, Karpaga Vinayaga College of Engineering and Technology, Chengalpattu, Tamil Nadu, India

D. J. Hemanth et al. (eds.), *Artificial Intelligence Techniques for Advanced Computing Applications*, Lecture Notes in Networks and Systems 130,
https://doi.org/10.1007/978-981-15-5329-5_12

1 Introduction

Numerous individuals like utilizing a marker to feature books while reading, particularly learners with course books close by. Research shows that appropriately featured substance indeed support understanding. Maybe this is the motivation behind why a considerable amount of book writers as of now feature the key ideas, highlights or equations in their books, and more are mentioned doing as such. For the most part, there are two kinds of highlighting: content highlighting and list of contents highlighting. The former for the most part underlines sentences, while the last works in a bigger scale, showing which area ought to be given special attention. Not only is a far-reaching practice with conventional paper books, the highlighting capacity also invited in the time of digital books. It is broadly executed in numerous e-book applications. In this case, while a marker is never again required, highlighting is yet dependent on the book-like literary materials.

The originality, circulation of participants, finding solution for the problem, investigation, association and analysis are in the place called Virtual learning environment. E-learning involves the use of data and communication technology in instructional process. Here, the accessibility is given to somebody, specifically who is in order to obtain wide-ranging data by exploiting the Web. According Berge et al., based on a fixed memory, the Internet or the intranet, the E-learning deals with instructional approaches occur on the system and planned to improve the intelligence and expertise associated to each one of the individuals or administrative targets. The inclusive concept of E-learning is seizing the attention of the users with variety of contents explicitly intended to fulfill the needs of the participants by the self-paced and contented setting. Web learning grants beginners to study by involving and performing, by getting immediate comments, and by letting them to track their growth with tests, quizzes and other similar activities. The application of e-learning is designed at lecturing the problem through the medium of Web as well as computer but parting among learners and teachers.

Nevertheless, in broadening the platform of learning there requires certain effective modifications and improvements to meet the needs and challenges faced by the everyday users. This paves the way for crowdsourcing, where a huge set of users raise queries regarding the clarifications in the courses being provided by the e-learning platform. Crowdsourcing generally is about seeking information from a huge set of people through the Internet generally toward completion of a specific goal. Here, crowdsourcing plays the role of improving the lecture quality by frequently generating queries by watching the videos. Crowdsourcing has several advantages in the current era. They are used in improving the quality of businesses by seeking the right information and by able to provide the right data for analysis.

Presently, YouTube is the world-wide methodology for sharing the video and queries are generated for every video in the comments section. It is having sure constrains such as having idleness in Web-based learning. In online investigation learners anticipating some additional rules from given asserts. Video annotations are used in order to navigate to similar sites and advertisement. In learning and instructions, making annotation plays a major role from cognitive perspective. Being the portion of guided observing, it comprises underscoring emphasizing, labeling, designating and provide feedback are the features of visual representation that supports deep learner's attention on visual intents.

The level of knowledge is enlarged in an infinite scale than our capacity to make use of it. Each participant provided with accessibility to variety of the latest books, feeds, articles, tweets, journals, Web sites coming out each day that creates high in dimensional and sophisticated organized knowledge. To preserve and investigate the large amount of information, the standard storage method doesn't seem to be enough. The methodology of massive knowledge and its investigation processes are typically accustomed to give the outline regarding larger datasets [1]. By differentiating ancient datasets and its processes, massive knowledge comprises semi structured and unstructured data that require a lot of real time analysis. Massive knowledge additionally obtains the information regarding new changes for determinative new values, supports to enhance the associate deep grasping the values that are out of sight, and additionally sustains new challenges. The amount of knowledge from numerous sources is increased in massive level, and it conjointly provides regarding some difficult problem's stern fast resolutions on analysis of this knowledge. This increase in data, challenges the sphere with the most issues of gathering and integration vast amount of information from cosmopolitan data sources like social media applications. As a conclusion, this survey will be providing a quick evaluation on the technologies employed in massive knowledge analytics and annotation systems so as reinforcing E-learning platform.

2 Related Works

2.1 E-Learning

Kimberly E. Arnold et al. designed a system which is useful to provide feedback to a learner in real time by the learner analytics using course signal which is developed by this system. The performance of the learners is not only predicted based on the grades, it is also based on history of the learners and their interaction regarding the subject by using the Course Signals. The e-mail will be sent to the individual learner based on the course signal and there will be some color indication for learner performance.

Paul R. Pint rich et al. proposed a system in which the learner can self-analysis of their learning strategy, self-efficiency, test anxiety, self-regulation and the performance in class room assignments performance is also measured [2]. Cognitive strategy the learner's effort for understanding the topic is measured as intrinsic value and self-efficacy [3, 4]. This system is the best predicting in the performance based on the test anxiety, self-efficacy and self-regulation. Intrinsic value is not directly related to performance, but it is related to cognitive strategy and self-regulation.

Mary McCaslin et al. studies about the past event and ineffectiveness of educational psychology's in-house fights over mission and contests are conducted for theoretical dominance supported in the name of unity. This proposed system has been recommended instead of advisability of collaboration between various participants and theoretical incorporation for the development of educational practices [5, 6].

Barry J. Zimmerman proposed a model based on triadic [7], it consists of three different relations of the personal, environmental and behavioral for analyzing the self-regulated concept. The additional hypothesis of this model is 3 self-regulatory processes: self-reactions, self-regulatory and self-judgment and construct a central role for educational self-efficacy. The psychological feature formulation is supported by research, and it is used to rise tutorial accomplishment and learner learning [8].

Daniel C. Moos proposed a model where the cognitive will be loaded with hypermedia while learning hypermedia with SRL strategy [9]. The undergraduate learners are taking notes while learning with hypermedia. The survey has been made with 53 UG learners while studying about a science topic over a period of 30 min with hypermedia learning regarding self-report, think-aloud, pretest [10]. The pretest performance is used to predict the prior knowledge about the domain [11]. The participant's notes and self-report questionnaire have the significant relationship.

Steven Lonn et al. detailed academic mentors of EWS about the data in mining learning management system data and translating those data [12]. The learning analytics importance is elaborated by advisors and mentors for higher education. To increase the academic success of M-STEM for getting successful in highly competitive exam, the academy is mainly aimed at the learners who are reasons for the status in socioeconomic and college status [13, 14].

2.2 *Crowdsourcing*

The main aim of crowdsourcing is to collaborate with people, society and improve the connectivity among them. Previously the open call is used for the undetermined group of people by employees for the function outsourcing. In this online fabrication model, the information is obtained from the group of people called crowd and the wisdom of crowds is the concept of this model. The group of people can give a better solution, decision making and predicting than the individual person [15], this is known as wisdom of crowds. Governments, communities, institutions, global societies and work groups are the major works carried out in crowdsourcing. Crowdsourcing learning is natural, and the best educational experiences are not obtained of necessity. Crowdsourcing should support these three reasons: unexplored experience in online education, existing techniques should be ready to apply and education in certain fields with a good quality.

Drew Paulin et al. addressed the openness of whom, with, where and how we learn [2]. The learning environment challenges are also addressed here. They decenter the classes provided by institution and the open virtual communities are created, discussed in broaden and the open operation for the online crowd [16]. To make the crowd sourcing as the curriculum, they adapt emerging educational practice, learning science and discuss how collaboration and peer production makes the learning more effective.

The main goal of this crowdsourcing is to explore the participatory practices lead to a new way of learning, the potential of creating and managing large twenty-six SLE is the main advantage of crowdsourcing. In this area, the prime focus for the large-scale online education is MOOCs and it is the increased production result of a culmination [17]. Here the open content and resources available in online are shared which increases the curiosity in establishing collaborative learning in vast scale and necessary to lay out learning possibility to all who want it without cost [18]. With the shared motivation of learning, the result has been obtained in unparalleled and massive scale.

Carlos Eduardo Barbosa et al. discussed the basic process in crowd sourcing tools in e-learning and analyzes ways to classify them, for development of software conceptual model tailored to each type [19, 20]. The new crowd sourcing tools for e-learning is built by using the conceptual models. There are two groups in crowd sourcing tools: peer-learning and traditional crowd sourcing e-learning. This model analysis each process model for software development and crowd sourcing learning [21].

3 Proposed Work

The micro lecture E-learning system (MELS) has three parts. They are the learner terminal, the teacher terminal, and the cloud platform. Here, cloud platforms are used by teachers and learners for the log in which hold on to the conventional learning platform but also includes the new E-learning platform. The Web site is based on AWS-EBS. To upgrade the cognitive activity, there are various ways for the learners who involved in this task. This proposed technique uses the active learning strategy with three tasks:

1. Teachers create the learning source videos;
2. Then teachers share their videos to the experts in cloud platform and get posted in the Web after getting the approval from the experts;
3. Learners learn from the videos, raise their queries and get clearance from the experts. Here, the new learners get benefited by observing the reviews of published video on the bottom of the page. Here, the role of new learners is categorized into two, as a reflective and as a social actor.

This proposed technique has a general function like uploading video and playback. In addition to this, some features are designed on the cloud servers which include annotation of video and video along with subtitles that are correlated with each other and use the Lucene full-text search technology. Learners can access micro lecture resources via Web, make annotation on it, analysis it and then raise question while learning. At the instance, the notification is received by the expert members, and the answers are given through the interactive medium where both the experts and the learners can communicate with each other and offers valuable advices. To calculate the quality of the video resource, clustering analysis algorithm is used in terminating platform and cleverly sends the result to the leaner's interface. And to enhance the learner's recall ratio and precision of penetrating, Lucene technology is used to guide micro lecture resource by the central Web server.

Figure 1 denotes the proposed system architecture diagram. The author of the video can upload the video along with relevant descriptions. The learners access the e-learning environment on their desired interest. When the learner wants to clarify their doubts, the learner clicks over the video at the specific portion of doubt. The click time of the learner is noted down and the query is typed in by the learner. This query reaches the crowd sourced expert's page along with an e-mail trigger. The author verifies the video along with the query and answers for the relevant questions. The answer for the query is annotated in the specific learner's video the next time he plays it. This makes it efficient for the learner to clarify his queries on time. Also, the other queries asked by some other learners are also annotated in the current video. The questions asked by other learners for the specific video is displayed at the bottom of the page. This enables crowd wisdom model, where a group of learners generate queries regarding the video uploaded and the author can check the quality of their video. The following algorithm will give the working process.

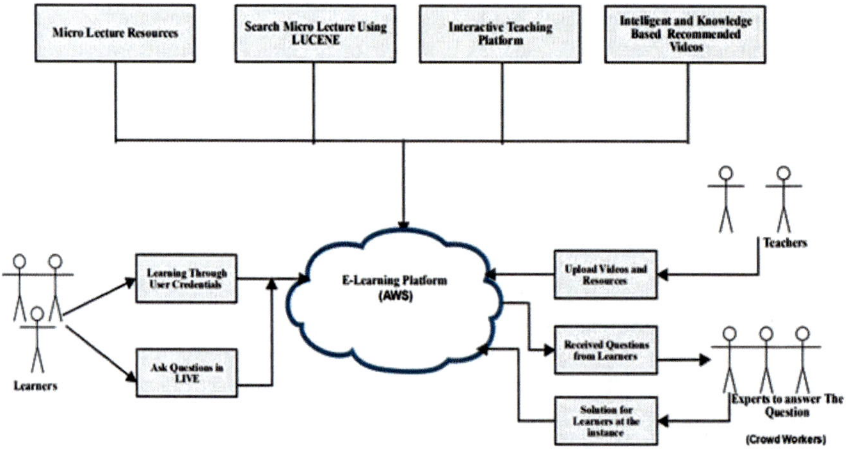

Fig. 1 Proposed system architecture diagram

```
INPUT: Search course, Question from learners.
OUTPUT: Knowledge based recommended resources, Answering the question at the time without delay(crowd of workers).
PARAMETER: L-Learners T-Teacher, EW-Expert workers, EP–Learning platform, C-Courses,
R-Resource
PROCESS:
L: request(C);
EP: received reques(c);
{ ack(c)–>L;

   List_courses();

}
L<–Choose the course
T–>upload(R)–>EP
EW<– receive questions from EP; & Answer to learners at the time through EP
EP–>recommend similar course to L
end;
```

4 Implementation

The experimental setup is being implemented using the Windows 10 OS. The Web page of e-learning platform was designed using JAVA and JSP program. The database relates to Amazon Web service cloud environment. Especially S3 (Simple Storage Service). The Web application is linked with S3 and the Web service is tested. The learner and the teacher's portal are both considered to be the crowd, where group of learners raise questions and the group of expert's members answer for the questions. The click time of the user over the video is acquired and the annotation process is carried over. The annotation of the video is performed using Web application by making use of the click time. The text is annotated to the video based on the time of the user query generation. The test results provided an efficient query answering system in e-learning.

5 Experimental Result

As we early talked about the concept and importance of micro lecture and E-learning, this paper prefers the new scheme of collaborating micro lecture and E-learning and particularized the MELS design which encourages the multi-platform learning. The first user interface (UI) displays the user's page where the query is raised. This query along with the portion of the video where the doubt is raised is sent to the teacher's

Attribute	Full Data (55.0)	0 (32.0)	1 (23.0)
S.no	28	27.25	29.0435
Learner ID	LID1	LID1	LID2
Ease of use	3.4	3.4375	3.3478
Innovation	3.6	3.8125	3.3043
Usefulness	3.5636	3.25	4
The quality of learning	3.7273	4	3.3478
Enjoyment	3.3273	3.8125	2.6522

Fig. 2 Outcome of the proposed model

Attribute	Full Data (55.0)	0 (23.0)	1 (32.0)
S.no	28	27.6087	28.2813
Learner ID	LID1	LID2	LID1
Ease of use	2.8364	1.6087	3.7188
Innovation	2.2545	2.6087	2
Usefulness	3.1636	2.3913	3.7188
The quality of learning	3.2545	4	2.7188
Enjoyment	2.5818	3	2.2813

Fig. 3 Outcome of the existing model

page. The second UI displays the teacher's page where the answers are provided. The final UI displays the annotated output of the proposed system with expert solution. When establishing the new technique for learning, the accommodation period and advancement are required by E-learning.

As a result, we got the 53 student's ratings on both existing and the proposed model based on the parameters such as ease of use, innovation, usefulness, the quality of learning and enjoyment. Having these datasets, we use a tool called WEKA to compare the results obtained on both proposed and existing method with the help of clustering algorithm. As a conclusion, the proposed model is more effective than the existing one [22] (Figs. 2 and 3).

6 Conclusion and Future Work

E-learning methodology has been introduced as the substitute for the regular classroom approach. The predominant technology used in e-learning system is typical Web sites and forums where information from various resources are congregated. In this platform, the queries of the users are answered efficiently by the crowd of experts. Thus, this paper provides an alternative solution to the existing system. The future work of the system involves handling the huge dataset effectively using the Hadoop ecosystem and generating a machine learning algorithm for the annotation process.

References

1. Dhinakaran K, Silviya Nancy J, Duraimurugan N (2015) Video analytics using HDVFS in cloud environment. ARPN J Eng Appl Sci 10(13)
2. Paulin D, Haythornthwaite C (2016) Crowdsourcing the curriculum: redefining E-learning practices through peer-generated approaches. School of Information Studies, Faculty Scholarship
3. Azevedo R, Moos DC, Johnson AM, Chauncey AD (2010) Measuring cognitive and metacognitive regulatory processes during hypermedia learning: issues and challenges. Educ Psychol 45(4)
4. Greene JA, Azevedo R (2010) The measurement of learners' self-regulated cognitive and metacognitive processes while using computer based learning environments. Educ Psychol 45(4)
5. McCaslin M, Hickey DT (2001) Educational psychology, social constructivism, and educational practice: a case of emergent identity. Educ Psychol 36(2)
6. Song L, Tekin C, van der Schaar M (2014) Clustering based online learning in recommender systems: a Bandit approach. In: IEEE international conference on acoustic, speech and signal processing (ICASSP)
7. Zimmerman BJ (2009) A social cognitive view of self-regulated academic learning. J Educ Psychol 81(3)
8. Murad H, Yang L (2018) Personalized E-learning recommender system using multimedia data. Int J Adv Comput Sci Appl (IJACSA) 9(9)
9. Moos DC (2009) Note-taking while learning hypermedia: cognitive and motivational considerations. Comput. Human Behav 25(5)
10. Aher SB, Lobo LMRJ (2013) Combination of machine learning algorithms for recommendation of courses in E-learning system based on historical data. Elsevier
11. Lykourentzou I, Giannoukos I, Nikolopoulos V, Mpardis G, Loumos V (2009) Dropout prediction in e-learning courses through the combination of machine learning techniques. Elsevier
12. Lonn S, Krumm AE, Waddington RJ, Teasley SD (2012) Bridging the gap from knowledge to action: putting analytics in the hands of academic advisors. In: International conference on Learning Analytics and Knowledge, Vancouver, Canada
13. Joksimović S, Gasevic D, Kovanović V, Riecke BE, Hatala M (2015) Social presence in online discussion as a process predictor of academic performance. J Comput Assist Learn
14. Pireva K, Kefalas P (2017) A recommender system based on hierarchical clustering for cloud e-learning
15. Ayodele T, Shoniregun CA (2010) Towards E-learning security a machine learning approach
16. Pardo A, Ellis RA, Calvo RA (2015) Combining observational and experiential data to inform the redesign of learning activities. In: International conference on Learning Analytics and Knowledge, Poughkeepsie, NY, USA
17. Aher SB, Lobo LMRJ (2012) Course recommender system in E-learning. Int J Comput Sci Commun 3(1)
18. Salehian H, Howell P, Lee C () Matching restaurant menus to crowdsourced food data: a scalable machine learning approach. In: KDD 2017 applied data science paper
19. Dhinakaran K, Silviya Nancy J, Saranya R, Classification and prediction of earthquake and tsunami using big data analytics. Online issue May 2015

20. Sivasakthi M (2017) Classification and prediction based data mining algorithms to predict learners' introductory programming performance. In: Proceedings of the international conference on inventive computing and informatics (ICICI 2017)
21. Barbosa CE, Epelbaum VJ, Antelio M, Oliveira J, Rodrigues JA, Medeiros SP, de Souza JM (2016) Conceptual crowdsourcing models for E-Learning. In: IEEE international conference on systems, man, and cybernetics
22. Nkoana TH (2016) E-learning: crowdsourcing as an alternative model to traditional learning. IEEE

Security for Data in IOT Using a New APS Elliptic Curve Light Weight Cryptography Algorithm

Ravi Sridharan and Thangakumar Jeyaprakash

Abstract A thought process on my mind says that IOT is expanding every minute and second due to many newer devices who join in the IOT chain every time. Necessity is the mother of invention, as per this when private and confidential data are moving across in the IOT environment across homogeneous and heterogeneous devices then in this case we should think about the data care. In this paper, a strong cryptographic algorithm is coined which will secure the data getting transformed over the IOT devices. A New APS Elliptic Curve Light Weight Cryptography algorithm is used which satisfies all the parameters in ensuring security across the IOT for data. If a text is given the user will have the private key and the hashing function is applied over the text and then the cipher text which moves across the signed hash checkers and after that the other user at the recipient side will be able to decrypt the cipher text using his public key and can read the message. This is actually done in the case of a Elliptic Curve Cryptography. The lightweight cipher text is mainly to handle the low resource device on the IOT chain. By considering the naturality of all the devices over the IOT chain and to satisfy all those parameters which are very essential to meet data security on IOT devices, the New APS Elliptic Curve Light Weight Cryptography algorithm is used.

Keywords IOT · Cryptography · Private and confidential

1 Introduction

The growth of IOT keeps on increasing day after day and this results in an increase in the number of data that moves across the connected devices in the IOT chain. As there are a lot of parameters to be concerned while thinking about data movement in IOT chain, the security which is an important issue to be thought off. Between

R. Sridharan (✉) · T. Jeyaprakash
Department of Computer Science and Engineering,
Hindustan Institute of Technology and Science, No.1, Rajiv Gandhi Salai,
Padur, Chennai, Tamil Nadu, India
e-mail: sravi@hindustanuniv.ac.in

D. J. Hemanth et al. (eds.), *Artificial Intelligence Techniques for Advanced Computing Applications*, Lecture Notes in Networks and Systems 130,
https://doi.org/10.1007/978-981-15-5329-5_13

peer to peer communicating devices, there must be a strong communication protocol using which the data could move safer to the other connected devices in the chain. Any devices in the IOT chain through which the data transfer would happen must be certified before it is linked onto the IOT chain. If any device which is already added into the link shall be updated on a regular basis. Self-initiating process of any devices on the IOT chain made that possible for them to communicate with the other devices on the chain even without human interventions and hence a strong cryptography algorithm should be coined that protects the data across its traversal through all the communicating devices on the IOT chain.

2 Related Works

Most of the researchers describe at least one or more of the complexities that exist in IOT environments.

Article on https://www.nabto.com/security-in-nabto-p2p-iot-solutions/: Security in IOT and especially in-home security solutions [1] where Security in IOT and mainly in-home security solutions were discussed. Security principle which was applied in the case of p2p communicating devices is discussed. Camera must turn on which communicates with central service that must know all about camera network configuration was discussed. Communications in terms of nap to communicating devices were discussed. Nap to client has DNA address that is connected to nap to base station and from where thru base station it connects with the dedicated client. Cryptography while sending data across the p2p devices is discussed.

Huichen Lin et al. proposed a system [2], where the discussion was about how security could be enhanced in IOT by upgrading the system configuration automatically every time and also by updating the software and the firmware to provide security in case of the online secure system operations. The backlogs understood in this work is that the untrained people due to their mistakes in installing and configuring the system the security is affected in depth. The Webservices nourished by Smarthouse architecture in case of certain network configuration and the automated devices, its auto-updating of the system was a greater strength in this work.

Suk Kyu Lee Mungyu et al. proposed a system [3], where mobile devices that are connected across the globe are well explained and how people get information in any nomadic environments using such a communicating device are well explained. Specialty of IOT network and how it was built and its role in IOT was explained well in the article. IOT technology and its capabilities are broader which is well understood in this article. But lack of strong standard protocol in IOT network is mentioned. A storing gateway or an IOT device connecting heterogeneous devices should be provided.

Kubrakalkan et al. proposed a system [4], where the smart control is being used and the devices on the IOT chain automatically sense the data and pass it through the gateway and the role-based software-defined network was discussed.

Vijayanand Thangavelu et al. proposed a system [5], where observing IOT network-connected devices and its advantages are discussed. Deployment of certain anomalies like behaviour and the vulnerability patching of specific devices and the dynamic mitigation attacks—to deploy at the centralized network using supervised machine learning algorithm–finds difficult to locate any new devices on the network so DEFT Controller-Addresses common devices connected to IOT chain-maintaining while gateways are nearer to the IOT devices, the classifier for Fingerprinting. DEFT is scalable, dynamic, and controlled through SDN.

George David Dehoz Diego et al. proposed a system [6], where Security to IOT devices are mainly concerned in this work. The devices before connected to the network must be certified for security. Already certified devices must be updated and newly added devices should be certified. The security protocols are upgraded using the Transport layer security but this leads to difficulty that the entire application to be upgraded and this shortens the life of IOT devices. An alternative to this is SSH (Secure socket shell), discretion to secure communication in IOT application based on certain principles. When a comparison was made it was identified that HTTP-SSH than HTTP-TLS consumes low power and also shows good throughput in terms of performance and energy consumption.

Kazim Rifat Ozyilmaz et al. proposed a system [7], where discussion about IOT and the mobile devices which are connected which help the system to be secured from onto that IOT chain and the vigorous increment of these devices day by day was the concern. The way how the data are stored on those devices and extracted in a secured manner was to be taken care of. Block chain technology was adopted which secured the system again distributed denial of service attack and fault tolerance. They used Lora, Swarm and Ethereum as the emerging networking technology, a distributed data platform for storage and block chain platforms.

Bestoun S. Ahmed et al. proposed a system [7], where IOT plays a vital role. IOT had promising power to change the future. But it was mentioned that due to high demand for IOT system to be released without proper testing it was released. But without proper quality testing, it should not be released. Hence starting from 2009, during the early period of IOT till today, a comparison was made regarding how to increase the Quality in IOT.

Sheng Ding et al. proposed a system [8], where attribute-based encryption for ciphertext was proposed (CP-ABE). This algorithm had a problem with bilinear pairing because of the device/processors low power supply and less computational resources. The authors proposed certain innovative methods and the balancing pairing is replaced with calcining multiplication on the elliptic curves thus reducing the total computational overheads.

Bassam J. Mohd et al. proposed a system [9], where low resource devices on the IOT chain were discussed. Actually, the low resource devices may not be able to attend sensitive information like the one which is more confidential. Conventional Cryptography methods are not suitable for low resource devices so instead light weight block ciphers are used to encrypt data on such devices and this normalize the security benchmarks and the energy consumptions on IOT devices. So, a strong energy management algorithm is used to improve IOT data against denial of service attacks in the form of battery winding off. A model for light weight cipher performance metrics is coined, and the results are compared with the published application-specific integrated circuit (ASIC) and field programable gate array (FPGA) designs. When block size is between 48 and 96 bits, optimum energy is achieved. When the number of rounds is less than 16, then optimum energy is obtained. A novel algorithm to manage cipher energy consumption is used.

Mirza Abdur Razzaq et al. proposed a system [10], where the different types of networks are being discussed and the various problem related to data security, security loss in case of insurance concerns, lack of certain standards, technical issues and the security attacks and certain vulnerabilities of the system are discussed. The importance of emphasis more security like disabling certain features that are not used to the users and avoiding the default passwords, etc. ate pointed out. So, a strong cryptography algorithm which prevents the data from being attacked by intruders is highly suggested.

Vignesh et al. proposed a system [11], where the three-layer architecture on IOT is taken into consideration the perception, network and application layer. The authors felt that the IOT framework is vulnerable to attack at each layer and hence a strong level of security is raised at certain protocols but due to fast-growing technology, it has to be consolidated with fewer more new protocols like 5G protocol and IPV6 along with the authorization protocol. Security in terms of private and confidential, access controls and the end points security management, Management of trust, Global policies/standards are very much lagged and hence a strong algorithm to overtake all these should be brought onto the light.

3 Comparison of Various Techniques, Methods and Models

Huge data are conquered over IOT chain. The data are taken from various sources. So, it is our duty to find out where the data is available and then applying data mining to extract knowledge from the data base. We can also adopt some statistical/Mathematical patterns for doing analysis of the data being captured on the system. When newer data are getting added every second intelligently the system should be able to monitor the data, process the data and also to take up decision making at all times. Many newer ideas are given by researchers regarding how to secure these data. By examining closely, the research papers which are published in recent years. Many new techniques, Models and algorithms are identified. Now a thought process is that there are several conventional cryptography algorithms like RSA, DES, etc. are being utilized in the previous research papers but the methods could not be applied onto low resource device on the IOT chain and hence a strong cryptography algorithm which can be used even in low resource device thereby saving energy and to increase the performance to be considered. Examples are shown in Table 1.

Table 1 Comparison of methods, models and techniques

Model/methods/techniques	Comments	Example
Conventional cryptography	To help the low resource devices while communication conventional cryptography methods are not suitable and hence light weight block ciphers are used for encryption and to balance energy as well as security issues	FIDES, SKINNY
Cyber-text policy attribute-based encryption (CP-ABE) method	This method had a problem with bilinear pairing because of the device/processors very low computational resources and its energy. This method is based on CP-ABE elliptic curve cryptography	Elliptic Curve Diffie–Hellman (ECDH) Elliptic curve digital signature algorithm (ECDSA)
Block chain technology	A question of how shall the data on IOT could be kept secured arised and then the thought process emerged of using the block chaining process where the data are divided into several pieces and stored in multiple locations and hence easy hacking of them could not be possible and hence the security was ensured	Bit Coin Block
Secured socket shell protocol	SSL is a standby to secure communication in Internet of Things application based on Hyper Text Transfer Protocol and Hypertext Transfer Protocol/2. On comparison, it was identified that HTTP-SSH than HTTP/TLS consumes low power and shows more throughput in terms of energy consumption and performance	The SSL server on a default assumption will listen to the standard transmission control protocol (TCP) PORT22, through which the graphical sessions on the X Window system could be run securely

(continued)

Table 1 (continued)

Model/methods/techniques	Comments	Example
Detection based on attitudes thereby finding the vulnerabilities of certain devices and by using supervised machine learning in the centralized network with an awareness of certain non-constant attack mitigations	This method finds that difficult to identify a new device on the network. So DEFT controller is used which can easily address any new device on IOT chain and also supports device classification	DEFT-JS
Smart control and the use of software-defined network	An efficient framework that is not constant and that can be easily managed, cost-effective and more flexible which gives more comfort that it is suitable for high bandwidth and the non-steady nature of today's applications	2011-SDN 2012-OPEN FLOW 2014-ONOS 2017-CORD
Smart home gateway architecture	Helps in automatic device, network configuration and system updates	IOT Industrial gateway application on to smart homes, smart health care, asset tracking, etc.
Napto devices and the use of Napto base stations	With the help of clients DNA address, mapped with the help of Napto base stations to the clients having the same DNA address	Base station is one side and the other devices/vehicles, etc. at the other side who communicates with its peer device through the base station. For example, Walkie Talkie

4 Proposed Model

The radio frequency identification devices, smart cards, debit cards, credit cards, wireless sensor nodes all have high probability of getting attacked by the attacker thereby losing all its information which are the most vital. The cost and the constrained resource of the high-volume consumer devices are taken into consideration. So, to reduce this, light weight and the specialized cryptography primitives for security applications are used. The light weight cryptographic algorithm possesses some common characteristics like Minimum Charge and its consumption, Requirement of only minimum energy, consuming only minimum cost and the reduced processing time. Humming bird cryptographic algorithm was being used which increased the throughput about 4.7 times bigger than the light weight cryptography algorithm. Humming bird 2 algorithms were also generated then which increased the throughput of the system comparatively with each other. Further extension to this New APS Elliptic Curve Light Weight Cryptography algorithm shall be generated which is the upgraded version to Elliptic Curve Cryptography algorithm, which shall be handled data very smartly satisfying all the parameters like the credential checking of the user, interface of the data over the IOT chain, Constant update of the information, Beware about the IOT device supplies, Keeping a correct load checking habits and the most important thing is that smarter handling of the low resource devices on the IOT chain and shall consume very lesser power than the predecessor algorithms (Fig. 1).

Parameters types	Credential based on password	Data interface	Constant updates	Beware of IOT device supplies	Load checking process	Low resource devices
Conventional Cryptography algorithms	YES	YES	NO	NO	NO	NO
Elliptic curve cryptography	YES	YES	NO	NO	NO	YES
Algorithms						
Block chain technology	YES	YES	YES	NO	YES	NO
Secured socket shell protocol	YES	YES	NO	NO	NO	NO
New APS elliptic curve light weight cryptography algorithm	YES	YES	YES	YES	YES	YES

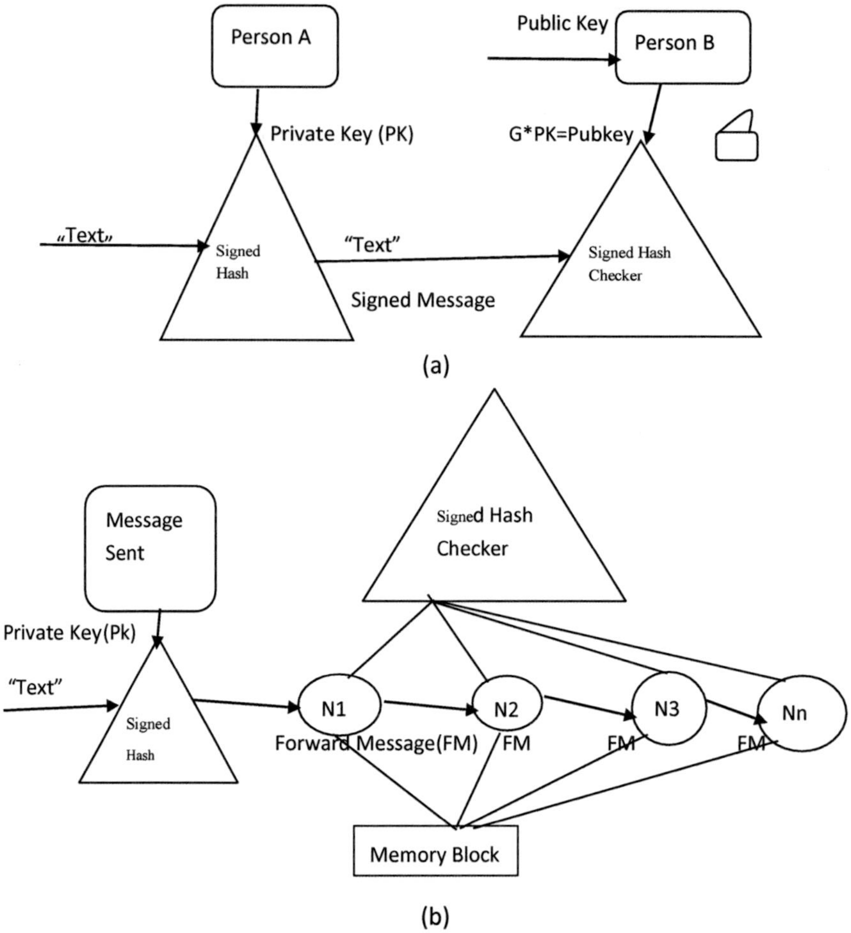

Fig. 1 **a** Earlier method. **b** Proposed method for security of data in IoT

5 Conclusion

The Internet of things which also ensures many low resources and the constraint devices to communicate with each other. These devices will be able to do computational processes and also will be in the position to make up a confirmed idea in the chain of networking. In the homo/heterogeneous setup, there could be many challenges starting from energy and its consumption, memory space, cost of performance and security in ICT as well. The lightweight cryptography primitive which has lightweight block/stream ciphers and hashing function as well for low resource devices on IOT chain. The light weight cryptographic algorithm based on several factors consideration not limited to its Key/ Block sizes and the structures thereby supporting the position aspects of the system. This ensures security with respect to power and the perfection of the system. New APS Elliptic Curve Light Weight Cryptography algorithm will be coined which shall handle smaller micro blocks and thereby increasing the throughput and saves energy consumption and increase the performance when compared with the predecessor algorithms. The cipher text which is intruded by an intruder shall be stored in these micro blocks and then the other stream texts could be forwarded on the IOT chain so that the recipient node could decrypt and see the message. Meanwhile, the embedded OS on the micro block could trace whether the data stored onto them is valid or invalid. If it is true information then this can also be forwarded on the IOT chain or otherwise the particular cipher text stored on that micro block could be sent via other nodes to the recipient with a flag set off and hence the recipient could not read that message or decrypt it. In this way, security is provided to the data across the IOT chain. From the above comparisons made on the proposed model, it is observed that all the parameters could be well satisfied in the New APS Elliptic Curve Light Weight Cryptography algorithm.

References

1. Gammelby U (2016). https://www.nabto.com/security-in-nabto-p2p-iot-solutions
2. Lin H, Bergmann NW (2016) IOT privacy and security challenges for smart home environment
3. Lee SK, Bae MY, Kim H (2017) Future of IOT networks: a survey:1–25
4. Kalkan K, Zeadally S (2018) Securing internet of things with software defined networking. IEEE Commun Mag:186–192
5. Thangavelu V, Divakaran DM, Sairam R, Bhunia SS, Gurusamy M (2019) DEFT: a distributed IOT fingerprinting technique. IEEE Internet Things J 6(1):940952
6. De HozDiego JD, Fernandez-Navajas J, Ruiz-Mas J (2019) IOTSafe, decoupling security from applications for a safer IOT. IEEE Transl Content Mining 7:29942–29962
7. Ozyilmaz KR, Yurdakul A (2019) Designing a blockchain—based IOT with Ethereum, Swarm and Lora. IEEE Consum Electron Mag
8. Ahmed BS, Bures M, Frajtak K, Cerny T (2019) Aspects of quality in IOT solutions: a systematic mapping study. IEEE Access J 7:13758–13780
9. Ding S, Li C, Li H (2018) A novel efficient pairing free CP-ABE based on elliptic curve cryptography for IOT. IEEE Access 6:27336–27345
10. Bassam JM, Hayajneh T () Light weight block ciphers for IOT: energy optimization and survivability techniques. IEEE Access 6:35966–35978
11. Razzag MA, Qureshi MA, Gill SH, Ullah S (2017) Security issues in the internet of things: a comprehensive study. Int J Adv Comput Sci Appl 8(6)

Ravi Sridharan is a member of ACM, IAENG, Toast Master's Club International. His qualifications include MCA from SRM Engineering College in the year 2000, Chennai, M.Phil from Periyar University in the year 2008, Salem, M.E., Computer Science & Engineering from Sree Sastha Institute of Engineering & Technology in the year 2010, Chennai. Currently, doing Research at Hindustan Institute of Technology &Science, Chennai. His area of interest includes Internet of Things, Software Engineering, and Mobile Adhoc Networks.

Dr. Thangakumar Jeyaprakash has received his B.E. Degree in Electrical & Electronics Engineering from Sivanthi Aditanar College of Engineering, Tamil Nadu in 2003. He obtained his M.Tech in Computer Science & Engineering, SRM University, Chennai. He obtained his Ph.D. Degree in Computer Science & Engineering from Hindustan Institute of Technology & Science, 2017. His area of interest include Mobile Adhoc Networks, Vehicular Adhoc Networks, Cryptography and Network Security, data Mining and Software Engineering.

Feature Selection Strategy for Academic Datasets Using Correlation Analysis

V. Sathya Durga and Thangakumar Jeyaprakash

Abstract In this work, we implement a correlation-based feature selection method on the academic dataset. A lot of research works are found on students' academic datasets. Increase in the number of data mining works on academic datasets brings the need to build mining models which are more accurate and precise. To build a good data mining model with high accuracy, the right subset of features must be selected on which the mining models can be built. There are many techniques for selecting the right features in data mining. A new methodology for selecting features based on correlation is proposed in this paper. Results prove that the features selected by correlation analysis increase the accuracy of the data mining models built.

Keywords Correlation analysis · Data mining · Feature selection · Performance prediction

1 Introduction

Data mining applications have found its way into our daily life inseparable. Almost all industry uses data mining to make smarter decisions at the right time. Importance of data mining applications has increased day by day. There are applications, which predict whether a person will have cancer or not from health datasets [1]. There are applications which predict grades of students with academic datasets. Whatever the industry may be, data forms the base of any mining application. The data which is collected from the real world is noisy and inconsistent. Moreover, it contains many features which do not correlate with the dataset and which reduce the efficiency of the data mining tasks. So, it is very important to identify those unimportant features and remove them before building a data mining model. The process of removing

V. S. Durga (✉) · T. Jeyaprakash
Department of CSE, Hindustan Institute of Technology and Science, Padur, Chennai, Tamil Nadu, India
e-mail: sathyadurga.v@gmail.com

T. Jeyaprakash
e-mail: tkumar@hindustanuniv.ac.in

D. J. Hemanth et al. (eds.), *Artificial Intelligence Techniques for Advanced Computing Applications*, Lecture Notes in Networks and Systems 130,
https://doi.org/10.1007/978-981-15-5329-5_14

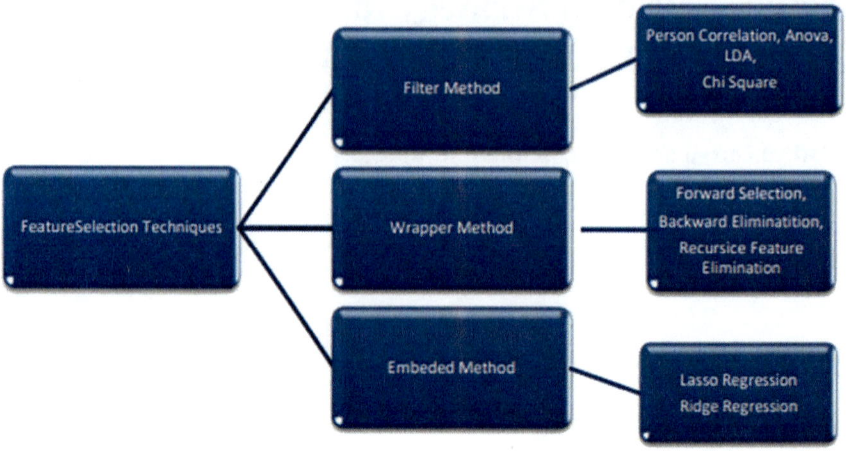

Fig. 1 Feature selection techniques

unwanted features or picking only the essential attributes for further processing is called feature selection. There are many methods for selecting features. These methods are broadly classified into filter, wrapper and embedded methods. Figure 1 shows the three types of feature selection techniques. Filter method works by selecting features based on scores obtained through statistical tests. Wrapper method works by selecting a subset of features and using those features to train the model. Based on the results, it is decided whether to retain the feature or to discard them. Embedded feature selection method is the third type of feature selection class which works by combing both filter method and wrapper method [2].

2 Review of the Literature

Shivakumar et al. (2016) use correlation-based feature selection algorithm (filter method) in their work to classify students based on their performance. The accuracy of their work is 90% [3]. Osmanbegović et al. (2015) use a gain ratio approach (filter method) for performance prediction. Random classifier achieved greater results than other algorithms [4]. Stephen et al. (2016) use chi-square-based feature selection method in their work to predict enrollment in respective courses [5]. Principal component analysis was used as a feature selection method in many works to predict the grades of students.

Huda et al. (2017) use correlation-based feature selection in their work to build two prediction models based on support vector machine and KNN. Students' final exam grades are predicted in this work [6]. Srinivas and Ramaraju (2017) build a prediction model using a Naïve Bayes classifier. Dataset consisted of 257 academic records. Two feature selection techniques, namely CfsSubsetEval and GainRatioAttributeEval, were used. Performance of both the algorithms is compared. Results reveal CfsSubsetEval performs better with 84% accuracy [7].

3 Materials and Methods

The material and methodology used in this research work are as follows.

3.1 Materials

The dataset used in this work is students' performance prediction dataset which was downloaded from the UCI repository [8]. It contains 33 attributes of students from Portuguese school.

3.2 Methods

Step 1: First, correlation values for all attributes are calculated. Table 1 shows the correlation values of all the features.
Step 2: Features with correlation value greater than zero are selected to build the prediction model.
Step 3: Based on the correlation values, we build two students' performance prediction models: one prediction model with all attributes and one prediction model with 18 attributes whose correlation values are greater than zero.
Step 4: Accuracy of the prediction model is determined, and results were analyzed. Results of this experimental study are discussed in the result section.

4 Results

Table 2 displays the accuracy of both the models. Model with no feature selection achieved 87.3% accuracy, whereas the prediction model built with features selected by correlation analysis achieved an higher accuracy of 89.7%. Thus, the second model performed with feature selection performed well than the first model.

Table 1 Correlation values of all the features

S. No.	Attributes	Correlation values
1.	Age	0.640939
2.	School	0.402857
3.	Schoolsup	0.200769
4.	Guardian	0.168603
5.	Studytime	0.16042
6.	G1	0.147762
7.	Activities	0.097226
8.	Traveltime	0.076904
9.	Nursery	0.069622
10.	G2	0.068726
11.	Absences	0.067116
12.	Famsup	0.064741
13.	Famsize	0.038131
14.	Goout	0.032334
15.	Higher	0.030654
16.	Famrel	0.017346
17.	Internet	0.010441
18.	Reason	0.006324
19.	Dalc	−0.005847
20.	Pstatus	−0.008269
21.	Freetime	−0.012036
22.	Walc	−0.023817
23.	Fjob	−0.042844
24.	Failures	−0.045507
25.	Mjob	−0.05279
26.	Medu	−0.060386
27.	Fedu	−0.067965
28.	Health	−0.071219
29.	Paid	−0.072014
30.	Sex	−0.111318
31.	Address	−0.140908
32.	Romantic	−0.145226

Table 2 Accuracy comparison of the prediction models

S. No	Attributes	Accuracy (%)
1.	Students' performance prediction model with all attributes	87.3
2.	Students' performance prediction model with 18 attributes	89.7

5 Discussion and Conclusion

From the previous section, it can be inferred that prediction model with features selected using correlation analysis yielded high accuracy than model built without any feature selection method. So, it can be concluded that the correlation-based feature selection method is more effective in picking the right features for building academic prediction models.

References

1. Anand A, Shakti D (2015) Prediction of diabetes based on personal lifestyle indicators. In: 1st international conference on next generation computing technologies, pp 673–676
2. Why, how and when to apply feature selection. https://towardsdatascience.com/why-how-and-when-to-apply-feature-selection-e9c69adfabf2
3. Sivakumar S, Venkataraman S, Selvaraj R (2016) Predictive modeling of student dropout indicators in educational data mining using improved decision tree. Indian J Sci Technol 9:1–5
4. Osmanbegović E, Suljić M, Agić H (2015) Determining dominant factor for students performance prediction by using data mining classification algorithms. Tranzicija 16:147–158
5. Stephen KW (2016) Data mining model for predicting student enrolment in STEM courses. IJCATR 5:683–724
6. Al-Shehri H, Al-Qarni A, Al-Saati L, Batoaq A (2017) Student performance prediction using support vector machine and K-nearest neighbor. In: IEEE 30th Canadian conference on electrical and computer engineering
7. Srinivas A, Ramaraju V (2018) Detection of failures by analysis of student academic performance using Naïve Bayes classifier. Int J Comput Math Sci 7(3):277–283
8. Student performance data set. https://archive.ics.uci.edu/ml/datasets/student+performance

Exploration of Magnetic Resonance Imaging for Prognosis of Alzheimer's Disease Using Convolutional Neural Network

M. S. Roobini and M. Lakshmi

Abstract Alzheimer's Disease (AD) is significant among the various dementia influencing endless senior individuals the world over which is the basic wellspring of dementia and memory hardship. Advancement causes shrinkage in hippocampus and cerebral cortex and it builds up the ventricles in the mind. Support vector machines have been utilized and a fragment of these techniques has been emitted an impression of being incredibly persuading in diagnosing AD from neuroimages, a part of the time on a very basic level more reasonable than human radiologists. X-beam uncovers the data of AD in any case decay regions are contrasting for various individuals which makes the discovering genuinely trickier. By utilizing the algorithm convolutional neural networks, the issue can be settled with insignificant error. This paper proposes a critical convolutional neural network (CNN) for Alzheimer's Disease finding utilizing mind MRI information appraisal. The tally was organized and tried utilizing the MRI information from Alzheimer's Disease Brain image.

Keywords Machine learning · Convolutional neural network · Alzheimer · Dementia

1 Introduction

Developing general well-being danger presented by Alzheimer's illness has raised the desperation to find and survey markers for the early discovery of the ailment. In this regard, an extraordinary arrangement of exertion has been committed to build models for foreseeing AD dependent on a solitary marker, or a mix of different markers, which catches the heterogeneity among subjects and identifies the sickness movement of subjects in danger. Since gentle subjective weakness is regularly

M. S. Roobini (✉)
Sathyabama Institute of Science and Technology, Chennai, India
e-mail: roobinims@gmail.com

M. Lakshmi
Saveetha School of Engineering, Chennai, India
e-mail: laks@icadsindia.com

D. J. Hemanth et al. (eds.), *Artificial Intelligence Techniques for Advanced Computing Applications*, Lecture Notes in Networks and Systems 130,
https://doi.org/10.1007/978-981-15-5329-5_15

considered as a transitional stage to AD, MCI patients are typically selected as the objective populace for early guess and assessing mediations. Existing exploration has recognized various biomarkers in anticipating a person's probability of changing over to AD, as well as contrasts in biomarker esteems among MCI and AD individual. It is broadly recognized that attractive reverberation imaging-based proportions of atrophy in key cerebrum districts, for example, the hippocampus, are prescient of movement from MCI to AD. Although a large portion of the ebb and flow thinks about measure local decay utilizing a solitary volume-based value, some analysts showed that the surface-based morphology investigation offers more favorable circumstances since this strategy ponders examples of subarea decay and creates nitty gritty point insightful relationship among decay and intellectual decline. An epic PC helped finding framework that utilizations include positioning and a hereditary calculation to break down basic attractive reverberation imaging information; utilizing this framework, we can anticipate change of mellow psychological impedance-to-Alzheimer's malady at somewhere in the range of one and three years before clinical determination. The CAD framework was created in four phases. Initially, we utilized a voxel-based morphometry method to examine worldwide and neighborhood dim issue decay in an AD bunch contrasted and sound controls. Districts with significant GM volume decrease were divided as volumes of intrigue. Second, these VOIs were utilized to remove voxel values from the individual decay districts in AD, HC, stable MCI, and dynamic MCI tolerant gatherings.

2 Data Preprocessing

The principle objective of the challenge was to assess how well classification calculations could gain from MRI information from different destinations, conventions, scanners, and advancements. As depicted in the challenge's site, MRIs were chosen from the ADNI database. Rivalry coordinators arbitrarily and consequently chose subjects with a static seed by utilizing the information examination stage Konstanz Information Miner. Subjects from ADNI were chosen by filtering content files downloaded from the site. Specifically, they utilized the file containing 52 transformation of finding for first picking HC, AD patients, and mild cognitive impairment patients who did not change over their conclusion in the development. In any case, its past point where it is possible to treat the infection in definite stages in this way, prior finding is fundamental. Hearty AI calculation, for example, CNN, which can arrange Alzheimer's malady, will help radiologists in diagnosing and will likewise help in the precise and auspicious conclusion of Alzheimer's patients.

3 Literature Survey

Prediction model is developed in this paper for prediction of AD by designing a decision tree model [1]. Neuroimaging procedures give an approach to clinicians to look at the basic and practical changes in the cerebrum related with the advancement of sicknesses [2]. Usually, utilized modalities incorporate attractive reverberation imaging, utilitarian attractive reverberation imaging, positron discharge tomography (PET), single photon emanation figured tomography, and dissemination tensor imaging. Owning to its simple access in clinical settings, MRI gets the most consideration of specialists contrasted and different modalities. The basic changes in the mind related with AD can be non-obtrusively evaluated utilizing MRI. AD patients regularly have proof of cortical decay and broadened ventricles in examination with wellbeing controls. Average worldly flap (MTL) decay, evaluated utilizing MRI, has demonstrated to be a successful clinical guide in the early determination [2]. The information from this investigation has suggestions for the utilization of imaging as surrogate markers of illness movement in remedial preliminaries for AD. In the first place, there was no clear favorable position of average fleeting over hemispheric rate measures for affectability to clinical transformation right off the bat in the illness procedure and MCI gatherings. Also, a pattern was available recommending better connection between change on MRI and change on psychological test/rating-scale execution for hemispheric decay rates contrasted and average transient projection decay rates. Accordingly, the more computerized and less work serious hemispheric technique may be best as a proportion of ailment movement in certain clinical preliminaries. Second, temperamental appraisal of sickness movement was essentially increasingly visits with psychological tests/rating scales than with the MRI measures. The more prominent exactness of MRI converts into a lot of littler assessed test size necessities for clinical preliminaries in MCI. This information bolsters the utilization of imaging notwithstanding standard clinical/psychometric measures as surrogate markers of ailment movement in AD restorative preliminaries. A differentiation ought to be made between approval of a surrogate proportion of restorative adequacy and approval of a surrogate marker of infection movement. Without a positive infection altering remedial that included imaging, one cannot profess to have approved imaging as a surrogate marker of helpful viability. In any case, one can profess to approve an imaging marker of ailment movement in a characteristic history concentrate, for example, this where the information shows obvious and reproducible relationships between various proportions of clinical sickness movement and proportions of progress on basic MRI [3]. To extract patterns from neuroimaging data, various techniques, including statistical methods and machine learning algorithms, have been explored to ultimately aid in Alzheimer's disease diagnosis of older adults in both clinical and research applications. However, identifying the distinctions between Alzheimer's brain data and healthy brain data in older adults whose age is less than 76 is challenging due to highly similar brain patterns and image intensities. Recently, cutting-edge deep learning technologies have been rapidly expanding into numerous fields, including medical image analysis. This work outlines state-of-the-art deep learning-based pipelines employed to distinguish Alzheimer's magnetic resonance imaging and functional MRI data from normal healthy control data for the same age group. Using these pipelines, which were executed on a GPU-based high-performance computing platform, the data were strictly and carefully preprocessed. Next, scale and shift invariant low- to high-level features were obtained from a high volume of training images using convolutional neural network architecture. In this study, functional MRI data were used for the first time in deep learning applications for the purposes of medical image analysis and Alzheimer's disease prediction. These proposed and implemented pipelines, which demonstrate a significant improvement in classification output when compared to other studies, resulted in high and reproducible accuracy rates of 88.8% and 99.66% for the fMRI and MRI pipelines, respectively [4]. Interclass connection coefficients for interrater dependability were determined in two different ways. Interclass connection coefficients for each ratter pair were reliably more prominent than 0.82 and shifted between 0.82 and 0.86. Second, all the ratters were looked at as a gathering. There was moderate to generous agreement; 32 with interclass connection coefficients of 0.81, 0.78, and 0.82 for right, left, and mean MTA scores, separately. Ratters infrequently varied by more than 1 classification of MTA score, and there were no precise contrasts among the ratters [5]. In this paper, we structured and tried an example grouping framework that joins inadequate autoencoders and convolutional neural systems [6]. All in all, our investigation has demonstrated that the proposed ICA-based strategy might be valuable for grouping AD and MCI subjects from typical controls. Nonetheless, the accomplished characterization exactness is as yet not ideal because of a few elements. To begin with, the ADNI is a multicenter database and we did not consider scanner or focus impacts [7]. So as to recognize minds influenced by Alzheimer's sickness from ordinary sound cerebrums in more established grown-ups, this investigation introduced two vigorous pipelines, including broad pre-processing modules and profound learning-based classifiers, utilizing auxiliary and utilitarian MRI information. This investigation additionally showed that the created pipelines filled in as productive calculations in portraying multimodal MRI biomarkers. Taking everything into account, the proposed strategies exhibit solid potential for foreseeing the phases of the movement of Alzheimer's ailment and ordering the impacts of maturing in the ordinary mind [8]. Alzheimer's Disease (AD), the most widely recognized type of dementia, is a hopeless neurological condition that outcomes in a dynamic mental decay. Albeit conclusive determination of AD is troublesome, by and by, AD finding is to a great extent dependent on clinical history and neuropsychological information including attractive asset imaging (MRI). Expanding research has been accounted for on applying AI to AD acknowledgment lately. This paper presents most recent commitment to the development. It depicts a programmed AD acknowledgment calculation that depends on profound learning on 3D mind MRI. The calculation utilizes a convolutional neural system (CNN) to satisfy AD acknowledgment. The three-dimensional topology of mind is considered all-in all-in AD acknowledgment, bringing about an exact acknowledgment [9]. Alzheimer's disease is an incurable, progressive neurological brain disorder. Earlier detection of Alzheimer's disease can help with proper treatment and prevents brain tissue damage. Several statistical and machine learning models have been exploited by researchers for Alzheimer's disease diagnosis. Analyzing magnetic resonance imaging (MRI) is a common practice for Alzheimer's disease diagnosis in clinical research. Detection of Alzheimer's disease is exacting due to the similarity in Alzheimer's disease MRI data and standard healthy MRI data of older people. Recently, advanced deep learning techniques have successfully demonstrated human-level performance in numerous fields including medical image analysis. They propose a deep convolutional neural network for Alzheimer's disease diagnosis using brain MRI data analysis. While most of the existing approaches perform binary classification, our model can identify different stages of Alzheimer's disease and obtains superior performance for early-stage diagnosis. Moreover, the proposed approach has strong potential to be used for applying CNN into other areas with a limited dataset. In the future, we plan to evaluate the proposed model for different AD datasets and other brain disease diagnosis.

4 Classification Frameworks for Prediction of Alzheimer's Disease and its Progress

Over the previous decade, Classification systems have been utilized effectively to investigate complex examples in neuroimaging information with a view to the classification of AD and MCI subjects. A classification system is involved four significant segments: highlight extraction, include determination, dimensionality decrease, and highlight-based classification calculation. The instances of such inferred measures incorporate provincial tissue densities, territorial cortical thickness, and so forth. The highlights separated from different modalities can be utilized in disengagement or joined to utilize the integral data gave by a few modalities. A classification calculation is then prepared on the separated highlights to give symptomatic help in foreseeing subjectively typical and unhealthy subjects. Along these lines, we gave more subtleties on highlight extraction for the remembered investigations for this audit. Generally speaking, the paper is isolated into different areas, where each segment centers around highlights removed from one specific imaging methodology.
Framework Used:
Framework which is used in this paper is PyTorch Deep Learning Library. The dataset has four unbalanced classes, namely mild demented, moderate demented, non-demented, and very mild demented input image from the dataset: [Fig: 1]
LBP Parameters:

$$\text{LBP energy} = 0.32325068209898705$$
$$\text{LBP entropy} = 2.1782987228452058$$

GLCM Feature Extraction:
The necessary parameters for Gray Level Co-occurrence Matrix are:

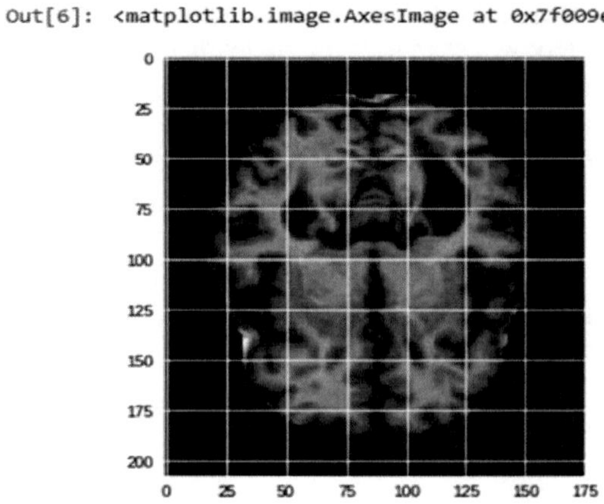

Fig. 1 Input image from the dataset

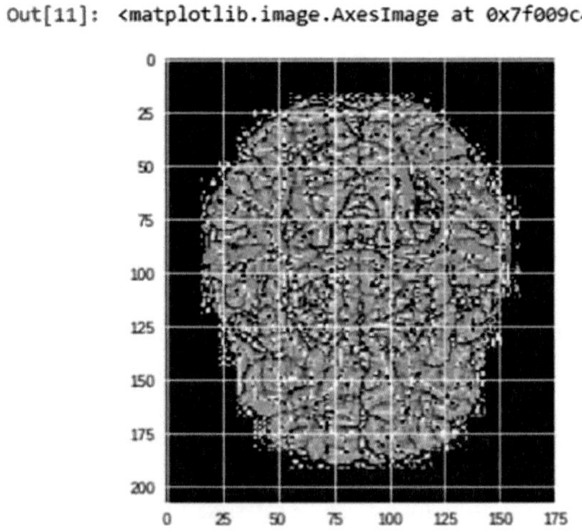

Fig. 2 Finding the local binary pattern for feature extraction

Contrast = 345.56283156498677
Dissimilarity = 9.781222369584437
Homogeneity = 0.48409272116505836
Energy = 0.4372697858687803
Correlation = 0.9424142926079172

The necessary parameters while applying the Gabor Filter are:

Gabor energy = 0.541411325269191
Gabor entropy = 1.3535283133672198

Applying a Convolutional Neural Network on the Dataset:

```
<matplotlib.image.AxesImage at 0x7f009c34c2e8>
```

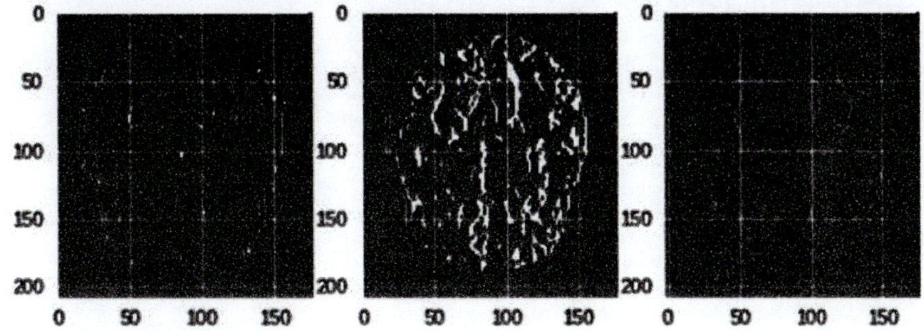

Fig. 3 Applying the Gabor filter

Network Architecture: ***Conv → Max pool → Conv → Max pool → FC***

The first convolutional layer consists of 96 kernels of size 11 × 11 applied with a stride of 4 and padding of 0. The max pool layer following Conv-1 consists of pooling size of 3 × 3 and stride 2. The second conv layer consists of 256 kernels of size 5 × 5 applied with a stride of 1 and padding of 2. The max pool layer following Conv-2 consists of pooling size of 3 × 3 and a stride of 2. The first fully connected (FC) layer has 4096 neurons.

The accuracy achieved after 2500 Iterations:

Loss: 0.7782031297683716
Accuracy: 54.26114151681001

```
/usr/local/lib/python3.6/dist-packages/ipykernel_launcher.py:43: UserWarning: invalid index of a 0-dim ten
sor. This will be an error in PyTorch 0.5. Use tensor.item() to convert a 0-dim tensor to a Python number
Iteration: 500, Loss: 0.9553308486938477, Accuracy:51.83737294761532
Iteration: 1000, Loss: 0.9312585592269897, Accuracy:53.87021110242377
Iteration: 1500, Loss: 0.7760425209999084, Accuracy:55.12118842845973
Iteration: 2000, Loss: 0.8502954244613647, Accuracy:55.90304925723221
Iteration: 2500, Loss: 0.7782031297683716, Accuracy:54.26114151681001
```

Applying a Neural Network on the Dataset:

Network Architecture: ***ogits → non-linear-op → logits → softmax → labels***

Input dimension: 784 (28 * 28)
Output dimension: 10
Hidden dimension: 100
Learning rate: 0.1

Accuracy Achieved on a Three-Layered Neural Network:
Loss: 0.06957100331783295, Accuracy: 97.13.

```
/usr/local/lib/python3.6/dist-packages/ipykernel_launcher.py:115: UserWarning: invalid index of a 0-dim te
nsor. This will be an error in PyTorch 0.5. Use tensor.item() to convert a 0-dim tensor to a Python number
Iteration: 500, Loss: 0.3420164883136749, Accuracy:89.59
Iteration: 1000, Loss: 0.15970270335674286, Accuracy:94.39
Iteration: 1500, Loss: 0.24343451857566833, Accuracy:95.01
Iteration: 2000, Loss: 0.055785685777664185, Accuracy:96.16
Iteration: 2500, Loss: 0.03965865448117256, Accuracy:97.13
Iteration: 3000, Loss: 0.06957100331783295, Accuracy:97.13
Time required is 43.638954877853394
```

Accuracy Achieved on a Two-Layered Neural Network:
Loss: 0.03204762563109398, Accuracy: 96.98.

```
/usr/local/lib/python3.6/dist-packages/ipykernel_launcher.py:114: UserWarning: invalid index of a 0-dim te
nsor. This will be an error in PyTorch 0.5. Use tensor.item() to convert a 0-dim tensor to a Python number
Iteration: 500, Loss: 0.28217756748199463, Accuracy:90.66
Iteration: 1000, Loss: 0.20606881380081177, Accuracy:93.82
Iteration: 1500, Loss: 0.2810133099555969, Accuracy:95.27
Iteration: 2000, Loss: 0.20160506665706635, Accuracy:95.56
Iteration: 2500, Loss: 0.11775699257850647, Accuracy:96.4
Iteration: 3000, Loss: 0.03204762563109398, Accuracy:96.98
Time required is 40.914836406707764
```

Applying Transfer Learning on the Dataset
Instead of random initialization, we initialize the network with a pretrained network, like the one that is trained on imagenet 1000 dataset. Rest of the training looks as usual. Here, we will freeze the weights for all of the networks except that of the final fully connected layer. This last fully connected layer is replaced with a new one with random weights and only this layer is trained. Uses Resnet-18 Pretrained Model and the model are trained for over 30 Epochs:

```
Epoch 23/29
----------
train Loss: 0.7750 Acc: 0.6430
test Loss: 0.8341 Acc: 0.6122

Epoch 24/29
----------
train Loss: 0.7836 Acc: 0.6284
test Loss: 0.8443 Acc: 0.6052

Epoch 25/29
----------
train Loss: 0.7739 Acc: 0.6399
test Loss: 0.8394 Acc: 0.6020

Epoch 26/29
----------
train Loss: 0.7819 Acc: 0.6251
test Loss: 0.8357 Acc: 0.5973

Epoch 27/29
----------
train Loss: 0.7798 Acc: 0.6419
test Loss: 0.8338 Acc: 0.5966

Epoch 28/29
----------
train Loss: 0.7714 Acc: 0.6384
test Loss: 0.8385 Acc: 0.5981

Epoch 29/29
----------
train Loss: 0.7746 Acc: 0.6389
test Loss: 0.8325 Acc: 0.6013

Training complete in 29m 20s
Best val Acc: 0.613761
```

Final Accuracy Achieved: 93.43%

Then, use ConvNet(VGG16) as fixed feature extractor which freezes all the network except the final layer. The model is trained for 14 Epochs and the accuracy achieved is 67.0310%.

```
Epoch 7/14
----------
train Loss: 0.5327 Acc: 0.7156
test Loss: 0.6042 Acc: 0.6530

Epoch 8/14
----------
train Loss: 0.5344 Acc: 0.7236
test Loss: 0.6048 Acc: 0.6539

Epoch 9/14
----------
train Loss: 0.5314 Acc: 0.7195
test Loss: 0.6011 Acc: 0.6530

Epoch 10/14
----------
train Loss: 0.5312 Acc: 0.7158
test Loss: 0.6004 Acc: 0.6667

Epoch 11/14
----------
train Loss: 0.5331 Acc: 0.7181
test Loss: 0.6003 Acc: 0.6485

Epoch 12/14
----------
train Loss: 0.5322 Acc: 0.7193
test Loss: 0.6000 Acc: 0.6667

Epoch 13/14
----------
train Loss: 0.5331 Acc: 0.7158
test Loss: 0.6016 Acc: 0.6548

Epoch 14/14
----------
train Loss: 0.5326 Acc: 0.7161
test Loss: 0.5994 Acc: 0.6576

Training complete in 8m 54s
Best val Acc: 0.670310
```

Implementing Transfer Learning using fast.ai
The dataset available with us:

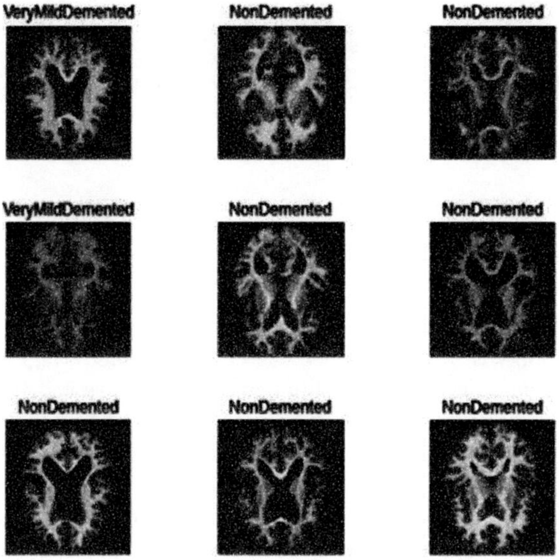

Using Resnet-34, the model is trained for 10 Epochs:

Total time: 06:40

epoch	train_loss	valid_loss	accuracy	
1	0.021454	1.995235	0.683346	(00:40)
2	0.072192	2.009616	0.595778	(00:40)
3	0.103748	5.072022	0.511337	(00:39)
4	0.087938	1.719478	0.652072	(00:39)
5	0.058591	1.404937	0.678655	(00:40)
6	0.042578	0.757219	0.759969	(00:40)
7	0.020298	1.019235	0.741204	(00:39)
8	0.007972	1.011817	0.749023	(00:40)
9	0.003368	1.209249	0.739640	(00:39)
10	0.001742	1.088688	0.750586	(00:39)

Final Accuracy achieved: 97.25%

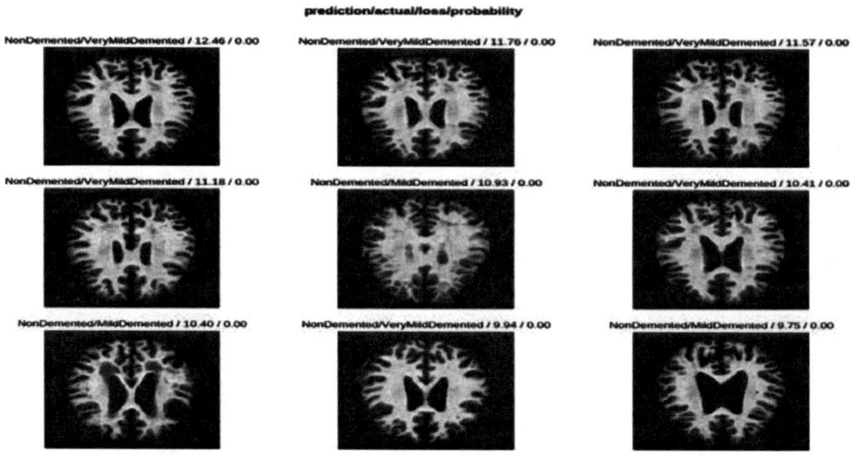

5 Diagnosis and Monitoring

Considering that the hardware expected to execute a pitch disturbance characteristic mechanical assembly is by and by open in PDAs and tablet PCs, this is a promising philosophy toward working up a simplicity, non-invasive gadget that may help perceive individuals requiring further evaluation (Fig. 4).

As another AI worldview, profound learning has been progressively investigated in the advancement of innovation for large information and man-made reasoning. The system was based upon a 3D convolutional autoencoder, which is pre-prepared to catch anatomical shape varieties in auxiliary cerebrum MRI examines. Trials on the received MRI dataset with no skull-stripping pre-processing had indicated

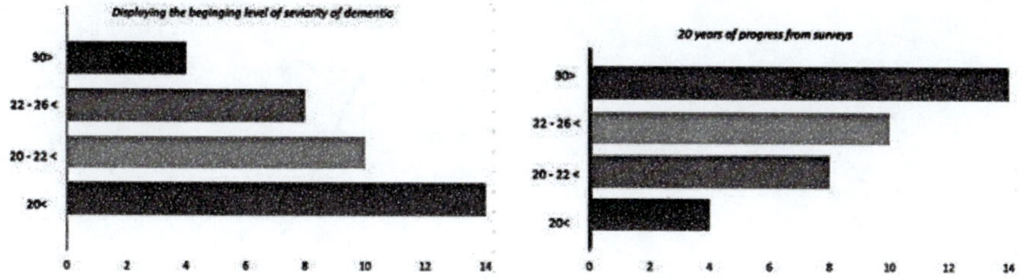

Fig. 4 Progress of disease

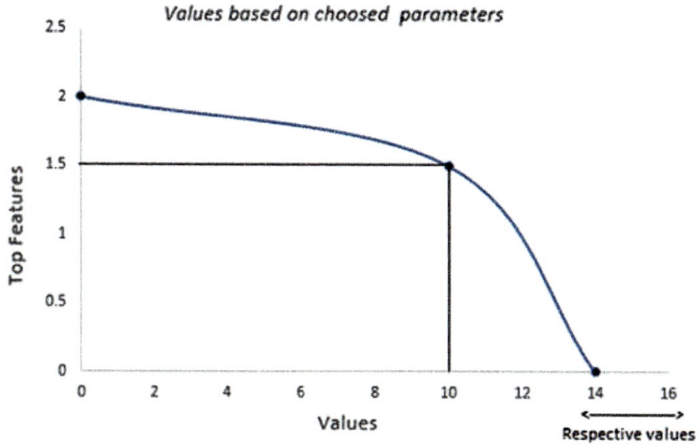

Fig. 5 Levels indicating the parameters

that it outflanked a few traditional classifiers by accuracy. This paper presents a programmed AD acknowledgment calculation that depends on profound learning on 3D mind MRI. We depict the aftereffects of the trials and the related information. It is inferred that outlining the methodology and calling attention to conceivable future interests in the region (Fig. 5).

6 Discussion

This investigation confirms the effectiveness of CNN models for PC supported identification frameworks. Indeed, the classification results got by CNN permitted the BMPG gathering to get one of the most exact expectation in the members' roaster. By and by, the multi-class classification precision is still a long way from arriving at acceptable outcomes for clinical materialness. Anyway, these findings contrast well and best in class results; for instance, in the best performing technique arrived at a 53% exactness for a three-class classification issue. Notwithstanding, different works

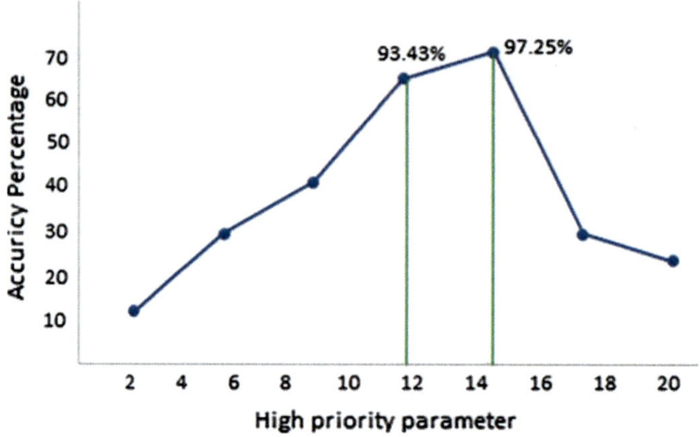

Fig. 6 Accuracy percentage

consider this as a naturally four-class classification issue, obviously for this situation revealed exact nesses, 54.6%, can arrive at lower esteems. This investigation has enabled us to measure and break down the influence and utility of the MMSEs on the 190 final classification and furthermore delighted that the significance of morphological highlights was monitored in the two cases. In this paper, we exhibited a top to bottom examination on the utilization of AI procedures in the recognition of AD and related states by utilizing just MRI basic data paying little heed to utilized MRI convention or MRI machine. Taking into account that the MMSEs were the psychological measurement present in the dataset, subjective information was restricted. In spite of the fact that the confinements and possibilities of the MMSEs have just been broadly examined, in this paper, we indicated the effect on disposing of this score from the information, which actuates a misfortune in exactness of roughly 10%. Regardless of this, the significance of the 255 principle morphometric highlights with or without the MMSEs was rationed when utilizing every one of the subjects as one single gathering. To take out the MMSEs from the classification procedure without losing its data, we chose to isolate the train information as per the members' psychological profile. This methodology enabled us to recoup the exactness lost by slicing of the MMSEs and to find unmistakable primary highlights for each gathering (Fig. 6).

7 Conclusion

In was concluded that, the proposed highlight determination technique chooses the top discriminative highlights as well as limits the dimensionality of the information vectors to low-dimensional space. The exploratory outcomes exhibit that a blend of highlight positioning and GA is a dependable method for MCI transformation forecast and early discovery of AD, particularly with respect to high-dimensional information design acknowledgment. Also, the trial results show that the GM decay design in AD subjects and HCs, as appeared in VBM investigation, can be valuable for removing highlights from sMCI and pMCI tests. The exploratory outcomes show that the exhibition of the proposed methodology can contend firmly with cutting-edge systems utilizing MRI information, as announced in the writing. This may bring about lackluster showing and absence of generalization, as well as in non-translate capacity of calculation forecasts. The accuracy prediction using CNN algorithm was done. On the off chance that example enhancement for momentary clinical preliminaries is required, different factors that foresee transient change must be incorporated alongside MRI. The prognosis of Alzheimer disease using CNN gives a prediction accuracy of 97.25%.

References

1. AL-Dlaeen D, Alashqur A (2014) Using decision tree classification to assist in the prediction of Alzheimer's Disease. IEEE Computer Society
2. Liu J, Li M, Lan W, Wu F-X, Pan Y, Wang J (2015) Classification of Alzheimer's Disease using whole brain hierarchical network
3. Jack CR, Petersen RC, Xu YC, Waring SC, O'Brien PC, Tangalos EG, Smith GE, Ivnik RJ, Kokmen E (1997) Medial temporal atrophy on MRI in normal aging and very mild Alzheimer's Disease. Neurology 49(3):786–794
4. Visser P, Verhey F, Hofman P, Scheltens P, Jolles J (2002) Medial temporal lobe atrophy predicts Alzheimer's Disease in patients with minor cognitive impairment. J Neurol Neurosurg Psychiatry 72(4):491–497
5. DeCarli C, Frisoni GB, Clark CM, Harvey D, Grundman M, Petersen RC, Thal LJ, Jin S, Jack CR, Scheltens P (2007) Qualitative estimates of medial temporal atrophy as a predictor of progression from mild cognitive impairment to dementia. Arch Neurol 64(1):108–115
6. Tong T, Wolz R, Ga Q, Guerrero R, Hajnal JV, Rueckert D, Alzheimer's Disease Neuroimaging Initiative (2014) Multiple instance learning for classification of dementia in brain MRI. Med Image Anal 18(5):808–818
7. Yang W, Lui RLM, Gao J-H, Chan TF, Yau S-T, Sperling RA, Huang X (2011) Independent component analysis-based classification of Alzheimer's disease MRI data. J Alzheimer's Dis 24(4):775–783
8. Sarraf S, Tofighi G, The Alzheimer's Disease Neuroimaging Initiative (2016) DeepAD: Alzheimer's disease classification via deep convolutional neural networks using MRI and fMRI
9. Luo S, Li X, Li J (2017) Automatic Alzheimer's Disease recognition from MRI data using deep learning method. J Appl Math Phys

Advanced Accident Avoiding, Tracking and SOS Alert System Using GPS Module and Raspberry Pi

Chaitanya Lakshmi Indukuri and Kottilingam Kottursamy

Abstract IoT connects different physical things or objects which can be integrated with several sensors which helps the objects to extract data and process on Internet. IoT is mostly used technology in present situation and it is thought that it is used mostly by the end of 2020 also. This aims at implementing an advanced and real-time system for monitoring and tracking the bus to ensure better safety of public on IoT platform. The system is about providing passengers safety about fatal crashes on winter fog/smoke and providing information of emergencies case such as accidents, break down, fire accidents, defilement by immediately sharing the location and images of the inside environment of the bus to the concern authorities by email alert. The system provided with user keys. In case of vehicle breakdown, siren is on through provided key. If there are any accidents and defilement an email alert will be sent to the authorities with the images of the inside environment of the bus along with GPS location of the bus. The complete system uses RPi build around using ARMII76JZF-S Microprocessor. This has interface devices such as camera, GPS, speakers to microprocessor. This basic goal of this is to provide information about location and images of the inside environment of the bus to the concern authorities by email alert. Location details obtained by GPS module.

Keywords Raspberry pi · TCP/IP protocol · SOS · Voice alert · AAU

1 Introduction

Day by day traffic is increasing rapidly in India. Most of the people travel by vehicle, as it is the safest mode of transportation. Transport system development has been increasing which is unbalanced. The burden on road transport has been rising by the Delhi metro. The need for an hour is to modernize and use the latest technology in

C. L. Indukuri (✉) · K. Kottursamy
SRM Institute of Science and Technology, Chennai, India
e-mail: Chaitanyaindukuri33@gmail.com

K. Kottursamy
e-mail: kottilik@srmist.edu.in

D. J. Hemanth et al. (eds.), *Artificial Intelligence Techniques for Advanced Computing Applications*, Lecture Notes in Networks and Systems 130,
https://doi.org/10.1007/978-981-15-5329-5_17

the transportation system. In the road transport system, we are making progress with old technologies. As the technologies are old numerous vehicles are abandoned by state road Transport Corporation. The main problems of existing vehicle transport system are irresponsible behavior of the driver, uncertain traffic conditions, vehicle equipment defects, and lack of passenger protection. The safety of children and students is important. Security camera monitoring in school buses is a great tool. Each day the control and monitoring system in the vehicles are becoming more and more complex. This leads to increased use of cluttered wiring difficulty which is difficult to install, service, high cost and hazardous. New technologies using in-vehicle networking methods cannot solve these problems. Safety of the students in the vehicle depends on the driver and also the physical conditions of that vehicle. A vehicle monitoring system is the combining usage of automatic vehicle location using GPS, speed control and live video streaming in the vehicle with specialized software helps to enhance the safety measurements in the school vehicles. GPS is a device which is more used in mobile phones to track the road maps. The whole processing of this system relays on ARM cortex board. ARM cortex board is interfaced to GPS.

Current records of accidents and crashes occur due to three major factors so-called weather events, i.e., rain, snow and fog/smoke. Figures 1 and 2 [1] show that the fatal crashes occur across Indian states in these weather conditions is clearly a toughest problem that needs to have some solution. Hence, it is a major problem which helps in traffic crashes mostly in some north states in the form of fog or smoke, as this is major these are reported.

The system is about providing information of emergencies case such as accidents, break down, fire accidents, defilement by immediately sharing the location and images of the inside environment of the vehicle to the concern authorities by email alert. The system provided with user SOS button in case of emergency, siren is on through provided SOS button and email alert sent to the concern authorities. An email alert will be sent to the authorities with GPS location of the bus. In case

Fig. 1 Statistics of four years fog-related road crashes across India

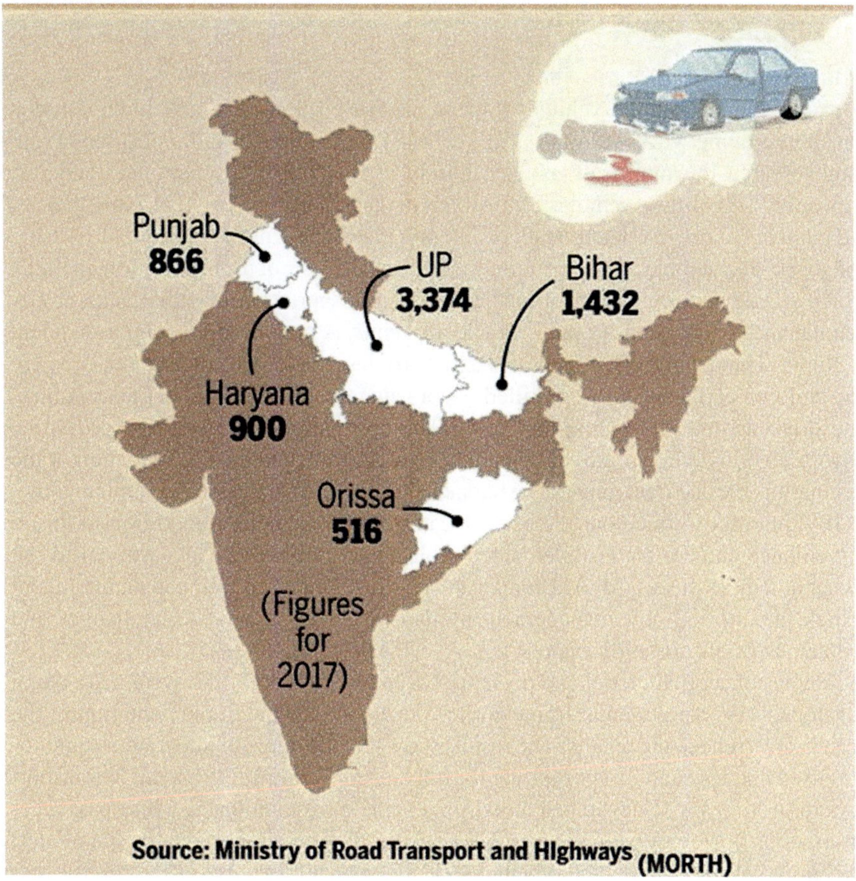

Fig. 2 Worst five states of India in fog-related fatal crashes

of winter accidents occurs due to dense fog are avoided by this system by using ultrasonic sensors, which give the alert to driver about distance of front and back vehicles approaching.

The second section explains related work in the technology field that is used to prevent accidents. The third section brings out the working model. The fourth section describes the system architecture. The fifth section gives importance of IoT. The sixth paragraph includes realistic understanding, performance, and discussion relevant to it. The final section discusses the results that explain the consequences and possible directions of work and progress.

2 Literature Survey

For most Indians, public vehicles are the most used transport, and India's public transportation networks are one of the world's most popular and well-used. From some research, the number of vehicles in India is small as per international standards, with only 24.85 million vehicles on the roads of the nation as of 2013 [2]. Nonetheless, the number of traffic-related deaths is among the highest in the world and is rising daily [3]. So, vehicle transportation is the basic mode of transportation in India which is safe and convenient. But the vehicle transportation in India lacks proper maintenance which leads to fatal vehicle accidents. A major school vehicle accident in Nupur, HimachalPradesh, in April 2018. To describe one of the incidents where few children and teachers were killed when vehicle feel from 328 feet near Punjab. The driver is unaware of the location of the destination that he has to go. So he drives the Vehicle in a route that which has gorge. If he had the GPS location of the destination, the accident may not be happened. This leads to the idea of implementing a GPS, in order to trace out the exact location of the vehicle. In addition to this, a surveillance camera also implemented which provides a live video stream to the owner at remote places. M. A. Hannan, A. M. Mustapha [4] developed an intelligent vehicle monitoring and management system that deals about the challenges and problems of transportation system.

A new conceptual model and decision-based algorithms for the system are being developed. An experimental framework is developed for the implementation of the design and to check the results. The results show that the range of integrated technologies used in the system is appropriate for tracking and controlling a vehicle transport system. Author [5, 6] developed a real-time vehicle for monitoring every passenger information for every vehicle. Vehicle tracking device can act as a practical alert system to help people decide whether to walk or wait. For this system, transmitter will be installed on vehicle, at vehicle stops receiver boards will be installed, using centralized controller and above methods best vehicle and transportation routes are taken using LED embedded map. For the user's precautions, the user information system software will also be installed at the stops of the vehicle and will provide user information on all vehicles. Author [7] designed a vehicle monitoring system using Polyline Algorithm to handle the vehicle's current location data, and using the data, the vehicle's real-time tracking will be detected and supplied to the vehicle's current status passenger. The system also includes mobile application, where users can track location of vehicle and can spot the nearby vehicle stop.

The system has enhanced GPS module which connects to a vehicle and finds out information at each interval and data is sent to database for storage and further processing. Our project "ADVANCED ACCIDENT AVOID, TRACKING AND SOS ALERTSYSTEM USING GPS MODULE ANDRASPBERRYPI" provides so many applications which are very useful for better vehicle transportation system. This enhances the ease of traveling even more secure. It is more helpful in panic situations by accessing the location using GPS and sends information to registered email.

3 System Architecture

Smart home security system consists of two parts, accident avoiding unit (AAU) is part of smart home where security system is implemented and remote unit (RU) is a device implemented on users' smart mobile phones. Vehicle monitoring system consists of two parts, accident avoiding unit (AAU) is part of vehicle monitoring where security system is applied and remote unit (RU) is a device implemented on smart mobile phone users.

3.1 Accident Avoiding Unit (AAU)

AAU is an effective, low-power consumption and low-cost embedded access control device for smart home protection and enables users to have remote monitoring and notification before they occur in pathetic situations. AAU is made up of Raspberry Pi built on an installed SD card with Raspbian Operating System which serves as a basic OS. The few modules that are used are user keys and Pi Camera that are connected to Raspberry Pi to track the exact location of the vehicle. Captured images are saved on SD card using the incident camera before or after with time and date. Raspberry Pi has been configured to enable the camera. AAU also includes a Wi-Fi module for transmitting the captured image or video via email alert to the remote-control unit.

3.2 Remote Unit

RCU is a software tool that can be applied on mobile phone users. One should provide Graphical User Interface (GUI) to send predefined Linux Terminal Commands to the Accident Evasion Unit via Secure Shell. SSH which is a basic protocol is used in communications that need to be secured (Fig. 3).

Accident avoiding unit (AAU) is efficient, low-power consumption, low-cost snored system to determine objects for safe driving in front of the vehicle and allows the user to monitor and control. AAA has a Raspberry Pi setup installed on a SD card with Raspbian Jessi operating system. Sensors are interfaced and integrated with Raspberry pi to detect the values of temperature, humidity, vibration, respectively, in front object values.

3.3 Raspberry Pi3

The RPi is an SBC which is developed to encourage teaching and guiding knowledge development in programming and computing [19]. It is mostly supportive for the

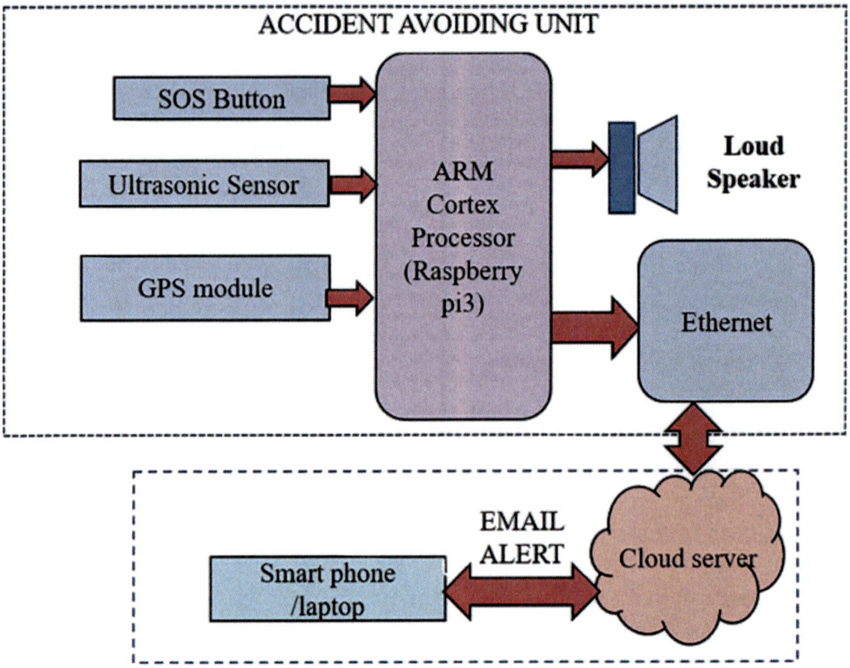

Fig. 3 Block diagram

IoT projects [8, 9]. Pi is the most commonly used experimental tool where you can use it without restrictions as a desktop computer, media center, database or monitoring/security device in your home [10]. Most programs and operating systems like Linux-based operating systems can work with Pi where you can have plenty of free software access and download [17, 18]. New Pi three has more features that can be introduced, such as computing power and on-board access to the outer parts, saving you time with the software (Fig. 4).

Features of the Raspberry Pi

- A 1.4Ghzprocessor.
- 1 GB RAM.
- Gigabit Ethernet Port.
- Power-over-Ethernet.
- Bluetooth4.2.
- 4 USB2.0.
- Camera port.
- Micro HDMI Video Output.
- SD Card Slot.
- 40-pinGPIO.

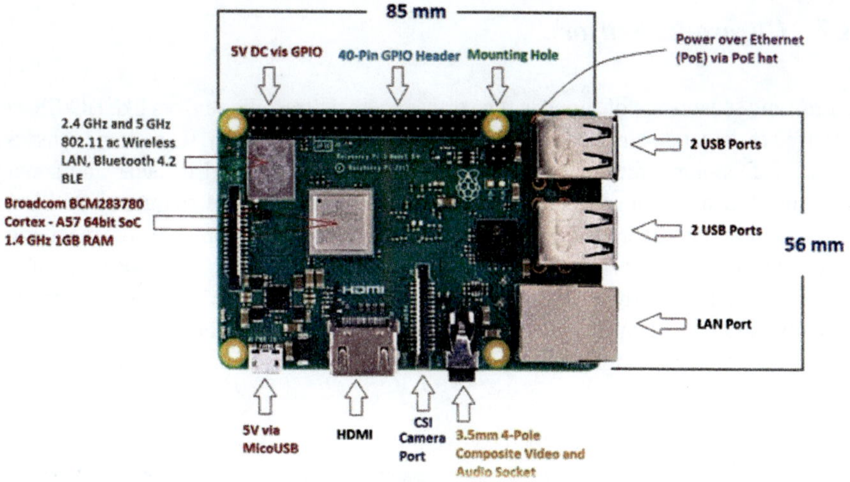

Fig. 4 Raspberry Pi3 [11]

3.4 Light Sensor (LDR)

An LDR is a component of sensors which has variable resistance changes when it intact with light when it falls on that sensor. These allow them to use in light sensing circuits.

3.5 Temperature and Humidity Sensor (DHT11)

DHT11 digital temperature and humidity sensor is a composite sensor [12] contain a calibrated digital signal output of the temperature and humidity [24]. This sensor which uses three pins out of four pins forms its module helps in determining temperature 0–50 °C and humidity about 20–90%.

3.6 Vibration Sensor (SW420)

The vibration sensor SW420 is used to detect vibrations. There will be certain threshold value, when the threshold value exceeds, then it will be detected [23]. This particular threshold is taken and fixed for the vehicle on the on-board potentiometer.

3.7 Ultrasonic Sensor

Ultrasonic sensors are of acoustic type transducers. Ultrasonic sensors like HC-SR04 are used to detect distance of the object in front of the vehicle [13]. Ultrasonic sensors work on Doppler effect that is similar to the transducers used in radar and sonar systems. It uses a transducer which sends ultrasonic sounds of frequency 20 kHz; these sound waves get reflected by object and received at ultrasonic sensors receiving antenna [20, 21]. The velocity of sounds known and the HC-SR04 circuitry measure time duration for echo received and thereby the algorithm written to calculate distance as shown in Fig. 6. The HC-SR04 Ultrasonic sensor [14] used for demonstration has maximum effective length of 400 cm with precision about 3 mm at angle of 15° (Fig. 5).

Fig. 5 Ultrasonic sensor [15]

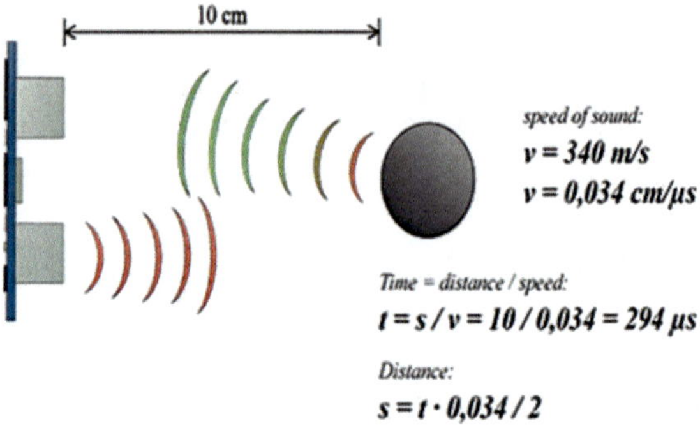

Fig. 6 Distance measurements [16]

3.8 *GPS Module*

GPS is general and primary navigation system, so this background navigation will give information and insights as to which extent the Global Positioning System is enabled [22]. It is the people first used module for positioning in differential dimensions such as latitude, altitude and longitude.

Class D Amplifier LM384

A Class D audio amplifier is usually a PWM toggle amplifier or amplifier. This code is this amplifier's main component. The switches are either entirely on or off in this type of amplifier, which helps greatly to reduce the power losses in the output devices. This method helps to achieve 90–95% efficiencies.

3.9 *User Keys*

There are two types of user keys used here. We considered push buttons as user keys. This is used to send an email alert to destination in case panic situations. We have given importance to this button in case of emergency where there is no other option to call or to leave a message. If the user can just press the SOS button which is connected to Pi, an email alert will be sent with the location of the vehicle.

3.10 *Database*

A database is a collection of information or data, which is organized that helps any person or application to get easy access to data, manageable and updated information. Databases support large storage and manipulation of data across many distributions. Databases make data management easy using different methodologies. In this, we are using a memory card as database.

4 Installing Raspbian OS

Raspberry Pis loaded Debian Linux as the operating system. With Raspberry Pis, it is easy to get an operating system (OS) installed since the OS is installed on a micro-SD card. Once you get an OS installed and configured it using certain specifications of your choice, we can copy the OS on the micro-SD card and load it on other micro-SD cards for the other Raspberry Pis for any other future works or application purpose.

5 Methodology

Breadboard acts as a major component to explain the technique, which helps to communicate with other modules such as sensors. First, along with the required connections, there should be a connection of ultrasonic sensor to the breadboard. For the ultrasonic sensor, different GPIO pin will be given. Raspberry Pi should be linked to the network and should be on. GPS module has four pins where two are connected to RPiGPIO pins and remaining are connected to breadboard for the power supply. Add speaker to the RPi for allowing voice alert notifications to the user. For the modules to work, connect them to RPi's GPIO pins. After the RPi has been granted with Wi-Fi connectivity, then you can see that the module will be up and running. There are three buttons like SOS, reset and shutdown which are used in this method. Specific GPIO pins are assigned for each button for the connectivity.

6 Evaluation

The results produced by the system of IoT-based accident avoiding system and emergency alert through Raspberry Pi. This sends the email alert to the authorized users helping in the emergency situation. Run the Python script and trigger the respective tools. This gives the vehicle's location in the alert which is being sent. The positioning of the vehicle can be easily tracked and monitored using GPS.

Figure 7 gives the evaluated results through graph where there are two constraints such as distance and time. Based on the distance calculated by the ultrasonic sensor along with particular time, the graph will be evaluated. The range in terms of distance for evaluation is given between 10 and 60 cm. Thereby distances of the object are known during fog conditions.

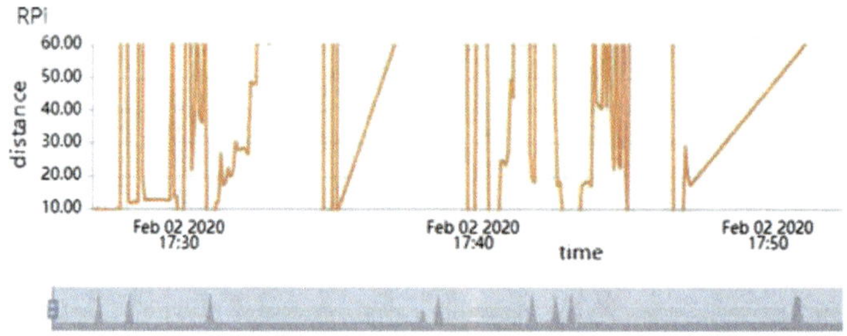

Fig. 7 Evaluation

7 Conclusion

This paper discusses smart vehicle monitoring system design and implementation which is advancement in commercial vehicle transportation system. The software consists of components which makes the vehicle journey safe and convenient. Vehicle alerts the owner of the remote location in case of panic situations through an email alert. By live alert, reports can be monitored such as over-speed, extreme failure, ideal vehicle time, halt time, on/off ignition to control enforcement closely and driver behavior. This monitoring helps to identify the current location of the car. Today, vehicle tracking system made use of GPS helps to reduce crime rates. By using GPS, many commercial vehicle transportation systems are increasing their performance, competitiveness and safety.

References

1. https://timesofindia.indiatimes.com/india/over-10000-lives-lost-in-fog-related-road-crashes/articleshow/67391588.cms
2. Golden R (2013) Raspberry Pi networking cookbook. Packt Publishing
3. CSU N et al (2004) A wireless sensor network for structural monitoring. In: Proceedings of ACM Symposium in networked embedded sensing (Sensys). ACM Press, pp 13–24
4. Chintalapudi K et al. (2006) Structures with wireless sensor network. In: Sensor-network application, IEEE internet computing. www.computer.org/internet/1089-7801/06/IEEE
5. Kottursamy K, Raja G, Padmanabhan J, Srinivasan V (2017) An improved database synchronization mechanism for mobile data using software-defined networking control. Comput Electr Eng 57:93–103
6. Robinson A, Cook M (2013) Raspberry Pi projects, 1st edn. Wiley
7. Molloy D (2014) Exploring BeagleBone: tools and techniques for building with embedded Linux, 1st edn. Wiley
8. Anwar S, Kishore D (2016) IOT Based smart home security system with alert and door access control using smart phone. Int J Eng Res Technol (IJERT)
9. Vakaavinash R, Venkatesh K (2020) Role of software-defined network in industry 4.0. In: EAI/Springer innovations in communication and computing-internet of things for industry 4.0 design, challenges and solutions, pp 197–218
10. Changsha G, Rice AJAP, Changzhi Li A (2012) Wireless smart sensor network based on multi-function interferometric radar sensors for strut. IEEE Trans Struct Health Monitor, 978-1-4577-1238-8/12 2012
11. https://www.theengineeringprojects.com/2018/07/introduction-to-raspberry-pi-3-b-plus.html
12. Feltrin G, Saukh O, Meyerand J, Motavalli M (2011) Structural monitoring with WSN: experiences from field deployments first middles east conference on smart monitoring, pp 8–10. http://www.mdpi.com/journal/sensors
13. Ampatzis T et al (2005) A survey of security issues in wireless sensors network. In: Proceeding of the IEEE international symposium on, mediterrean conference on control and automation in intelligent control, pp 719–724
14. Jayavardhana G, Rajkumar B, Marusic S, Palaniswami M (2013) Internet of things: a vision, architectural elements, and future directions. Future Gener Comput Syst
15. https://www.elecparts101.com/hc-sr04-ultrasonic-sensor-noncontact-range-detection-datasheet-and-pinout/
16. http://electrobist.com/product/hc-sr04-hc-sr04-ultrasonic-sensor/

17. Naveen Chandar B, Arivazhagan N, Venkatesh NK (2019) Improving the network performance using mp-olsr protocol for wireless ad hoc network (MANET). Int J Recent Technol Eng 8(3):5700–5707
18. Venkatesh K, Srinivas LNB, Mukesh Krishnan MB, Shanthini A (2018) QoS improvisation of delay sensitive communication using SDN based multipath routing for medical applications. Future Gener Comput Syst 93(2019):256–265
19. Gigli M, Koo S (2011) Internet of things: services and applications categorization. Adv Internet Things 1:27–31
20. Pelegri J, Alberola J, Llario V (2002) Vehicle detection and car speed monitoring system using GMR magnetic sensors. IEEE Conf Ind Electr (2)
21. Megalingam RK, Mohan V, Ajay M, Leons P, Shooja R (2010) Wireless sensor network for vehicle speed monitoring and traffic routing system. In: IEEE international conference on mechanical and electrical technology
22. Kirankumar G, Samsuresh J, Balaji G (2012) Vehicle speed monitoring system [VSM] (using Rubee protocol). IACSIT Int J Eng Technol (1)
23. Lavanya G, PreethyW, Shameen A, Sushmita R (2013) Passenger bus alert system for easy navigation of blind. In: IEEE international conference on circuits, power and computing technologies
24. Chatterjee ,S Timande B (2012) Public transport system ticketing system using RFID and ARM processor Perspective Mumbai bus facilityB.E.S.T. Int J Electron Comput Sci Eng

Review on Traffic Engineering and Load Balancing Techniques in Software Defined Networking

Nidhi Kawale, L. N. B. Srinivas, and K. Venkatesh

Abstract Traffic Engineering-TE deals with the major issue in the arena of the computing and network. There will be so many failures in and out of the network as traffic demand increases. Thus there are many techniques available for countering this problem. Many researchers also proposed a number of traffic engineering techniques to be successfully deal with the issue. But all these tactics and techniques that have been suggested for traditional methods. Such methods are difficult to implement or there is no effective way to implement. Research communities have been working extensively on methods for traditional networking structures that allow the network to adapt to changes in traffic patterns. So here comes new ideology in which there is good manageability of network. Each time network stages will be varied and also these changes should be well managed for the good traffic outcomes. These are done through the modern age of networking SDN (Software Defined Networking), as it makes optimum use of resources in the network. We can have a quick glance where the techniques of traffic engineering (TE) perform for the traditional networks and even the new concept of networking known as SDN.

Keywords TE-Traffic engineering · SDN · Load balancing · Link-Failure recovery · Optimal rerouting

N. Kawale · L. N. B. Srinivas (✉) · K. Venkatesh
SRM Institute of Science and Technology, Chennai, India
e-mail: srilnb@gmail.com

N. Kawale
e-mail: niddhikawale@gmail.com

K. Venkatesh
e-mail: venkatek2@srmist.edu.in

D. J. Hemanth et al. (eds.), *Artificial Intelligence Techniques for Advanced Computing Applications*, Lecture Notes in Networks and Systems 130,
https://doi.org/10.1007/978-981-15-5329-5_18

1 Introduction

TE is just how network operators manage large amounts of data that stream across their network traffic. Reconfiguring the network in response to changing traffic loads in order to achieve functional objectives by putting the essential technique for modularizing the performance of the data network. It increases performance by regulating & managing the data flow in the system, also by predicting the dynamic nature of transmission it is easy to analyse the network traffic. MPLS-Multiprotocol Label Switching & ATM is a growing implementation in today's networks. It helps in representing the current and past data networks. These networks are better adapted for today's network traffic but as taking into consideration of increasing data flows in the TE solutions & for future demanding these networks will not be best suited. Mainly because the architecture of network should be capable of identifying the variety of data traffic coming from variety of applications. It should be able to provide the proper solution for this much of data flowing through TE in short span of time. As all the data in today's world is migrating into cloud computing for data storage. Because of this rapidly growing world of cloud computing the network should able to handle massive-loads of data traffic for better system performance. This is not the case in architecture of above networks. For the above reasons there is a need of improved network which is reliable, dynamic & intelligent in nature.

For maintaining the high performance of network, varies techniques are used irrespective of the failures in networks. But for quality service the performance of networks must not be affected by the failures in the networks itself [1–3]. These failures can be overcome by using the backup routes. Even though these backup routes rerouted the traffic but this rerouting of paths cause the sufficient amount of delay to affect the performance of distributed network application. This happens only when the readily available backup routes are not generated properly.

To address network failures, most approaches were developed to enhance network performance as a result of increasing network traffic demand. The important feature of any network for network management is its Fault Tolerance. Fault tolerance in a network is a service that allows the network to resume its operations even if a failure or malfunction occurs. Instead of failing completely, the network will continue operations at a reduced level. In computer network there different devices used of networking for e.g. Routers, bridges, switches, firewalls, etc. Figures 1 and 2 shows the basic network architecture as well as SDN architecture. It contains the control plane & data plane which are the part of Software Defined Networking (SDN) architecture. In which control plane performs data forwarding & decision making according to data traffic in the network. Whereas data plane manages the user traffic connected to network. Infrastructure layer is a collection of network switches and routers used for network traffic forwarding. The SDN controllers are included in the monitor layer. The controller uses the business logic in this layer to control and maintain data, topologies and statistics on the network. Northbound interface is the intermediate API between the SDN controller & application layer. Southbound interface is communicator between the infrastructure layer & control layer which

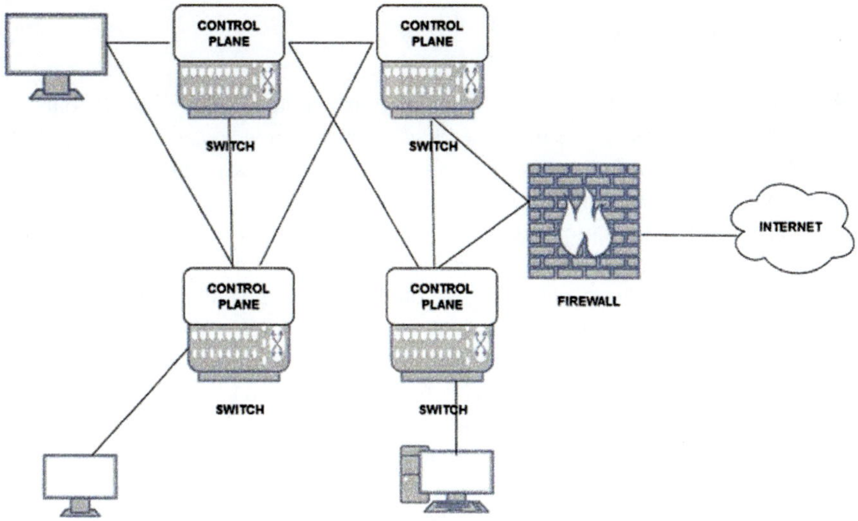

Fig. 1 Traditional network architecture [21]

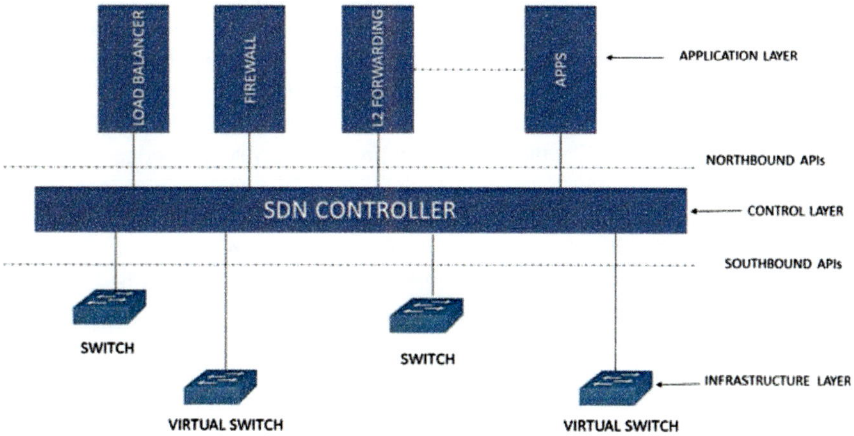

Fig. 2 SDN architecture

uses southbound protocols. Application layer is the end-to-end solution to real world entity & can various applications related to network monitoring, configuration & management, troubleshooting, etc.

Various TE procedures are created to improve the performance of SDN-based networks to be ideal. Even though these procedures are doing well as per traffic engineering perspective but has some drawbacks to itself. In this paper we are trying to analyse these procedures & present the review on some techniques used for TE in Software Defined Network.

2 Literature Review

The researchers in this paper [4] shows that there is a reliable solution to all double-edge disillusionment as long as the network graph is 3-related and heuristic. Another way to manage disillusionment control is to reconfigure the ways for pre-configured service paths. Such a theory is discussed and this text describes the preface of our arrangements. Reconfiguration provides an excellent structure in view of how the choice of the fortification route for the second failed edge is subject to the right learning of what has already tumbled (instead of coping with any potential dissatisfaction). Reconfiguration generates full to higher-assortment problems in the same way up to more viable. Nonetheless, not in the least as our proven most pessimistic condition guarantees, due to a resolved veiled network, the focus is on analysis and reconfiguration layout structure. The research in also takes into account the re-provisioning of service paths, yet to the point of association all the way. There is an enormous category of "p-cycles" function. A p-cycle is a pre-configured unit limit cycle that is built from within the network as far as possible. This offers snappy restoration of loop dissatisfactions close to the ranges of "straddling" with both end canters around the loop.

This paper [5] author proposes the mathematical formation for MPLS fast reroute layout & different protection mechanisms. Depending on FRR mechanism there are mainly two mechanisms. 1. One-to-one backup & 2. Many-to-one backup. It provides an ILP which provides a simple stream errand (with stream part) and a "flow redistribution" plot in situations that ensures no blocking of k frustrations: all streams can be directed to F without stopping, provided any set of F edges with full scale usage up to k.

In recent years, focusing on the traditional computer network, this paper [6] has been favoured. SDN networks. Although the SDN network is superior to the conventional network, it has the load balancing problem. The author classifies the load balancing schemes for SDN networks by evaluating & providing benefits & drawbacks of balancing schemes to affect network performance. Based on Load Balancing studies in data plane, this paper provides the proper guidance for other experiments by balancing the reference load. This article advocates SDN architecture & analysis advantages for further progress in the SDN network.

The router is the primary component responsible for all of the network's data traffic. This includes the collection of IP addresses that can be forwarded to a specific destination. The router manages to find new paths by using the Transmission Control Protocol (TCP). This mechanism is not perfect to work efficiently yet, all the flow works properly. This article [7] is focused on the Local Area Network (LAN) router stream of different data. Using OSPF & IS-IS Intradomain routing protocol, they approach the flow of traffic in this area by monitoring traffic & topology. In other words, certain protocols use the pinning to transfer the data from one path to another without disrupting the other paths. Additionally, some protocols use data flow backup routes. Depending on increasing traffic requirements, modifying static connection weights to IP networks is acceptable or beneficial.

The paper [8] provides a brief discussion on the services used by MPLS/DiffServ in IP networking. MPLS is now a day's most frequently used service. But this is not enough because of the system's complexity, i.e. it can not distinguish the variety of data flow from different network applications. It provides another server's need to distinguish the network's data flow. The architecture of DiffServ offers the assurance and optimizes capital. DiffServ is the mechanism of TE for all service classes. Combining MPLS & DiffServ is the most promising mechanism for ensuring reliability, compatibility, and service quality. The researchers proposed a new algorithm based on integrating both MPLS/DiffServ to boost network capacity use.

Authors address the idea of circuit switching and network packet switching in this paper [9]. Once transmitting the data to the network if any connection fails due to any circumstances, the packet will not be transmitted to the destination unless and until it identifies the alternate origin to be delivered for the same packet. Because of this issue, the network performance will be degraded. The researchers proposed a structure for Internet Engineering Task Force-(IETF) and IP Quick Reroute-(IPFRR) in order to overcome this problem. There are plenty of protocols available for network recovery from failures like the IPFRR method. The main function of this is to reduce the recovery time after a link fails and to allow secondary paths to escape network congestion. In this paper creators used the Dijkstra and Disjoint spanning tree algorithm, which provides an effective way to redirect data from source to destination in the event of a multi-link failure. This two algorithm stated in the paper used to recover the network's failure link and to balance the network infrastructure's load traffic.

Authors in this paper [10] address few methods in a network to protect multiple connection failures (MLF). Few current techniques such as MPLS Quick Re-route (FRR) link-based restoration (LBR). Such methods are fast, but by overloading edges they still create congestion in a network. The enhancement for FRR restore for multiple failures in a network design was proposed in this authors. We define different network topology here, which adds a number of edges to an existing topology layout of the network. Several protocols are stated for each single network topology, such as a distributed algorithm, which helps to distribute information among the network nodes and also helps reconfigure backup paths after each state information failure.

Researchers in this paper [11] focus on the technique called voice over internet protocols-(VoIP) that provides end-to-end communication guarantee. Here authors evaluate the quality of VoIP traffic using the BGP-Border Gateway Protocol. Some of the other protocols, namely Multiprotocol Label Switching (MPLS), Virtual Private Network (VPN), Open Shortest Path first-(OSPF), are also used in this article. The transmission system and overall network performance was improved through the combination of MPLS with traditional routing. OPNET model 14.5 is used for situations and metrics like delay testing.

The SDN controller implementations are open source for networks of tenants such as Floodlight, Trema, Open Daylight, POX, ONOS, and Ryu, etc. Every VTN has a dedicated host that includes an independently operating SDN controller. It is therefore important at the time of each dynamic VTN job to physically deploy and configure the SDN controller at the dedicated host. This SDN controller execution requires postponements of a few days in the provision of necessary assistance. Virtualization of the SDN controller works by NFV worldview methods is assumed to be an increasingly refined methodology for the use of system capabilities including burden adjustment, steering and sending, firewall and traffic construction. NFV provides the ability to virtualize the SDN controller and transfer it to the cloud for adaptive delivery, creating a separate SDN controller template network within minutes. Therefore, whenever another VTN is powerfully sent, the utility of the whole process can be cultivated in two or three minutes [12]. Additionally, this method also offers beneficial preferences such as a decrease in interruption of retention of equipment and improved performance of recovery time in situation of disaster or disappointment. A virtualized SDN controller [13–15] can be quickly and easily moved between physical servers in a server farm haze when equipment retention is needed (less equipment retention stoppage), depictions and improvements of virtualized SDN controller conditions can be shared from one server farm to another in a cloud for brisk reconfiguration after failure (quinquency).

There are some relevant chips away from the SDN controller's burden modification, some of which are referenced here. The change structure like stream table passages can be modified in OpenFlow portrayals only through the ace c-hub suggested in [16]. This ace c-hub is responsible for adjusting the progression of approaching and active messages to build versatility at the various number of changes. In [17] for burden adjustment in SDN-enabled systems, a technique called BalanceFlow was suggested that a super controller be distributed among reasonable controllers to deal with uneven traffic load challenges. A chief controller hub assembles nearly all other controller hubs and then address the problem of heap shift by taking into account the heap varieties of all considerations. Impediments of this technique include I execution bargains due to trade in daily control messages and restricted resources such as memory, transfer speed, and CPU power (ii) load information is acquired with defers that do not reflect genuine load conditions due to two system transmissions (sending directions and collecting burdens) and (iii) Total load adjustment operation may be down if focal controller breakdown.

Dynamic and scalable calculation (DALB) proposed in [18, 19] enabled all slave SDN controllers simply as an ace controller for neighbourhood choices. This equation enables the adaptability and usability of distributed SDN controllers and only one transmission system for charging on social occasions. Therefore, choice postpone decreased in light of the fact that all controllers don't gather the heap data too as often as possible. While thinking about the system assets, joining of SDN and NFV acquainted in [14] with upgrade the system convention and capacities programmability. NFV worldview bolsters the dynamic change of system assets and gives the idea of virtualized system control capacities for inhabitant systems. Along these lines, control work programming occasions can be progressively sent and moved if the requirement for proficient usage of accessible assets.

In this paper [20], the authors mentioned the decentralized SDN controller architecture which helps to handle various customer applications which work together with existing load balancing servers. In order to optimize the weight parameter and dynamically reduce the load balancing traffic, Creators suggested one of the algorithms. This algorithm assumes that among all types of applications the loads of processing a request are the same. Through implementing SDN in actual environments, it can be avoided.

Past work on burden adjustment depends on whether the cantered or distributed SDN controller is considered to be physical SDN resources. Increasing asset can be virtualized via NFV and if an increased abnormal traffic load occurs in vSDN-enabled systems, additional vSDN controllers can be included for burden adjustment. A dual vSDN will increasingly be built to share the heap and make the same mistakes as a single vSDN. All in all, the first question here is the stage we need to duplicate the vSDN controller and the other question is how are hubs going to think about the presence of an additional controller.

Our workbooks could be said to enhance the utility of the SDN/NFV combination and present IP burden by using NFV philosophy to change the usefulness of digital SDN controller-empowered systems so that resources can be saved with better execution.

3 Proposed System

The unpredictability of the existing system is by one way or another high because of the devouring of numerous controllers. It might increment with the number of controllers. Another impediment of existing work is, it doesn't specify the difficulties of vitality utilization and carbon discharge. In this way, to defeat the disadvantages of burden adjusting instrument as far as vitality utilization and carbon discharge we proposed Hybrid Load adjusting calculation which uses Least Response Load adjusting and most limited way first calculation and to lessen the carbon emanation adaptable sensor data framework is utilized which change the transmission of information and recurrence of information gathering.

NFV provides effective application structures for the efficient management of VNFs and associated administrations. We have this opening when using digital SDN as a (Virtual Network Function) VNF, we can also include increasingly indistinguishable VNFs for a similar errand in order to share traffic through the basic system. If increased irregular traffic load may occur, an auxiliary vSDN can be used to change the burden in digital SDN-enabled systems. Since each resource (controls, switches, and association, etc.) is used mainly under NFV, we must allow/include equipment resources according to the precondition. The need for an auxiliary vSDN (Virtual Software Defined Networking) controller is overcome and a duplicate of vSDN (Virtual Software Defined Networking) controller is rendered using exactly the same arrangements as original vSDN (Virtual Software Defined Networking) that function precisely and provide traffic load. Both vSDN controllers are freely put in the cloud with simplicity to ensure that the newly made auxiliary vSDN controller is

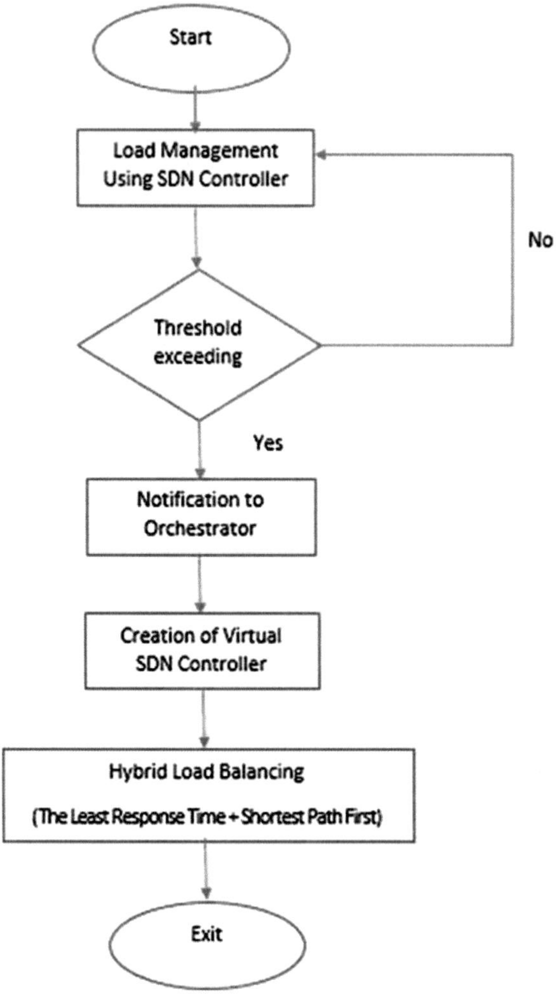

Fig. 3 Flow chart of proposed work

accessible to every user in the network. We present the proposed traffic load balancing model in occupant systems using the SDN controller as VNF in this region as shown in Fig. 3.

Proposed Algorithm:

Hybrid Load Balancing Algorithm

(Least Response Load Balancing + Shortest Path First)

(1) Find information about hosts connected.
(2) Using shortest route concept finding the information.
(3) Find total link cost for all the routes.
(4) Get current transmission rate.

(5) Selecting the best path.
(6) Push the traffic into each switch in the current best path and go to step 2.

Distribution of load among the various servers is been balanced by load balancing as shown in Fig. 4 using the algorithms. When an operator chooses load balancer it means it uses least response time method. The methodology of Least Response Time (LRT) incorporates fewer efficient connections and less average response time (LART). Time-to- First-Byte (TTFB) is known for sending a request packet to a server and receiving the first response packet back. Thus the load balancing algorithm helps operator to distribute all the request send by the client to various servers.

One of the biggest advantage of this method is to improve the n user response time and server fault tolerance. Hence this method is resourceful optimized method for any operator. In a technical era were to many operations depend upon the single server uses create performance degradation. Hence the load balancing method gives the best solution.

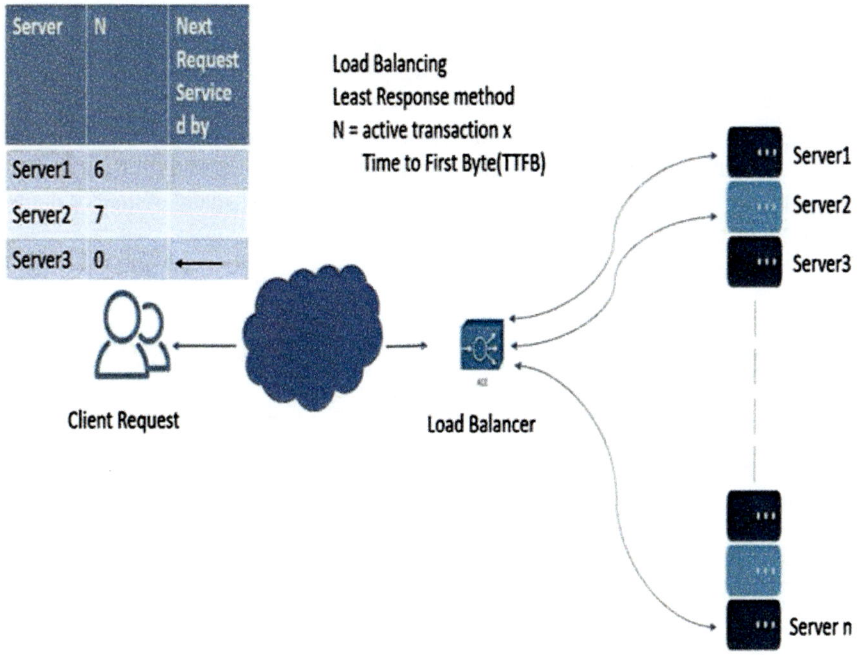

Fig. 4 Least response method [22]

4 Experimental Validation

The proposed approach is being experimentally tested with Mininet's support. A realistic virtual topology can be created in a matter of seconds using a mininet with SDN support. Much attention will be paid to the data-based network infrastructure. Fat-Tree topology will be used, as at every bisection it comes with the same bandwidth. We will use Iperf to measure performance, data rate and use of connections, as it provides accurate measurement of the defined parameters. In addition to the Wireshark, the remote control plane will be used for use in bandwidth, packet capture, and load balancing to avoid the output disruption. Open Daylight as a SDN controller will be used during this entire experiment.

5 Conclusion

The study discussed here is a summary of the existing traffic engineering literature and research conducted. Many techniques are ideal for conventional network structure although few are related to SDN techniques. Often addressed are the risks associated with the implementation and potential changes required to upgrade the systems. SDN is a popular solution in the model of networking. This separates the network's control plane from the plane used to forward data. It addresses many of the problems of modern network architecture and yields the result.

References

1. Elhourani T, Gopalan A, Ramasubramanian S (2014) IP fast rerouting for multi-link failures. In: IEEE INFOCOM 2014—IEEE conference on computer communications, pp 2148–2156
2. She Q, Huang X, Jue JP (2007) Survivable routing for segment protection under multiple failures. In: OFC/NFOEC 2007, pp 1–3
3. Kim H, Schlansker M, Santos, JR, Tourrilhes J, Turner Y, Feamster N (2012) Coronet: fault tolerance for software defined networks. In: ICNP '12, pp 1–2
4. Todimala A, Ramakrishnan KK, Sinha RK (2009) Cross-layer reconfiguration for surviving multiple-link failures in backbone networks. IEEE
5. Zukowski C, Tomaszewski A, Pióro M, Hock D, Hartmann M, Menth M (2011) Compact node-link formulations for the optimal single path MPLS fast reroute layout. Adv Electron Telecommun 2(3)
6. Qilin M, Weikang S (2015) A load balancing method based on SDN. In: 2015 seventh international conference on measuring technology and mechatronics automation Year: 2015|Conference Paper|Publisher: IEEE
7. Fortz B, Rexford J, Thorup M (2002) Traffic engineering with traditional IP routing protocols. Commun Mag 40(10):118–124
8. Akyildiz IF, Anjali T, Chen LC (2003) A new traffic engineering manager for DiffServ/MPLS networks: design and implementation on an IP QoS testbed. Comput Commun 26(4):388–403
9. Bhor M, Karia DC (2018) Network recovery using IP fast rerouting for multi link failures. In: International conference on intelligent computing and control (I2C2). IEEE Xplore

10. Sinha RK, Ergun F, Oikonomou KN, Ramakrishnan KK (2014) Network design for tolerating multiple link failures using Fast Re-Route (FRR). IEEE
11. Yunos R, Ahmad SA, Noor NM, Saidi RM, Zaino Z (2013) Analysis of routing protocols of VoIP VPN over MPLS Network. In: 2013 IEEE conference on systems, process & control (ICSPC2013)
12. Muñoz R, Vilalta R, Casellas R, Martinez R, Szyrkowiec T, Autenrieth A, López V, López D (2015) Integrated SDN/NFV management and orchestration architecture for dynamic deployment of virtual SDN control instances for virtual tenant networks. J Optical Commun Netw 7(11):B62–B70
13. Vilalta R, Mayoral A, Munoz R, Casellas R, Martínez R (2016) Multitenant transport networks with SDN/NFV. J Lightw Technol 34(6):1509–1515
14. Naveen Chandar B, Arivazhagan N, Venkatesh K (2019) Improving the network performance using MP-OLSR protocol for wireless ad hoc network (MANET). Int J Recent Technol Eng 8(3):5700–5707
15. Venkatesh K, Srinivas LNB, Mukesh Krishnan MB, Shanthini A (2018) QoS improvisation of delay sensitive communication using SDN based multipath routing for medical applications. Future Gener Comput Syst 93(2019):256–265
16. Hadjiona M, Georgiou C, Vassiliou V, A hybrid fault-tolerant algorithm for MPLS networks, Department of Computer Science University of Cyprus
17. Hu Y, Wang W, Gong X, Que X, Cheng S (2012) Balanceflow: controller load balancing for openflow networks. In: 2012 IEEE 2nd international conference on cloud computing and intelligence systems
18. Vakaavinash R, Venkatesh K (2020) Role of software-defined network in industry 4.0. In: EAI/Springer innovations in communication and computing-internet of things for industry 4.0 design, challenges and solutions, pp 197–218
19. Hikichi K, Soumiya T, Yamada A (2016) Dynamic application load balancing in distributed SDN controller. In: Network operations and management symposium (APNOMS), 2016 18th Asia-Pacific. IEEE, pp 1–6
20. Dixit A, Hao F, Mukherjee S, Lakshman T, Kompella R (2013) Towards an elastic distributed SDN controller
21. Abbasi MR, Guleria A, Devi MS (2016) Traffic engineering in software defined networks: a survey. J Telecommun Inf Technol
22. https://www.franken.pro/sites/default/files/styles/portfolio_image_slide/public/LB_LeastResp_900x495.png?itok=PHDbzJ1y

Malware Classification Using CNN-XGBoost Model

Sumaya Saadat and V. Joseph Raymond

Abstract This paper attempted to introduce a deep learning-based model for the classification of malicious software (Malware). Malware is growing exponentially every year and malware writers try to evade the antivirus software by producing polymorphic and metamorphic malware. Most antiviruses are based on signature detection which is not sufficient against the new generation of malware. For a solution against malicious software, antivirus vendors started to use Machine Learning approaches which had a positive impact on malware detection and classification. Recently, Deep Learning algorithms and specifically Convolutional Neural Networks (CNN) caught more attraction for malware classification and it is the best deep learning algorithm for extracting features from images. By integrating the CNN with Gradient Boosting (XG-Boost) algorithm we can have a powerful model to classify malware images into their classes or families. The input source for the model is the Malimg dataset [1] which is an open collection of already converted malware to a grayscale image. There are many papers used CNN-SVM, CNN-Softmax and other models for malware image classification and they got good accuracies, but this paper proposed to used CNN-XGBoost model and achieve more accuracy than previously used algorithms for malware classification.

Keywords Deep learning · CNN · Malware classification · XGBoost · CNN-XGBoost

1 Introduction

We are living in the digital age and the internet plays a fundamental role in this fast movement toward digitalization. These days the internet is an important part of

S. Saadat (✉) · V. Joseph Raymond
Department of Information Security and Cyber Forensics, SRM Institute of Science and Technology, Chennai, India
e-mail: sumaya.saadat@gmail.com

V. Joseph Raymond
e-mail: josephrv@srmist.edu.in

D. J. Hemanth et al. (eds.), *Artificial Intelligence Techniques for Advanced Computing Applications*, Lecture Notes in Networks and Systems 130,
https://doi.org/10.1007/978-981-15-5329-5_19

our life and cyber space become a platform for online activities like communication and online financial transactions. This wide usage of internet and computer systems attracted malware writers to infect this environment with their malicious code and reach their sinister intentions.

Malware, stand for malicious software, refers to a software program developed for malicious purpose, such as unauthorized access to personal and organizational sensitive information, disrupting computer operations, bypass access controls and display unwanted advertisements. According to the AV-TEST report, which has been illustrated in Fig. 1, there is a huge growth in the total number of malware over the years and more than 969 million malware registered in 2019 which shows an increase of 100 million from 2018 [2]. Symantec reports 246 million of new variant malware in 2018 [3], we can say the main reason behind this fast increase of new malware is use of encryption and obfuscation techniques by malware writers to design a different look malicious file. For handling this rapid evolution in new generation of malware, it is required to develop powerful malware classification techniques that be able to classify every variants of malware into families that they are originally belong.

Previous researches indicate that all malware belong to a family share the same malicious behavior, for this reason building a model that can efficiently classify malware based on their families even if they have appeared in a variant, is a fruitful task and frustrates the increase of malware variation. After machine learning, these days, deep learning has shown state-of-the-art performance for different tasks and areas such as speech recognition, natural language processing, and malware classification. Cybersecurity is one of the fields which benefit significantly by using and advancing deep learning. Convolutional Neural Networks (CNN) is a multi-layer architecture

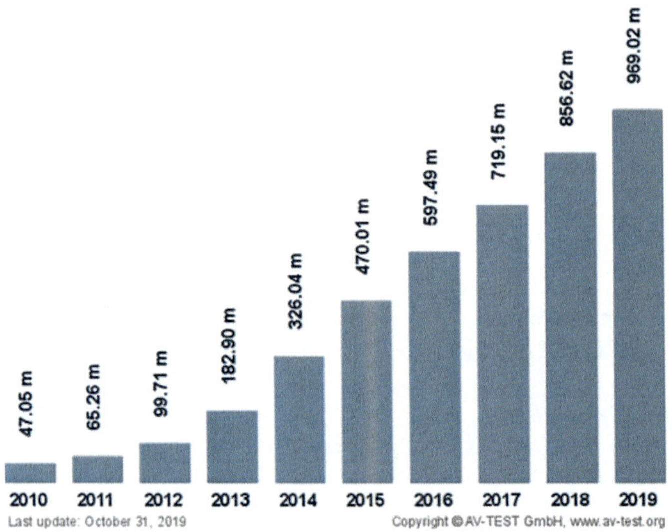

Fig. 1 AV-Test report (*source* https://www.av-test.org/en/statistics/malware, accessed 31 October 2019)

of deep learning that experiments and researches prove its ability and success in classification problems of the huge number of images. Inspiring the power of CNN and the novel image classification model of Ren et al. [4], we want to convert the malware classification problem to an image classification problem and suggest the possibility of classifying malware with the CNN-XGBoost model. This model is combination of CNN for automatic feature extraction and XGBoost for classification, as XG-Boost makes a strong classifer using weak classifiers we expect to get the best performance and higher accuracy than any other method for classification of malware.

2 Related Work

Past and existing methods of detection and classification of malware can be divided into two categories: traditional method and machine learning/deep learning-based methods.

2.1 Traditional/Classical Method

Previously, the malware were detected by static or dynamic analyzing and signature-based antiviruses, for malware detection in the static method, without executing the code we observe the structural properties, operation code or syntax of the program. On another hand, malware writers uses various encryption, obfuscation and polymorphic techniques to bypass these static detection algorithms and the main drawback is its insufficiency for unseen malware. For detection of malware, there is another way to run the file in a test or virtual environment and monitor the behavior of application in order to find any abnormality or malicious behavior, this method is called dynamic analyzing. However dynamic analysis is a promising method, still, it is very complex and not sufficient for polymorphic malware. Because these kinds of malware will not show their main operation and look like a benign file, if recognize that are run in a test environment. By growing Artificial Intelligent and machine learning, new approach for malware detection came on the scene.

2.2 Machine Learning and Deep Learning Based Methods

To overcome the limitations of the previous methods and understanding the fact that malware in a family share the same behavior [5], anti-malware organizations started to get benefit from machine learning and data mining classification techniques [6] for feature engineering and malware classification and design smarter anti-malware systems. Machien learning algorithms have this drawback that they are not scalable for these fast growth of variety in malware and they are depend on already known features of malware. Regarding this problem scientist, used deep learning architecture, for its ability in automatic feature extraction and and its functionality to work with huge amount of data.

As a successful implementation of deep learning and machine learning for malware detection and classification we can mention the experiment of Drew et al. [7], on detection of polymorphic malware on the Microsoft Malware Dataset [8] and gaining above 0.95 of accuracy using machine learning algorithms. Among other machine learning techniques the ensemble method of eXtreme Gradient Boosting grabbed more attention and the winner team of Microsoft Malware Classification Challenge (BIG 2015) [8] used a complex combination of features extractor and classifier based on the XGBoost and he was able to achieve the best accuracy of 0.998 [9]. Ahmadi et al. [10] also used a machine learning classifier based on forward stepwise feature selection and XGBoost and got a wonderful accuracy of ~0.998.

Inspiring the novel method of converting malware to an image by Nataraj et al. [5], experiments focused on malware image classification and using powerful image classifier models for this purpose. Gibert [11], used Microsoft Malware Challenge dataset and CNN architecture, to find 98.56% accuracy for image classification. Cui and Xue [12], also used the CNN model for malware image classification and their proposed approach achieved 94.5% accuracy. Kabanga and Kim [13], used CNN for image classification and resulted in 98% accuracy. Elleuch et al. [14], used the CNN-SVM model and the classification result on HACDB (Arabic handwriting dataset) shown 99.15% accuracy.

Recently a novel method for image classification proposed by Ren et al. [4] using CNN-XGBoost model, the experiment implemented on MNIST (a large database of handwritten digits) and CIFAR-10 dataset (dataset of 60,000 image in different classes) and the result proved that the new method can result in 99.22% accuracy, that compare with accuracy of other methods on the same datasets it performed more better. According to latest paper named "Performance Comparison of Hybrid CNN-SVM and CNN-XGBoost models in Concrete Crack Detection" [18] done by Sahana Thiyagarajan at 2019, CNN-XGBoost model had 99.8% accuracy and CNN-SVM had 98.8% accuracy on concrete crack dataset. The result shows integration of CNN with XGBoost produce more accuracy for this classification comparison.

3 Proposed System

Deep learning is a member of artificial intelligence family that mimics the brain processing of humans for analyzing data, creating patterns and making the decisions. Deep learning is a subset of machine learning with supervised and unsupervised learning capability. It learns from large quantities of data, which usually people can understand and process for decades. Deep learning architectures like recurrent neural networks, deep neural network, convolutional neural networks, and deep belief network have been implemented to many areas like speech recognition, natural language processing, machine translation, computer vision, social network filtering, drug design, analyzing medical image for diagnosis and malware classification where they have produced results that are something superior to human experts.

The malware classification problem has become the focus of research in many years and is one of the most important research directions for cybersecurity. Machine learning and deep learning both have shown their best performance regard malware detection and classification, as this paper focuses on classification only, deep learning as the current world's salient topic selected for the core function of our malware classification algorithm. The CNN architecture of deep learning is an outstanding method for image detection and classification and can extract the features automatically from the given dataset with no need to features be known or listed by man or other parts of the program previously.

For malware classification purpose this paper suggest to use the CNN-XGBoost model (Fig. 2) which is one of excellent image classification models recently, that for the fist time introduced and experimented by Ren et al. [4]. He implemented this model on the MNIST and CIFAR-10 datasets and got the accuracies of 99.22 and 80.77 respectively and these results were higher than the accuracies of other models like CNN, CNN-SVM, SAE + CNN, Linear SVM and K-means. As this model is based on CNN, and as CNN mostly targeted for image classification, then we need to first convert all targeted malware to images, and collect samples of every malware family. Afterward we convert the malware binaries to an 8-bit binary vector and then transform to a matrix that visualized as a grayscale image with values in the range of [0, 255], that 0 represents black color and 1 represents white (Fig. 3). Nataraj et al. [5] for the first time suggested malware to image conversion and he created the Malimg dataset. For our malware classification demo program, we are supposed to use the Malimg dataset as input for our model that provides malware images that are already classified in their families (Fig. 4).

After converting the malware bytes into the image, every malware from a family produced lots of similarity with other malware in the same family, and this close similarity of malware images strongly help the image classifier to easily distinguish between malware of each family (Fig. 4).

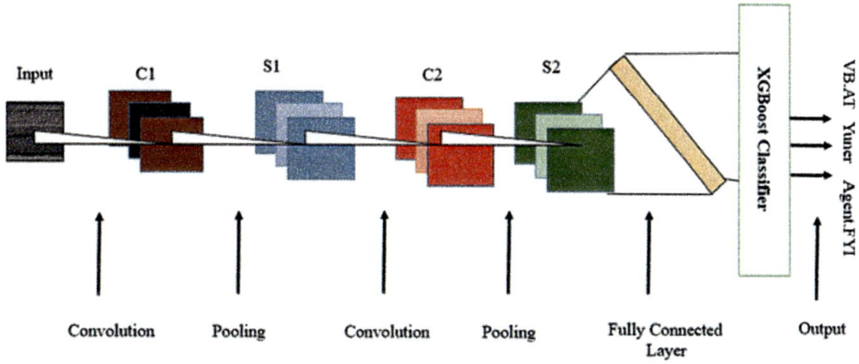

Fig. 2 CNN-XGBoost model

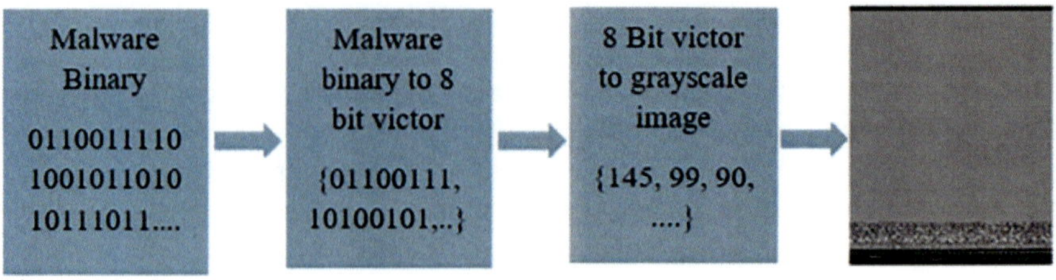

Fig. 3 Converting malware to grayscale image

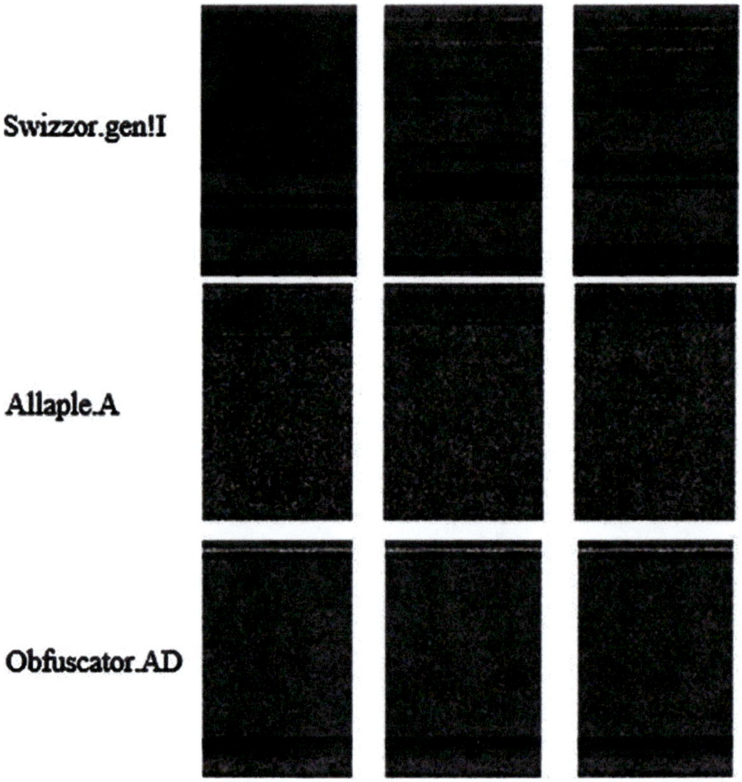

Fig. 4 Malware images

After preparing the input we are going to use it to feed CNN-XGBoost model, the model asks CNN to read every image of all classes from dataset and automatically find the features of every malware image in families, CNN with its own three layers: Convolution layer, pooling layer, and Fully-connected layer can extract the high-quality features or characteristics of images. After completing feature engineering we need to use these features to do classification. CNN is one of the best feature extractors but it cannot produce good accuracy in classification until it integrates with a powerful classifier. Instead of CNN's most used, soft-max classifier, we add the XGBoost layer to classify every image to its corresponding family. XGBoost is one of the famous classifiers based on Gradient Boosting and has the capability of repeatedly computing for providing better accuracy for weak classifiers. CNN-GX-Boost model consist of these components:

1. CNN
    1. Convolutional layer
    2. Pooling layer
    3. Fully connected layer
2. XGBoost layer

3.1 Convolutional Neural Network (CNN)

A convolutional neural is a highly neutralized category of deep learning, which is mostly used for the study of visual imaging. Mostly used in applications such as photo and video recognition, identification of objects, natural language processing, and recommendation system. Convolution Neural Network (CNN) is a neural, biologically inspired feed-forward network specifically based on an analysis of animal visual cortex organization. CNN is the new neuro web technology, which addresses the problem of image classification. Neurons with learning weights and biases are included in CNN. These three parts are the main components of CNN [15].

Convolutional layers: These layers have applied for every sequential image with a certain number of convolution (linear filtering). This layer usually filters and extracts the input image information from the edge, form, and color. Filters are used in each sub-region of a picture and computational inputs for each sub-region to generate a single value. The output of this layer (assume x) is transferred to a non-linear function, which can be defined as ReLU Activation.

$$f(x) = \max(0, x)$$

Pooling layers: The layer guarantees the information generated by convolution layers to decrease the processing time (i.e. minimize the spatial resolution of the imagery in the output layer), so the data size can be managed by computing tools. This leads to a reduction in the number of learning parameters in the following network layers. Max pooling is a commonly used pooling process that retains the maximum value and discards the remaining values from 2×2 non-overlapping data regions. The dimension of input image decrease after implementing convolution on it. If we continue applying convolution on the input image, the dimension of the input image would decrease faster. To preserve as much information, we can add zeros on the boundary of the image after convolution operator so that we can maintain the dimension as same as the origin and this process is called padding. Besides, applying padding and pooling are not only preserver of the information but also reduce the computation and avoid overfitting [16].

Fully connected layers: In a standard CNN model the classification of the output produced from convolution and pooling layers is performed in this layer. All neuron in this layer is associated with each relating neuron in the past layer. All values from previous layer's matrix transform to a one dimensional vector and ready for classification. In CNN-XGBoost architecture, the classification is not done in a fully connected layer, perhaps we have the next layer (XGBoost) to do this.

3.2 EXtreme Gradient Boosting

It is a library of modern data-science problem and tools which stands on sequential ensemble learning. Boosting helps many weak classifiers to work strongly in classification purposes. It leverages the techniques mentioned with boosting and comes with an easy to use library. The main advantages of XGBoost are high scalability or parallelizability, speed and outperforming than other algorithms. XGBoost is a well-known algorithm, developed by Tianqi Chen to win classification challenge and used for prediction tasks like regression or classification. XGBoost is already well known for its performances in various Kaggle competitions [8]. These are major positive points in XGBoost:

Performance and speed: Originally written in C++ and compare to other ensemble classifiers it is faster than others.

Core algorithm is parallelizable: The core XGBoost algorithm is parallel and allows the capacity of multi-core processors to be used. It is also parallel to GPUs in computer networks and allows to be trained by a large amount of data.

Consistently outperforms other algorithm methods: It has shown better performance on benchmark datasets of machine learning and deep learning.

Wide variety of tuning parameters: XGBoost has outstanding parameters for regularization, cross-validation, missing values, user-defined objective functions, and tree parameters.

Comparison of different classifiers: There is a result of a comparison that has been recently done by Nag [17], for comparing the accuracy and performance of different classifiers for high dimensional data classification challenge. The experiment did on dataset from Kaggle, for all classifiers.

1. *Random-Forest Classifier*: Random-Forest is a tree and bagging approach-based ensemble classifier, and it will automatically reduce the number of features by its probabilistic entropy calculation approach. Result is:

   Accuracy for RandomForest: 0.97793
   CPU time: user 3.25 s, sys: 40.4 ms, total: 3.29 s
   Wall time: 3.29 s.

2. *Logistic Regression Classifier*: Logistic Regression is one of linear classifiers and works in the same way as linear regression.

   Accuracy for Logistic Regression: 0.931977
   CPU time: user 2 min 11 s, sys: 1.02 s, total: 2 min 12 s
   Wall time: 2 min 7 s.

3. *Artificial Neural Network Classifier (ANN)*: ANN classifier is non-linear classifier with automatic feature extraction and dimensional reduction techniques.

   Accuracy for ANN: 0.97559701
   CPU time: user 29 min 28 s, sys: 4 min 47 s, total: 34 min 15 s
   Wall time: 5 min 50 s.

4. *Linear Support Vector Machines Classifier (SVM)*: We will now apply 'Linear SVM' on the data and see how accuracy is coming along. Here also scaling is required as a preprocessing stage.

   Accuracy for Lienear SVM: 0.9643467
   CPU time: user 59.4 s, sys: 1.04 s, total: 1 min
   Wall time: 55 s.

5. *Extreme Gradient Boosting Classifier (XGBoost)*: XGBoost is a boosted tree-based ensemble classifier. Like 'Random-Forest', it will also automatically reduce the features and it use a separate 'XGBoost' library which does not come with scikit-learn.

   Accuracy for XGBoost Classifier: 0.9943615
   CPU time: user 15 min 3 s, sys: 1.49 s, total: 15 min 5 s
   Wall time: 15 min 6 s.

Among all of the classifiers, it is clear that, in accuracy prespective, 'XGBoost' is the winner by the accuracy of 0.99. [17] By considering the power of XGBoost for classification and CNN as a lead architecture for feature engineering, the combination of these algorithms cause another powerful model for image classification which in this paper specifically used for classifying malware into their families.

4 Dataset

The deep learning-based model which has been considered in this paper evaluated on the Malimg dataset, which is one of the well-known datasets, created by Nataraj et al. [5] and it is containing 9339 examples of malware belong to 25 different families. The number of malware variants in every family has been mentioned in Table 1.

Table 1 Malimg dataset structure

No.	Family	Family name	No. of variants
01	Backdoor	Agent.FYI	116
02	Backdoor	Rbot!gen	158
03	Dialer	Adialer.C	122
04	Dialer	Instantacces	431
05	Dialer	Dialplatform.B	177
06	PWS	Lolyda.AA 1	213
07	PWS	Lolyda.AA 2	184
08	PWS	Lolyda.AA 3	123
09	PWS	Lolyda.AT	159
10	Rogue	Fakerean	381
11	Trojan	Alueron.gen!J	198
12	Trojan	C2Lop.P	146
13	Trojan	C2Lop.gen!G	200
14	Trojan	Malex.gen!J	136
15	Trojan	Skintrim.N	80
16	Trojan downloader	Swizzor.gen!E	128
17	Trojan downloader	Swizzor.gen!I	132
18	Trojan downloader	Obfuscator.AD	142
19	Trojan downloader	Wintrim.BX	97
20	Trojan downloader	Dontovo.A	162
21	Worm	Allaple.A	2949
22	Worm	Allaple.L	1591
23	Worm	VB.AT	408
24	Worm	Yunner.A	800
25	Worm: AutoIT	Autorun.K	106

5 Expected Result

The CNN and XGBoost individually and their combination in the CNN-XGBoost model proved their best performance in many classification experiments and challenges, and that's why we are expecting a good result from CNN-XGBoost model by implementing on Malimg dataset. We can use this model for classification of those kinds of malware that are unseen for antiviruses, and existing malware classification methods fail to recognize them. We believe the CNN-XGBoost model with proper tuning and parameters, achieves higher accuracy than other methods for malware classification on the mentioned dataset. Adding layers and having a large amount of input help the CNN to extract more features and learn more characteristic of images and produce much better accuracy after classification, but it requires more cost and equipment. We implement this model using Malimg dataset and we got 98.7% accuracy, which is an outstanding outcome for malware classification problem.

6 Conclusion

The digital world is always exposed to affecting malware and malware writers are trying to make new malware that easily bypasses antiviruses. Malware who originated from a family shares the most characteristic and behavior, however, it may differ in appearance to trick the users and evade antiviruses. By recognizing the family of malware we can guess the behavior or target of malware and it helps us to understand how to protect ourselves and safeguard against them, and even after infection, know how to care and handle. For malware classification, we propose CNN-XGBoost model and we believe this model performs more fast, efficient and accurate than other existing methods and hope this research be effective to reduce malware harms and damages.

References

1. Malimg Dataset. https://www.kaggle.com/c/malware_classification/discussion/73433. Accessed 2019
2. Malware statistic. https://www.av-test.org/en/statistics/malware/. Accessed 31 Oct 2019
3. Internet security threat.https://resource.elq.symantec.com/e/f2ISTR_24_2019_April_en.pdf, report 2019, volume 24
4. Ren X et al. (2017) A novel image classification method with CNN-XGBoost model. In: IWDW
5. Nataraj L, Karthikeyan S, Jacob G, Manjunath B (2011) Malware images: visualization and automatic classification. In: Proceedings of the 8th international symposium on visualization for cyber security. ACM, p 4
6. Siddiqui M, Wang MC, Lee J (2008) A survey of data mining techniques for malware detection using file features. In: Proceedings of the 46th annual southeast regional conference on XX. ACM, pp 509–510
7. Drew J, Moore T, Hahsler M (2016) Polymorphic malware detection using sequence classification methods. In: Security and privacy workshops. IEEE, pp 81–87
8. Microsoft malware classification challenge (big 2015) (2017) https://www.kaggle.com/c/malware-classification. Accessed 30 Sept 2019
9. Microsoft malware classification challenge (big 2015) first place team: Say no to overfitting. http://blog.kaggle.com/2015/05/26/. Accessed 20 Nov 2019
10. Ahmadi M, Ulyanov D, Semenov S, Trofimov M, Giacinto G Novel feature extraction, selection and fusion for effective malware family classification. In: Proceedings of the sixth ACM conference
11. Gibert D (2016) Convolutional neural networks for malware classification. Universitat de Barcelona

12. Cui Z, Xue F (2018) Detection of malicious code variants based on deep learning. IEEE Trans Ind Informat 14(7)
13. Kabanga EK, Kim CH (2018) Malware images classification using convolutional neural network. J Comput Commun 6:153–158. https://doi.org/10.4236/jcc.2018.61016
14. Elleuch M, Maalej R, Kherallah M (2016) A new design based-SVM of the CNN classifier architecture with dropout for offline Arabic handwritten recognition. Proc Comput Sci 80:1712–1723
15. Intro to convolutional neural networks (2019) https://web.stanford.edu/class/cs231a/lectures/intro_cnn
16. Lin M, Chen Q, Yan S (2014) Network in network. In: ICLR
17. A comparison of different classifiers' accuracy & performance for high-dimensional data. https://www.freecodecamp.org/news/multi-class-classification-with-sci-kit-learn-xgboost-a-case-study-using-brainwave-data-363d7fca5f69/. Published on 9 May 2019, accessed on 20 Oct 2019
18. A comprehensive guide to boosting machine learning algorithms. https://www.edureka.co/blog/boosting-machine-learning/, retrieved at April 2020

Design of E-Water Application to Maintain the Flow of Water from Common Faucets by Enabling GSM

P. Baskaran, Kaaviya Baskaran, and V. Rajaram

Abstract Water Scarcity is a major problem faced by all sections of the society. Though people are informed about the water crisis, they still fail to utilise water in the wisest possible way. This happens because people do not receive direct data and reminders about the amount of water that is being utilised. Thus this proposed work aims at providing a solution to the above mentioned problem. It creates awareness allowing the users to about the amount of water that they are consuming by enabling direct tracking of water consumption levels. Furthermore, this work drafts an ideal water consumption level that households should attempt to stick to on a monthly basis and provides reminders for the same. The users can regulate the flow of water from water faucets directly using the application and data about the water used by the people can be stored in a database.

Keywords Solenoidal valve · Water motor · Firebase · Android app · Water conservation · Water controller · Storage tank

1 Introduction

Water is the most basic commodity that proves quintessential in order to ensure the sustenance of life on Earth. The drastic tip in the levels of potable water has taken a major hit on people from all walks of life regardless of their social status [1]. In India, the fight and chaos at the water faucets have been a typical daily morning scenario. Spurious fights erupt at water collection especially over the contention of

P. Baskaran (✉) · K. Baskaran · V. Rajaram
Information Technology, Sri Venkateswara College of Engineering, Sriperumbudur, Chennai, India
e-mail: baskpann@gmail.com

K. Baskaran
e-mail: kaaviyabas@gmail.com

V. Rajaram
e-mail: vrajaram@svce.ac.in

D. J. Hemanth et al. (eds.), *Artificial Intelligence Techniques for Advanced Computing Applications*, Lecture Notes in Networks and Systems 130,
https://doi.org/10.1007/978-981-15-5329-5_20

the order of filling the buckets resulting in wastage of time and water. This proposed work aims at tackling this hubbub and regulating the usage of water.

Taking into account the current situation, this work will greatly increase awareness amidst people about water consumption levels as direct data about water consumption [2] every time they switch on the faucet can be viewed from the mobile application along with the monthly water consumption levels.

2 Statistics

According to official sources, the total water demand in the city is 950 million liters per day (MLD), whereas the supply is 750 MLD, including 200 MLD from private tankers. That leaves a dangerous deficit of 200 MLD. During the period of an acute water crisis in the summer of 2019, MTC water tankers charged Rs. 700–800 for 9000 litres of water while private water tankers forked up to Rs. 4000–5000 for the same.

3 Existing Solution

As of present times, there are no technological applications or solutions available which can enable people to collect water from the water faucets and lorries without any inconvenience. People try to fight with their buckets in line fighting for water, which usually results in wastage of water many times. Inefficient filing of buckets also results in overflowing buckets and spillage of water along the way.

4 Hardware Components

The work utilizes the following hardware components to enable the user to regulate the flow [3] of water from the water faucets into the containers at the click of a button.

4.1 Solenoidal Electro Valve

The solenoidal valve as shown in Fig. 1 can be electromechanically operated. Electric current is passed through the solenoid regulated by a valve. Operating voltage is set to 12 V for the solenoidal valve. It allows around 3 L/minute of water flow. The flow of water will be interrupted until the fast-on conductors receive 12 V supply [4]. The connection is established between ON-OFF switches and the solenoidal valve. When

Fig. 1 Solenoidal valve

the switch is open, the switch is activated permitting liquid flow whereas when the switch is closed, the solenoidal valve gets energized prohibiting further liquid flow.

4.2 Arduino UNO

Arduino allows the implementation of interactive environments accessible using an individual microcontroller. There are several types available notably the Arduino UNO REV3 model. Arduino UNO specified in Fig. 2 is based on ATmega328 with 14 IO pins, 6 analogs I/P, a USB connection, power jack, etc. [5]. Arduino consists of all the components necessary to establish a connection with the computer via the USB cable. Arduino can be programmed using an Arduino IDE which is installed in a Windows PC. With the aid of serial cable, the program is written and uploaded to the IDE. Data from the flow meter will be received and converted into digital pulses.

4.3 Gsm

A GSM module shown in Fig. 3 is a chip for establishing communication between a mobile device or a computing machine. GSM (Global System for Mobile Communications, originally Groupe Spécial Mobile), was developed by the European Telecommunications Standards Institute (ETSI). It describes the protocols for

Fig. 2 Arduino UNO

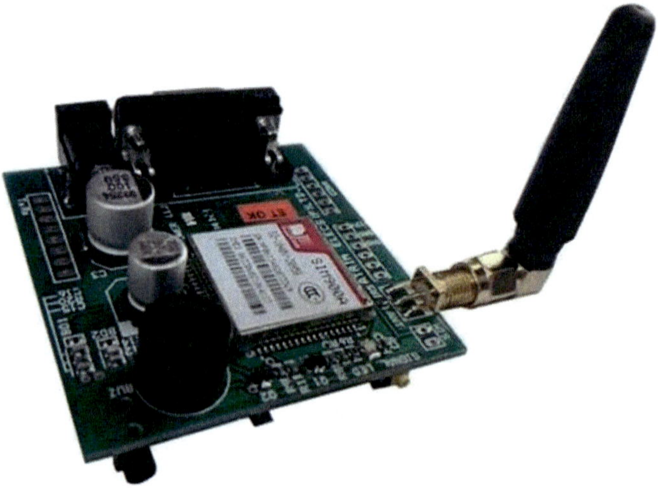

Fig. 3 GSM module

second-generation (2G) digital cellular networks utilized in over 219 countries and territories.

4.4 Relay

Fig. 4 Relay

A relay is an electrically operated switch that can be turned on or off, letting the current go through or not, and can be controlled with low voltages, like the 5 V provided by the Arduino pins. As shown in Fig. 4, it consists of three pins namely COM, NC and NO. COM is the common pin [6]. NC refers to the normally closed configuration which is used when the relay has to be closed by default. NO refers to the normally open configuration which works the other way around where a signal has to be sent from the Arduino to close the circuit.A nozzle is fitted to ensure that only a steady stream of water flows out from the tap in all conditions.

5 Software Components

5.1 Android Application

Each individual user, identified by a unique mobile number will be granted access to a fixed amount of water per day. Once the user logs into the portal with his/her mobile number, data about the user can be retrieved from the database. The user can then control the flow of water from the common faucet/tap with the click of a button on his/her app which will trigger a timer to run [7]. The amount of usage and the remaining balance of water for the day will be sent as an SMS to the user's registered mobile number.

A registration portal as shown in Fig. 5 will be available for new users. This will enable a provision for users to provide a unique mobile number and their name. These details will be stored in the database. The mobile number will be the primary key in

Fig. 5 User login and register

identifying the user. Users who have already registered can log into their portal with their mobile number.

For each mobile number, a stipulated amount of water for collection will be allocated, the common water faucet/water lorries. Once the user clicks a button, it will enable the water flow from the faucet and trigger the operation of a timer. Once the timer as shown in Fig. 6 is switched off, the application [8] immediately calculates the amount of water that has been used up within the time and sends an SMS to the user.

The SMS as displayed in Fig. 7 which is sent to the user contains details of the water consumption by the user and the balance amount of water left for collection based on the set limit for the particular day. This enables the user to monitor their daily water usage and use it wisely as well [9]. Once the set limit for the day is reached, this will prevent further water from being sent out as recorded in Fig. 8.

6 Overall Process

The overall process involves user registering and logging into the portal designed to regulate and monitor water usage. Once the user starts off the timer, the water flow through the valve will also be initiated. As soon as the user switches off the timer, the water flow will be stopped. The application will calculate the water consumption and balance for that day and send the data as an SMS to the user.

Fig. 6 Timer display

7 Conclusion

This proposed work will enable every individual user to be mindful of their water usage and to use it wisely. It will also reduce the amount of water that goes wasted since a set limit is provided. It will also ensure orderliness in the collection of water since only if the number is provided access will they be able to collect water from the common faucets/water lorries.

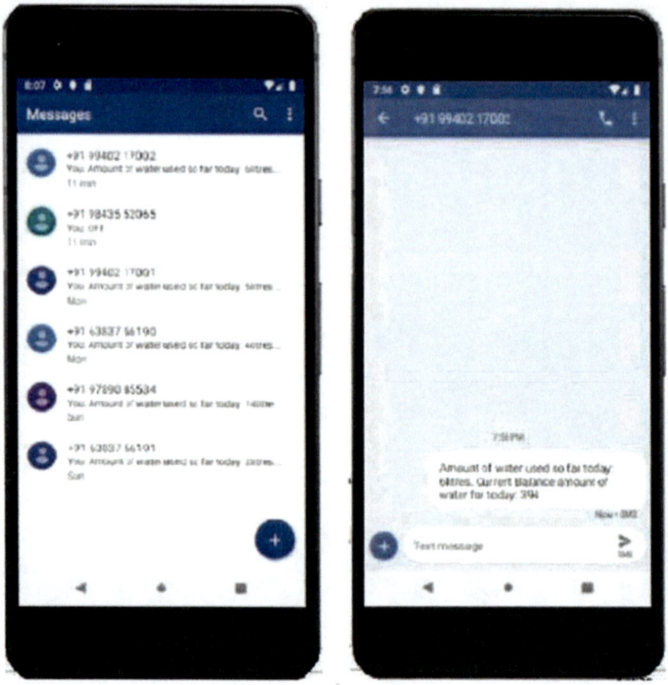

Fig. 7 SMS received

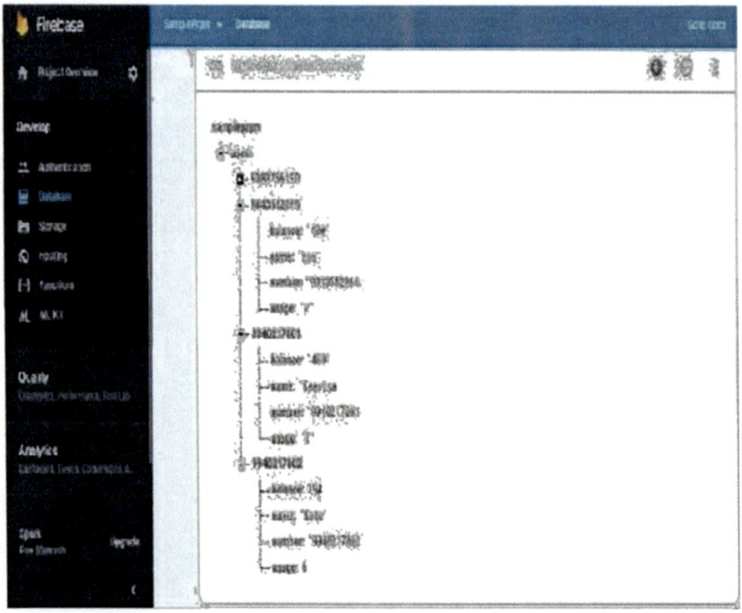

Fig. 8 Application Database

References

1. Karmoker S, Arefin KS, Momtaz ASZ (2011) Controlling water pump machine using cellular mobile telephony. In: 2011 IJCIT. ISSN 2078-5828 (Print)
2. Madhavireddy V, Koteswarrao B (2018) Smart water monitoring system using IoT. Gowthamy J, Int J Eng Technol 7(4.36):636–639
3. Pagare N (2017) Water pump control using GSM. Int J Electr Electron Res 5(3). ISSN 2348-6988
4. Selvaraj JDF, Paul PM, Jingle IDJ (2019) Automatic wireless water management system (AWWMS) for smart vineyard irrigation using IoT technology. Int J Oceans Oceanogr 13(1):211–218. ISSN 0973-2667
5. Robles T, Alcarria R, Martin D, Navarro M, Calero R, Iglesias S, Lopez M (2017) An IoT based reference architecture for smart water management processes. In: Piresa A, Moratoa J, Peixotob H, Boteroc V, Zuluaga L, Figueroad CA (eds) Sustainability assessment of indicators for integrated water resources management, vol 578, pp 139–147. https://doi.org/10.1016/j.scitotenv.2016.10.217
6. Vijaia P, Sivakumar PB (2016) Design of IoT systems and analytics in the context of smart city initiatives in India. In: 2nd international conference on intelligent computing, communication & convergence (ICCC-2016)
7. Yang S-H, Chen X, Chen X A case study of internet of things: a wireless household water consumption monitoring system
8. Vörösmarty CJ, Hoekstra AY, Bunn SE, Conway D, Gupta J Freshwater goes global. Science 349 (6247):478–479. ISSN 0036-8075
9. Perumal T, Sulaiman MN, Leong CY (2015) Internet of things (IoT) enabled water monitoring system. In: 2015 IEEE 4th global conference on consumer electronics (GCCE)

Secure Public Cloud Storage Using Collobrative Access Control and Privacy Aware Data Deduplication

A. Aarthi, G. M. Karthik, and M. Sayeekumar

Abstract Attribute Based Encryption (ABE) proved to be a powerful tool of cryptography able to express more quality access policies, which can provide fine-grained control, adaptable, and secure access to the data outsourcing. In any case, existing ABE-based access control plans don't back clients to pick up the access authorization by collaboration, In this paper, Explore the attribute-based access control in situations where a lot of users who have the qualities typical set can collaborate to gain access authorization if the owner of the data enable their collaboration in getting the policy. Meanwhile, collaboration is not defined in getting to the approach should be esteem as collusion and get to ask would be rejected. Propose collaborative access control attribute-based flow control by assigning a translation hub in the access structure. Then create ZEUS and ZEUS +, indicates the aware of two privacy with the deduplication protocols, the guarantees with the weaker protection in productive communication cost, ZEUS + provides the stronger safety profile with the cost communication costs increases. This is the high priority way to security addresses two sides with one or the other does not utilize additional hardware or depending on the parameters chosen heuristics used by existing solutions, in this way reducing the cost and complexity of cloud storage.

Keywords Attribute based encryption (ABE) · Cryptography · Collaboration access control · ZEUS · ZEUS+ · Deduplication

A. Aarthi · G. M. Karthik (✉) · M. Sayeekumar
Department of Information Technology, SRM Institute of Science and Technology, Chennai, India
e-mail: profgmkarthik16@gmail.com

A. Aarthi
e-mail: aarthi2324@gmail.com

M. Sayeekumar
e-mail: sayeekum@srmist.edu.in

D. J. Hemanth et al. (eds.), *Artificial Intelligence Techniques for Advanced Computing Applications*, Lecture Notes in Networks and Systems 130,
https://doi.org/10.1007/978-981-15-5329-5_21

1 Introduction

Cloud computing when relates with the conventional computing methods it shares the devices assets instead of using individual device. Computing over the cloud gives a simple way adaptable and cost-effective way to get information in a multi-platform client at any time [1]. The sharing of assets incorporates capacity, program and equipment. The basic concept behind the cloud is Virtualization. The secrecy, availability, security, protection, execution, uprightness is the significant problem of cloud. The services of cloud provide various kinds of cloud development models like community, Isolated, Distributed computing is a rising innovation as the quantity of cloud specialist co-ops and the cloud clients are expanded as of late. The income of cloud specialist co-ops expanded through year by year. Income from computing over the cloud in 2009 was nearer 60 billion dollars. The profit increases year by the year. Now applications managed cloud buyers and private business needs as opposed to business-critical applications. Effect of security breaks for large scale business and key application will be incredibly high when compare to little scale business [2]. The compensation made by the scattered handling relies upon the Quality of Service offered by the cloud. The fundamental property of Quality of Service is security and the cloud authority focus needs to give full certification of security with respect to question, availability, confirmation and respectability. Among the issue's protection might be an essential and uncompromisable factor of security. Encoding of information is much needed to protect the data from untrusted access [3]. Especially in Public cloud storage data will be manage by unknown third party therefore their no confidentiality to our data this made a high security breach therefore to bring high confidentiality to our data and to secure our data from access of unknown strong encryption of data is much needed, So here try out to bring strong encryption methodology by using attributes that is strong encryption take place by taking attributes into the consideration [4], As well as here consider the situation of need of access in collaboration manner so here implement collaboration access control by ABE (Attribute Based Encryption) methodology at the same time to provide high level of security during collaborative access implemented deduplication method ZEUS +.

2 Related Works

Rui Guo et al., displayed a paper named Certificateless Public Key Encryption Conspire with Hybrid Issues and Its Application to Web of Things [5]. This paper manages the idea of certificateless open key encryption plan to dodge the key escrow issue. Certificateless cryptography plot goes for joining the benefits of open key cryptography and personality-based cryptography to stay away from the certificate management and the key escrow issue.

Al-Riyami et al., manages certificateless open key cryptography [6]. In this paper certificateless open key cryptography (CL-PKE) conspire is utilized to take care of the key escrow issue and certificate revocation issue. It doesn't require authentications and it additionally conquers the issue of implicit key escrow. The downside of the over two plans is that it depends on matching activity. Pairing based protocols are used in a variety of protocols and it is found that many applications uses pairing based protocols as the solution where ID-based cryptographic schemes and the short signature schemes are employed.

To overcome this impediments Y. Sun et al., proposed a paper called Strongly secure guards less key for public purpose encoding with no blending. It gives the upside of character based open key cryptography with no input escrow issue. This work has been the essential certificate less encryption method without utilizing mixing movement. In this paper they have illustrated the security against versatile chosen cipher content assault within the self-assertive demonstrates [6].

In identity based and security concerned systems [7, 8], a third party is needed to generate the key which is termed as key escrow problem. This becomes the major disadvantage of identity-based system. To eradicate this problem, certificate less cryptosystem is proposed. Every entity will possess a public key but no certificate [9]. Identity strings can be used to assure that only correct entity can have the private key with respective to its public key. Then instant revocation cannot be obtained in this case.

When considering identity-based encryption method, it does not depend on security intermediator. It has predefined keys and there are chances of key escrow and certificate issue. Certificateless encryption methodology has removed the key escrow issue but the certificate revocation problem still prevails. In Identity based safety mediate method there is no chance for getting a certificate revocation problem. Therefore, security mediated certificate less method can overcome both key escrow and certificate revocation. Also, it uses bilinear method which is highly expensive. So, there emerges a need for method without using pairing technique [4]. Thus, security mediated certificate less encryption can be used to preserve the data in public cloud which does not employ any pairing technique.

Cyber-attack also a major considered issue in cloud through network hackers able to steal the data from cloud therefore it is necessary to secure network protocol [10, 11], By use of open daylight controller in operating system open stack cloud achieved transport layer security [12].

2.1 Issues in Cloud Computing

Cyber crime's impact is felt all over the Web, and cloud computing is striking targets for a variety of reasons. An existing infrastructure to avoid and survive a cyber-attack, but not every cloud has such abilities. On the off chance that a cybercriminal can recognize provider's easiest vulnerability to abuse, then these substances will be highly visible targets. Otherwise, all cloud providers to supply adequate safety measures, at the time this will turn into a cloud of high-needs focus for cyber criminals. By nature of their intrinsic design, the cloud offers an open door to strike simultaneously to a variety of sites, and without appropriate security [13], some sites may be disturbed by evil movement solitary. Distributed computing security incorporates a variety of issues such as multi-occupancy, misfortune information and spills, openness simple cloud, personable executive, dangerous API, rate of administration and understand the deviation, improving executive, danger interiors etc. It is not easy to apply everyone security measures need involve in the security of large number of owners, arguing that the various client may have security requests vary depending on the target they take advantage of cloud administration [14].

3 Problem Statement

Some safety measures is sensitive subject in computing over the cloud. The information is beginning from cloud using open structure (web) there are openings to hack the information. There have been bit of work done on security issues and difficulties anyway simultaneously there isn't 100% full confirmation course of action. There are different physical and a couple of different attacks on the data that smashes the data on server. One answer for that is dispersed the information on more than one server instead of one server. By and by, this not manages issue totally considering the way that informational index away in blended mode utilizing encryption key. The aggressor's assault on key and might be hacking the information.

4 Problem Solution

To effectively address the problem of security in the cloud, Need to understand the security of the compound. The challenge in every respect. Primarily, Must: (1) investigate various properties compute cloud security vulnerability, danger, danger, and models of assault; (2) identify security requirements calculate secrecy, sharpness, accessibility, candor, etc.; (3) distinguish the gatherings in question (customer, administration giving, pariahs, insiders) and the job of each gathering in a cycle of assault protection; (4) comprehend the effect of security on the different cloud sending models (open, open, private, cross breed). The significant commitment of this paper is that, a safe encryption plan to share scrambled information among a lot of approved clients and to accomplish effective client disavowal for inconsistent mists, The collaboration access allow under the policy the policy determine the number of authority persons can access and in which combination the persons can access data combination determined by setting attribute his/her position, priority. when data get access by a group the group must consider high priority authorized person of data should be present under his/her accountability access get permit to that respective group therefore through this able to achieve secure collaboratives access. In distributed computing security is a significant part of nature of administration. A few encryption methods are utilized to keep the sensitive client information secret against untrusted servers. ABE encryption plot dependent on the believed authority has been suggested that use the presentation required for encryption errands inside the cloud itself. A believed authority is liable for key managing, resulting in a more efficient and scalable security. Propose a cloud-based verified information structure, which engages control to safely store their riddle information on the semi-accepted cloud ace associations, and especially share their mystery information with a wide extent of information gatherer, to diminish the key organization complexity of intensity owners and data recipients. Detachment from past cloud-based information frameworks, Data proprietors encodes their mystery information for the information recipients utilizing ABE ENCRYPTION SCHEME [15] used to scramble the information at long last put away in the cloud (server farms). Another moved detail is, if any data recipient needs some record to download, the data beneficiary will send the auctions to the power (information proprietor). The position proprietor has the Access Control. On the off chance that the Proprietor needs to impart the underlying record to the information recipient, he recognizes the ask something different data owner rot the inquire. After recognizes ask the data beneficiary to download the keys [16] and this key are fundamentally for endorsement and to download the data in the first configuration (deciphered design).

4.1 System Model

5 Data Owner

The data owner will only have the rights to upload the data in the format of excel sheet. While uploading the data is read from the excel sheet and inserted into the database of the respective cloud server and before encryption, the data get encrypted and stored in the cloud server.

6 Data Consumer

The data consumer will have all the rights to search the data stored by the data owner on various cloud server, but they have to know the respective security credentials provided by the cloud service provided. If the credentials are does not match the data consumer will not have an option to view the data stored on the respective cloud server.

7 Cloud Server

The data uploaded by the data owner will get uploaded in the form of encrypted format. For data storage had used Amazon ec2 cloud server, which is an open source cloud server. The algorithm used to encrypt and decrypt the data is Blow Fish algorithm. There are so many algorithms for encrypt and decrypt the data But had selected the Blow Fish algorithm because of the maximum accuracy provided by it.

8 Checksum Generation

The checksum generation logic is a unique process for each and every application which is being deployed in the various servers. When the files are uploaded, before encryption the check sum is generated with the same name of the particular uploaded file name. Capturing the same checksum in database against each and every uploaded file.

9 Find Out Duplicate Files

When the files are being uploaded the check, sum got stored in the database. When a new file is uploaded the check, which is generated to the file is being compared with the old one which got stored in the database. If the checksum is already present in the database, the files are not allowed to get uploaded and duplicate files which is being uploaded got restricted (Fig. 1).

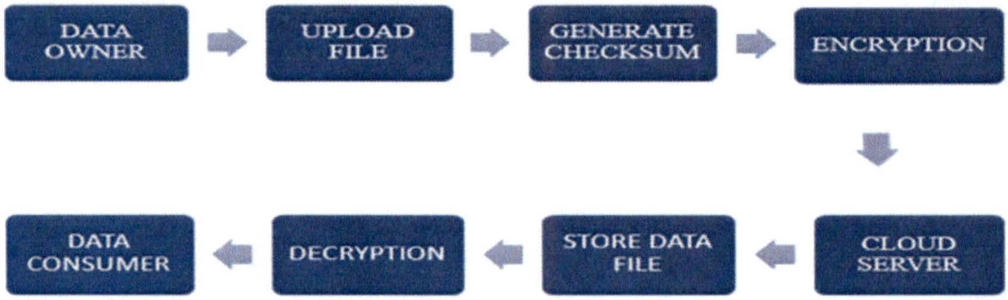

Fig. 1 System model flow diagram

9.1 Initialization Processes

cipher index file size and the classes get initialize and then through cipher index class and attributes public key and another key generate, Iteration take place to calculate string by using cipher index class name and random integers this integer should converted to binary string by using interger to binary string function.

Public_ key = Random generation _index (cipher index class, attributes)

Other_ key = Generate random index key (cipher index class, attributes)

9.2 Encryption of File

Get the plain text and first encryption take place through using symmetric key by using blow fish symmetric key algorithm the person who is the authority of documents of data carry the encryption process of symmetric key according to the collaborative access policy to allow collaboration access under the collaborative access policy condition. Collaborative access policy generated by first having some nodes declared by authority to get the access in collaboration manner, here collaborative access policy determines by using tree structure relationship attributes, Tree determining in top down approach from root to non-leaf node degree get reduce by value one than the threshold value, leaf node degree remain as 0 according to the node value priority will be set and collaborative access policy get generate.

9.3 Key Generation

To generate key summation of key take place, follow by mask operation for public key and the byte format and private key and byte format after this iteration take place.

Set j= Bytes12 [i]; (j=0 to j < byptes12.length)

Str_3= Int. Binary string(j)

Temp_string= Temp+ Int. parse_Int(S3)

S3= converted to binary String(temp)

9.4 File Deduplication Using ZEUS+

ZEUS + provide none knowledge response to the one who is checking for presence of data to maintain confidentiality of the data, In this method first uploaded document split into some parts then the first two parts get upload into ZEUS, ZEUS check the size of uploaded parts and then check hash value if it is determined the content is their it response take place in 2 ways. 1. if part 1 is available and part 2 s not available it returns 1 indicate 1 part is missing does not give knowledge which part is available and which part is not available, 2. If 2 parts are not available return 2. 3. If no parts are available return 0 In existing the problem here is the attacker can get knowledge about presence of content by duplicate check so to avoid duplicate check here dirty parts check take place dirty parts already uploaded and checked part if dirty part is determined ZEUS + always return 2 as well as the uploaded person not able to check the existence of their own self uploaded part they sure that is already uploaded part is in cloud because sometime through this unauthorized one can able to get knowledge which part is available by self-uploading the one part previously. Therefore, here non-independent check take place to achieve secure deduplication.

9.5 Decryption of File

Get the encrypted text from database then decrypt by using secret key and private key, Decryption take place character by character and then store decrypted one in list have iteration to check whether received text after decryption is equal to text before encryption to determine it have check of size whether the size is matching or not and check character by character to determine the secure communication (data manage) in cloud.

10 Experiment and Results

To evaluate the performance of the system, the encryption time, the time needed for generate Key and time of decryption took for consideration. The processing time required by ABE redistributing is resolved on Integrated Development Environment (IDE) JAVA [17] while the database is kept in MYSQL.

Execution investigation is finished by as pertaining Encryption processes and transfer time, Decryption processes and downloads time and time to generate key. performance analysis performed by utilizing the typical size of the document and the results obtained by observing the time utilized for encrypt those various sized documents, the time required for decoding the records and the time required for the generation of keys (Fig. 2).

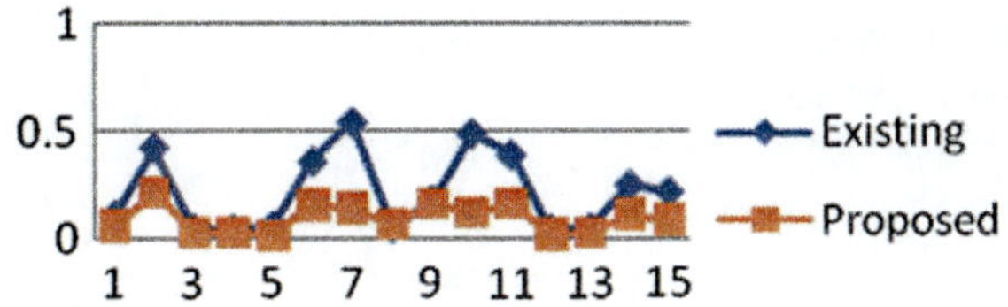

Fig. 2 Shows the comparison graph for encryption time

To evaluate ABE performance, time taken to encrypt and upload the file on the MYSQL Database by using different file sizes-1, 2, 5, 50, 100 KB is considered. The time taken to encrypt they data is high in prevailed solution [18, 19] when compare with the time taken to encrypt the data by proposed solution, the main lead of the planned solution is size of in taking plaintext is self-determining so can able to choose our own different various sizes plain text even though having various sizes plain text able to encrypt quickly and efficiently in planned solution when compare to prevailed solution [19] and also independent to get different size of ciphertext. The above graph showing that the time taken by prevailing solution [20] to encrypt the intake plaintext is increased twice than the time taken by planned solution to encrypt the same type of data and then according to the size of block of data of various sized documents increase the time taken to encryption also varying in mentioned prevailed solution therefore the prevailed solution determining as having more blocks of data of plaintext efficient of encryption will be getting down as the solution to this issue in the planned solution efficient of encryption better than the prevailed solution as the planned solution endure neutral for 30 ms therefore by utilizing the planned solution able to encrypt efficiently and quickly the multiple block of data of multiple size documents and also quickly able to upload in MySQL database (Fig. 3).

Key production time here is to produce key while downloading file from the system. Here the file sizes used are 1, 2, 5, 50, and 100 KB. And by using the same file have to note the time for generating the secret key which is used for the file downloading process. In prevailed solution [21] to produce secret key for above mentioned size of cipher text contained file took time twice more than the planned solution it is determined in above graph (Fig. 4).

The operations which are performed while downloading file are: the encrypted file is taken from Storage Service Provider, and the file is decrypted by using the

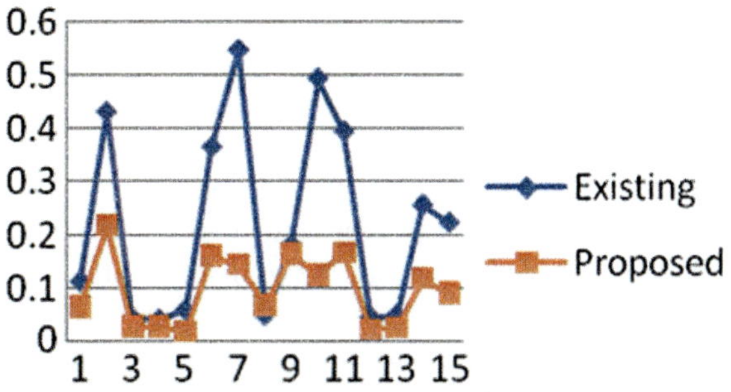

Fig. 3 Shows the comparison graph for key generation time

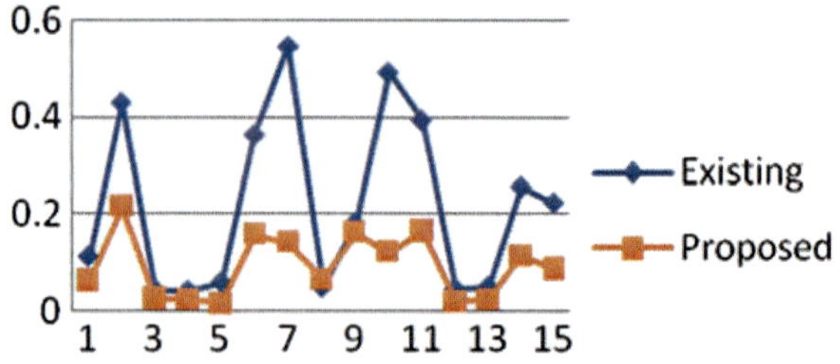

Fig. 4 Shows the comparison graph for decryption time

Table 1 Shows the comparison table for encryption algorithm

Algorithm	Block-cipher & key length	Flexibility/modification	Data center and data tenant	Block size	Effectiveness	Level of security	Encrypt speed
Existing	Binary, 128 bits, 192 bits & 256 bits	YES, extended from 56 to 168 bits	No	128 bits	Slow in software	Adequate security	Slow
proposed	Binary, 1–4096 set of integers	YES, 256-key size is multiple of 64	Yes	1024 bits	Efficient in software	Highly secure	fast

user's private key. If the file size increases, then decryption time also increases. In prevailed solution [18, 19] as the downloaded encrypted file size increased level the time taken to decrypt that encrypted file also twice increased in prevailed solution when compare to the time taken by planned solution to encrypt those downloaded different size encryption file of ciphertext and also here free to have the size of downloading encryption file to decrypt and also free to get the various size(as equal to size of file before encryption) of decrypted file after decryption of ciphertext as well as here too decryption process is efficient than prevailed solution [20] by endure neutral for 30 ms therefore by utilizing the planned solution able to decrypt efficiently and quickly the multiple block of data of multiple size encrypted documents (Table 1).

11 Conclusions

Cloud computing with the nature of Distributed computing is said to be a combined form of key strategies that have advanced and created over a long time. Distributed computing contains a potential for cost undertaking assets to the under taking yet the security chance is also huge. Undertaking researching distributed computing advancement as a way to deal with critical harm and expansion proficiency should separate the security plausibility of distributed computing. Intensity of distributed computing in the organization of the possibility of information ability to regulate even more compelling possibility. In spite of the way that the preparing of the Cloud can be viewed as another phenomenon that is set to agitate the manner by which utilize the Internet, there is abundance to be watchful about There are numerous new advances expanded at a fast rate, each with a mechanical movement and the capacity to make human lives more straightforward. Be that as it may, one must be extremely mindful so as to comprehend the wellbeing risks and difficulties acted to exploit this development. Distributed computing is no exception. In this paper the key security contemplations and difficulties right now looked in the Cloud are featured. Distributed computing can possibly get the opportunity to be a developer in advancing arrangements that safe, virtual and monetarily later on.

References

1. Buyya R, Yeo CS, Venugopal S, Broberg J, Bandic (2009) Cloud computing and emerging IT platforms: vision, hype, and reality for delivering computing as the 5th utility. Future Gener Comput Syst 25(6):599–616
2. Hough A (2010) Google engineer fired for privacy breach after staking and harassing teenagers. The telegraph
3. Yang JF, Chen ZB (2010) Cloud computing research and security issues. In: 2010 IEEE international conference on computational intelligence and software engineering (CiSE), Wuhan, pp 1–3
4. Kaur PJ, Kaushal K (2011) Security concerns in cloud computing. In: Accepted for international conference on high performance architecture and grid computing
5. Guo R, Wen Q, Shi H, Jin Z, Zhang H (2014) Certificateless public key encryption scheme with hybrid problems and its application to internet of things. Math Probl Eng 2014
6. Al-Riyami S, Paterson K (2003) Certificate less public key cryptography. In: Proceedings of ASIACRYPT, Springer, LNCS 2894, pp 452–473
7. Yuen TH, Zhang Y, Yiu SM, Liu JK (2014) Identity-based encryption with post-challenge auxiliary inputs for secure cloud applications and sensor networks. In: Proceedings of 19th ESORICS, vol. 8712, pp 130–147
8. Liang K, Liu JK, Wong DS, Susilo W (2014) An efficient cloud-based revocable identity based proxy re-encryption scheme for public clouds data sharing, vol. 8712, pp 257–272
9. Takabi H, Joshi JBD (2010) Security and privacy challenges in cloud computing environment. IEEE J Secur Priv 8(6)
10. Naveen Chandar B, Arivazhagan N, Venkatesh K (2019) Improving the network performance using MP-OLSR protocol for wireless ad hoc network (MANET). Int J Recent Technol Eng 8(3):5700–5707
11. Vakaavinash V, Venkatesh K (2020) Role of software-defined network in industry 4.0". In: EAI/springer innovations in communication and computing-internet of things for industry 4.0 design, challenges and solutions, pp 197–218
12. Venkatesh K, Srinivas LNB, Mukesh Krishnan MB, Shanthini A (2018) QoS improvisation of delay sensitive communication using SDN based multipath routing for medical applications. Future Gener Comput Syst 93(2019):256–265
13. Jiang T, Chen X, Li J, Wong DS, Ma J, Liu J (2014) TIMER: secure and reliable cloud storage against data re-outsourcing. In: Proceedings of 10th international conference information security practice and experience, vol. 8434, pp 346–358
14. Yang J, Chen Z (2010) Cloud computing research and security issues. In: The proceeding of IEEE international conference on computational intelligence and software engineering, pp 1–3
15. Hohenberger S, Waters B (2014) Online/offline attribute-based encryption. In: Proceedings of the 17th international conference on practice and theory in public-key cryptography (PKC). Springer, pp 293–310

16. Lenstra AK, Hughes JP, Augier M, Bos JW, Kleinjung T, Wachter (2012) "Public keys", volume 7417 of LNCS. Springer, pages 626–642
17. Michaelis K, Meyer C, Schwenk J (2013) Randomly failed the state of randomness in current java implementations, volume 7779 of LNCS. Springer, pages 129–144
18. Rifki S, Park Y, Moon S (2015) A fully secure cipher text-policy attribute-based encryption with a tree-based access structure, pp 247–265
19. Yang Y, Liu JK, Liang K, Choo K-KR, Zhou J (2015) Extended proxy-assisted approach: achieving revocable fine-grained encryption of cloud data. In: Proceedings of 20th ESORICS, vol. 9327, pp 146–166
20. Bethencourt J, Sahai A, Waters B (2007) Ciphertext-policy attribute-based encryption. In: Proceedings of the 28th IEEE symposium on security and privacy (Oak-land). IEEE, pp 321–334
21. Liang K et al (2014) A DFA-based functional proxy re-encryption scheme for secure public cloud data sharing. IEEE Trans Inf Forensics Secur 9(10):1667–1680

A Comparative Study of Techniques, Datasets and Performances for Intrusion Detection Systems in IoT

Arathi Boyanapalli and A. Shanthini

Abstract IoT Security is the area concerned with safeguarding connected systems. IoT involves the set-up of various integrated devices. Devices are identified with a unique identifier, and provided with the ability to transfer data over the network opens them up to several serious vulnerabilities, if not appropriately protected. An Intrusion Detection and Prevention System (IDPS) plays a crucial part in discovering and preventing numerous attacks entering the network and provide an uncompromised secure system. The sheer volume of the sensors in a system comes with limitations such as interoperability, scalability, and storage, where security algorithms like IDS couldn't perform well as it requires a huge amount of labelled data for training, to detect intrusions and ascertain new attacks. Fog computing plays a major role with a decentralized architecture allows IoT devices to compute, make decisions, take actions, and push only relevant information to the cloud. Data availability is closer and can act immediately for the sensitive information, which in turn helps the IDS to perform well using Artificial Intelligence algorithm to detect and prevent various attacks. This paper categorizes the existing recent researches in IoT Intrusion Detection systems using artificial intelligence and fog computing architecture in terms of technical constraints.

Keywords Intrusion detection system (IDS) · Internet of things (IoT) · Machine learning · Deep learning · Anomaly detection

1 Introduction

Emerging developments in electronic devices and communication systems lead to the concept of the Internet of Things (IoT). IoT is a group of interconnected devices that can sense, connect, and exchange the data between the devices over the network.

A. Boyanapalli · A. Shanthini (✉)
SRM Institute of Science and Technology, Chennai, India
e-mail: shanthia@srmist.edu.in

A. Boyanapalli
e-mail: ab1114@srmist.edu.in

D. J. Hemanth et al. (eds.), *Artificial Intelligence Techniques for Advanced Computing Applications*, Lecture Notes in Networks and Systems 130,
https://doi.org/10.1007/978-981-15-5329-5_22

IoT devices lead to the development of smart cities, smart agriculture, healthcare, and intelligent appliances. According to Gartner reports, Twenty billion internet-connected things are expected by 2020 [1]. Massive data collected from interconnected devices not only create awareness of the environment where the sensor is situated but can also be used for analysis and predict the future condition of the environment. These scenarios can automate device administration, improve proficiencies, and reduce functioning costs while improving the customer experience. Security challenges have increased with the increase of IoT devices. The security challenges in IoT devices differ from traditional cyber security as real-time IoT networks can have far-reaching effects on security and safety.

1.1 Relevant Terms

Attacks existed around for a very long time, but the way they act in the IoT environment differs. Many of the devices connected can be household appliances or vehicles act as end-node that are utilized as entry points for attacks. And therefore posing increasing security and privacy risk. Various attacks that can be categorized as

Botnets: A botnet takes remote control of devices in a network and distributes malware. These botnets are used by criminals to steal private information.

Man in the Middle attack (MITM): In MITM, the attacker or hacker secretly listens between the communications of two parties or transmits messages directly to the recipient in disguise and can trick the other party as they are getting the legitimate message.

Data and Identity Theft: The data collected from social media, smart watches, smart fridges, fitness trackers can give a lot of personal information, which makes it easier to aim an identity theft.

Social Engineering: Social Engineering is an act to influence people and get their confidential information such as passwords, banking information, or install malicious software that gives control to their computers and access personal information.

Denial of Service (DoS): Malicious systems attack one target machine by sending requests continuously due to which the target becomes unavailable for service. In Distributed Denial of Service (DDoS), a large number of malicious systems attack one target.

Intrusion Detection Systems (IDS): Intrusion detection systems monitor network traffic to detect intrusion by unauthorized entities. The functions of IDS are

- IDS can be used to monitor a single system or large networks.
- Monitors routers, firewalls, servers to detect, prevent, and recover from attacks.
- Helps administrators to perform audit trails on Operating systems and System log files.
- Any alteration in data files can be recognized and reported.

- Notifies and generates an alarm when a security breach is detected.

IDS mainly classifies into two categories:

- **Network intrusion detection systems (NIDS)**: Monitors intrusions entering the whole network.
- **Host-based intrusion detection systems (HIDS)**: Monitors intrusions entering a particular host.

IDS variants into two types depending on the detection methods, detection methods can be applied either on NIDS or HIDS

- **Signature-based detection**: This approach considers attack patterns as signatures stored as a database and compares the traffic with the stored signatures. It works well with known attacks.
- **Anomaly-based detection**: This approach alarms when the traffic deviates from a model. Deep Learning and Machine Learning algorithms can be used to sense variations from regular traffic. It works well even with zero-day attacks or unknown attacks.

Limitations of IDS:

- Noise can affect intrusion detection systems and can raise the alarm.
- An outdated signature database can make IDS vulnerable to newer attacks.
- Fake IP addresses can be more challenging to detect.
- False Positives are frequent and cannot be ignored, as real attacks also raise a false positive.
- They do not process encrypted packets.
- Protocol-based attacks are susceptible.

2 IDSs Designed for IoT

A summary of recent studies on network-based intrusion detection systems with fog computing architecture and techniques utilized for maintaining security in IoT devices discussed with an overview of datasets and features used by the researchers.

N. Moustafa et al. (2019) [2] Proposed a statistical contrivance for anomaly detection, Outlier Dirichlet Mixture-based ADS (ODM-ADS). The method is adapted to detect malicious nodes and proposed instances in the training phase, and creates a statistical profile, checks for dissimilarities in the starting point of the pattern. This method is suited for large scale networks.

M. Hasan et al. (2019) [3] several machine learning models are compared based on performances to predict irregularities on the IoT systems exactly. The machine learning (ML) algorithms used are Logistic Regression (LR), Support Vector Machine (SVM), Decision Tree (DT), Random Forest (RF), and Artificial Neural Network (ANN).

T. Nguyen et al. (2019) [4] Proposed new SDN-based NIDS architecture that improved system resources and path selection optimizations. The architecture generated exceptional performance in anomaly discovery, mitigation and management of bottleneck problem in the SDN-based cloud IoT networks.

E. Anthi et al. (2019) [5] Proposed a three-layered supervised approach for IDS. The system classifies the attack type. Identify the malicious packets and classify the nature of the attacks. The system is assessed on a smart home test bed with eight devices and 12 attacks from four main network-based attack categories and against four scenarios of attacks with a multifarious chain of events.

B. Sudqi Khater et al. (2019) [6] ADFA-LD dataset is used on MLP single hidden layer using Raspberry Pi, which performances as the Fog device. The MLP single hidden layer is tested with two nodes, three nodes, four nodes, five nodes, and six nodes. The experiment conducted found out that the MLP single hidden layer has three nodes with 120 2-gram & 80 1-gram achieved noticeable results for accuracy, recall rate, F1 measure.

H. Alaiz-Moreton et al. (2019) [7] Deep Learning and ensemble methods are used to classify attacks, especially on MQTT protocol using dataset feed into IDS and achieved better results using deep learning models.

N. Chaabouni et al. (2019) [8] discussed about traditional and existing NIDS implementation tools using machine learning techniques and datasets. A comparison is done on the architectures, detection methods, and deploying aspects of learning techniques for IoT.

S. Jan et al. (2019) [9] Support Vector Machine (SVM) is used to identify data that compromises the IoT network. Simulation results are used to evaluate the performance using simplex features that determine accuracy and detection time.

Y. Wang et al. (2018) [10] A signature-based intrusion detection method is proposed to preserve privacy, reduce the load on cloud side with a quick response to the detected attacks. This method is demonstrated on simulation as well as real-time.

A. Abeshu et al. (2018) [11] A distributed Deep Learning scheme is proposed for detecting cyber-attacks in fog- to-things computing consists of a master node and worker node. Training data is local to the worker node that gets initialized parameters from master node. The gradients of Deep Learning and updating of parameters using optimizers are done on worker node and aggregated to master node. The malicious events are detected locally and updated to the master node.

X. An et al. (2018) [12] Sample Selected Extreme Learning Machine was suggested. The data is calculated and sampled by cloud servers. The limitation of storage capacity in IoT devices only selected samples are given to fog nodes for training and detected attacks on the fog nodes. Performance metrics are evaluated using accuracy, training time, and ROC value.

A. Diro et al. (2018) [13] LSTM network is used for detecting distributed cyber-attack in fog-to-things communication. The main aim of the system is to analyze Critical attacks and threats targeting IoT devices and identify attacks that are vulnerabilities of wireless communications. Two scenarios are taken into consideration to exhibit the effectiveness and proficiency of deeper models.

S. Prabavathy et al. (2018) [14] A distributed architecture is proposed to detect attacks at a faster rate using Online Sequential Extreme Learning Machine (OS-ELM) to understand attacks from the IoT flow. The distributed architecture enabled scalability, flexibility, and interoperability.

Y. Meidan et al. (2018) [15] MIRAI and BASHLITE botnets are used to initiate attacks and N-BaIoT; the proposed anomaly detection method is employed using auto encoders to detect anomalous traffic from compromised IoT devices. The evaluation results demonstrated the ability to accurately and instantly detect the attacked devices that were part of a botnet.

C. McDermott et al. (2018) [16] developed the BLSTM-RNN detection model that is related with LSTM-RNN using Mirai botnet for discovering four attack vectors and evaluated for accuracy and loss. The model demonstrated to be progressive while putting an overhead of computation. This research created a labelled dataset as a part of their work.

Deng L et al. (2018) [17] Proposed a method that combines the FCM algorithm and PCA algorithms with Intrusion Detection scheme to provide an efficient way to lower false-positive rate.

Amouri A et al. (2018) [18] A two-stage is proposed by providing functions that prohibit direct access to data. CCI is collected for every transmission to check the variations in network behavior, which helps to determine between normal and malicious node after AMoF sample.

Liu L et al. (2018) [19] The proposed method used suppressed fuzzy clustering algorithm and the principal component analysis algorithm to classify data into high-risk data and low-risk data, showed better adaptability when compared to the traditional method.

R. Stephen et al. (2017) [20] proposed IDS detects sinkhole attack in the network, which uses the RPL as a routing protocol and detection metrics. This technique identifies the malicious node using the IR value. If the IDS system detects the malicious node, it sends the alert message to the leaf nodes to isolate the malicious node in the next data transmission.

Bostani H et al. (2017) [21] Selective forwarding and sinkhole attacks are detected using an actual hybrid intrusion detection structure that decides suspicious conduct by using a elective mechanism. The hybrid model, which is anomaly-based and specification-based, is deployed and investigated in a smart-city scenario.

Fu Y et al. (2017) [22] An automata model is proposed heterogeneous IoT networks based on that detects and reports the possible IoT attacks. Jam-attack, false-attack, and reply-attack are the three attacks that can be automatically detected.

Khan ZA et al. (2017) [23] the proposed method is dedicated to healthcare applications, designed a trust management mechanism that allows managing information about the neighbors periodically, which 0helps to identify malicious nodes in the neighborhood.

M. Lopez-Martin et al. (2107) [24] Conditional variational encoders are utilized to provide a less complicated method that performs feature reconstruction, which recovers missing features of the dataset. The reconstruction mechanism has higher accuracy. The unique algorithm helps to reconstruct the dataset that would be useful for instruction detection and improves the efficiency in detecting attacks.

H. Pajouh et al. (2016) [25] Proposed an Anomaly-based two-tier dimension reduction and classification model to detect attacks utilizing Naive Bayes and Certainty Factor version of K-Nearest Neighbor. NSL-KDD dataset is utilized to identify suspicious behavior.

Le A et al. (2016) [26] Proposed a method that records the RPL Information to eliminate delay in transmitting the report. This method detects RPL topology attacks with high accuracy by minimizing the overhead and enabling scalability.

3 Comparative Study

Table 1 categorizes the machine learning, Deep Learning and data mining approaches that can be detected by the various authors. Numerous methods of machine learning approaches with dissimilar datasets are given below and Table 2. Categorizes the performances of the IDs's.

4 Discussion

The exploration of machine learning approaches for intrusion detection has continued from the beginning of this decade [27]. A lot of simulation research is continuing to improve the security and computational overhead on the IoT devices. The substantial characteristics of the IoT system and massive data labelling for machine learning recommends intrusion detection mechanisms at the device level and at the edge routers that gain effective performance. The functionalities at the device level and router level of IDS are separated to reduce the computational overhead caused due to intrusion detection mechanisms. Alerts are generated to the edge router by device-level components on detection of intrusions and edge router report to the administrators about the patterns and mechanisms of threat.

5 Conclusion

An Intrusion Detection System (IDS), as well as Intrusion Prevention System (IPS), is required that not only detects various existing attacks but also prevent attacks by mitigating and involving very less computational overhead. The system should be adaptable to multiple protocols used in WSNs. The placement of IDS also plays an essential role in improving efficiency and scalability. The distributed technique is advantageous than the centralized procedure as the load is distributed and reduces the amount of traffic monitored by increasing the processing speed.

Table 1 Summary of some data mining and machine learning approaches

Reference	Method	Dataset	Highlight
Moustafa et al. [2]	Outlier Dirichlet mixture (ODM)	(1) NSL-KDD (2) UNSW-NB15	(1) Identify zero-day attacks (2) Can self-adapt against data poisoning attacks
Hasan et al. [3]	(1) Logistic Regression (2) Support Vector Machine (3) Decision Tree- (4) Random Forest- (5) Artificial Neural	(1) Open source dataset from kaggle	(1) Identifies DoS, Probing, Malicious Control, Malicious Operation, Scan Spying, Wrong Setup
Nguyen et al. [4]	(1) Support Vector Machine, (2) Self-Organizing Map (3) Stacked Auto Encoder	(1) CAIDA (2) KDD-CUP 99 (3) UNSW-NB15	(1) Identified DDoS attack using CAIDA (2) Identified Probe, DoS, U2R, R2L using KDD-CUP99 (3) Detected Fuzzers, Backdoor, Dos, Exploits, Generic, Reconnaissance, shellcode, worms using UNSW-NB15
Anthi et al. [5]	(1) Naïve Bayes (2) Bayesian network, (3) J-48 (4) Zero R (5) One R, (6) Simple Logistic (7) Support Vector Machine (8) Multi-Layer Perception (9) Random Forest	(1) Generated from 4 IoT devices	(1) Detects and classifies the attack
Sudqi Khater et al. [6]	(1) Multilayer Perceptron (MLP) model	(1) ADFA-LD (2) ADFA-WD	(1) Latest datasets (2) Realtime Evaluation

(continued)

Table 1 (continued)

Reference	Method	Dataset	Highlight
Abeshu et al. [11]	(1) Stacked Auto Encoders as unsupervised Deep Learning	(1) NSL-KDD	(1) Improved efficiency than shallow models
Diro et al. [13]	(1) LSTM-RNN(Long Short Term Memory-Recurrent Neural Network)	(1) ISCX (2) AWID	(1) Discriminates algorithmically generated domains from normal URLs (2) Resilient to attacks (3) Improved scalability (4) Better results with ISCX dataset
Prabavathy et al. [14]	(1) Online Sequential Extreme Learning Machine (OS-ELM)	(1) NSL-KDD	(1) Detects DoS, Probe, R2L and U2R attacks (2) Improved response time and detection accuracy
Meidan et al. [15]	(1) 4 layers of Auto encoders	(1) Generated by IoT devices	(1) Used Bashlite and Mirai botnets (2) Detects Scan, TCP flooding, UDP flooding, SYN flooding, Junk
McDermott et al. [16]	(1) BLSTM-RNN	(1) Generated by IoT devices	(1) Detects UDP flood, ACK flood, DNS flood, SYN flood
Deng et al. [17]	(1) K-means clustering algorithm at $k = 12$	(1) KDD-CUP 99	(1) Matlab simulation platform (2) Lightweight (3) Finding efficiency with a low FPR

(continued)

Table 1 (continued)

Reference	Method	Dataset	Highlight
Liu et al. [19]	(1) suppressed fuzzy clustering (SFC) algorithm (2) Principal Component Analysis (PCA) algorithm	(1) Generated	(1) Adaptability to high dimension spaces (2) Reduces detection time
Hosseinpour et al. [28]	(1) Artificial Immune System (AIS)	(1) KDD-CUP 99 (2) SSH Brute force	(1) Detect zero-day attacks and existing Dos, DDos, Probe, U2R (2) Real-time detection (3) Light-weight IDS
Pajouh et al. [25]	(1) Naive Bayes (2) CF-KNN classifier	(1) NSL-KDD	(1) Detects Dos, Probe, R2L, U2R attacks (2) Best performance on R2L and U2R attacks (3) Two-tier classification

Table 2 A comparison of surveyed IDSs performances

Reference	Detection rate (%)	FPR (%)	Accuracy (%)	STD	Precision (%)	Recall (%)	F1 (%)	Energy (J)	Processing time (ms)
Moustafa et al. [2]	99.89	0.21							
Hasan et al. [3]			99.4	0.014	0.99	0.99	0.99		
Nguyen et al. [4]	95.5		95.5						7.0 ms
Anthi et al. [5]					99.0	99.0	99.0		0.2 ms
Sudqi Khater et al. [6]			94			95	94	0.0014	
Abeshu et al. [11]	99.27	0.85							
An et al. [12]			99.7						
Diro et al. [13]			99.91		99.85	99			
Prabavathy et al. [14]	97.72	0.37	97.36						

References

1. Gartner.com (2019) [Online]. Available: https://www.gartner.com/imagesrv/books/iot/iotEbook_digital.pdf. Accessed 29 Oct 2019
2. Moustafa N, Choo K, Radwan I, Camtepe S (2019) Outlier Dirichlet mixture mechanism: adversarial statistical learning for anomaly Detection in the fog. IEEE Trans Inf Forensics Secur 14:1975–1987
3. Hasan M, Islam M, Zarif M, Hashem M (2019) Attack and anomaly detection in IoT sensors in IoT sites using machine learning approaches. Internet Things 7:100059
4. Nguyen T, Phan T, Nguyen B, So-In C, Baig Z, Sanguanpong S (2019) SeArch: a collaborative and intelligent NIDS architecture for SDN-based cloud IoT networks. IEEE Access 7:107678–107694
5. Anthi E, Williams L, Slowinska M, Theodorakopoulos G, Burnap P (2019) A supervised intrusion detection system for smart home IoT devices. IEEE Internet Things J 6:9042–9053
6. B. Sudqi Khater, A. Abdul Wahab, M. Idris, M. Abdulla Hussain, A. Ahmed Ibrahim, A lightweight perceptron-based intrusion detection system for fog computing. Appl Sci 9 (2019) 178
7. Alaiz-Moreton H, Aveleira-Mata J, Ondicol-Garcia J, Muñoz-Castañeda A, García I, Benavides C (2019) Multiclass classification procedure for detecting attacks on MQTT-IoT protocol. Complexity 2019:1–11
8. Chaabouni N, Mosbah M, Zemmari A, Sauvignac C, Faruki P (2019) Network intrusion detection for IoT security based on learning techniques. IEEE Commun Surv Tutor 21:2671–2701
9. Jan S, Ahmed S, Shakhov V, Koo I (2019) Toward a lightweight intrusion detection system for the internet of things. IEEE Access 7:42450–42471
10. Wang Y, Meng W, Li W, Li J, Liu W, Xiang Y (2018) A fog-based privacy-preserving approach for distributed signature-based intrusion detection. J Parallel Distrib Comput 122:26–35
11. Abeshu A, Chilamkurti N (2018) Deep learning: the frontier for distributed attack detection in fog-to-things computing. IEEE Commun Mag 56:169–175
12. An X, Zhou X, Lü X, Lin F, Yang L (2018) Sample selected extreme learning machine based intrusion detection in fog computing and MEC. Wirel Commun Mobile Comput 2018:1–10
13. Diro A, Chilamkurti N (2018) Leveraging LSTM networks for attack detection in fog-to-things communications. IEEE Commun Mag 56:124–130
14. Prabavathy S, Sundarakantham K, Shalinie S (2018) Design of cognitive fog computing for intrusion detection in internet of things. J Commun Netw 20:291–298
15. Meidan Y, Bohadana M, Mathov Y, Mirsky Y, Shabtai A, Breitenbacher D et al (2018) N-BaIoT—Network-based detection of IoT botnet attacks using deep autoencoders. IEEE Pervasive Comput 17:12–22
16. McDermott C, Majdani F, Petrovski A (2018) Botnet detection in the internet of things using deep learning approaches. In: 2018 international joint conference on neural networks (IJCNN), pp 1–8
17. Deng L, Li D, Yao X, Cox D, Wang H (2018) Mobile network intrusion detection for IoT system based on transfer learning algorithm. Clust Comput 21:1–16
18. Amouri A, Alaparthy VT, Morgera SD (2018) Cross layer-based intrusion detection based on network behavior for IoT. In: IEEE 19th wireless and microwave technology conference (WAMICON). IEEE, Sand Key, pp 1–4
19. Liu L, Xu B, Zhang X, Wu X (2018) An intrusion detection method for internet of things based on suppressed fuzzy clustering. EURASIP J Wirel Commun Netw 1:113
20. Stephen R, Arockiam L (2017) Intrusion detection system to detect sinkhole attack on RPL protocol in internet of things. Int J Electr Electron Comput Sci 4(4):16–20
21. Bostani H, Sheikhan M (2017) Hybrid of anomaly-based and specification-based IDS for internet of things using unsupervised OPF based on MapReduce approach. Comput Commun 98:52–71

22. Fu Y, Yan Z, Cao J, Ousmane K, Cao X (2017) An automata based intrusion detection method for internet of things. Mob Inf Syst 2017:13
23. Khan ZA, Herrmann P (2017) A trust based distributed intrusion detection mechanism for internet of things. In: 2017 IEEE 31st international conference on advanced information networking and applications (AINA). IEEE, Taipei, pp 1169–1176
24. Lopez-Martin M, Carro B, Sanchez-Esguevillas A, Lloret J (2017) Conditional variational autoencoder for prediction and feature recovery applied to intrusion detection in IoT. Sensors 17:1967
25. Pajouh H, Javidan R, Khayami R, Dehghantanha A, Choo K (2019) A two-layer dimension reduction and two-tier classification model for anomaly-based intrusion detection in IoT backbone networks. IEEE Trans Emerg Topics Comput 7:314–323
26. Le A, Loo J, Chai KK, Aiash M (2016) A specification-based IDS for detecting attacks on RPL-based network topology. Information 7(2):1–19
27. Bace R, Mell P (2001) Intrusion detections systems, National Institute of Standards and Technology
28. Hosseinpour F, Vahdani Amoli P, Plosila J, Hämäläinen T, Tenhunen H (2016) An intrusion detection system for fog computing and IoT based logistic systems using a smart data approach. Int J Digital Content Technol Appl 10(5):34–46
29. Cha H, Yang H, Song Y (2018) A study on the design of fog computing architecture using Sensor networks. Sensors 18(11):3633

Unusual Behavior Analysis of Virtual Machine in Cloud Environment

S. Nivetha, M. Saravanan, and V. Lavanya

Abstract Cloud computing is a model for enabling the resources to access over the internet, rather than managing files on local storage. Virtual machine is the abstraction of hardware of a distinct Personal Computer into numerous diverse implementing environments. Virtual machines are also known as virtual servers, full-fledged computers, but it is all virtual or simulated, not physical. They have operating systems, files and applications. Cloud service provider offers virtual machine to be used by any user. So, any user can gain control of virtual servers. Unusual behavior of virtual machine may occur in the cloud environment. The proposed system identifies the virtual machine theft attack, which discusses about Spectral Analysis and Energy Calculation which avoids the attackers to stay in the Virtual Machine for a long Period and paying less amount for the service time providers. Hence the total estimation time and Energy might be calculated to overcome the loss for actual time. We proposed and implemented using JAVA software with an API, which identifies and predicts the virtual machine theft attack by Spectral Analysis of Virtual Machine Activities and Energy Calculation.

Keywords Cloud computing · Virtual machine theft · K-means clustering · Spectral analysis of VM · Energy calculation

S. Nivetha · M. Saravanan (✉) · V. Lavanya
SRM Institute of Science and Technology, Chennai, India
e-mail: saran84gct@gmail.com

S. Nivetha
e-mail: Shreenivi1016@gmail.com

V. Lavanya
e-mail: lavanyav@srmist.edu.in

D. J. Hemanth et al. (eds.), *Artificial Intelligence Techniques for Advanced Computing Applications*, Lecture Notes in Networks and Systems 130,
https://doi.org/10.1007/978-981-15-5329-5_23

1 Introduction

1.1 Cloud Computing

Cloud computing [1] is the distribution of on-demand computing facilities over the Internet on a wage as you go basis, e.g., Webmail. Somebody's server is used by the cloud for hosting, processing and for collecting the data. It is storage, computation, network-based and widely distributed. Cloud is nothing but we can access at anytime, anywhere, with any device and any service can be accessed. The evolution of cloud computing is as follows, i.e., grid computing solves great complications by parallel computing. It is completed with mainstream through Globus Alliance. Utility computing is introduced in the late 1990s. It offers computing assets as a metered facility. Software as a service is Web-based payments toward applications. SaaS gained momentum in 2001. After 2001, cloud enters the market. Cloud computing is the next-generation Internet computing as well as next-generation [2] datacenters. Internet computing is for accessing data based on subscriptions, usage and requirement and datacenters are cast-off toward storing different organizations records in mist with the help of datacenters. Cloud computing is utilized because it is paid for what you use by scale up and scale down process, no server space is required, automatic software updates, no experts are required for hardware and software maintenance, high flexibility, data can be accessed and shared anywhere over the internet, rapid implementation and better data security.

Cloud computing provides both deployment models and service models [3]. Deployment models consist of public, private, hybrid and community cloud. A private cloud is a cloud infrastructure created and operated by a single organization. Although some might think that the private clouds are not really cloud, in the same sense, several companies run their own private cloud to get all the accomplishes while maintaining full control over their infrastructure. Public cloud is common model of doing cloud computing provided with open access to everyone. Consumers can specify the geographic region in which it resides to aid with data access latency based on the end users location. Providers like Amazon, Google and Microsoft are the important players in the field and their solutions are widely used and appreciated. Hybrid cloud is a mixture of both private and public cloud, e.g., SRM University is providing materials through email to students. Here, SRM wants to store its data to a dedicated storage is called private cloud, and the students accessing through email is called public cloud. Community cloud is a cloud infrastructure created and operated by a group of communities. Cloud applications can be run according to three different service models under SaaS, PaaS and IaaS. The top-level cloud computing model is SaaS. G-mail, salesforce are the example of software as a service. We get a complete software package ready to use on demand and a month of our yearly fee. However, we cannot customize it more than the provider allows. The second level is PaaS. Services like Cloud Foundry or Google App Engine are two common examples of PaaS which is platform as a service. With a PaaS, we can deploy our own application and take care of it while our provider will manage all the underlying infrastructure aspects. PaaS is an intermediate level between SaaS. IaaS or infrastructure [4] as a service is the most basic service model. In this scenario, we have full control over our infrastructure and we are provided with resources such as power supply, network connections, load balancers, firewalls, blocker and object storage. e.g. Amazon web service [5], Rackspace. Benefits of cloud computing [6] include reduced investments, increased scalability and increased reliability and availability.

1.2 Virtual Machine

Virtual machine [7] is a software computer that is stored on a physical hard drive. Virtualization is a capability of running several operational systems taking place in a particular physical system. It is a method in which a single computer multitudes the presence of the several computer. It makes a life of a software developers much easier by streamlining the process to deploy software. They have operating systems, files and applications. They do not necessarily live inside a physical server on a one-to-one ratio. In fact, several virtual machines can live within a single physical server. To the outside world, there is really no difference between a physical server and a virtual server. They both can serve up websites, host applications, and contain data. You can spin up a virtual machine on a desktop PC, on a server, or in the cloud. Both servers and virtual machines run operating systems and applications, serve up websites and respond to pings, and they can be networked. In contrast to servers, virtual machines do not have a one-to-one relationship to hardware. You can run several virtual machines on a single piece of equipment. They allow you to add hard drive space or memory much more easily. You do not have to buy a new server when you run out of space. They boot faster and easier to move. They require more processing power and memory. This is especially true for hosted virtual machines. They are easier to troubleshoot, and they have different requirements for security. Hypervisor is a software which manages the virtual device and permits for running an OS inside the additional OS (Fig. 1).

1.3 Threats

Remote access enables users to access the system through Internet [8] connection. So, anyone can misuse the virtual machine by stealing the data. A virtual atmosphere inside a virtual machine VM hosts is generated when a hacker proceeds mischievous control [9] above the hypervisor, results in hyperjacking attack. Targeting operating systems which is under the virtual device is the aim of a Bout. Accordingly, intruders can run their database and the requests on the VMs overhead will be entirely ignorant to its existence. The complete server method is managed, when hyperjacking implicates mounting a mischievous false hypervisor [3]. Hypervisor exactly functions in secrecy that goes underneath the device; hence, it is further challenging for

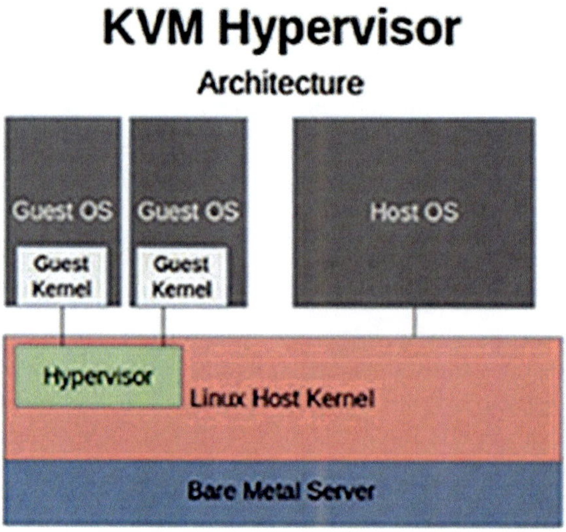

Fig. 1 KVM-QEMU architecture

detection and achieving right to use the server system which results in disturbing the complete institution process. In VM escape [10], the guest OS will escape from VM and interact with hypervisor in unauthorized manner, which makes the attackers to easily access all VMs. Server configuration changes and comprised hypervisor results with resource exhaustion.

1.4 Characteristics of Virtual Machine

1. Security Isolation: It occurs in shared environment relied on physical separation to enforce security. Access control mechanism and encryption play a key part in security isolation. Existing security policies may need to be re-written.
2. Resource Isolation: Virtualized environment which uses dedicated hardware to guarantee access to resource [11]. Shared environments allow better use of separate capacity. Workloads should still be guaranteed minimum resource level.
3. Workload Efficiency: Virtualization usually introduces overhead, i.e., hypervisor overhead, virtualized resource overhead, o/overhead. Virtualization is sometimes used to address scalability issues either at the hardwires or application level.
4. Availability: Consolidated workloads suffer a higher impact due to infrastructure failure. An individual work within the pool may not require high availability (HA), but the combination of several workloads does.
5. Serviceability: Here, consolidated workloads suffer a higher impact due to infrastructure maintenance. Negotiating a joint outage (system power off) window

across multiple business owners can be extremely difficult. The number of service events that require that require an outage should be as low as possible. In case, where an outage is unavoidable, workloads should be able to move with minimum of disruption.
6. Flexibility: Workloads typically have resource utilization patterns that vary overtime. They may also grow or shrink over a longer periods of time. It is therefore important to be able to alter the resource allocation of these workloads. Ideally, these should be automatic, dynamic and immediate, so that idle workloads do not reserve resources, and busy workloads can get them.
7. Agility: In a consolidated infrastructure, it is often useful to move workloads between servers within a resource pool. Agility is a measure of workload migration between physical servers and takes into account as ease/simplicity of migration, impact of migration, speed of migration.

1.5 Unusual Behavior of Virtual Machine

Unusual characteristics are as follows, i.e., control may take over the virtual machine. When your values are modified, we can notice that our integrity of data is at risk. Integrity is all about no unauthorized change to data or information/no modification/validity.

Web server attacks [12] targets networks, OS, also the Web server software and the applications that are used on website. Web servers have the added vulnerability of being accessible from the internet. Brutal trying of other systems in network (server hacking) may happen.

Malware (malicious scripts, Trojans, etc.) may be used to infect the server and its applications. System attacks may take advantage of weak passwords, missing patches, or other built-in OS flaws and application vulnerabilities. Network attacks [13] could be launched against machine, resulting in a DOS condition. Application attacks target the website and database services. Its goal is to access or alter the data. Attackers will try to enter into server side.

Intrusion exists that attempts for intruding into confidentiality of the network. Masquerader, misfeasor and clandestine are different types of intruders. In masquerader, consumer without authority will access the system. It penetrates the security system as a legitimate user. Misfeasor user is legitimate user but misuses the privileges. Clandestine user attempts for stealing and using the IDs of their supervisor. Intrusion aims to compromise the security ghost of an organization.

2 Literature Review

T Safrir et al. [14] created a demonstrate for "Cheat attack" with a rate for the CPU required for the organization. A few tests for a few forms speak to the utilization of non-usage of central processing unit.

Zhou et al. [15] revealed the implementation for the attackers with shares of CPU in VM. Some implementation required for an attacker, uses the share of schedule process, with an end never schedules for CPU services. Some theft of services attacks might be implemented in the smart phones with the different values resources in the knowledge for the users. Some several cases might be studied required for the provider the comparative analysis for KVM with respect to the hypervision for different activities.

Zhou et al. [15] developed with the implementation for the theft of services attacks in Xen and the implementation for KVM in the studies. The work might be maintained for the theft of the attackers in modifying the CPU clock. Our proposed advance of activity is arresting in such an address area it requires no about-face about at CPU's active associated analogously as can be accomplished on any billow with no change at compassionate level.

Azeem Ahmad et al. [16] proposed a Documentation and Preclusion of Burglary-of-Service Paper in fog security, proposes the kernel of virtual machine with QEMU emulator, which identifies the VM theft attack by calculating the API-based power consumption of the user. When the values are modified in Virtual Machine it is noticed that our integrity of data is at risk. Integrity is all about no unauthorized change to data or Information/no Modification/validity. L.N.B.Srinivas and Subburaj Ramasamy [17] proposed a method for Missing of data by Wireless Sensor Network.

3 Proposed Methodology

Proposed method is conducted with hundreds of VMs in primary cloud. We considered VM theft as an example to identify and predict the unusual behavior of virtual machine. In this paper, we implemented a spectral analysis of virtual machine activities and energy calculation method to avoid the attacker from using the VM a long while and paying very less amount to the service provider. The cost average value and time average value of each user of VM are considered, by which average time of virtual machine will be calculated. Every task which is processed by the particular VM will log the running time to that particular user and the total average time will be calculated to overcome the loss of actual time. The proposed clustering and MEWMA algorithm will be used to find out the service theft which is done by the attacker by grouping the VM into separate gangs. The formed gangs are responsible for the service theft and each and every gang is well formulated to find out the service theft. If the average time of vm exits, service theft is detected and rectified by setting maximum and minimum time limit for user.

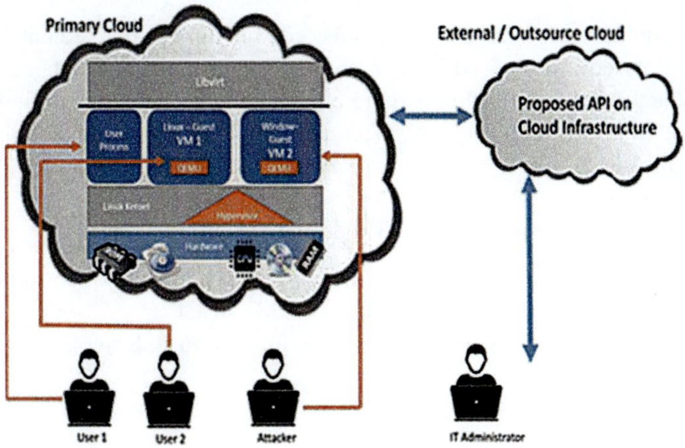

Fig. 2 Represents vulnerable cloud infrastructure

3.1 Proposed Architecture

The proposed architecture discusses the kernel-based virtual machine (KVM [18]); it is a type1 hypervisor which is built into kernel of Linux. Quick EMUlator virtual machine format (QEMU) is a type 2 hypervisor that runs within user space and performs virtual hardware emulation. The KVM module creates a bare metal hypervisor on the Linux kernel. Virtual machines can be loaded onto this hypervisor, running separate OSes.

Figure 2 represents vulnerable cloud infrastructure which consists of primary cloud and out-source cloud. VM 1 is used by users and VM 2 is compromised by the attacker. Attackers are using the VM a long while and paying very less amount to the service provider. The proposed solution is based on two steps, (1) spectral analysis of virtual machine activities and (2) energy calculation of virtual machine user.

3.2 Proposed Algorithm

K-means clustering and MEWMA algorithm is used in this project. Sorting of an items obsessed by different groups is called as clustering. Separate dataset into subgroups, therefore collective information of subgroups prefers for sharing some mutual characteristic according to exact distance measure. Grouping objects created by its features obsessed by K number of group. K is a positive integer number [19]. Combination is through reducing the amount of squares of distances among data and the equivalent cluster centroid.

MEWMA [20] is also a clustering-based algorithm. We used this algorithm in our project to find out the VMs utilized average time beyond the max time limit.

While execution of task in all the VMs is recorded in our pre-defined database and further the dataset is processed by our proposed algorithm. The obtained results show that the proposed algorithm is more efficient and accurate one to find out the VMTHEFT. Henceforth, the values are plotted into a graph notation and displayed to the administrated.

Steps of algorithm

Step 1: Spectral Analysis of Virtual Machine (Execution Time)

- **Input**: A dataset of items P = {P1, P2,………,Pn}, a number of clusters k.
- **Output**: Find time limit exceeds, to predict VMTHEFT
- Centers {C1, C2,………,Cn} implicitly dividing P into k clusters.
- Initialize the dataset
- Divide the dataset into "**k**" subgroups
- K← is positive integer
- Choose k initial centers **$C = \{C_1, \ldots\ldots, C_n\}$**
- **While** stopping criterion has not been met
- **do**→ assignment step:
- **for** i = 1,……..,N
- **do** find closet center c_k $\dot{\varepsilon}$ C to instance p_i
- Assign instance p_i to set C_k.
- → update step:
- **End for**
- **for** i = 1,……..,k
- **do** set allocation job execution time
- **Average Time VM_i= Tmin – Tmax**
- **Cost x_i= Tavg**
- **Cost y_i= Tavg**
- **End for**
- Predict→VMTHEFT based on maximum and minimum time limit exceeds.
- **return** clusters

Step 2: Energy Consumption

- **Input**: Average time of virtual machine.
- **Output**: To calculate the energy consumption of virtual machines.
- Initialize the virtual machine job(**j**)
- **While do→** load job:
- **for** i = 1,……..,N
- **do** find closet execution time
- Execution Time E_i = Start time – End Time.
- Energy = W * S;
- W← Amount of work Done and S← Slab Capacity
- **End for**
- Calculate→Energy consumption of virtual machine.
- **End**.

4 Experiment and Results

An host device, functioning by means of private cloud, comprises of Windows 95/98/2000/NT4.0 using Linux kernel 4.1.0 as well as QEMU emulation software on Pentium—III CPU with Partial RAM 256 MB(min). Cloud simulation module has been used for simulation. In this experiment, hundred different virtual machines are conducted in primary cloud to monitor the energy consumption and spectral analysis of virtualized and host machine. We simulated VM theft occurrence as well as manipulate variables that stored in virtual machine energy depletion indicators, hence invader claim of no usage of VM. Average time of each VM is fixed by calculating the average cost period and average time period of each VM. Minimum and maximum time limit has been considered for users to identify attackers. Analyze time utilization by calculating the work of past span of time in virtual machine.

If user exits maximum time limits, then VM theft has been detected. Utilization of energy can also be calculated by the amount of work done multiplied by slab capacity.

(1) Virtual machine time and energy calculation:
    Execution Time = Start Time – End Time;
    Energy = Amount of Work Done * Slab Capacity;
    Max = 10000 unit.
(2) Average time details:
    VM1 = Tmin - Tmax
    VM2 = Tmin – Tmax
    Cost x = Tavg
    Time y = Tavg
    MAX time = 50(min)
    MIN time = 20(min)
    VMTHEFT is detected, when execution time of VMs differs with allocated job execution time. VMTHEFT is rectified, by setting maximum and minimum time limit for execution of job (Figs. 3 and 4).

5 Conclusion

Cloud computing enables us to utilize high end resources so we build great applications without worrying about the infrastructure. Virtual machine is the traditional physical architecture model in which operating systems and applications are installed, i.e., may be a Web server, email service. Attacker takes malicious control of virtual machine for his own personal gain. We explored limitations of KVM hypervisor with QEMU emulation method to prevent stealing-of-VM occurrence that permits

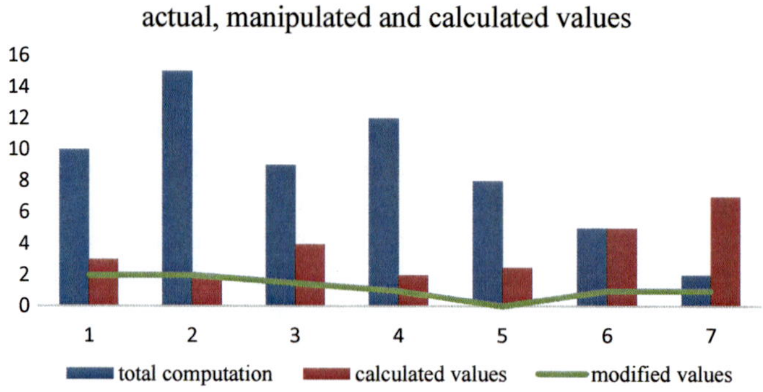

Fig. 3 Spectral analysis

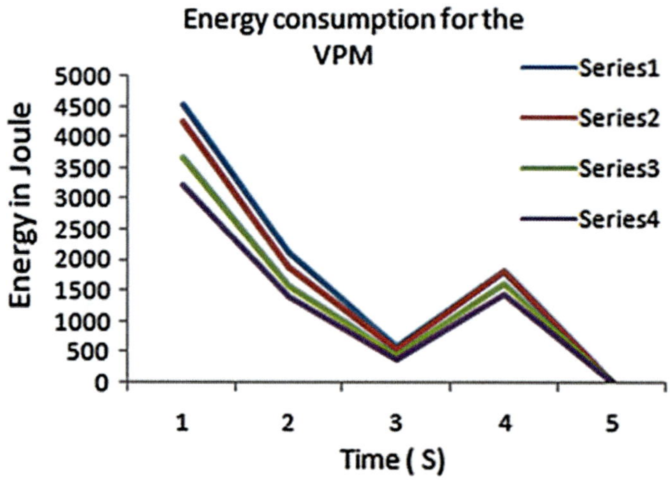

Fig. 4 Energy calculation

the operator for using VM for an extended period. We planned and executed an API, which can find and prevent theft-of-service attack by Spectral Analysis of Virtual Machine Activities and Energy Calculation as well as alert the administrator during attack.

References

1. Nycz M, Polkowski Z (2015) Cloud computing in government units. In: 2015 fifth international conference on advanced computing communication technologies (ACCT), 2015, pp 513–520
2. Zissis D, Lekkas D (2012) Addressing cloud computing security issues. Future Generation Comput Systems 28(3):583–592

3. Bazargan, Yeun F, Zemerly C, Jamal (2013) State-of-the-Art of virtualization, its security threats and deployment models. Int J Inf Security Res 3, https://doi.org/10.20533/ijisr.2042.4639.2013.0039
4. Althobaiti AFS (2017) Analyzing security threats to virtual machines monitor in cloud computing environment. J inf Security 08 (01):1–7
5. AWS| Amazon Elastic Compute Cloud (EC2)—Scalable Cloud Hosting.[Online]. http://aws.amazon.com/ec2/. Accessed 21 Sept 2015
6. Srivastava JP, Verma VK (2015) Cloud computing in libraries: its needs, applications, issues and best practices. In: 2015 4th international symposium on emerging trends and technologies in libraries and information services (ETTLIS), 2015, pp 33–38
7. Kim H, Lim H, Jeong J, Jo H, Lee J (2009) Task-aware virtual machine scheduling for I/O performance. In: ACM VEE
8. Choubey R, Dubey R, Bhattacharjee J (2011) A survey on cloud computing security, challenges and threats. Int J Comput Sci Eng (IJCSE) 3(3):1227–1231
9. Talbot D (2009) Vulnerability seen in amazon's cloud computing. MIT Tech Review, October 2009
10. Tank D, Aggarwal A & Chaubey N (2019) Virtualization vulnerabilities, security issues, and solutions: a critical study and comparison. Int j inf tecnol https://doi.org/10.1007/s41870-019-00294
11. Cherkasova L, Gupta D, Vahdat A (2007) When virtual is harder than real: resource allocation challenges in virtual machine based IT environments
12. Seo J, Kim H, Cho s and Cha s (2004) "Web server attack categorization based on root causes and their locations," International Conference on Information Technology: Coding and Computing, 2004. Proceedings. ITCC 2004. Las Vegas, NV, USA, pp. 90–96, https://doi.org/10.1109/ITCC.2004.1286431
13. Qinquan W, ZaiLin P (2010) "Research on Network Attack and Detection Methods," 2010 Second International Workshop on Education Technology and Computer Science, Wuhan, pp. 630–633, https://doi.org/10.1109/ETCS.2010.196
14. Tsafrir D, Etsion Y, Feitelson DG (2007) "Secretly Monopolizing the CPU Without Superuser Privileges," in Pro-ceedings of 16th USENIX Security Symposium on USENIX Security Symposium, Berkeley, CA,USA, pp. 17:1–17:18
15. Zhou F, Goel M, Desnoyers P, Sundaram R (2010) Scheduler vulnerabilities and coordinated attacks in cloud computing. In: Northeastern Technical Report, 2010
16. Ahmad A, Nasser N, Anan M (2016) "An Identification and Prevention of Theft-of-Service Attack on Cloud Computing," in 2016 International Conference on Selected Topics in Mobile & Wireless Net-working (MoWNeT)
17. Srinivas LNB, Ramasamy S (2017) "An Improvized Missing Data Estimation Algorithm for Wireless Sensor Network Applications". J Adv Res Dynamical and Control Systems 9(18):pp. 913–918
18. "What is KVM | Open Virtualization Alliance." [Online]. Available: https://openvirtualizationalliance.org/what-kvm. [Accessed: 22-Sep-2015]
19. Na S, Xumin L, Yong G (2010) "Research on k-means Clustering Algorithm: An Improved k-means Clustering Algorithm," 2010 Third International Symposium on Intelligent Information Technology and Security Informatics, Jinggangshan, pp. 63–67, https://doi.org/10.1109/IITSI.2010.74
20. Huda, Abawajy S, Alrubaiey J, Pan B, Hassan L, Mohammad (2019) Automatic extraction and integration of behavioural indicators of malware for protection of cyber-physical networks. Future Generation Comput Syst https://doi.org/101.10.1016/j.future.2019.07.005

Feature Selection Techniques for Disease Diagnosis System: A Survey

G. Saranya and A. Pravin

Abstract Reducing dimensionality as a preprocessing phase towards artificial intelligence effectively eliminates unnecessary and repetitive information, raises the efficiency of learning accuracy, and improving output understandability. Nonetheless, the rapid surge in data dimension presents a real challenge to several current collections of features and extraction methods as regards efficiency and effectiveness. Many researchers build and use plenty of feature selection algorithms. But an emerging field of machine learning is still to be based on data mining and method of analysis for information processing. Due to the recent increase in data variation and speed, many feature selection algorithms face serious efficiency and performance problems. Different kinds of feature selection algorithms are available in research such as algorithms based on filters, algorithms based on wrappers, and algorithms based on hybrids. Additionally, a literature study analyzes some of the existing popular feature selection algorithms and also addresses the strong points and difficulties of those algorithms.

Keywords Feature selection · Machine learning · Dimensionality reduction · Feature optimization

1 Introduction

Reduction of dimensionality is a wide-spread preprocessing in the study, visualization, and modeling of high-dimensional data. Feature Selection is one of the easiest ways of reducing dimensionality; only those input parameters containing the necessary information are selected to solve the specific issue. Feature Mining is a more

G. Saranya (✉) · A. Pravin
Department of Computer Science and Engineering, Sathyabama Institute of Science and Technology, Chennai, India
e-mail: saranyag3@srmist.edu.in

A. Pravin
e-mail: pravin_ane@rediffmail.com

D. J. Hemanth et al. (eds.), *Artificial Intelligence Techniques for Advanced Computing Applications*, Lecture Notes in Networks and Systems 130,
https://doi.org/10.1007/978-981-15-5329-5_24

particular strategy of transforming the input space into a low-dimensional subspace that retains most of the information required [1].

Retrieval methods and feature selection methods are used in isolation or in tandem with the goal of improving efficiencies, such as approximate precision, representation, and understandability of the studied information [2]. The aim of selecting a feature is to decrease the number of parameters in the set of data so that the feature chosen integrates as much data as possible from the whole set of data. Limiting the number of features can have several positive effects, such as deleting unnecessary or obsolete features, reducing production time, and enhancing the efficiency of learning systems [3].

The advantage of feature selection is that there is no loss of valuable information related to a particular feature, but if a limited number of parameters are required and design parameters are very complex, there is a possibility of losing information as some of the features must be skipped away. And from the other side, with fractal dimension elimination also known as characteristic mining, the length of the attribute space can be reduced periodically before information about the initial space is lost. One downside to extracting the function is the fact that the spatial structure of the unique features is usually incomprehensible and the information about how much an original function impact is sometimes missed [4].

A brief survey of these strategies is conducted depends on a review of the literature to examine the competency of several feature selection techniques and feature mining techniques in some situations depends on researchers' studies to analyze how these strategies help increase the statistical precision of the classification method.

In this analysis, we bring to the attention of interested readers different techniques for minimizing dimensions. The paper is structured in the following way. Section 2 consists of background details relating to the selection of features, its approaches, and optimization strategies to examine how efficiently these methods can be used to obtain higher efficiency that ultimately enhances the statistical precision of classifiers. Section 3 provides similar studies conducted to assess the efficiency in the medical field of the various function selection techniques. Section 4: an attempt to compare and analyze feature selection approaches briefly with the advantages and disadvantages of some widely used dimensional reduction methods.

2 Background

Introduction to Feature Selection

High-dimensional information comprises characteristics that might be obsolete, disappointing, or meaningless, increasing the capacity of the search space ensuing in difficulties in information handing, thus not contributing to the learning process. Choosing a subset of features is the method of choosing the good features among all of the parameters that are beneficial for class discrimination [5].

Feature selection process

The feature selection process involves four important steps such as the generation of subset features, the evaluation of subsets, the criterion stoppage, and the validation of tests. The generation of the subset function allows for evaluation in the candidate selection subset. In fact, a heuristic approach follows. A progressive, comprehensive, and random search of features is the search approaches it follows for producing subsets.

The efficiency of the subset being built is measured and use as an assessment parameter. The new subset is compared with the preceding subset and the best subset was found. More use of the first-rated subset for the next comparison. This process of comparison is repeated until the criterion of the stop is reached and the best subset is generated. The final best subset is further validated either through various tests or with prior knowledge. Figure 1 illustrates the feature selection of pro-cess.

Feature selection approaches

Differentiation of feature selection strategies into three divisions: wrappers, filters, and embedded/hybrid method as shown in Fig. 2.

Filter Method: The filter-based method is driven by rating techniques. The variables allocated to a score with an appropriate ranking criterion and the variables

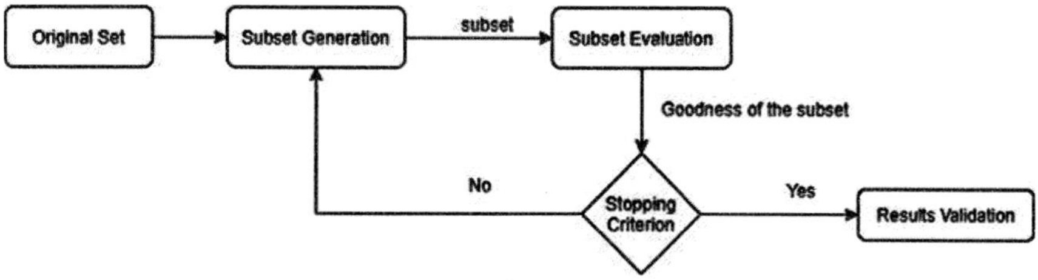

Fig. 1 General feature selection process [6]

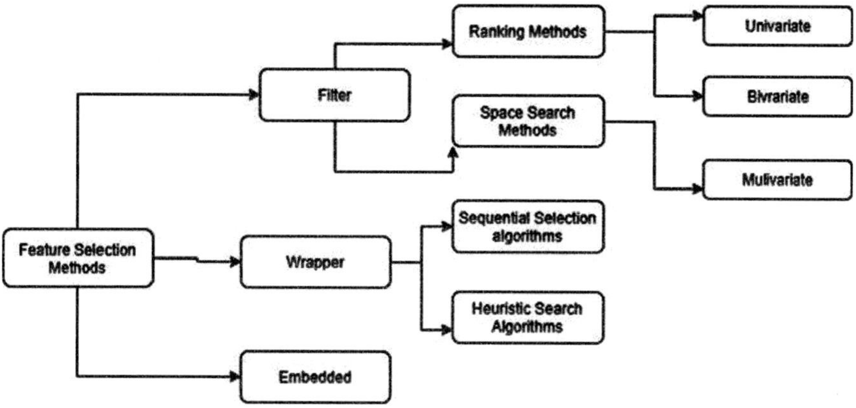

Fig. 2 Hierarchical structure of feature selection approaches [3]

with a score below the threshold value are excluded. As they are independent of the supervised learning algorithm, the filter-based approaches offer more generality [7, 8].

Wrapper method: Wrapper methods find the selection of a set of characteristics as an issue of query where various types are designed, checked, and matched with others. The model uses reverse elimination to delete the irrelevant features from the subset [7, 8].

Embedded method: The embedded approach can be divided into three groups, namely method of pruning, built-in process, and models of regularization. In the pruning system, first, all the features are processed into the classification model building training phase and features with a lower correlation coefficient value are recurrently removed using the support vector machine (SVM)[r]. In the built-in mechanism-based feature selection method, features are selected using a part of the C4.5 and ID3 supervised training phase. The regularization method uses objective functions to minimize the fitting error and excludes the features with near-zero regression coefficients [7, 8].

Wrappers methods work well compare to the filter methods, as the review process is optimized with the discriminator for use. Nonetheless, due to high computational costs, wrapper approaches have to be used expensively for broad feature space and each feature set has to be tested with the qualified discriminator, which eventually slows down the process of selecting. Filter methods have small computing costs and faster relative to wrapper methods but with poor categorization efficiency and are best suited to large-dimensional data sets. Hybrid approaches that utilize both filter and wrapper approaches advantages are being developed recently. A hybrid approach uses the feature subset's function as both an independent analysis and a performance assessment [9].

Heuristic methods for feature selection

Ideally, check feature selection approaches through the feature subsets and attempt to find the good one among the participating 2 m subsets of candidates according to certain evaluation functions. Nevertheless, this approach is comprehensive, since it attempts to find just the right one.

It may be very expensive and almost unfeasible, even for a scale-sized function of a fixed size. Certain approaches depend on deterministic or conditional search methods to try to reduce the computational difficulty by reducing efficiency. Such methods require a stop criterion to stop the search for subsets being comprehensive.

A binary cuckoo search optimization

Binary cuckoo search methodology was introduced based on cuckoo bird behavior, where the search area is designed as a b-cube, where b represents the number of parameters. In traditional CS, the solutions are updated to continuously valued search space positions. But in BCS the search area is modeled as an m-dimensional Boolean structure, in which the remedies are modified to pick features around the edges of a hypercube. It is based on several transition functions that link static answers to discrete binary solutions.

A binary vector solution is used to select the set of features or not in which one correlates to how a parameter will be selected to form the new set of data and 0 otherwise the Ideal Path Classifier accuracy will be used as a fitness component. We test the power of algorithm compared to the binary versions of the Bat Algorithm, Firefly algorithm, and Particle Swarm Optimization to accomplish the selection function task. The experiments and assessment provided over four public datasets, use a navigate verification strategy to verify how the techniques work for the purposes of selecting features [10].

Bat algorithm and collection of features based on a rough set

A new classification strategy is being introduced, relying on the hard-sets and bat method. Bat Algorithm (BA) is a classification method that helps bats to fly within the space subset of function and to discover the best combinations of functions. Bat algorithm requires only un-composite operators of basic and easy numerical operators such as hybrid and mutation. In terms of both memory and run time, it is algorithmically cheap. Fitness combines both categorization precision and the number of features chosen, and therefore checks categorization reliability and decrease size. The used sample-set fitness model guarantees better-classified outcomes while also maintaining limited feature-length. The random set theory offers a statistical methodology that uses purely functional approaches to define information constraints and decrease the number of functions included in the sample. Sample sets were a significant tool for clinical use. The complete solution for detecting marginal reductions is to generate all feasible reductions and pick one with marginal eigenvalues which can be done by building and simplifying the function from the set of data [11].

Multi-objective genetic algorithm method for a subset of features

The Setback function selection is multiple objectives in general and therefore, there is no need to refine parameter subgroups with respect to any particular assessment criteria. Multi-objective genetic algorithm uses multi-objective consistency subset optimization to put multiple key criteria for applications. The results demonstrate that the current system is capable of determining varied optimal subsets of features well distributed across the overall parameter space, and the classified precision of subsequent parameter subgroups is relatively high. Various solutions reach the normal scale of computational complexity with regard to function subsets scientific value, un-redundancy, and cardinality [12].

Optimization of binary ant colony for feature selection

Techniques for optimizing the ant colony were commonly used to solve the problems of combinatorial optimization. Bearing this in mind, ABACO was introduced to solve the problems of the selection of apps. A graph model is generated by treating the features as graph nodes and is connected to each other completely. Each node has two sub-nodes in the graph which represent selecting and unselecting features. To pick the nodes, ant will visit all the features in the ACO algorithm. In the end, each ant carries a binary function of the similar size as function, where one implies selection and zero implies deletion of the related characteristic. The observational analysis checks that the methodology provides good classification accuracy as compared to another known feature selection method based on ACO [13], using a small feature set [13].

3 Literature Review

Dimensionality reduction techniques in the medicine sector become a clear necessity. Today, the health domain produces a large amount of information. It includes the symptoms a patient might have, as well as numerous cases of potential medical tests. The function is correlated with parameters and control parameters.

Soliz [14] has suggested a method for acquiring image-based characteristics for categorizing optical retinal pictures of AMD. An ophthalmologist classified 100 images into 12 categories based on the disease's unique features. Independent Component Analysis (ICA) is used to identify traits and classifier feedback was used. ICA is shown to be able to strongly identify and classify features in fund pictures and to implicitly extract the computational attributes from each image to describe the phenotype.

Ladha et al. was provided in [15] following feature specification advantages:

- Minimizes the complexity of the space reduces computing needs and increases the implementation rate.
- It suppresses old, obsolete, or disruptive data.
- The immediate consequences of the data processing activities increase the run time of the learning algorithms.
- Improving the quality of the results.
- Making the resulting model more accurate.
- Reduction of the set function, to save resources during or after the next round of data collection.

Zheng et al. [16] suggested the diabetes diagnosis method using ANN and a set of features developed by implementing Singular Variable Decomposition and the Principle Component Assessment approach. Findings indicate that the composition of ANN, PCA, and SVD is a valid method of diabetes diagnosis, with smaller operational expenses and high precision. Because of noisy data, ophthalmologists found attribute extraction strategies to be more suitable for artificial condition detection than attribute extraction strategies. Since most biomedical data sets do not contain obsolete or redundant data, but noisy data.

Tan et al. [17] have proposed a hybrid model composed of methodologies for artificial intelligence. GA and SVM in which they both proposed methodologies for artificial intelligence and were effectively combined using a wrapper method. DC Irvine natural language repository received five data sets which were managed by the hybrid method GA and SVM. The GA and SVM solution included adding a comparative test to boost the mean fitness of a chromosome. When eliminating the unnecessary attributes, the study checked the GA-SVM approach as a good classification. The precision achieved by the GA-SVM integrated system is 84.67%, which was very high.

Yan et al. [18] introduced a true-coded genetic algorithm system appropriate for selecting important and significant characteristics and neglecting the insignificant and redundant characteristics. In the treatment of five major cardiac conditions, this approach has been used. Data from heart disease with 351 cases were utilized in this system. And every case has 41 diagnostic features. Of the 351 cases of heart disease data, the 23 key diagnostic features have been reported.

Batla et al. [19] proposed research which included various methods for predicting cardiovascular disease in data analysis. The automated cardiovascular disease prediction programs can use many forms of statistical analysis. This research comprises of many techniques and classifiers used for data analysis to treat cardiovascular disease in an excellently organized and vibrant way. The article shows that the use of 16 attributes by Neural Network has provided the maximum precision. Decision Tree has achieved good accuracy by using 16 attributes of 99.72%. The Decision Tree and GA combination reduced the parameters count 16 to 6 attributes and has 99.22% accuracy.

Quinlan was using the decision tree C4.5 categorization system for breast cancer diagnosis and obtained 94.74% accuracy with cross-validation [20]. Hamilton achieved 96.12% precision use the RIAC technique while using Linear Discrete Analysis, Dobnikar reached 96.9% [21, 22]. In another research work, Peña and Sipper obtained 97.37% precision with the Fuzzy genetic algorithm methodology [23].

Akay et al. utilized a model-based classification feature score while Chen utilized sample-set FS-based SVM for the diagnosis of breast cancer [24–26]. Rathore and Agarwal have used an integrated method to calculate the killing power of cancer patients [27]. Polat developed a method for the diagnosis of breast cancer utilizing minimum-square SVM [28]. Karabatak incorporated the principles of the association and the method to the algorithm to create a professional framework for cancer diagnoses.

4 Comparative Analysis of Feature Selection Algorithms

The benefits of filter-based approaches are high statistical effectiveness and generality. The wrapper-based approach provides better results but for a large data set, it is computationally expensive.

The pros of both methods will be derived from embedded or hybrid approaches. Anyhow all of these approaches have been commonly used for classification problems by many researchers. If a dataset's dimensionality is different, the same algorithm for selecting the feature may not be appropriate. New feature selection algorithms approaches are always in need. Table 1 summarizes many of the three types of feature selection methodologies, namely filter-based, Wrapper-based, and hybrid. Each algorithm has its own merits, its own demerits [29, 30].

Table 1 Merits and demerits of prediction techniques

Algorithm/Techniques	Type	Factors/approaches used	Result/Inference	Limitation/s
Novel hybrid feature selection algorithm	Hybrid	Rough conditional mutual information. Bayesian classifier	Reduces computational complexity Irrelevant features are eliminated. Enhances statistical accuracy	Accuracy can be improved
Incremental feature selection (IFS) with analysis of variance (ANOVA)	Filter	ANOVA	Elevated statistical importance	Other validations can be done
Fuzzy rough set feature selection algorithm	Filter	Fuzzy based\Greedy forward algorithm	Works best in highly overlapping datasets	Does not work for small stack datasets
Class dependent density based feature elimination	Filter	Feature ranking feature elimination selection	Works better for binary data in high-dimensional Functions for classifier as well	Other data types can be verified
Unsupervised and multivariate filter-based feature selection method3	Filter-based	Ant colony optimization	The algorithm performance improves	New state transmission rule to control the randomness can be developed
Affinity propagation-sequential feature selection algorithm8	Wrapper-based	Cluster based	High-dimensional data quicker	Accuracy is comparable

5 Conclusion

This paper summarizes different available feature selection methods and optimization techniques. A large proportion of attributes in a set of data may result in lower categorization accuracy with higher computing costs and the threat of overfitting. The survey provides a comprehensive overview of different diseases such as cardiovascular disease, pulmonary disease, diabetes, and breast cancer that can be successfully treated using machine learning feature selection methods. And from this study, it is clear that an efficient unified framework is required which should provide feature selection without incomplete data, low computational complexity, and highest precision for any size of the dataset.

References

1. Chumerin N, Hulle MMV (2006) Comparison of two feature extraction methods based on maximization of mutual information. In: Proceedings of the 16th IEEE signal processing society workshop on machine learning for signal processing, pp 343–348
2. Motoda H, Liu H (2002) Feature selection, extraction and construction. In: Towards the foundation of data mining workshop, sixth Pacific-Asia conference on knowledge discovery and data mining (PAKDD2002), Taipei, Taiwan, pp 67–72
3. Mhamdi H, Mhamdi F (2014) Feature selection methods on biological knowledge discovery and data mining: a survey. In: 2014 25th international workshop on database and expert systems applications, pp 46–50. IEEE
4. Janecek AGK, Gansterer GF et al (2008) On the relationship between feature selection and classification accuracy. In: Proceeding of new challenges for feature selection, pp 40–105
5. Khalid S, Khalil T, Nasreen S (2014) A survey of feature selection and feature extraction techniques in machine learning. In: 2014 science and information conference, pp 372–378. IEEE
6. Xue B, Zhang M, Browne WN, Yao X (2015) A survey on evolutionary computation approaches to feature selection. IEEE Trans Evol Comput 20(4):606–626
7. Liu X, Shang L (2013) A fast wrapper feature subset selection method based on binary particle swarm optimization. In: 2013 IEEE congress on evolutionary computation
8. Asir Antony Gnana Singh D, Appavu alias Balamurugan S, Jebamalar Leavline E (2016) Literature review on feature selection methods for high-dimensional data. Int J Comput Appl (0975–8887) February 2016
9. Veerabhadrappa L, Rangarajan (2010) Bi-level dimensionality reduction methods using feature selection and feature extraction. Int J Comput Appl 4(2):33–38
10. Pereira LAM, Rodrigues D, Almeida TNS, Ramos CO, Souza AN, Yang XS, Papa JP (2014) A binary cuckoo search for feature selection. Springer International Publishing Switzerland
11. Nakamura RYM, Pereira LAM, Costa KA, Rodrigues D, Papa JP, Yang X-S (2011) BBA: a binary bat algorithm for feature selection. In: Tangri N, Stevens LA, Griffith J, Tighiouart H, Djurdjev O, Naimark D, Levin A, Levey AS (eds) 2012 XXV SIBGRAPI conference on graphics, patterns and images. A predictive model for progression of chronic kidney disease to kidney failure. JAMA 305(15), 1553–1559 (2011)
12. Spolaor N, Carolina Lorena A, Diana Lee H (2010) Use of multi objective genetic algorithms in feature selection. In: 2010 eleventh Brazilian symposium on neural networks
13. James Mathai K, Agnihotri K (2017) Optimization techniques for feature selection in classification. 2017 IJEDR | Volume 5, Issue 3 | ISSN: 2321-9939
14. Soliz P et al (2008) Independent component analysis for vision-inspired classification of retinal images with age-related macular degeneration. In: Proceeding of IEEE international conference on image processing SSIAI, pp 65–68
15. Ladla L, Deepa T (2011) Feature selection methods and algorithms. Int J Comput Sci Eng (IJCSE) 3(5):1787–1797
16. Zheng Y et al (2013) An automated drusen detection system for classifying age-related macular degeneration with color fundus photographs. In: IEEE 10th international symposium on biomedical imaging, pp 1440–1443
17. Tan et al (2009) A hybrid evolutionary algorithm for attribute selection in data mining. J Expert Syst Appl 36:8616–8630
18. Yan H et al (2008) Selecting critical clinical features for heart diseases diagnosis with a real-coded genetic algorithm. J Appl Soft Comput 8:1105–1111
19. Bhatia N et al (2012) An analysis of heart disease prediction using different data mining techniques. Int J Eng Res Technol 1(8)
20. Quinlan JR (1996) Improved use of continuous attributes in C4.5. J Artif Intell Res 4:77–90
21. Hamilton HJ, Shan N, Cercone N (1996) RIAC: a rule induction algorithm based on approximate classification. Computer Science Department, University of Regina

22. Ster B, Dobnikar A (1996) Neural networks in medical diagnosis: comparison with other methods. In: Proceedings of the international conference on engineering applications of neural networks (EANN '96), pp 427–430
23. Peña-Reyes CA, Sipper M (1999) A fuzzy-genetic approach to breast cancer diagnosis. Artif Intell Med 17(2):131–155
24. Akay MF (2009) Support vector machines combined with feature selection for breast cancer diagnosis. Expert Syst Appl 36(2):3240–3247
25. Huang C-L, Liao H-C, Chen M-C (2008) Prediction model building and feature selection with support vector machines in breast cancer diagnosis. Expert Syst Appl 34(1):578–587
26. Chen H-L, Yang B, Liu J, Liu D-Y (2011) A support vector machine classifier with rough set-based feature selection for breast cancer diagnosis. Expert Syst Appl 38(7):9014–9022
27. Rathore N, Agarwal S (2014) Predicting the survivability of breast cancer patients using ensemble approach. In: Proceedings of the international conference on issues and challenges in intelligent computing techniques (ICICT '14), February 2014. IEEE, pp 459–464
28. Polat K, Güneş S (2007) Breast cancer diagnosis using least square support vector machine. Digit Signal Proc 17(4):694–701
29. Krishnaveni N, Radha V (2019) Feature selection algorithms for data mining classification: a survey. Indian J Sci Technol 12(6). https://doi.org/10.17485/ijst/2018/v12i6/139581, February 2019
30. Chandrashekar G, Sahin F (2014) A survey on feature selection methods. Comput Electr Eng 40(1):16–28

Survey on Real-Time Diabetic Patient's Monitoring Using Internet of Things

G. Geetha and K. Mohana Prasad

Abstract According to the statistics, diabetes is a life-threatening disease and also paves the way to other acute and chronic diseases. Diabetes management is a crucial task even after the technology improvement. In this paper, a study has been done for monitoring real-time patients using the Internet of Things (IoT). In a diabetes Healthcare application, a large amount of data will be generated continuously; hence there is a need to employ some efficient technique to convert this raw data into useful information. Combining IoT and Big Data will help us to address the issues of storage, processing, and analytics. Diabetes can be predicted based on the blood glucose level in the blood samples. Insulin is a hormone generated by the pancreas, if there is insufficient secretion of insulin or if it's not consumed by the body properly, it leads to a condition called diabetes. There are three different types of diabetes: (1) Type 1 diabetes is due to insufficient secretion of insulin (2) Type 2 diabetes occurs when the body does not use the insulin secreted (3) Gestational diabetes that comes during pregnancy and it goes off after giving birth.

Keywords Real-time monitoring · IoT · Big data analytics

1 Introduction

Technologies behind the real-time patient monitoring system

Cloud computing: The major advantage of cloud computing is that you can access information anywhere and anytime if you have an internet connection. This concept is extended to real-time patient monitoring to achieve reliability because it involves human life. It offers services on-demand and it is easy to deploy and access the services. Cloud computing seems to be a better choice in the Healthcare sector rather than using multiple servers in different locations. But two major concerns which arise while dealing with cloud computing are security and privacy of data. These can be overcome by getting connected with reliable service providers like Microsoft

G. Geetha (✉) · K. Mohana Prasad
Sathyabama Institute of Science and Technology, Chennai, India
e-mail: geethag@srmist.edu.in; muraligeetha0105@gmail.com

D. J. Hemanth et al. (eds.), *Artificial Intelligence Techniques for Advanced Computing Applications*, Lecture Notes in Networks and Systems 130,
https://doi.org/10.1007/978-981-15-5329-5_25

Azure because they rely on security policies set forth during agreement. One can opt for a hybrid cloud for storing patient's information. There are three ways to upload files into the cloud: (1) Web UI, (2) Command line and (3) programmatically. There are different ways in which the sensor data are collected and uploaded in the cloud, among which the best is the BLE (Bluetooth Low Energy) sensor that directly connects with Wi-Fi routers to send data to the cloud without using mobile phones as the interfacing unit. Another way is to use MIT APP Inverter to create a mobile app to receive the sensor data and eventually send it to cloud storage.

Sensor Data

Data used for real-time patient monitoring are mainly sensor data, which has a high voluminous unstructured or semi-structured format. It is very difficult to handle it with a relational database because it can handle only relational schemes and sensor data are schema-less. These kinds of data can be handled by SQL databases like MongoDB. MongoDBsupportingJSON (Java Scripts Object Notation) formats for document storage. It uses the sharding technique for distributed data storage to achieve parallel read/write or access. There are some needleless sensors to measure blood glucose levels without pricking fingertips like Freestyle Libre, GlucoTrack and ever sense, etc. These types of sensors measure blood glucose level by measuring interstitial fluid between the cells beneath the skin and send the readings directly to the mobile app of the patient. Based on the requirements, it can be customized to send a notification to the caretaker as well (Fig. 1).

Big Data Analytics on Healthcare systems

In the Healthcare system, there will be a huge volume of continuous data coming out of various sources like electronic devices, sensors, and other applications, and it is also in an unstructured, semi-structured or structured format. Hence traditional database systems are inefficient to store, process, and analyze the data. This can be overcome by Big Data because it has more enhanced tools for storage, processing, and visualization. Big Data analytics is the method of examining the large data sets of diverse data types to discover hidden patterns, concealed correlations, market trends, customer preference, and other practical business information. Existing different types of analytics systems are real-time analytics, off-line analytics, memory-level analytics, BI-analytics, and massive analytics (Fig. 2).

To fulfill the procedural needs of huge information analysis, a framework is important to style, implement, and manage the desired pipelines and algorithms.

The framework functions on the following four levels:

1. Data ingestion and acquisition level for gathering mixed data related to Healthcare operations,
2. Data storage and processing level such as filtering, analyzing and storing data to make decisions and events autonomously,
3. Data query level where active analytic processing takes place and
4. Visualization level initiating execution of the events corresponding to the received decisions.

Ingestion is collecting and storing data, particularly the unstructured data, from where it is initiated, into a system where it can be stored and investigated for drawing a

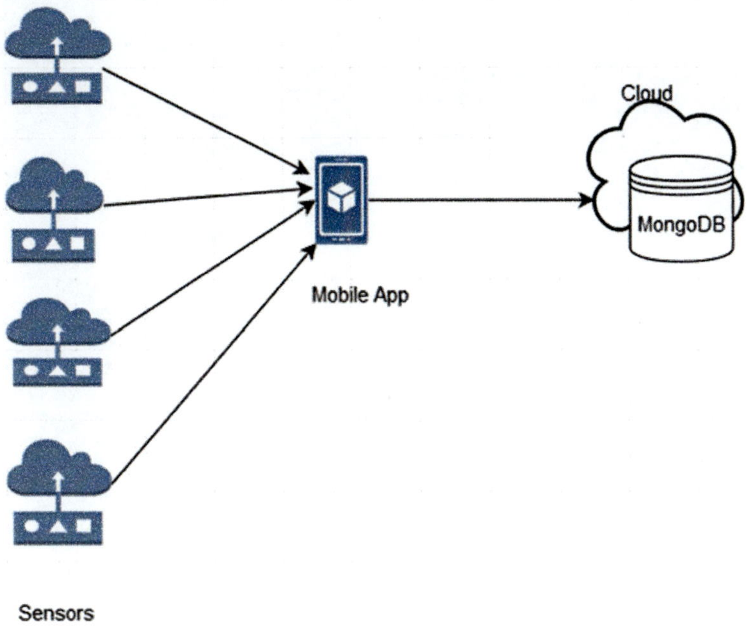

Fig. 1 Cloud architecture for IoT applications

conclusion. Data can be streamed in real-time or ingested in batches. Apache Flume is an ingestion tool to Ingest streaming data from multiple sources into Hadoop Distributed File System (HDFS) for storage and analysis. Data processing: Among all existing processing systems, we choose Apache Spark, because it will process both real-time data streams as well as batch processing. For data storage, HDFS is used because IoT devices generate structured, semi-structured, and unstructured data, and yet another reason is Apache Flume supports mainly HDFS. So we need schema-less query language for querying purposes hence HQL is preferred to use. For decision making, BigML is a scalable and programmable machine learning platform that consists of a range of tools to perform machine learning tasks such as classification, clustering, etc. For data visualization, certain visualization tools are available like Google Chats, Tableau Software, and D3.js. Among all, Tableau is the best suitable tool for Big Data (Fig. 3).

Big Data Analytics Methods

Clustering Analysis: Clustering is an unsupervised learning technique to put similar data in one cluster and dissimilar data in other clusters to recognize different patterns. Clustering algorithms can be generally classified into hierarchical, partition, and density-based clustering. Classification: Classification is supervised learning to classify data into groups based on the existing or prior knowledge of training data. Bayesian network, SVM (support vector machine), KNN (k-Nearest Neighbor) are

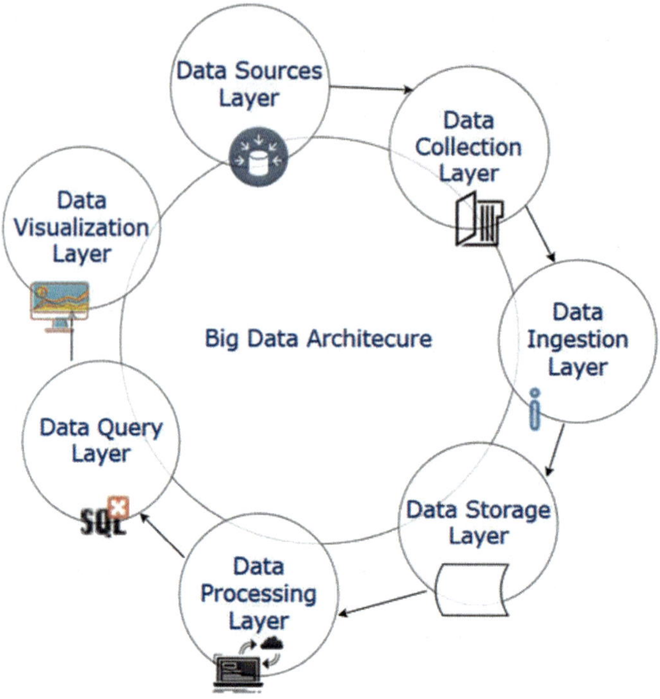

Fig. 2 Big data architecture

various classification methods. Association Rule Mining: Association rule mining identifies the relationship among the different groups, objects, and events to analyze customer behavior and unseen trends. Predictive Analysis: Predictive analysis uses existing training datasets to reveal unseen patterns and behavior of data. Challenges in the field of big IoT data analytics are privacy issues, data exploration, information extraction, data visualization, and Integrating diverse data types. Hadoop in Big-data: Hadoop is an open-source framework for storing and processing Big Data in a distributed manner. The two major components of Hadoop are HDFS and YARN. HDFS is for storing data in a distributed fashion on different DataNode and YARN is for reducing the processing time.

Real-Time Diabetics monitoring products

Continuous Glucose Monitoring

Continuous Glucose Monitoring, automatically and continuously, monitors the blood glucose level throughout the day and informs the patient or caretaker to balance the diet or physical activity. CGM fixes a sensor under the skin, then measures the blood glucose level using interstitial fluid between the cells, hence avoiding finger prick frequency, but in order to confirm the blood glucose level twice a day, fringer prick is necessary. It will also send an alarm signal in case the blood glucose level

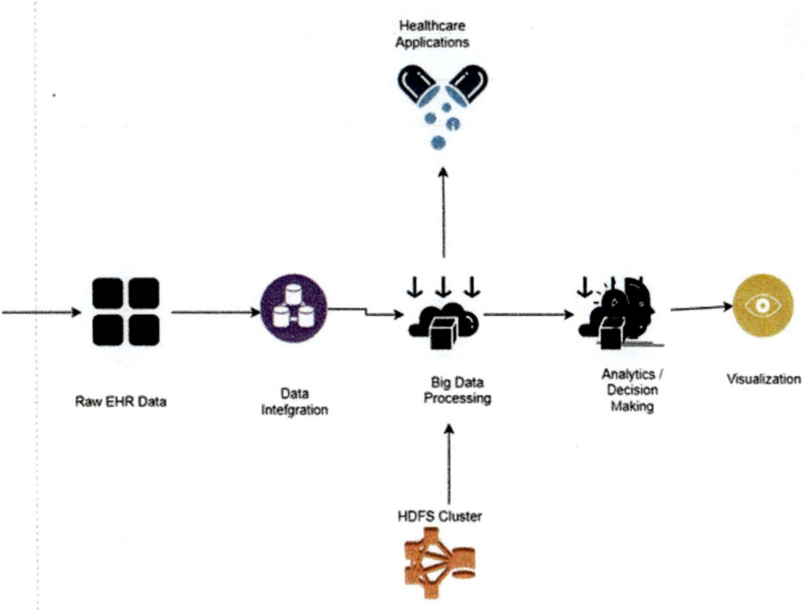

Fig. 3 Healthcare software stack

increases or decreases above or below the threshold limit. The main disadvantage is that accuracy is less compared to the finger prick method.

Artificial Pancreas

This device acts exactly like a real pancreas which continuously monitors the blood glucose level and releases the required quantity of insulin into the body when the blood sugar level increases. Artificial Pancreas uses continuous glucose monitoring (CGM) to measure the blood glucose level and communicate with an insulin pump to inject the right amount of Basel insulin automatically. When you want to adjust the amount of insulin to be injected, it can be done manually.

2 Literature Survey

"*A reliable IoT system for Personal Healthcare Devices*". A reliable oneM2M-based IoT system for Personal Healthcare Devices system uses protocol conversion between ISO/IEEE 11073 protocol messages and oneM2M protocol messages have been done in gateways located between Personal Healthcare Devices and the PHD management server [1] (Fig. 4).

They also implemented a fault-tolerant algorithm for the reliable IoT system in which gateways on the same layer in the system are connected to form a daisy chain

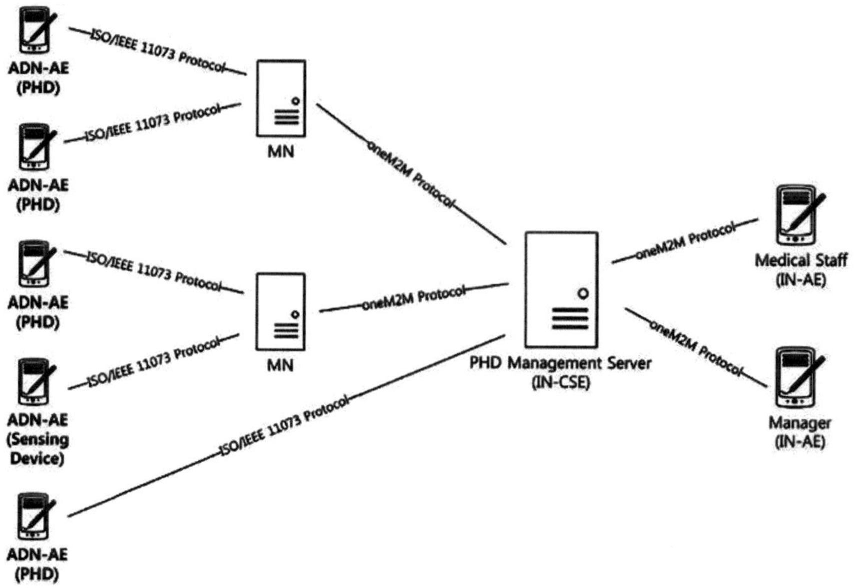

Fig. 4 Structure of the oneM2M IoT system for PHDs [1]

for fault tolerance at the level, and each gateway stores the copies of the previous gateway [1].

"*Fault-Tolerant and Scalable IoT-based Architecture for Health Monitoring [2]*". This paper focuses on scalability and fault tolerance for Healthcare. Backup routing between nodes has been done to achieve Fault tolerance and advanced service mechanisms to preserve communication links if there are any failures between system nodes. They have built a tailored tunneling gateway for routing packets from nodes to a server on the Internet (Fig. 5).

"*Secure Data Aggregation of Lightweight E-Healthcare IoT Devices With Fair Incentives" [3].* This paper focuses on security issues, privacy issues, and also fault tolerance. They have developed an architectural framework which poses three-layer (1) Data Collection Layer (2) Data aggregation layer and (3) service layer. In the data collection layer, the patient's health information is collected from the wearables and sent to the aggregation layer. In the aggregation layer, data are collected and noise data are annotated followed by encryption process taking place and stored in the nearby health centers whereby finally the data moves to the cloud storage for analysis. Encryption with a digital signature is combined used to maintain the privacy and confidentiality of user data. Then the cloud server aggregates the data from multiple health centers and responds to user queries (Fig. 6).

"*Secure Edge of Things for Smart Healthcare Surveillance Framework" [4].* In this, they combined edge computing and cloud computing platforms to perform local and global analytics respectively. Biosignal data are collected from smart community members and then transmitted from medical services gateway to edge computing

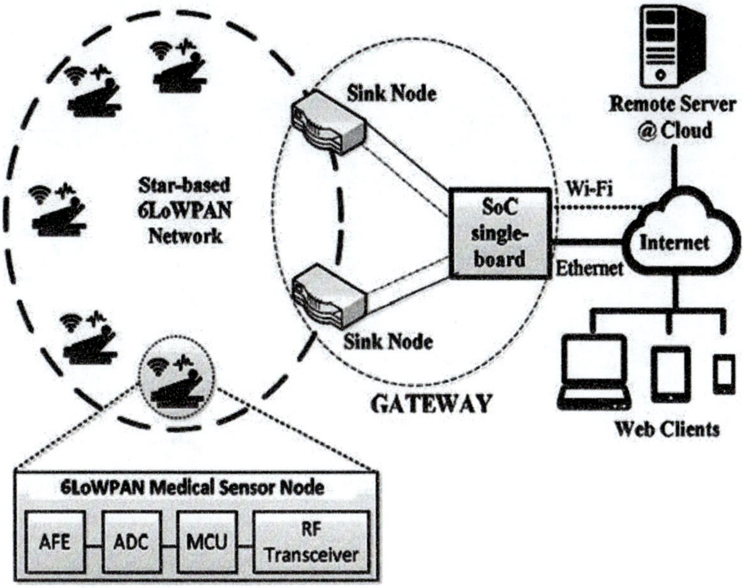

Fig. 5 IoT-based healthcare system architecture [2]

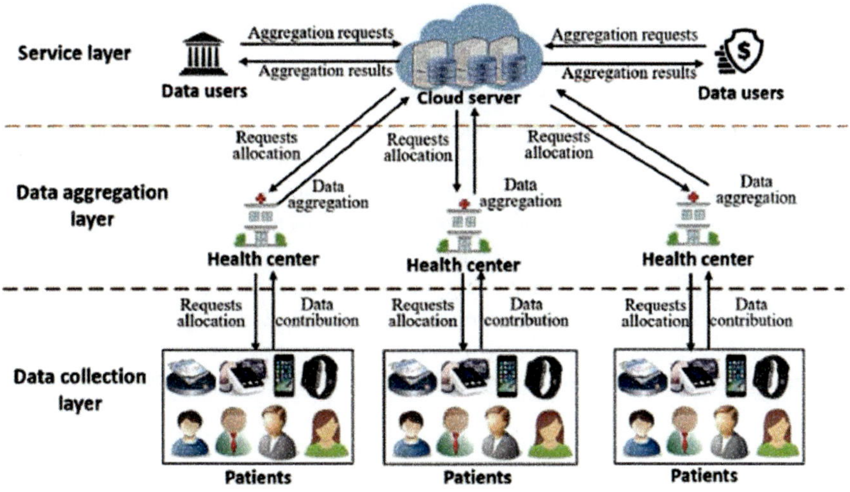

Fig. 6 Three-layer framework [3]

gateway where the encryption and local analysis takes place, hence enhancing the privacy protection and reducing the latency. EoT layer performs real-time data analytics and decision making locally through which early detection and treatment of diseases could be done locally. Machine learning techniques like k-means clustering

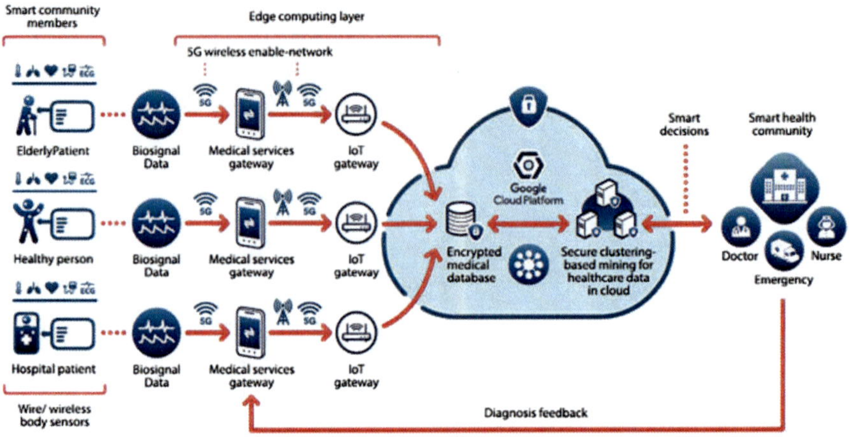

Fig. 7 System architecture showing different components of proposed privacy-preserving change detection and abnormality prediction model in the cloud [4]

and fuzzy c-means clustering has been used for decision making. 5G wireless or wired has been used to improve the efficiency of data transfer (Fig. 7).

"IoT-based continuous glucose monitoring system: A feasibility study" [5]. In this paper, continuous glucose monitoring IoT-based system architecture has been designed to cover up from front end data collection to back end presentation system in graphical and textual format to the doctors and caretakers.

They have implemented nRF protocol to accomplish a high level of energy efficiency and also focused on how to reduce the energy consumption of sensor devices. nRF communication module is responsible for transferring sensor data from microcontroller to the gateway. Additionally, this system sends push notification using the MQTT message broker in case of emergency situations to the patient caretaker and doctors. Within the sensor node, the energy harvesting unit and power management unit has been implemented to save energy consumption (Fig. 8).

"A context ware Interactive m-Health System for Diabetics" [6]. In this paper, the cloud server plays an important role which stores and processes the patient's information. It performs two major tasks: abnormal blood glucose level detection

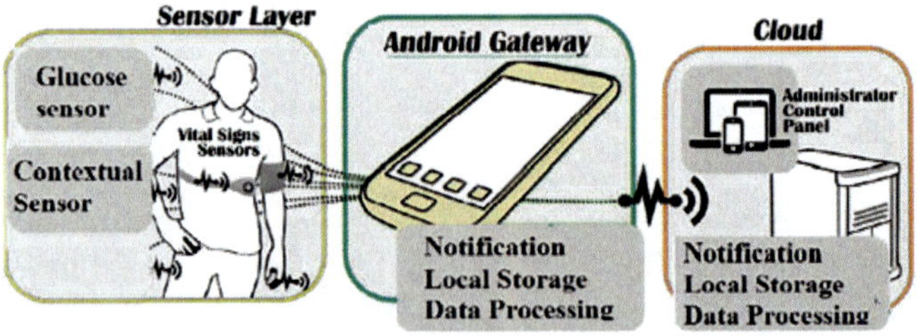

Fig. 8 Continuous glucose monitoring using IoT [5]

(ABLD) service and a proactive communication engine (PNE). This ImHS uses a rule-based system for Abnormal blood glucose level detection. ABLD module performs two functions: blood glucose abnormalities detection and the measured data are not received on the scheduled date. Blood Glucose abnormalities could be decided based on blood glucose level and if this value is above the threshold, an alert message will be sent to the doctors and caretakers by the PNE module using the MQTT message broker. If the ABLD module does not receive the measurement data on time, then the message will be sent to the patient to resend the data again and wait for some specific time but if the data is not received within the stipulated time then necessary action will be performed. GPRS CGM sensor is used to obtain the glucose level reading and it is automatically stored at the cloud for further processing (Fig. 9).

"A Modular approach to diabetic examination with an alert system based on IoT" [7]. They have designed the needleless method to ensure the blood glucose level. It uses light and body mass index to determine the blood glucose level, hence preventing the need of pricking every time. They have created an electronic nose like an IR pulse detector which works on the principle of the amount of light transmitted and amount of light received to measure pulse level. The electronic nose will be fixed in between the index and thumb finger since this is the thinnest region to measure the blood glucose level. Then the measured value will be sent to the patient mobile through Bluetooth along with the dietary plan. If the blood glucose level is abnormal then the message will be sent to the doctor as well to take follow up action.

"Mobility Based Self-Assessment monitoring System For Diabetes Patients using IoT" [8]. They have proposed self-assessment monitoring system to monitor diabetic patients with the help of a patient's body sensors data such as BP, Blood glucose, and

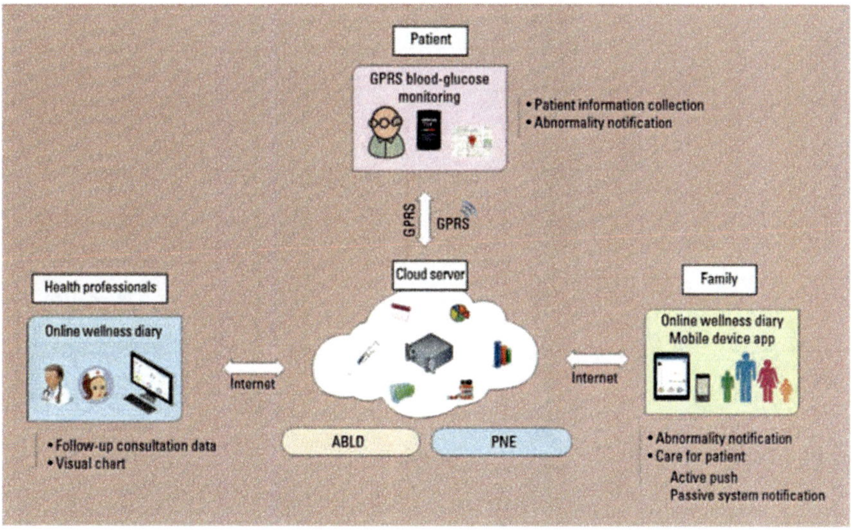

Fig. 9 Interactive mobile health service (ImHS) architecture [6]

ECG. They used 8051 microcontroller to collect sensor data through analog to digital converter ADC0808. Collected sensor values are transmitted through ESP8266 Wi-Fi module to cover wider range. KNN classifier has been implemented to classify and analyze the medical data and inform the patients about the risk factor, physical activity, diet, and amount of insulin intake.

"IoT-based Personal Health Care Monitoring Device for Diabetic Patients" [9]. This paper presented a non-invasive breath test method for diabetic patient monitoring. Usually, diabetics will be monitored based on urine tests or blood test but this method measures breath acetone since this is a convenient, non-invasive and accurate method for ketone level measurement. FIGARO TGS 822 gas sensor has been used to detect the amount of gas acetone in the patient's breath. In addition to that, temperature and humidity (DHT11) sensors also used to balance the decrease in resistance of gas sensors. These sensors values are read using Arduino and sent to database using ESP 8266 Wi-Fi Module. Patients can sign into the personal web site created for them to monitor their current conditions and medication, accordingly (Fig. 10).

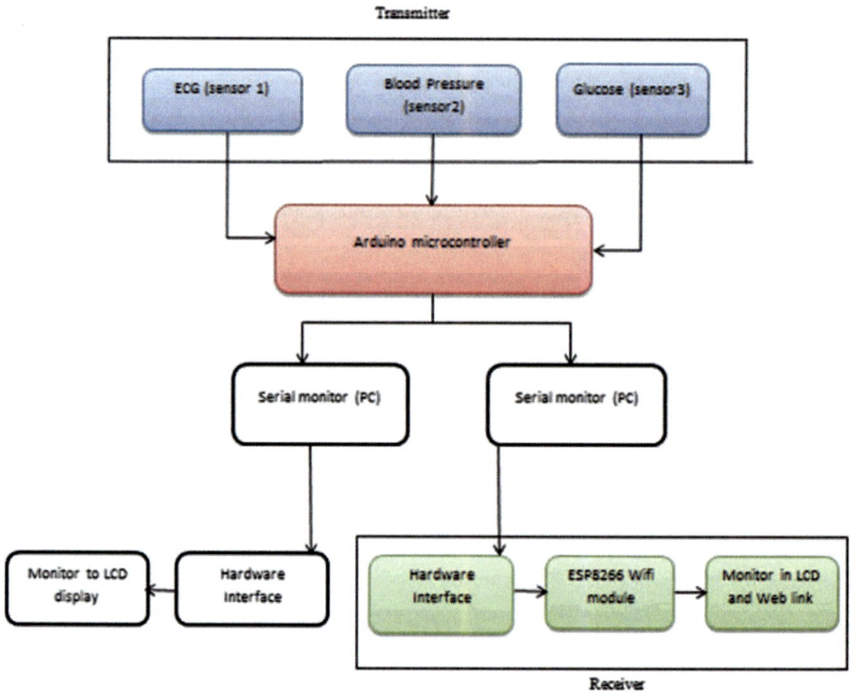

Fig. 10 System architecture [8]

Challenges

1. Security and Privacy:

Security and privacy are the major issues in IoT-based Healthcare applications because IoT devices do not follow data protocols and standards and also ambiguity in data ownership regulations. This may allow criminals to compromise the patient's details and hence they can create a fake ID, can claim Medical Insurance, etc.

2. Integrating Multiple devices:

Integrating multiple devices is yet another hurdle in the scalability of IoT in the health care sector, this is due to the lack of communication standards and protocols followed by vendors to make the IoT devices.

3. Data Overload and Accuracy:

A tremendous amount of data will be continuously generated by the IoT devices every minute, hence aggregating and processing this data for decision making is a highly tedious job which reduces the accuracy.

3 Conclusion

In this paper, we have done a survey on prediction and real-time monitoring of diabetic patients using Internet of Things. Through this real-time monitoring system, doctors and caretakers can easily monitor the patient health remotely using a smart phone or web application. In recent years, many refined e-Health applications have been recommended and effectively put into practice. We have presented the working and fundamental architecture of the most recent Healthcare applications based on Internet of Things used in diabetes management system. We also reviewed the issues and challenges faced by these latest applications.

References

1. Woo Woo M, Lee J, key him Park (2017) A reliable IoT system for personal health care system. Future Generation Computer System, Elsevier
2. Nguyen Gia1 T, Rahmani1 A, Westerlund1 T, Liljeberg1 P, Tenhunen1,2 H (2015) Fault-tolerant and scalable IoT-based architecture for health monitoring. IEEE Instrumentation and Measurement Society
3. Tang W, Member, IEEE, JuRen, Member, IEEE, Deng K, Zhang Y, Senior Member, IEEE (2019) Secure data aggregation of lightweight E-Healthcare IoT devices with fair incentives. IEEE Internet Things J 6(5)
4. Alabdulatif 1 A, Khalil2 I, Yi 2 X, Guizani 3 M (Fellow, IEEE) (2019) Secure edge of things for smart healthcare surveillance framework. IEEE Access, January 27, 2019
5. Nguyen Gia T, Ali M, Ben Dhaou I, Rahmani AM, Westerlund T, Liljeberg P, Tenhunen H (2017) IoT-based continuous glucose monitoring system: a feasibility study. In: 8th international conference on ambient systems, networks and technologies (ANT-2017), Elsevier

6. Maria Boncy1 JN, Shanthi2 D, Somasundaram SK (2018) Mobility based self-assessment monitoring system for diabetes patients using IoT. Int J Pure Appl Math 118(20)
7. Rabiul Alam1,2 G (Member, IEEE), Fakhrul Abedin1 S (Student Member, IEEE), Il Moon1 S, Talukder1 A (Member, IEEE), HONG1 C (Senior Member, IEEE) (2017) Healthcare IoT-based affective state mining using a deep convolutional neural network. IEEE Access
8. Vasanthakumar R, Darsini KD, Subbaiah S, K.Lakshmi,"IoT for monitoring diabetic patients"-Vasanthakumar. R et.al; International Journal of Advance Research, Ideas and Innovations in Technology, Volume 4, Issue 2,2018
9. Ab Rahman R, Abdul Aziz N, MurizahKassim, IkramYusof M (2017) IoT-based personal health care monitoring device for diabetic patients. IEEE
10. Riazul Islam1 SM (Member, IEEE), Kwak2 D, Kabir1 H, Hossain3 M, Kwak1 K-S (2015) The internet of things for health care: a comprehensive survey. IEEE Access
11. Usha Rani1 M, Goutham2 J, Monicka3 S, ANUPA ELIZABETH4 P (2013) Web based service for diabetes patient monitoring sensors. Int J Comput Sci Inf 3(2)
12. 1Bhat GM, 2Pradeep R, 3Praveen K, 4Sathvik BS, 5Yashas KR (2017) A modular approach to diabetic examination with alert system based on IoT. Int J Sci Eng Res
13. Ismail SF (2017) IOE solution for a diabetic patient monitoring. In: 2017 8th international conference on information technology (ICIT)
14. Suh1 M, Moin2 T, Woodbridge1 J, Lan1 M, Ghasemzadeh1 H, Bui3 A, Ahmadi4 S, Sarrafzadeh1 M (2012) Dynamic self-adaptive remote health monitoring system for diabetics. In: 34th annual international conference of the IEEE EMBS, September 2012
15. Wang N, Kang G (2012) A monitoring system for type 2 diabetes mellitus. In: IEEE 14 international conference on e-health networking, applications and services
16. Sood SK, Mahajan I (2019) IoT-fog-based healthcare framework to identify and control hypertension attack. IEEE Internet Things J 6(2)
17. Gómez J, Oviedo B, Zhuma E (2016) Patient monitoring system based on internet of things. In: The 7th international conference on ambient systems, networks and technologies (ANT 2016). Elsevier
18. Zeadallya S, Bello O (2019) Harnessing the power of internet of things based connectivity to improve healthcare. Internet of Things, Elsevier 2019
19. Chang S-H, Chiang R-D, Wu S-J, Chang W-T (2016) A context ware interactive m-health system for diabetics. IEEE Computer Society

Automatic Text Summarization of Article (NEWS) Using Lexical Chains and WordNet—A Review

K. Janaki Raman and K. Meenakshi

Abstract The process of selecting important information or extracting the same from the original text of large size and present that data in the form of smaller summaries for easy reading is text summarization. Text summarization has now become the need for numerous applications, like market review for analysts, search engine for phones or PCs, business analysis for businesses. The procedure conveyed for outline ranges from structured to linguistic. In this article, we propose a system where we center around the issue to distinguish the most significant piece of the record and produce an intelligent synopsis for them. Proposed system, we don't require total semantic interpretation for the substance present, rather, we just make a synopsis utilizing a model of point development in the substance shaped from lexical chains. We use NLP, WordNet, Lexical Chains and present a progressed and successful computation to deliver a Summary of the Text.

Keywords Summarization · Linguistic · Semantics · NLP · WordNet · Lexical chain

1 Introduction

With the growth of the amount of data, which became very tough for person to retrieve materials/information of private hobby, to gain an outline of influential, important information or to search effectively for particular content from relevant material. In vogue time of information, an assortment of individuals are endeavoring to discover educational records on the net, yet on each event, it isn't possible that they may get all important data in a solitary report or on a solitary net Web site page. All things considered the computerization of featuring the literary content can be useful

K. Janaki Raman (✉) · K. Meenakshi
Department of Information Technology, SRM Institute of Science and Technology, Chengalpattu, India
e-mail: kr4607@srmist.edu.in

K. Meenakshi
e-mail: meenaksk@srmist.edu.in

D. J. Hemanth et al. (eds.), *Artificial Intelligence Techniques for Advanced Computing Applications*, Lecture Notes in Networks and Systems 130,
https://doi.org/10.1007/978-981-15-5329-5_26

software for humans, such as academic college students, politicians, administrators or lawyers, who need to look at and survey numerous writings. The automatic content synopsis is as of now accessible; however, there is no correct execution for literary substance featuring. This examination is an endeavor to find a response to how to place into impact modernized text summarization as a book extraction-based and abstractive based methodology for incredible robotization.

Summarization of the data is a single process of extracting or amassing key records from the unique textual content and presents the ones key factors within the form of a summary. As of late, the requirement for summarization might be found in various thought process and in any region alongside data articles, e-mail, short message of news on portable devices, and records synopsis for representative, specialists authorities, scientists online inquiry through Web crawler to procure the summary of appropriate pages, therapeutic subject for following patient's medicinal history for furthermore treatment. On the Web, numerous such models are accessible like data article summarizers together with Microsoft News, Google or Yahoo. There are few biomedical summarization gears like BaseLine, FreqDist, SumBasic, and MEAD. Other online tools for summarizations are Text Compacter, Sumplify, Tools4Noobs, FreeSummarizer, WikiSummarizer and SummarizeTool. The urgency of getting certainties in an abridged structure on a tick on has quickened just as the need for computerized printed content rundown has additionally expanded in heaps of regions to be explicit, news stories outline, electronic mail outline, brief message news on cell and records for government officials look into, big business, online research engine to gain a synopsis. The first device which got here into existence become in the late 1950s, The automatic summarizer, in preferred, extracts essential phrases from the record, later joins these phrases on the whole, it expends considerably less time or efficient to perceive substance inside the large report. The intention of computerized textual content summarization is to transform a big document right into a shorter one and saves the basic substance material. The mechanized rundown of content is one of the perceived endeavors within the subject of Natural Language Processing (NLP).

Huge accomplishments in literary substance summarization were acquired the usage of statistical analysis and sentence extraction. Text summarization procedures may be widely divided into: Abstractive and Extractive summarization. In The extractive rundown, the extraction of basic words or terms from the first documents and consolidates these words to deliver a synopsis without changing the legitimate content is cultivated. The extractive substance diagram is proposed subject to POS marking by considering Hidden Markov Model (HMM) the usage of the corpus to select crucial phrases from it to form a suitable summary, whereas in Abstractive summarization, this includes a detailed understanding of the supply textual content by way of the usage of a linguistic technique to interpret and observe the textual content. This method needs a deeper evaluation of the textual content. These methods have the potential to create an advance statement. This can upgrade the point of interest of a summarized form, reduces its recurrence, and maintains a terrific compression fee.

Figure 1, portrays how the Synopsis of the content happens. First the information is been Pre-processed, at that point the things are been separated later on the Lexical Chain is been created and finally Sentence is been extricated and the yield is framed as content. This Current Framework is for Extractive Based Model.

The whole process carried out is described in detail below.

- **Pre-Processing**—This is a basic step where the input data has to be cleaned, so that the following stages doesn't affect with any misleading ways. The author used three kinds of process in this stage.
  - **Pronoun Resolution**—This procedure is otherwise called Anaphora Resolution, where it decides the antecedent of an anaphor. That is which one of the pronouns is connected to which nouns. As a sentence would have numerous quantities of things and pronouns, the whole process goes such that, the

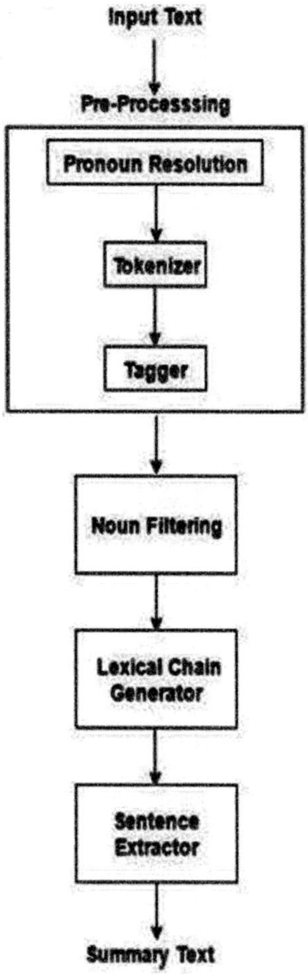

Fig. 1 Existing methodology

model distinguishes specific pronouns' (known as Anaphor) mapped to a noun (known as Antecedent).
  - **Tokenizer**—Tokenizing implies parting the content into negligible important units. It is a required advance before any sort of preparation. One can consider token parts like a word is a token in a sentence, and a sentence is a token in a section.
  - **Tagger**—Tags are useful for building parse trees, which are used in building NERs (most named entities are Nouns) and extracting relations between words. POS Tagging is also essential for building lemmatizes which are used to reduce a word to its root form.

  These are processes carried out on the raw data to extract the required data from it.

- **Noun Filtering**—From the past stage with the assistance of the NLTK toolbox, each word is mapped with their specific POS. Presently this stage channels every one of the things present in the information mapping to its position and dependent on the number of events. As indicated by the creator, Nouns are assumed to be a significant job and which thing has more happened, those lines are chosen.
- **Lexical Chain Generation**—A lexical chain is a succession of related words recorded in writing, spreading over short (nearby words or sentences) or long separations (whole content). A chain is autonomous of the syntactic structure of the content and essentially, it is a rundown of words that catches a part of the firm structure of the content. A lexical chain can give a setting to the goals of a vague term and enable identification of the concept that the term represents.
- **Sentence Extractor**—Based on the previous results of Lexical Chains, Sentences are been selected from the original document without any disrupting the meaning of the sentences. Overall the selection of important sentences from the whole document is combined together forming the summary of the document or file which is been provided as the input.

Well now we know what is text summarization, but let's see types of summarization for the text. It is said that there are two types of summarization.

1.1 Abstractive Text Summarization

Abstractive summarization communicates the contemplations inside the report in different expressions. Procedures use all the additional prevailing regular language preparing methods to interpret the message and make new framework content, rather than picking the most specialist existing choices to play out the rundown. In this procedure, information from source content is re-expressed. Be that as it may, it is increasingly hard to use as it gives unified issues that incorporate semantic portrayals. For example, Book Reviews: If we need a synopsis of any books, at that point by utilizing this technique we can make a synopsis from it. These strategies, for the most part, utilize propelled procedures, for example, NLP to produce a totally new outline. Now and again there are not many pieces of this synopsis that may not, in any case, show up in the first content.

1.2 Extractive Text Summarization

Extractive summarizer focuses on settling on the most appropriate sentences inside the report in the document while holding a limited repetition inside the framework. It is made by reusing segments (word, sentences and so on.) of info message verbatim. For example, Web indexes ordinarily produce extractive synopses from site pages. Generally, the methods include positioning the pertinent expressions to pick just the most important to the information from the source. These strategies depend on extricating a few sections, for example, expressions and sentences, from a bit of content and stack them together to make an outline. Along these lines, distinguishing the correct sentences for a synopsis is of the most extreme significance in an extractive strategy (Table 1).

The brief details of Introduction of text summarization and its types are explained in Sect. 1, whereas in Sect. 2, it contains the related work of text summarization frameworks (i.e., Literature Survey). Coming to Sect. 3, it mostly contains proposed methodology. And at last, but not the least Sect. 4 is concluded with Survey of Research Work with Future directions.

Table 1 Difference between abstractive and extractive text summarization

	Abstractive text summarization	Extractive text summarization
General definition	Involves in generating new phrases and sentences to capture the meaningful sentence	Involves in generating a summary by basis of selection of phrases and sentences from the source document
Method carried out	Method is based on semantic representation and then use Natural language processing techniques	Process where, it just selects a set of words, phrases, paragraph or sentences
Challenging	More challenging approach as the process of understanding and rewriting takes place	Less challenging approach, as the process is of ranking and selection
Grammatical issues	Generates the whole summary, so has to be in proper grammar	No issue with grammar

2 Related Work

Sethi et al. [1], proposed methods in this paper was found out to be working on news articles. It is said that the journalist or the pressman continuously pursue a specific example to distribute a news story. They start the article with "what occurred" and "when it occurred" and proceed in the accompanying sections with an "elaboration of what occurred" and "why it occurred". The Author gained this knowledge and utilizes this knowledge to score the sentence. This takes place by giving the main parts of Speech that is nouns which most of the time appears in the first lines of the document giving a high score. The author analyzed and found out that the first sentence present in any reports always has a high score. This is on the grounds that the report has Nouns were utilized rehashed a few times in the article. This is instinctively reliable since the primary sentence of the article consistently has things which are the names or things, that the article discusses, that is the theme of the article.

Lynn et al. [2] gave an improved framework for mechanization of summary for content fôr Web content utilizing lexical chain with semantic-related terms proposes an improved extractive book outline strategy for reports by refreshing the standard lexical chain technique to improve huge data. By then, the maker at first inquired about the approaches to manage separate sentences from the report(s) in view of the appropriation of lexical chains at that point fabricated a "Transition Probability Distribution Generator" (TPDG) for n-gram catchphrases that learn the characteristics of the apportioned watchwords from the preparation instructive file. Another procedure for customized watchword extraction additionally included in the framework dependent on the Markov chain process. Among the removed n-gram catchphrases, just unigrams are caught to gather the lexical chain.

Gupta et al. [3], stated that sentences from single archive rundowns are grouped and the most elevated sentences from each bundle are used for making a multi-report rundown. The model created by the creators contains the accompanying advances; pre-handling, commotion expulsion, tokenization, stop words evacuation, stemming of words, sentence parting and highlight extraction. As the procedure said is done the most noteworthy sentences are isolated from each bundle and assembled into sections and set up all together. None the less in the event that there are numerous records of a similar name and features, at that point those two documents are been contrasted word with word, sentence by sentence and passage by section and finally, the comparable things are spared into different spots for future references.

Bronselaer et al. [4], evaluated how to utilize information blending methods to merge a huge amount of co-referent records that have been consolidated, while utilizing computational techniques. The standard point of convergence in this paper lies in the f-ideal union capacity which uses the weighted symphonious (harmonic mean) plan to discover a congruity among review and exactness. The overall precision and review gauges referenced are portrayed by techniques for a triangular standard enduring near to exactness and review views as a dedication, so as to make a multi-set of key considerations that can use to make a synopsis.

Krishnaveni and Balasundaram [5] proposed that the automatic content rundown by close by scoring and arranging for improving the knowledge approach gives modified feature-based extractive heading clever substance summarizer to improve the insight in like way improving the understandability of the consolidated substance. It solidifies the given information record utilizing neighborhood/nearby scoring and closes by situating that is, it gives heading astute summation. Headings of a document give reasonable data and grant visual looking at the record to discover the interest substance. The delayed consequences of the evaluation certainly show that heading brisk summarizer gives better exactness, overview and f-measure over the essential summarizer, MS-word summarizer, free summarizer and Auto summarizer.

Saggion and Poibeau [6], stated that the prior methodologies in the content outline concentrated on getting content from lexical chains made during the subject development of the article. These frameworks were favored since they didn't require a full semantic elucidation of the article. The frameworks moreover blended a few in number data sources like a linguistic highlight like POS tagger, the shallow parser for the unmistakable confirmation of apparent gathering, a division estimation, and the WordNet.

Lawrence et al. [7] proposed the repeat of area explicit thoughts as a segment for perceiving outstanding sentences in biomedical works. We displayed an appraisal of a couple of existing outline structures to choose an introduction benchmark. We by then overviewed a top-level continue figuring utilizing the two terms and musings as thing units to show the use of the rehash of contemplations is as persuading, and every so often an improvement over, the utilization of rehash of terms. The makers grew new calculation that is subject to repeat spread showing and assess it utilizing terms comparatively as thoughts. The recurrence distribution algorithm beats a present top tier recurrence based calculation to the detriment of higher computational multifaceted nature. The usage of thoughts can be progressively important in making tweaked traces. An imagined structure enables a client to pick space unequivocal thoughts basic to the client, and in this way have the summarizer make a summary where those musings are more fundamentally weighted than the contemplations showing up in the source content.

Manne et al. [8], stated that the extractive automation of content synopsis approach by sentence extraction using an administered POS naming. A successive term-based substance rundown method with an HMM tagger is organized and executed in java. Positioned sentences are assembled by perceiving the component terms and content abstract is formed. This gave the benefit of seeing the most related sentences as added to the synopsis content. The framework delivered the most compressed synopsis with high caliber and great outcomes in contrast with manual summarization extraction.

Gnes and Radev [9], came up with another strategy for deciding the most significant sentences in a given corpus was talked about. The new strategy, named LexRank, is distinguished by the PageRank technique. This technique works right off the bat by producing a chart, made out of all sentences in the corpus. Each sentence speaks to one node/edge, and the edges are comparability connections between sentences in the corpus. In this research, the authors measure the closeness between sentences by considering each sentence as a bag-of-words model. That is the closeness or similarity measure between sentences is processed by the recurrence of word event in a sentence. The essential estimation is utilizing TF-IDF formulation. This is then utilized as an estimation for the similarity between sentences by utilizing it in this idf-modified cosine equation. This equation is basically estimating the 'distance' between two sentences x and y. the more comparative two sentences, the more 'closer' they are to one another.

Day et al. [10], gave the best execution of quantifiable model is the ROUGE-1 result of KS results, which is quarterly whimsically. In any case, the three inevitable results of RS results have a higher identicalness in one case. The computer-based intelligence insight model uses 1000 randomly processes to prepare and test. The prepared model gets the accuracy/precision which is 82.47% to foresee the token is the bit of the up-and-comer title or not. The result of the significant learning model shows that the two attributes diminished from 0.72 to 0.2 generally. The seq 2seq model was not blending.

Sun and Zhuge [11], proposed the technique which utilizes is Reinforcement Learning. Where the support positioning on the Semantic Link framework can be applied to any associate substance and the game plan of different outline administrations, for example, along these lines making the MindMap of Scientific paper, slides for a given paper and expanded hypothetical for a long consistent paper or book to give customers a catalyst impression of the center substance.

Pourvali and Abadeh [12], proposed calculation depends on WordNet which is hypothetically space autonomous that is domain independent, and furthermore the author has utilized Wikipedia for a segment of the words that don't exist in the WordNet. For synopsis, the creators intended to utilize more union hints than other lexical chain-based blueprint computations. Assessed results were commanding with various outline calculations and accomplished uncommon outcomes. Utilizing the co-occasion of lexical chain people. The calculation tries to fabricate the bond between subject terms and the article terms in the substance.

Paulus et al. [13], introduced a new model and training method that gets best in class brings about text summarization for the CNN/Daily Mail, improves the coherence of the produced summaries and is more qualified to since quite a long output sequence. The model likewise to run on abstractive model on the NYT dataset. The author stated that, in spite of their regular use for assessment, ROUGE scores have their deficiencies and ought not be the main measurement to advance on summarization model for large groupings. The intra-consideration decoder and consolidated training object could be applied to other succession to sequence tasks with long information sources and yields.

Zeng et al. [14], have proposed two straightforward systems to mitigate the issues of current encoder-decoder models. First commitment is a Read Again' model which doesn't frame a portrayal of the info word until the entire sentence is perused. Subsequent contribution is a duplicate system that can deal with out-of-vocabulary words in a principled way enabling us to lessen the decoder vocabulary size and significantly accelerate derivation. They had exhibited the viability of their methodology with regards to synopsis and demonstrated best in class execution. Later on, they intend to handle synopsis issues with huge information content.

Pal [15], proposed a methodology which depends on the semantic data of the concentrates in a text. In this way, various parameters like format, places of various units in the texts are not considered. Be that as it may, in scarcely any cases, there are ordering amounts of named components in a book. In those cases, hybridization of the proposed approach with some specific rules in regards to Named Entity Recognition should give increasingly powerful outcomes.

Hu and Wan [16], proposes a novel system called PPSGen to deliver presentation slides from academic/look into papers. The creator arranged a sentence scoring model that is subject to SVR and usages the ILP methodology to modify and evacuate key articulations and sentences for making the slides. Preliminary outcomes show that the proposed methodology can create numerous ideal slides over standard systems. In this paper, the creator just considers one a customary style of slides that beginners typically use.

Le et al. [17] approach uses a word chart to express source reports. This philosophy fuses two phases. The first stage is the decrease stage and the second is sentence mix. The primary stage depends upon explicit norms and syntactic prerequisites for clearing excess provisos and to finish the finish of the decreased sentence. Word chart is used for sentence mixes and to address word relations between compositions. New sentences are created from the word chart. In word diagram center points are used to address the information about words and their linguistic structure tagger and the proximity relations between word sets are addressed on edges. This strategy makes phonetically right sentences anyway couldn't think less about word meaning.

Gao et al. [18], proposed the algorithm named PASCAL which presents a novel improvement of the algorithm Apriori. This improvement depends on another system called design tallying induction that depends on the idea of key examples. Framework shows that the help of regular non-key examples can be induced from visit key examples without getting into the database. Analyses contrasting PASCAL with the three calculations Apriori, Close and Max-Miner show that PASCAL is among the most effective calculations for mining incessant examples.

Wei and Croft [19], expressed that issue with association rule mining is the over-abundance existing in the evacuated association rules which staggeringly impacts the convincing usage of the removed gauges in dealing with veritable issues. A smart response for the issue should be one that can maximally oust abundance yet doesn't hurt the reasoning furthest reaches of and the confidence in the expelled standards. Furthermore, a fitting measure to portray a point of confinement among abundance and non-redundancy is alluring. In this paper, the creators proposed a concise depiction of affiliation rules called Reliable reason was displayed which can ensure the clearing of the maximal proportion of overabundance without reducing the derivation limit of the remaining removed standards. Moreover, the creators proposed to use the conviction factor as the premise to measure the nature of the discovered association rules.

Farman et al. [20] approach chip away at clustered semantic diagram-based methodology for multiple records abstractive text summarization. This methodology is like genetic semantic chart-based methodology yet here they utilized the clustering algorithm to dispense with excess. In clustering algorithm PASs with highest similarity weight score from each cluster is picked and apply to language. At that point language age rules are utilized to create abstractive rundown sentences. For making group, they utilize hierarchical agglomerative clustering (HAC) calculation. HAC algorithm acknowledges the semantic similitude grid as information. The algorithm combines most comparable groups and updates the semantic similitude network to speak to likeness between the closest bunch and the first group. End-user will choose the compression rate of summary document. This procedure will be rehashed until the client characterized compression rate of rundown is reached.

3 Proposed Methodology

Based on Literature Survey, it is found that the summarization techniques are unique in their own way with respect to document processing, algorithms and final outputs. To overcome the limitations for existing systems, we suggest the following methods.

3.1 Use of Similarity Algorithm for Sentence Extraction

We investigated a few impediments of existing frameworks one of them was single area outline that is calculation just takes a shot at explicit records like News Reports, Scientific papers, Sports. To maintain a strategic distance from this we assume to utilize a cosine similitude calculation which gives better sentence extraction results paying little mind to the kind of report or size of the archive. While removing sentences, we will regard a heading as a general sentence so the framework will perform on archives with or without heading.

3.2 Generation of Cluster

To build a model with higher accuracy, we need to utilize clustering so we can keep away the irrelevant reports, over that the algorithms should have the least time complexity which will help to limit the execution time. Let's state that there are multi-archives which must be condensed into a solitary piece, here this clustering strategy is valuable. This technique will bring all the comparative sentences selected from past strategy and gathering it keeping the key-value sets unblemished, where the key is the keywords and the values are those of the sentences been rehashed. In that capacity, a gathering of groups is been shaped from multi-report, yet in addition from a solitary archive as well. This technique diminishes an opportunity to discover the keywords over and over and its position.

3.3 Position Score Algorithm to Rank the Sentences

To rank the separated sentences, we use position score algorithm. It amplifies the precision rate of the framework. This strategy is used on the earlier formed clustered sentences and ranks the priority of each sentences to be framed. As per the ranks of the sentences, a new summary is been formed. With the assistance of over three strategies we have proposed a system that performs extractive text summarization and for abstractive text summarization we implement the usage of NLP techniques that is semantic words. To do this we are utilizing various parts of text mining. That guides the summarizer with the draft format of the substance. It passes on the significant ideas of the content outline by outperforming the precision rate of existing framework with least time complexity.

4 Conclusion

Nowadays the development of information is expanded in an organized or unstructured form and we need a summarize/diagram from that information in less time. So, there is a requirement for a summarization tool. In this survey paper, we have examined different sorts of content synopsis methods. Furthermore, constraints are found all throughout the papers and to beat the disadvantage of existing models, here we have proposed another model: use of similarity algorithm for sentence extraction, generation of cluster, and position score algorithm to rank the sentences. The proposed system is a work in progress. The presently taken results are giving positive outcome from proposed system.

References

1. Sethi P, Sonawane S, Khanwalker S, Keskar RB (2017) Automatic text summarization of news articles. IEEE, Department of Computer Science Engineering, Visvesvaraya National Institute of Technology, India
2. Lynn H, Choi C, Kim P (2017) An improved method of automatic text summarization for web contents using lexical chain with semantic-related terms. Springer, Berlin, Heidelberg
3. Gupta VK, Siddiqui TJ (2012) Multidocument summarization using sentence clustering. In: 2012 4th international conference on intelligent human computer interaction (IHCI), IEEE, pp 1–5
4. Van Britsom D, Bronselaer A, De Tre G (2015) Using data merging techniques for generating multidocument summarizations. IEEE Trans Fuzzy Syst 23(3)
5. Krishnaveni P, Balasundaram SR (2017) Automatic text summarization by local scoring and ranking for improving coherence. In: Proceedings of the IEEE 2017 international conference on computing methodologies and communication
6. Saggion H, Poibeau T (2013) Automatic text summarization: past, present and future, multi-source, multilingual information extraction and summarization. Springer, pp 3–21

7. Lawrence HR, Hyoil H, Saya VN, Jonathan CY, Tamara AS, Ari DB (2006) Concept frequency distribution in biomedical text summarization. In: ACM 15th conference on information and knowledge management (CIKM), Arlington, VA
8. Manne S, Mohd ZPS, Fatima SS (2012) Extraction based automatic text summarization system with HMM tagger. In: Proceedings of the international conference on information systems design and intelligent applications, vol 132, pp 421–428
9. Gnes E, Radev DR (2004) Lexrank: graph-based lexical centrality as salience in text summarization. J Artif Intell Res 22:457–479
10. Day M-Y, Chen CY (2018) Artificial intelligence for automatic text summarization. In: 2018 IEEE international conference on information reuse and integration for data science, Department of Information Management, Tamkang University, New Taipei City
11. Sun X, Zhuge H Senior Member (2018) Summarization of scientific paper through reinforcement ranking on semantic link network, IEEE, Laboratory of Cyber-Physical-Social Intelligence, Guangzhou University, China Key Laboratory of Intelligent Information Processing, Institute of Computing Technology, University of Chinese Academy of Sciences, Chinese Academy of Sciences, China System Analytics Research Institute, Aston University UK
12. Pourvali M, Abadeh MS (2012) Automated text summarization base on Lexicales Chain and graph Using of WordNet and wikipedia knowledge base. IJCSI Int J Comput Sci Issues 9(1), no 3. Department of Electrical and Computer Qazvin Branch Islamic Azad University Qazvin, Iran Department of Electrical and Computer Engineering at Tarbiat Modares University Tehran, Iran
13. Paulus R, Xiong C, Socher R (2017) A deep reinforced model for abstractive summarization. arXiv:1705.04304v3[cs.CL]
14. Zeng W, Luo W, Fidler S, Urtasun R (2017) Summarization with read-again and copy mechanism, p 111
15. Pal A (2014) An approach to automatic text summarization using WordNet. IEEE, Department of Computer Science and Engineering College of Engineering and Management, Jadavpur University Kolkata
16. Hu Y, Wan X (2015) PPSGen: learning-based presentation slides generation for academic papers. IEEE Trans Knowl Data Eng 27(4)
17. Le HT, Le TM (2013) An approach to abstractive text summarization. Soft Comput Pattern Recognit (SoCPaR), IEEE
18. Gao Y, Xu Y, Fengli Y (2015) Pattern-based topics for document modeling in information filtering. IEEE Trans Knowl Data Eng 27(6)
19. Wei X, Croft WB (2006) LDA-based document models for ad-hoc retrieval. In: Proc. 29th Annu. Int. ACM SIGIR Conf. Res. Develop. Inform., pp 178–185
20. Khan A, Salim N, Farman H (2016) Clustered genetic semantic graph approach for multi-document abstractive summarization. Intell Syst Eng (ICISE), IEEE

Prediction of the Ship Collision Point—A Review

Neelima Roy

Abstract With the goal to improve marine traffic and evasion of any setback, the automotive industry is pushing towards intelligent vehicles. One of the significant difficulties is to identify dangerous situations and respond appropriately so as to stay away from or moderate mishaps. This requires foreseeing/predicting the feasible advancement of the present traffic circumstance and surveying how perilous that future circumstance may be. This paper is an overview of existing techniques for forecast and hazard appraisal for colossal ships on the ocean. A procedure of deciding if at least two bodies are reaching at least one focal point is impact recognition or collision detection. Crash discovery is an indivisible piece of computer graphics, stimulations and apply autonomy. This paper gives a complete characterization of a collision detection writing into the two stages, and we have endeavoured to clarify a portion of the current algorithm which is difficult to translate.

Keywords Marine traffic · Foreseeing · Hazard appraisal · Colossal ships · Collision detection · Focal point

1 Introduction

At the point when one uses any method of transport, mishaps will undoubtedly occur. Mishaps happen due to careless missteps; however, the impacts of the equivalent are enduring and waiting. There have risen and are rising such a significant number of mishap cases that it has become to monitor them. Street mishaps, rail mishaps and air ship crash arrivals are mishaps that everybody today has gotten acclimated catching wind of. Along these lines, even sea mishaps happen, setbacks are caused and harms must be borne. Nonetheless, dissimilar to in the previous three cases, there are a few potential sorts of sea mishaps. The maritime territory is exceptionally immense and thusly the varieties in mishaps are likewise various. The impacts of the events

N. Roy (✉)
Department of Information Technology, SRM Institute of Science and Technology, Chengalpattu, India
e-mail: nr6492@srmist.edu.in

D. J. Hemanth et al. (eds.), *Artificial Intelligence Techniques for Advanced Computing Applications*, Lecture Notes in Networks and Systems 130,
https://doi.org/10.1007/978-981-15-5329-5_27

of marine mishaps incorporate people as well as the marine animals and the marine condition and biological system.

Vessel crash is the name given to the physical impact that occurs between two vessels realizing a hurting setback. This particular accident can moreover occur between a vessel and a relentless or a floating structure like an offshore exhausting stage or an ice retire or even a port. In a great deal of such cases, the effect is pulverizing no doubt in the world. The damage that such a disaster causes cannot just be evaluated with respect to costing or money, in reality it goes past that. With the development in the busy time gridlock on the high seas and the mechanical movements in the marine structuring realizing the headway of overpowering and huge pontoons with unfathomable velocity, the threat of such accidents has extended a lot. Exactly when a vessel sway happens, it has boundless outcomes. Right off the bat, the death toll is constantly an unsalvageable harm and something that can never be made up for. Sadly, the potential outcomes of death toll in such cases are extremely high. Also, the ecological effect is exceptionally negative particularly if any of the vessels in the crash happens to convey any synthetics or some other hurtful material that could be hazardous for marine life.

We have brought to you 12 fundamental known kinds of sea mishaps can be recorded down as pursues:

(A) **Offshore Oil Rig Mishaps**:

Offshore oil rigs establish extraordinary threat as far as their overwhelming apparatus and the complexities of the procedures in question. Indeed, even a minor mistake by method for carelessness of a basic procedure or neglecting in the working of a hardware part can prompt gigantic harming results over the world.

(B) **Voyage Vessel Mishaps**:

Voyage vessels structure a significant part in the get-away schedule of individuals. In any case, a significant kind of oceanic mishap happens in voyage vessels. Voyage vessels could invert or confront intense climate conditions making the ship create serious problems. Another significant instance of mishaps in journey ships is a direct result of the carelessness with respect to labourers. According to factual information, about 75% of flames are caused on account of an insignificant mix-up by individuals chipping away at the voyage transport.

(C) **Commercial Fishing Mishaps**:

In any event, angling for business purposes can prompt lethal occurrences being caused. Unpractised anglers—in some cases even experienced ones—can fall over the edge. Unforgiving climate conditions can likewise could extremely harm to a business angling undertaking.

(D) **Mishaps on Towing Boats**:

Towing boats are those which help move enormous pontoons to enter docks. They are little in nature anyway are mind boggling to ensure that the colossal vessels are managed safely. Regardless, a portion of the time because of the blockage of the visibility of towing boats by the greater vessels, maritime disasters occur. Likewise, human blunder with respect to the pilot of the towing boat can likewise prompt undesirable and unforeseen towing boat incidents.

(E) **Mishaps on Crude Oil Tankers and Cargo Vessels**:

The huge explanation behind disasters on freight tankers is impacted. Since the general thought of the materials these tankers transport is dangerous and significantly burnable, even the most minor of impacts can cause tremendous disasters. As indicated by insights, one of the primary purposes behind oil tanker mishaps happening is a direct result of labourer's carelessness—almost 84–88%.

(F) **Establishing of Vessels**:

A vessel building up happens when the base of the ship's body scratches through the ocean bed. This sort of ocean accident has a lot of impact on the ship's structure and more on the general sea region where the building up has started to occur and has finally finished. The danger to workers on board the ship is another critical outcome in perspective on the misfortune.

(G) **Oceanic Accident in view of Drugs and liquor**:

Drug or substance abuse is a huge issue over the world. To be sure, even in the marine world, substance abuse can produce sad damage. In the event that the labourers of a specific ship participate in substance misuse or liquor, the fixation instigated free for all could make the specialist carry on sporadically and along these lines lead to an undesirable oceanic mishap on board delivers.

(H) **Crane Mishaps**:

Much equivalent to extend undertakings on the land, marine crane exercises on ports and on send is moreover perilous. The danger is furthermore centred on because of the sea assignments where the cranes are required. By virtue of defective wires or winches, crane workers can lose their life or in a most desperate result comprehensible, be alive yet with unsalvageable physical weakening. Then again, mishaps due to extend activities are likewise caused as a result of carelessness and naiveté with respect to the specialist.

(I) **Mishaps in Dockyards**:

The dockyard is where the vessel is collected and built completely. Installing and welding mishaps are regular in the dockyard which could spare the worker his life yet hamper the master's general working limits. Correspondingly steady breathing in of harmful exhaust additionally turns into another shipyard mishap cause.

(J) **Naval Mishaps on Diving Support Ships**:

People who need to examine the riddles of the remote sea use a hopping reinforce strength to wander out into the water. Regardless if the plunging support makes it unsuitable and if the group moreover happens to be genuinely unsuitable to oversee and move the whole action effectively, a critical incident can be generated.

(K) **Mishaps on Flatboat**:

Flatboat mishaps happen principally by virtue of the general type of the scows themselves, which grants them obliged improvement on the water and considering the issues of the cargo vessel towing kinds of rigging. These issues could be delivered in view of innocence regarding the person accountable for the towing vessels or as a result of utilization of broken towing joins.

(L) **Cargo Hauling Accidents**:

Freight pulling sea mishaps are those mishaps caused to labourers who fill in as load haulers. In any case, as indicated by a few sea mishap examinations, it has been accounted for that payload pulling labourers exaggerate their load pulling wounds. The sea mishap examination, importantly reports that along these lines, this calling has one of the most serious paces of workplace non-appearance.

It very well may be seen from the previously mentioned sorts of sea mishaps that slip-ups with respect to the labourers and administrators assume a significant job in the mishaps being caused. Be that as it may, so as to discover what the genuine reason for the marine mishap was, a sea mishap examination is essential. Sea mishap examination will limit on the genuine reason for the mishap which will help the damage inquirers to guarantee their legitimate due with total clearness.

The tracking and monitoring of ship's region and advancement for crash avoidance and control using vessel pioneer. It is essential to extend degree of ocean development blockage; the sea setbacks are one of the creation security stresses in maritime involved time gridlock condition care. To grasp the accident conditions can help the ocean blockage directors to improve the prosperity control of maritime clog. One of the huge challenges is to distinguish hazardous circumstances and react fittingly in order to avoid or direct incidents. This requires anticipating/foreseeing the attainable progression of the present traffic condition and looking over how risky that future situation might be (Fig. 1).

Fig. 1 Type of ship collision

2 Related Work

Considering the inadequacies of the present effect chance figure models, in this assessment, we develop an accident chance desire model reliant on the Classification and Regression Trees (CART) count to anticipate the accident threat of a ship under different conditions. Our proposed model in a general sense involves two areas: beginning, a soft intensive evaluation system to survey the threat of the accumulated models is used to collect an effect peril recognizing confirmation library that contains information subject to ace accident avoiding; second, six basic parts affecting the ship crash risk are considered as the commitment of the CART model, and the degree of danger is resolved using the cushioned total appraisal technique as the certifiable yield for the proposed CART model in order to build up the effect risk model reliant on the CART. Finally, the introduction of the model is differentiated and other effect chance desire models. Our test outcomes show that the proposed model is completely superior to various models the extent that the accident chance affirmation precision and desire speed [1].

The computation shows extraordinary potential for vessel course gauge for medium time horizons approaching around 30 min. Further, it can seek after methods for various recurring patterns. Regardless, the estimation is fragile to the choice of certain decision parameters. Likewise, the calculation cannot manage extending of sea ways and it does not yield any helplessness extent of the estimates. Despite these shortcomings, the figuring may regardless be suitable to proactively help sway avoidance structures [2].

This paper means to join the practical utilization of man-made thinking and present-day course crash avoidance, to deal with the problematic effect issue of vessels which encountering the confounding water an area. Considering the Analytic Hierarchy Process (AHP), Artificial Neural Network (ANN) and different techniques, this paper proposes the T minutes want check. In this figuring, the parameter T shifts with the developments of ship's particular condition, for example, the velocity, the bearing and the air conditions, etc. The paper makes a far-reaching want got together with the genuine needs of the unpredictable condition of the vessel, including the figure time T and the position P where the vessel may land after T minutes. By then anticipate the impact risk a catalyst at the point P as showed by the course information. In the event that there exists a mishap plausibility, the framework will make cautioned in time and gives an impact shirking need outline to the driver, with the target that they can take a dodging measure deliberately early. The clarification behind this examination is to offer functional help to a guaranteed course [3].

In this appraisal, a vessel traffic service/automatic identification system/marine geographic information system (VTS/AIS/MGIS) intertwined arranged structure is proposed to foresee the district and time of every single potential mishap. It ought to be seen that our work depends upon clear advancement acumen. The proposed warm guarding model offers another system to diminish crash episodes because of rash slip-ups, considering the way that impact mishaps typically happen when approach velocity is less in the inward harbour. In addition, the arranged structure gives a record, which can be utilized as check if there should be an occasion of oceanic mishap and play back the condition from start to finish with time stamps. This data can be considered as oceanic mishap confirmation if pilots who overlooked the counsel heading from the VTS official saw in the readied framework. The proposed VTS arranged framework can be applied to pass on to dispatch exchanges and working stages above water as an impact arranged structure [4].

In this paper, information-driven way to deal with oversee vessel course figure for time skylines of 5–30 min using chronicled AIS information is assessed. A social occasion-based Single Point closest Search Technique is asked about near to a novel Multiple Trajectory Extraction Method. Conjectures have been driven utilizing these strategies and separated and the constant velocity method. Moreover, the Multiple Trajectory Extraction Technique is used to review surveyed convey courses [5].

Nonlinear finite element method (FEM) is a necessary resource for separating ship sway issue. The work showed in this paper displays an assessment between FEM numerical results and lab test delayed consequences of a scaled twofold casing structure addressing convey to send crash/setting up circumstances. The general assistant responses (i.e. weight and imperativeness results) and noteworthy disillusionment modes chose from the FEM differentiated well and the exploration office tests. Some express FEM parameters were analyzed, including limit conditions, work size, segment shape and course, break strain and pounding. The middle is to best match FEM re-enactment to the best with test discernments. This activity supports the use of numerical diversion strategies to move crash issues and provides information and guidance into a bit of the key numerical showing procedures and controls required in the propagation of these mind-boggling fundamental cooperation issues [6].

This paper portrays a thought for an effect avoiding structure for ships, which relies upon model judicious control. A constrained game plan of elective control practices is made by moving two parameters: balances the bearing course guide taught towards the autopilot and alters to the motivation course stretching out from apparent speed to full transform. Using reproduced measures of the headings of the tangles and ship, consistence with the Convention on the International Regulations for Preventing Collisions at Sea and crash risks related with the entirety of the elective control practices are assessed on a constrained want skyline, and the ideal control lead is picked. Liberality to distinguishing mess up, predicted hindrance direct, and natural conditions can be guaranteed by assessing different conditions for each control lead. The procedure is reasonably and computationally fundamental yet completely versatile as it can address the parts of the vessel, the segments of the organizing and impetus structure, controls considering wind and sea to and fro development, and any number of tangles. Re-enactments show that the methodology is suitable and can direct complex circumstances with different one of a kind obstructions and weakness related with sensors and desires [7].

Given a lot of oceanic traffic information amassed from AIS structure, we attempt to find vessels' upgrades that have battle practices and these practices may obtain a potential impact the event that they don't make any dubious move. We propose a structure of clustering and detection to regularly find the social events of conflict heading from AIS bearing information in an exhibition way. So as to beat trademark and multifaceted nature of the sea AIS information, the proposed structure merges two stages, experience packaging and battle conspicuous verification. The experience pressing is made to look at a great deal of boats' direction are experiencing each other from gathered AIS oceanic headings. By at that point, conflict affirmation is proposed to survey the conceivable conflict for each heap of comprehension. At last, the bunches of sea conflict headings can be found. In context on genuine AIS information, the primer results show that the proposed clustering-and-detection system can practically find sets of holding on for battle circumstance from sea AIS traffic information [8].

In requesting to move sea security and decreasing the death toll and property, and accomplish a dynamically secure course, cleaner seas goal, and this article reviewed the vessel congestion control success from four bits of the vessel, to be express "vessels, course condition, individuals, the heads". By investigating the procedure for the flourishing evaluation, this article gives a basic reference of the vessel security, and a persuading hypothesis reason behind the significant association measures. Legitimately off the bat, considering the oceanic congestion control mishaps, the four bits of the vessel congestion control thriving were gotten. Furthermore, the model was required to review the vessel congestion control flourishing by methodologies for pro evaluation and utilizing the strategy for AHP and padded absolute examination. The objective layer of vessel congestion control security assessment model is vessel congestion control success. The model layers are the ship, course condition, work control and the managers. The 14 overview layers are associated with the optional pointers. They are the structure of the body, the particular state of the packaging, the payload, the atmosphere, the ocean course, the computing vessels, the gathering business authority level, the social affair security wisdom, the get-together mental quality, the dispersing of the get-together individuals, the voyager, the association of skipper, the related corporate association, the association of ship proprietor. By managing aces, the enormity of each record was made. At last, the examination model was applied to assess and break down the ship traffic control thriving. This model was exemplified in one vessel congestion. The outcomes show that the assessment arrangement of vessel congestion control flourishing is sensible astonishing and doable [9].

A moving boat target location and following strategy dependent on the mind-boggling foundation of the extension region is displayed. The foundation model dependent on a changed Gaussian Mixture Model. The computation redesigns the establishment model in certified hours foremost part range is evacuated by flexible edge system by indicating each operational region in the matched picture, the geometrical features variable for instance, width, stature, centroid location and velocity can be isolated, which build up the structure for the examination and track of moving. Vessel development condition calculation module subject to visual geometry model, and figures the improvement state information of spotlight on vessel's circumstance in space orchestrate, vessel's size, velocity, heading, etc. Impact hazard forecast and crash shirking moving basic leadership module joined ship on-going movement state data, and bridge region vessel's movement, acquainting the crash chance with rank term, and early cautioning and impact evasion moving choice making. Accuracy in the paper is not just chosen by the calculation yet in addition by the particular test variables and related data, such as sensor variables, bridge environment variables, condor, relevant judgment standard, etc. All of these are fundamental to the objectives recognizable proof and hazard level expectation [10].

This paper depicts a technique for creating Probability Density Functions (PDFS) portraying struck ship harm in dispatch impacts. Struck and striking boat speed, crash edge, striking boat type and striking boat removal are treated as free arbitrary factors in this issue. Other striking boat attributes are treated as needy factors got from the autonomous irregular factors dependent on connections created from overall ship information. A simplified collision model (SIMCOL) is utilized in a Monte Carlo reproduction to foresee probabilistic harm degrees. SIMCOL applies the situation factors legitimately in a period venturing concurrent arrangement of inside (auxiliary) twisting and outer (ship) elements. Results are exhibited for crashes with four notional tanker plans [11].

This paper proposes another methodology to enable the evading to route of a vessel inside a straight range by exhibiting the accident probability sea range to each vessel through a web system. This assessment focused on crash at the exit of a sea range. By then, the vessel's new range at the leave point was foreseen by the assistance vector machine, which examined oceanic congestion data. The accident probability sea area at the leave location was resolved for each vessel by considering the obstacle range through the proposed procedure and the data of the foreseen vessel's new route. The feasibility of the proposed philosophy was displayed by the re-institution outcome [12].

The objective of this work is to develop a before time notice structure for angling barges hapless to avoid crash with ships. A preservationist stunned point to many-point lengthy-go little-control offshore broad range correspondence orchestrates vessels has been made by our assessment centre. The framework is watched and regulated by a waterfront Network Operations Centre (NOC). The early advised structure gains the range information of vessels from the NOC and that of pontoons from the Automatic Identification System (AIS). It predicts crashes by following the bearings of vessels and ships and alerts the boats in hazard with the objective that they will get adequate chance to move out of the vessel's way. In this work, we have viably made and prototyped a reasonable and strong effect evading structure using the zone information gained from the angling vessels and the pontoons. The shrouded correspondence composes for vessels gives a wide incorporation of in excess of 32 nm into the sea and even past that innovatively. The zone information is conveyed remotely to the NOC which assesses the likelihood of an effect and passes an appropriate alert to the angling vessels at whatever point required. The structure passes an advancement alert to the angling vessel thusly giving enough time to the angling vessel to make appropriate preventive move. The exploratory course of action was gone after for a couple of circumstances and the results were dismembered and saw as tasteful. We can reliably anticipate the probability of effect of angling vessels and pontoons using this structure and alert the fishermen early with the objective that they will get enough time to avoid real incidents [13].

In this paper, we propose conflict-discoverer to give a system to oceanic traffic conflict mining. Not quite the same as existing techniques those emphasis on identifying the clashes between two vessels in a confined conduit, we find the clashes happened by many transports in untamed ocean. For investigation of oceanic congestion clashes, a model of conflict-discoverer is actualized which assists with increasing a superior comprehension of traffic conflicts found and can be applied to the betterment of sea congestion security assessment and the board. In light of an enormous AIS direction information gathered, we centre around mining the boats' development practices those may bring a potential crash in the event that they don't take any evasion, called oceanic congestion clashes. Despite the fact that the sea congestion clashes are a non-mishap occurrence, the development practices of sea congestion clashes may have the comparable practices of maritime impact for investigation [14].

Another hazard recreation model dependent on Markov Chain-Monte Carlo calculation (MCMC) is suggested for oceanic congestion chance investigation close to land. While showing Markov Chain-Monte Carlo figuring, we can acknowledge that the danger can walk indiscriminately in all possible complete designated territorial states. During reproduction, we can transform each factor in turn, yet the perception variable stays consistent. Consequently, when the arbitrary testing strategies are streamlined, the proficiency of reproduction is improved. Also, utilizing Markov model and MCMC recreation for oceanic congestion hazard, and producing hazard tests through the stochastic difference in the past condition of oceanic congestion chance in assigned waters, test age calculation is changed which creates test freely according to every occasion previously. Prototypes confirm that we can consider the pattern of hazard advancement under time arrangement by presenting MCMC calculation and using territorial state progress [15].

This paper shows an autonomous development orchestrating computation for unmanned surface vehicles (USVs) to investigate safely in one of a kind, muddled circumstances. The estimation not simply will in general hazard avoiding (HA) for stationary and moving risks, yet moreover put in the International Regulations for Stopping Mishaps at Sea (known as COLREGS, for Collision Regulations). The COLREGS rules show, for example, which vessel is liable for offering course to the following and to which side of the "stay on" vessel to move. Three basic COLREGS rules are considered in this paper: convergence, overpowering, and head-on situations. We furthermore show a use of this development coordinator to a target trailing task, where a key coordinator headings USV waypoints reliant on raised level goals, and the close by development coordinator ensures chance evading and consistence with COLREGS during a cross [16].

This paper proposes an unaided and gradual learning way to deal with remove the authentic traffic designs from AIS information. The introduced technique called Traffic Route Extraction for Anomaly Detection (TREAD) successfully forms crude AIS information to derive various degrees of logical data, spreading over from the recognizable proof of ports and seaward stages to spatial and fleeting disseminations of traffic courses. Besides, the exact comprehension of the verifiable traffic empowers the characterization and forecast of vessel practices just as the discovery of low-probability practices, or irregularities. A definitive objective is to furnish administrators with a configurable information system supporting step by step basic leadership and general attention to vessel example of life movement. The strategy is exhibited by means of a genuine contextual investigation, which can be utilized as a source of perspective informational collection for further examination [17].

In this paper, the fuzzy surmising framework joined with a specialist framework is applied to impact shirking framework. Particularly, figuring technique for the crash chance by utilizing neural system is proposed. From the start, the participation elements of Distance to the Closest Point of Approach (DCPA) and Time to the Closest Point of Approach (TCPA) are resolved based on reproduction results utilizing the KT conditions. Also, a while later, the inducing table is redesigned by using the Adaptive Network-based Fuzzy Inference System (ANFIS) count. Moreover, additional elements, the vessel territory, topological traits and bound visibility, which can impact pilot's thinking about the accident chance other than DCPA and TCPA, are considered. Finally, Multilayer Perceptron (MLP) neural framework to the effect avoiding system is petitioned to make up for fuzzy rationale [18].

By taking Istanbul Strait as a model for this examination, we expected to build up a choice emotionally supportive network as well as a guidance method to help correspondingly passing vessels. The fundamental reason for existing is to build up an Artificial Neural Network (ANN) that utilizes the information of physically controlled vessels to produce forecasts about the future areas of those vessels. On the off chance that there is any plausibility of crash, this framework is expected to caution the administrators in the vessel traffic services (VTS) focus and to direct the workforce of the vessels. In this investigation, physically controlled and equally passing vessels' information was utilized (counting directions, speed, and natural conditions), neural systems were prepared, and forecasts were made about the areas of vessels three minutes after the underlying purpose of assessment (this length was dictated by thinking about the states of Istanbul Strait). With this reason, we utilized information assembled from vessels and demonstrated the accomplishment of the framework, particularly concerning forecasts made during turnings, by deciding the plausibility of impact between two vessels three minutes after the information was accumulated [19].

This paper features how the programmed identification system (AIS) information can be utilized in the investigations of ship impact in occupied conduits. In this unique circumstance, AIS information off Rotterdam Port in Europe is gathered as a pilot concentrate to create straight relapse models. Furthermore, in light of SAMSON, a dynamic chance technique is created for checking traffic and it is received and tried in the European undertaking MarNIS. The investigation appears: (1) AIS gives exact continuous information, which are important in traffic research and hazard examination in occupied conduits; (2) the intricacy of traffic qualities and CPA is found by factual investigation and a relapse model; (3) SAMSON, which depends on the long haul introduction information and reached out with dynamic parameters, for example, multipliers of TCPA, CPA, and experiencing point, is a pragmatic apparatus in real-time impact hazard examination [20] (Table 1).

Table 1 Algorithms/techniques and drawback of existing system

S. no.	Paper title	Algorithms/techniques	Drawback
1	Prediction of ship collision risk based on CART	Fuzzy comprehensive evaluation method, Classification and Regression Trees (CART)	The feature dimension is low and the sample size is small
2	AIS-based Vessel Trajectory Prediction	Single Point Neighbour Search (SPNS) method	The calculation is delicate to the decision of certain choice parameters. Likewise, the calculation cannot deal with stretching of ocean paths and it doesn't yield any vulnerability proportion of the expectations
3	An Intelligent collision avoidance algorithm research	The Analytic Hierarchy Process (AHP), Artificial Neural Network (ANN)	The algorithm gives a list of collision point to the user and then the user check it manually the collision point and then the user will notify the captain of the ship. It is a time-consuming process
4	A new method of collision avoidance for vessel traffic service	Fuzzy logic method, AHP method (Analytic Hierarchy Process)	The framework can't ascertain the timing and edge of rudder required for restorative activity, which can be conveyed from the VTS administrator to the guide
5	A Data-Driven Approach to Vessel Trajectory Prediction for Safe Autonomous Ship Operations	Single Point closest Search Technique, Multiple Trajectory Extraction Method	It is anyway unequipped for dealing with traffic division, where such cases bring about poor course over ground gauges and all things considered incorrect position gauges
6	Using Numerical Simulation to Analyze Ship Collision	Nonlinear finite element method (FEM)	It supports the use of numerical diversion strategies to move crash issues and provides information and guidance into a bit of the key numerical showing procedures and controls required in the propagation of these mind-boggling fundamental cooperation issues. It doesn't give more accuracy

(continued)

Table 1 (continued)

S. no.	Paper title	Algorithms/techniques	Drawback
7	Ship Collision Avoidance and COLREGS Compliance Using Simulation-Based Control Behaviour Selection With Predictive Hazard Assessment	Collision hazard avoidance method based on simulation and optimization is based on brute force evaluation	The calculation is inconsequential to actualize with parallel preparing and the ship elements are generally moderate, this isn't viewed as a significant useful constraint
8	A Framework for Discovering Maritime Traffic Conflict from AIS Network	Clustering algorithm using Euclidean distance	For every ship movement, there is an update of the values and such this algorithm is not appropriate for recursive updates
9	Evaluation of Ship Traffic Control Safety Based on Analytic Hierarchy Process	Analytic Hierarchy Process, fuzzy comprehensive evaluation	So as to precisely anticipate the security of the ship traffic control, the components of the ship traffic control wellbeing must be investigated all-a roundly. It takes long time for analyzing all the components
10	Intelligent Ship-Bridge Collision Avoidance Algorithm Research Based on A Modified Gaussian Mixture Model	Gaussian Mixture Model	To find the level of risk involved in collision between the ships uses minimal distance and not for the wide region
11	Probabilistic method for predicting ship collision damage	Probability Density Functions (PDFS), A simplified collision model (SIMCOL)	Takes times to process each iteration and predict the damage based on the kinetic energy of the ship
12	Feasibility Study for Predicting Collision Possibility Sea Area for Each Ship by Using Support Vector Machine	Support Vector Machine	The legitimacy of the anticipated ship's behaviour was not guarantee safe route
13	Harnessing Low-Cost Marine Internet for Collision Avoidance of Vessels at Sea	Distance of Closest Approach, Time of Closest Approach	In the event that the separation of nearest approach is from 20 to 50 m, a data alert is given. For separations more prominent than 50 m, no alarm is given

(continued)

Table 1 (continued)

S. no.	Paper title	Algorithms/techniques	Drawback
14	ConflictFinder: Mining Maritime Traffic Conflict from Massive Ship Trajectories	Clustering algorithm	Mining the ships' movements is proposed in this research. It does not give navigation collision analysis
15	Spatial Markov Chain Simulation Model of Accident Risk for Marine Traffic	Markov model, Markov Chain-Monte Carlo algorithm (MCMC)	On the off chance that the information isn't inside the extent of the autocorrelation test, the information would then be able to be made a decision as invalid, for example simulation results are mistaken
16	Safe Maritime Autonomous Navigation With COLREGS, Using Velocity Obstacles	Maritime navigation algorithm, Velocity obstacles (VO) method	More accurate solution cannot be found
17	Traffic knowledge discovery from AIS data	Traffic Route Extraction for Anomaly Detection (TREAD), Unsupervised and Incremental Learning Approach	If the knowledge discovery is inaccurate then vessel pattern will also be inaccurate
18	A study on the collision avoidance of a ship using neural networks and fuzzy logic	Adaptive Network-based Fuzzy Inference System (ANFIS), Multilayer Perceptron (MLP)	Implementing Fuzzy Logic and Neural Network to work together is time consuming
19	Decision support system for collision avoidance of vessels	Artificial Neural Network (ANN), Levenberg-Marquardt Learning Algorithm	The framework is delicate to contravention
20	Study on collision avoidance in busy waterways by using AIS data	Linear Regression model, SAMSON a dynamic risk method	The procedure can be alluded to in other safe zones; however, the utilization of the hazard model may make deviation from zone territory and tune ought to be altered to meet the genuine circumstance

3 Conclusion

In recent days, the development of sharing of locations of ships is feasible through the network has increased tremendously, and due to which we are able to locate the position of the ships. And with this, we are going to implement a model which can predict the collision points of the ships. This is a survey paper, where we have examined different sorts of clustering algorithms and predictions of collision points taken place until now. Considering the pros and cons in the list of reviewed papers, we are going to build a model that is going to predict the collision points with higher accuracy.

References

1. Li Y, Guo Z, Yang J, Fang H, Hu Y (2018) Prediction of ship collision risk based on CART. IET Intell Transp Syst 12(10):12
2. Hexeberg S, Flaten AL, Eriksen B-OH, Brekke EF (2017) AIS-based vessel trajectory prediction. In: 20th International conference on information fusion, Xi'an, 10–13 July 2017
3. Wang X, Zheng R, Simsir U, Xiao Y (2016) An intelligent collision avoidance algorithm research. In: 9th International congress on image and signal processing, BioMedical Engineering and Informatics (CISP-BMEI)
4. Kao S-L, Su C-M, Cheng C-Y, Chang K-Y (2007) A new method of collision avoidance for vessel traffic service. In: International conference maritime technology
5. Murray B, Perera LP (2018) A data-driven approach to vessel trajectory prediction for safe autonomous ship operations. In: 13th International conference on digital information management (ICDIM)
6. Wu F, Spong R, Wang G (2004) Using numerical simulation to analyze ship collision. In: 3rd International conference on collision and grounding of ships (ICCGS), Izu, 25–27 Oct 2004
7. Johansen TA, Perez T, Cristofaro A (2016) Ship collision avoidance and COLREGS compliance using simulation-based control behaviour selection with predictive hazard assessment. IEEE Trans Intell Transp Syst
8. Lei P-R, Tsai T-H, Wen Y-T, Peng W-C (2017) A framework for discovering maritime traffic conflict from AIS network. In: 19th Asia-Pacific network operations and management symposium (APNOMS)
9. S Ren (2010) Evaluation of ship traffic control safety based on analytic hierarchy process. In: International conference on intelligent computation technology and automation
10. Zhang W, Zheng Y (2011) Intelligent ship-bridge collision avoidance algorithm research based on a modified gaussian mixture model. In: International conference on multimedia technology, Hangzhou, 26–28 July 2011
11. Brown A, Chen D (2002) Probabilistic method for predicting ship collision damage. Oceanic Eng Int 6(1):54–65
12. Okazaki T, Terayama M, Nishizaki C (2018) Feasibility study for predicting collision possibility sea area for each ship by using support vector machine. In: IEEE international conference on systems, man, and cybernetics (SMC)
13. Rao SN, Balakrishnan A (2017) Harnessing low-cost marine internet for collision avoidance of vessels at sea. In: International conference on wireless communications, signal processing and networking (WiSPNET), Chennai, 22–24 Mar 2017
14. Lei P-R, Tsai T-H, Wen Y-T, Peng W-C (2017) "ConflictFinder: mining maritime traffic conflict from massive ship trajectories. In: 18th International conference on mobile data management

15. Xuan S, Xi Y, Huang C, Hu S, Zhang L (2017) Spatial markov chain simulation model of accident risk for marine traffic. In: 4th International conference on transportation information and safety (ICTIS), Banff, 8–10 Aug 2017
16. Kuwata Y, Wolf MT, Zarzhitsky D, Huntsberger TL (2014) Safe maritime autonomous navigation with COLREGS, using velocity obstacles. IEEE J Oceanic Eng 39(1)
17. Pallotta G, Vespe M, Bryan K (2013) Traffic knowledge discovery from AIS data. In: 16th International conference on information fusion, Istanbul, 9–12 July 2013
18. Ahn J-H, Rhee K-P, You Y-J (2012) A study on the collision avoidance of a ship using neural networks and fuzzy logic. Appl Ocean Res 37:162–173
19. Simsir U, Amasyah MF, Bal M, Celebi UB, Ertugrul S (2014) Decision support system for collision avoidance of vessels. Appl Soft Comput 25:369–378
20. Mou JM, van der Tak C, Ligteringen H (2010) Study on collision avoidance in busy waterways by using AIS data. IEEE Ocean Eng 37(5):483–490

Prediction of Cardiovascular Diseases in Diabetic Patients Using Machine Learning Techniques

K. Hemanth Reddy and G. Saranya

Abstract Diabetes and its related consequences are some of the most alarming health problems the world is currently facing. Recent studies have shown that one-fourth of adults all over the world are overweight and more than one-tenth are obese. Diabetes is identified as a major risk factor for determining the severity of coronary artery disease (CAD). Diagnosing a heart disease is a very tedious task. In this paper, several attributes like age, BMI, glycated haemoglobin, triglyceride, total cholesterol, HDL, LDL, insulin resistance, smoker, alcoholic, and family diabetic history are identified as major causes for chronic diseases. We have explained how different attributes are related to each other and how various machine learning models are implicated and applied to them.

Keywords Lipoproteins · Diabetes · Triglycerides · Cardiovascular · hs-CRP · HomoIR

1 Introduction

Diabetes mellitus also known as diabetes is a chronic disease which occurs when the glucose% in your blood is very high. Glucose is one of the main sources of energy which comes from the food we eat. Insulin is a hormone which regulates the level of blood sugar in our body. Diabetes is a common disease which mostly occurs when the pancreas does not produce enough insulin. It also arises when the body does not use the insulin provided. Hyperglycaemia called as raised blood sugar is a common effect of uncontrolled diabetes. Overtime of hyperglycaemia often causes vital damage in the body, it affects the nerves, eyes which may result in eye cataract, and it also results in kidney failure and damage in blood vessels.

K. Hemanth Reddy (✉) · G. Saranya
Department of Information Technology, SRM Institute of Science and Technology, Chengalpattu, India
e-mail: hk6027@srmist.edu.in

G. Saranya
e-mail: saranyag3@srmist.edu.in

D. J. Hemanth et al. (eds.), *Artificial Intelligence Techniques for Advanced Computing Applications*, Lecture Notes in Networks and Systems 130,
https://doi.org/10.1007/978-981-15-5329-5_28

The different types of diabetes are as follows:

(a) **Type-1 Diabetes**: This occurs when the body does not produce enough insulin. This kind of diabetes is usually seen in children and teenagers; it may occur at any age. It shows an impact on the immune system of the body and destroys all the cells which make pancreas.
(b) **Type-2 Diabetes**: Mostly seen in people of middle and old age category. It is the most common disease in diabetes seen in many parts of the world. It occurs when the body does not make insulin or make use of it efficiently.
(c) **Gestational Diabetes**: It is if usually found in women when they are pregnant. Generally, gestational diabetes will be cured when a child is born. People having this diabetes have a greater chance of developing type-2 diabetes.

People having diabetes may get affected by various diseases such as heart disease, stroke, kidney problem, vision loss, dental disease, nerve damage, and foot problems.

Of all these diseases, cardiovascular diseases are more likely to occur for patients who are having diabetes. According to statistics released American Heart Association, it is seen that there is a strong correlation between diabetes and cardiovascular diseases.

Around 70% of people having age 60 and above and having diabetes usually die from some heart disease. Adults with diabetes are highly prone to die from cardiovascular diseases than adults without diabetes. The American Heart Association finds diabetes to be one of the main controllable risk factors for cardiovascular disease.

According to the Framingham heart study, in the prediction of heart diseases, they observed that there are various predictors such as age, gender, fasting glucose, body mass index (BMI), high-density lipoprotein, triglycerides, blood pressure, parental history of diabetes which may contribute to the cardiovascular diseases and heart attacks.

1.1 Glycated Haemoglobin

Glycated haemoglobin test has various names like haemoglobin A1C, HbA1C, HbgA1c, glycosylated haemoglobin test, and A1C test. The A1C test is a blood test in which it gives the average levels of blood glucose over the past 3 months. If haemoglobin A1c levels lie between 5.7 and 6.4%, then they have a greater prone of getting diabetes. Having levels above 6.5% or higher indicates that the person has diabetes. (The higher the A1c, the higher the risk or diseases).

1.2 Fasting Plasma Glucose

It is a test that indicates the blood glucose level in the body. Increase In FPG results in hyper glamea which may result in kidney damage, heart attack—or other cardiovascular complications, nerve damage, and stroke.

1.3 Cholesterol

Cholesterol is a kind of fat produced by the liver and found in your blood. If the levels of cholesterol are less than 200, then it is desirable; else, if it is between 200 and 239, it is high, and if it is more than 240, then there is a greater risk of getting a heart stroke.

The two types of cholesterol in the blood are:

(1) **Low-Density Lipoprotein**: It is also called "bad" cholesterol, and it builds up and clogs blood vessels in the arteries. If the level of LDL is >190, then the risk is very high.
(2) **High-Density Lipoprotein**: It is a lipoprotein having high density and is also called good cholesterol. If the level of HDL is less than 40 in the body, then there is a greater risk for the heart.

1.4 Triglycerides

Having a high amount of triglycerides in your blood can increase your risk of heart disease. Highly triglycerides may often lead to the hardening of the arteries and thickening of the artery walls which can increase the chance of getting a stroke or a heart attack and further leads to heart disease. If the triglyceride level is more than 500, then there is a greater chance of getting a stroke.

1.5 hs-CRP

hs-CRP is a test used to find if there is any chance of developing a coronary artery disease, it is a disorder in which the arteries of the heart are compressed, and further it may lead to a heart attack.

The chance of getting a stroke is lower if the hs-CRP level is less than 2, and the risk is higher if the hs-CRP level is greater than 2. If the CRP level is higher, then the doctor may also request for the cholesterol test.

1.6 Homa_Ir

IR stands for insulin resistance. It is called an homeostasis model assessment-estimated insulin, which is also an indicator for predicting the heart disease in type 2 diabetes.

All these attributes that are demonstrated above are the factors that contribute to heart disease.

2 Related Work

Parthiban et al. [1] has proposed this system that comes under the category of machine learning. The system detects the major cause of disease in diabetic patients using support vector machines. Support vector machines have a classifier called SVM classifier. It divides the data into hyperplanes based on the support vectors. In this paper, they have also given brief information regarding data mining in which it helps in understanding the data patterns in heart data.

Parveen and Pattekari [2] has proposed this system using Naïve Bayes method. Naïve Bayes uses the Bayesian method for classifying the data. They made a web-based system which user answers the predefined questions. Naive Bayes is a classification technique used in machine learning. Naive Bayes works on the Bayesian approach. They have made this to assist healthcare practitioners and to make intelligent clinical decisions.

Patil [3] has proposed this system using Naïve Bayes and they also have a smoothening technique called Jelinek-mercer smoothening which is a smoothening technique for the attributes in naïve Bayes. Smoothening is an important technique used in capturing the important patterns in the data to avoid noise and other structures. They have made a web-based prediction system where the user enters the data manually. This draws hidden knowledge from the heart disease database. They used 13 attributes from the dataset for the calculation and predictions.

Folsom et al. [4] has proposed a system for predicting the ten-year probability of chronic heart disease in middle-aged adults of diabetes. They measured risk factors of nearly 15,000 adults and identified various parameters affect the heart disease. Based on their observation from the data they have found various risk factors that contribute to cardiovascular diseases.

Dinesh Kumar [5] has proposed a system that explains about the prediction of cardiovascular disease using machine learning techniques. In this paper, they have chosen the data from the VA heart disease database in the UCI machine learning repository for predicting the heart disease. Various machine learning techniques such as Logistic Regression, Support vector machines and Naive Bayes are used for analysing the data and estimate the occurrence of disease.

Prasad et al. [6] has also proposed a system that predicts heart disease using Logistic regression. Logistic regression is a classifier that classifies whether the person has a chance of getting heart disease. They have used various algorithms such as Logistic regression, Naïve Bayes, Comparing and confusing matrices.

Repaka et al. [7] has implemented a naïve Bayesian model in designing and implementing heart disease prediction. They built an effective cost-cutting approach for applying data mining techniques so that they can enhance the decision support system. Predicting the heart disease as per the numerical attributes and symptoms is quite complicated so, they have used Naïve–Bayesian data mining classification technique. The prediction is made on the UCI repository data. And the results generated from the system built by them have predicted the level of risk associated with heart disease.

Jabbar and Samreen [8] has proposed a hidden naïve Bayes classifier for the prediction of the heart disease. The hidden Naïve Bayes model generates more accurate results when compared with the Naïve Bayes. The paper describes that it takes influence from all the features and avoids complexity in the system. The Hidden Naïve Bayes creates a parent for each of the feature and integrates the influences from other features. Hidden Naïve Bayes algorithm requires more training time because of its parent creation from each feature. In this paper, they have used Weka 6.4 tool to apply hidden Naïve Bayes classifier. They used heart dataset which is downloaded from the UCI Repository to run their model.

Purushottam et al. [9] proposed a system for heart disease prediction. It was said that it would enhance medical care and reduces medical costs. They designed a system that defines the rules for predicting the patient's risk level on the basis of the parameters of their health. Rules can also be prioritized based on the parameters. In this paper, the data is taken from the Cleveland database which contains 76 attributes. After doing various features extraction techniques they have taken 14 attributes of them. The dataset used in this paper has multiple important parameters like cholesterol, chest pain, heartbeat rate and other attributes. In this paper, they used Knowledge Extraction based on Evolutionary Learning tool is a java tool used in accessing the algorithms for the data mining problems.

Kohli and Arora [10] presented the application of various classification algorithms on different diseases like Heart, Breast cancer, Diabetes. All these disease data is taken from the UCI machine learning repository for disease prediction. Using P-test, the selection of features for each data set was achieved through a technique called backward modelling. In this paper, the results of this study gave us an idea of the use of machine learning in the early detection of various cardiovascular diseases. Various machine learning algorithms such as Random Forest, Decision Tree, Logistic Regression, Support Vector Machine, Adaptive boosting are used in the prediction and detecting the early occurrence of various diseases.

Ambekar and Phalnikar [11] proposed a system on heart disease prediction using KNN algorithm and Naïve Bayes. In this paper, they applied the Naïve Bayes algorithm to predict the occurrence of disease. To extend their work, they have proposed a model for disease risk prediction using structured data. They have also used CNN algorithm for the prediction of unimodal disease risk. In deep learning, a convolutional neural network is an important technique in which the features are automatically extracted to get the appropriate result. CNNUDRP is an algorithm mostly used in extracting the values and required features from the dataset. In this paper, to predict the heart disease they have used the dataset extracted from the UCI Repository which consists of 12 attributes. By using CNN-UDRP algorithm it was given a 64% accuracy.

Meenakshi and Niranjana Murthy [12] has proposed a model for predicting coronary heart disease based on various risk factors. They have explained about the dimensionality reduction using a model called neurogenetic approach for early detection of CHD. They have used multivariate datasets in order to get good percentage in testing the accuracy. They have applied neural networks for prediction of heart disease on Ucl machine learning repository.

3 Conclusion

This paper gives an idea about the analysis and applications of machine learning techniques such as Regression, Classification and various types to identify the heart disease risk in diabetic patients. It also explained various feature selection and extraction techniques in the study. We have given a clear understanding of the attributes and how cardiovascular diseases are related to them. Our future scope includes predicting other diseases in diabetic patients using the latest technologies and advanced machine learning concepts. In this paper, we have examined various classification algorithms used in heart disease prediction. Considering the pros and cons in the list of reviewed papers, we are going to build a model that is going to predict the heart diseases in diabetic patients with better accuracy.

References

1. Parthiban G, Rajesh A, Srivatsa SK (2012) Diagnosing vulnerability of diabetic patients to heart diseases using support vector machines. Int J Comput Appl 48(2):888–975
2. Praveen A, Pattekari SA (2012) Prediction system for heart disease using naive Bayes. Int J Adv Comput Math Sci 3(3):290–294. ISSN 2230-9624
3. Patil RR (2014) Heart disease prediction system using Naive Bayes and Jelinek-mercer smoothing. Int J Adv Res Comput Commun Eng 3(5
4. Folsom AR, Duncan BB, Gilbert AC (2003) Prediction of coronary heart disease in middle-aged adults with diabetes. Diabet Care
5. Dinesh Kumar G (2018) Prediction of cardiovascular disease using machine learning algorithms. In: Proceeding of 2018 IEEE international conference on current trends toward converging technologies, Coimbatore
6. Prasad R, Anjali P, Adil S, Deepa N (2019) Heart disease prediction using logistic regression algorithm using machine learning. Int J Eng Adv Technol (IJEAT), 8. ISSN 2249-8958
7. Repaka AN, Ravikanti SD, Ramya GF (2019) Design and implementing heart disease prediction using Naive Bayesian. In: Proceedings of the third international conference on trends in electronics and informatics (ICOEI 2019), IEEE

8. Jabbar MA, Samreen S (2016) Heart disease prediction system based on hidden Naïve Bayes classifier. In: International conference on circuits, controls, communications and computing (I4C)
9. Purushottam, Saxena K, Sharma R (2015) efficient heart disease prediction system using decision tree. In: International conference on computing, communication and the automation (ICCCA 2015)
10. Kohli PS, Arora S (2018) Application of machine learning in disease prediction. In: 2018 4th International conference on computing communication and automation (ICCCA)
11. Ambekar S, Phalnikar R (2018) Disease risk prediction by using convolutional neural network. In: 2018 Fourth international conference on computing communication control and automation (ICCUBEA)
12. Meenakshi M, Niranjana Murthy HS (2014) Dimensionality reduction using neuro-genetic approach for early prediction of coronary heart disease. In: Proceedings of international conference on circuits, communication, control and computing (I4C 2014)

Optimization for Lung Cancer Prediction Using Support Vector Machine with Artificial Neural Networks—A Review

Aditya Katari and A. Shanthini

Abstract Lung cancer is not an accidental death nowadays. After undergoing many researches, it is observed that death rate is increased. So, to reduce the death rate, people should go through some quicker diagnosis. Generally, lung cancer detection is done by using many techniques of different domains. Currently, there are many algorithms in use to detect the lung cancer for feature extraction and selection in the domain of machine learning. For segmentation purpose, super-pixel segmentation is highly used. In this paper, we are going to propose the algorithm that is used to show accurate results in the detection of lung cancer. To find a algorithm or optimizing the existing algorithms like genetic optimization algorithms, PSO, SVM and comparing these algorithms with the existing ones then finalizing the standard algorithm that gives accuracy in formation of lung cancer detection system.

Keywords Machine learning · Super-pixel segmentation · Genetic optimization · Partial swarm optimization

1 Introduction

Several trends are opening in lung cancer detecting, which has become uppermost cause of death worldwide. Basically, cancer is nothing but a group of cells will be slowly stopping its growth in a right way which starts effecting the formation of malignant tumors. Because of this uncontrollable growth of cells, surrounding tissues are invaded. Classification of cancer is of two types namely non-small cell lung cancer and small cell lung cancer. In these two types of lung cancer, non-small cell lung cancer is prevalent by undergoing some workouts. It is also said that diagnosis and treatment vary between non-small cell lung cancer and small cell lung cancer.

A. Katari · A. Shanthini (✉)
Department of Information Technology, SRM Institute of Science and Technology, Chengalpattu, India
e-mail: shanthia@srmist.edu.in

A. Katari
e-mail: ka4393@srmist.edu.in

D. J. Hemanth et al. (eds.), *Artificial Intelligence Techniques for Advanced Computing Applications*, Lecture Notes in Networks and Systems 130,
https://doi.org/10.1007/978-981-15-5329-5_29

The lung cancer detection is proved to be second most common cancer in human society. Worldwide people are suffering from lung cancer. About 224,300 new cases of lung cancer and 158,060 deaths of lung cancer is the approximate estimation by American lung cancer society in the USA for year 2016. Clinical decision support system is another methodology that detects lung cancer in an accurate way. Diagnosis regarding health has become very tough job nowadays. But pre-diagnosis by analyzing data may become easy to do. Many health checking systems are developed for accurate results to increase the patient's life span. There are convolutional neural networks in the field of machine learning techniques that greatly contribute to the speed and accuracy of the images captured. 3D CNN is a neural network that is used for image classification in few tasks. It is also used in medical imaging that is developed for accuracy results in medical treatment (Fig. 1).

Development in computer-aided diagnosis is done in several ways like pattern recognition, neural networks, etc. CT scan images, ultra sounds, and MRI scanning are several methodologies to detect the cancer cells in the body of men or women. Previously, these three methodologies are used for detecting cancerous cells effectively. In this paper, the used algorithms are efficient for accuracy is proved (Fig. 2).

Moreover, the most sensitive modality is CT scan image that produces cross-sectional images in the specific areas of the objects that are scanned. Some X-ray images are taken by observing that formation of lung cancer detection is done. Lung cancer detection system in machine learning is used to explain the presence of cancer caused through CT images and blood samples. Every year death rate is increasing due to lung cancer which is said to be dangerous disease not only in the case of men but also in Women. Usually, for the person who is suffering from lung cancer, life

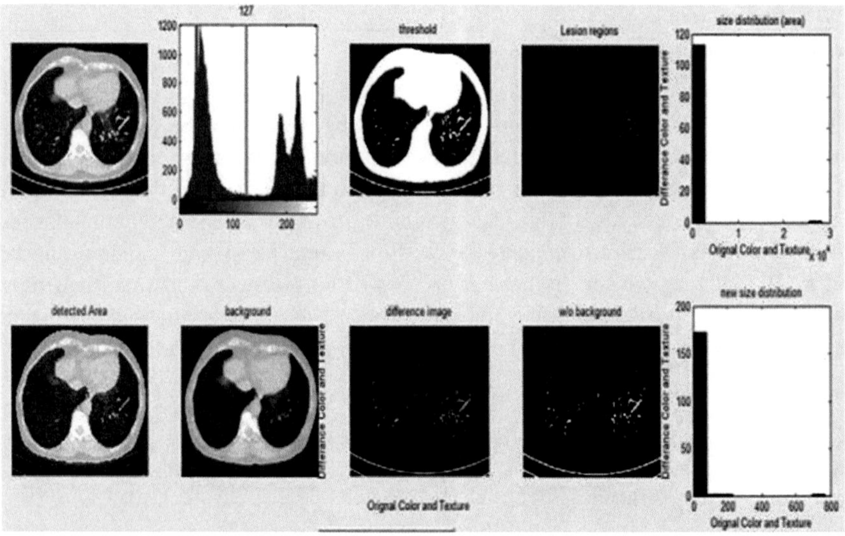

Fig. 1 CT scan image for lung cancer detection

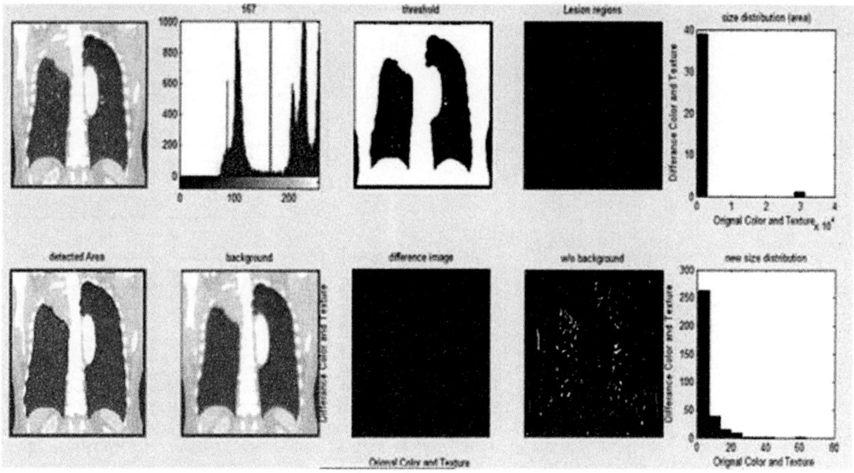

Fig. 2 MRI modality of lung cancer detection

span is very less. But in several detection, it is observed that if a patient undergoes quick diagnosis in early stages life span may increase.

1.1 Lung Cancer Detection System

- **Image Enhancement**:
  Here, various enhancement techniques are involved after capturing the medical bits of patient's tumor regarding cancer. For different images, different techniques of enhancements are used (Fig. 3).
- **Image Preprocessing**:
  Image preprocessing plays a major role in reducing the noise and also prepares the images captured for further steps like segmentation. Colorimetric, image enhancement, and smoothing are the steps used in image preprocessing.
- **Feature Extraction**:
  Feature extraction is used to provide final outputs and uses different techniques and algorithms that determine image is normal or abnormal. Algorithms and techniques which are used examine the image and eliminate the non-desirable portions of images. Segmentation is an important step that is carried out on the region of lung and then followed by feature extraction for further features.

Moreover, some diagnosis rules are followed to finally detect the lung cancer. Those diagnosis rules which are followed will eliminate the portions that detected wrongly of cancer nodules. This is done through segmentation.

Proposed work focuses on finding the accurate results of lung cancer detection. For this purpose, optimization of existing algorithms and comparing those algorithms

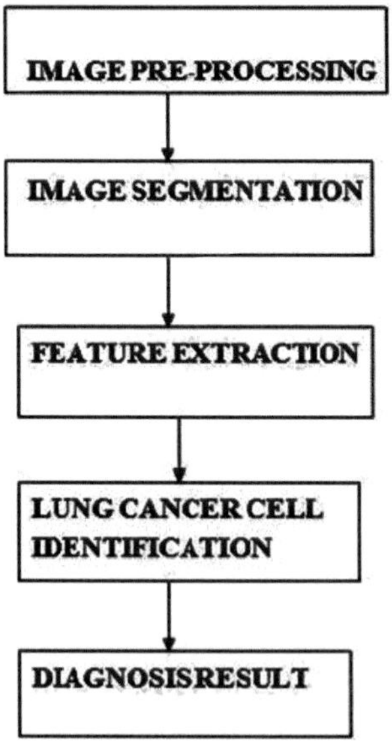

Fig. 3 Lung cancer detection system

take place. Then, the standard algorithm is chosen that is efficient in accurate detection in lung cancer and that increases the patient survival rate. In previous works, few algorithms are used so taking those algorithms into considerations comparison is done.

2 Related Work

One of the novel approaches to detect lung cancer undergone many techniques and data analysis. In many cases, it also has been proved that lung cancer detection is impossible to prevent, thus early detection of cancer is the only way to cure. Basically, lung cancer detection by using machine learning involves taking CT scanning images and blood samples of a patient. CT scan images are classified as normal and abnormal images. If patient has normal report, then it says that there is no cancer tumor. If it is an abnormal report, then it says that patient is suffering from cancer. This abnormal report is classified based on segmentation. An efficient way is taken to detect lung cancer to find accurate results by algorithms like SVM, etc. [1].

Previously, sensor systems brought up few applications like personal and easy-to-use health checking system. The gas chromatography and mass spectrometer generate the human urine into numerical data. Patient's urine contains some chemical substances called biomakers that detect the lung cancer. Human urine is of three-dimensional feature extraction [2].

Deep neural network is the effective one that classifies lung cancer patients from GC-MS data of human urine. By this, it is proved that deep learning is more effective and achieved 90% of accuracy in detection of lung cancer [2].

Another important approach is made to detect formation of lung cancer is low-dose computed tomography (LDCT). This modality includes the detection of indeterminate pulmonary nodules. Low-dose computed tomography plays a vital role in the early detection of lung cancer. Demonstration of combining utility of radiomics features to improve lung cancer detection modalities performance screening is done. Extraction and analyzing the features to improve and develop tools support in the lung cancer detection screening [3].

Nucleus segmentation is another modality to the sign of early lung cancer detection. The process first collects the sputum samples from the patient and labels it. The labeling is done by tetrakis carboxy phenyl porphine (TCPP) which assists in lung cancer cells by increasing low-density lipoproteins coating on surface of cancer. Biomoda is another important platform where well-known machine learning techniques are exploited for accurate detection of lung cancer. Approximately 80% of accuracy is obtained in the previous works. By adding the nucleus segmented features, it increased up to 89% of accuracy. But there are some problems in the potentiality of detecting lung cancer [4].

In the recent observation local energy-based shape histogram (LESH) feature extraction technique is used to detect breast cancer diagnosis. Further, it is applied to detect the lung cancer by mammogram images. In this accuracy, performance measure is proposed by using LESH technique in prior to radiography images. And also, enhancement of diagnosis outcome is involved [5].

The watershed segmentation by approaching different algorithms for performance measurements is presented. In this segmentation, background objects are taken into consideration. That is, it indicates background object markers of some specific images. The experimental methods are improving the performance by remote sensing and granulometry [6].

This paper presents the authentication errors by using techniques in the trend. CT and MRI images are scanned by regressions methods. Calculation of signal-to-noise ratio for absolute errors undergoes the improvisation of those errors to increase the accuracy rate [6].

In the years of 2008–2010, several cases are found like people suffering from pulmonary nodules problem in china. In the examination of health diagnosis, approximately up to 50% patients were found with 30–31 nodules that affect their life. Development of computer-aided diagnosis system is involved [7].

In 1988, it has been observed that the major problem of medical data analysis is image segmentation, as choosing large data set has become an expensive matter in the view of networking and deep neural networks. Finally, a standard data set is fixed that showed better results in the performance of accurate results [7].

The proposal of neural fuzzy model in the year 2002 played a vital role in the detection of pulmonary modules. The topmost features extracted are size, circularity, and brightness. The techniques included are segmentation, thresholding in the area of effected parts of lungs [8].

In previous approaches, neural information processing is one of the main approaches in the detection of lung cancer formation. Detecting the tumor and quick diagnosis in the morphological way took place for accurate results [8].

This paper explains about detecting the opacities by chest radiography. And the other major proposed work is to detect the opacities around large lung areas by using iris filter. Different filters for performance and accurate results in the detection of cancer tumor in the human lung area are compared [9].

Automation is the major key role in the detection of nodules in the lungs. Application of computer-aided diagnosis is developed for CT scanning images. Appropriate techniques are used in the process of detection. Completely automated CAD systems are used for the better performance in the presence of visual interpretation [10, 11].

Lung cancer stands as a dangerous disease that is decreasing person's survival rate in the world. CT image is used for scanning and detecting the tumor in the complex area of human lungs, which is used for quick diagnosis to improve the survival rate of person [12].

Extraction of lung cancer segmentation by CT images is enhanced by method Hopfield neural networks model. The diagnostics rules are verified by normal and abnormal points. Proposed model explains negative and positive status of scanned images [13, 14].

In the era of lung cancer, screening detection of lung nodules plays a vital role. The proposed system is all about the early detection and diagnosis of lung cancer to increase the life span of a person. Luna 16 is a challenging one in this paper. Luna 16 is an expensive data set for quicker diagnosis [15].

Previously, it has observed that pulmonary nodule is the early signature of lung disease. Performance of lung cancer disease is called as nodules which explains us the nature of the disease. A standard algorithm is used in the diagnosis which is simple, by using an effective method to improve early detection [16, 17].

3 Conclusion

In this survey of lung cancer detection system, major methodologies have been studied through different domains. Exact identification of tumor regarding cancer is done by observing necessary feature extractions. Efficient algorithms are exploited to detect the formation of lung cancer. In this paper, finding an algorithm or optimizing the existing algorithms and comparing the chosen algorithms with the existing ones is done. Then, a standardized algorithm is fixed which gives accurate results in the formation of lung cancer detection system.

References

1. Chaudhary A, Singh SS (2005) A Cancer J Clinic, 55:10–30. Int Trans Comput Sci, 4, 2012. American Cancer Society, Cancer Statistics, CA
2. Sharma D, Jindal G (2011) Identifying lung cancer using image processing techniques. In: International conference on computational techniques and artificial intelligence (ICCTAI'2011), vol 17(1), 872–880
3. El-Baz A, Farag AA, Falk R, Rocco RL (2002) Information Conference on Biomedical Engineering, Egypt
4. Ginneken BV, Romeny BM, Viergever MA (2001) Computer-aided diagnosis in chest radiography: a survey. IEEE Trans Med Imaging 20(12):1228–1241
5. Ginneken BV, Romeny BM, Viergever MA (2001) IEEE Trans Med Imaging 20(12)
6. Levner I, Zhangm H (2007) Classification driven watershed segmentation. IEEE Trans Image Process 16(5):1437–1445
7. Kanazawa GK, Kawata Y, Niki N, Satoh H, Ohmatsu H, Kakinuma R, Kaneko M, Moriyama N, Eguchi K (1998) Computer-aided diagnosis for pulmonary nodules based on helical CT images. Comput Med Image Graph 22(2):157–167
8. Lin D, Yan C (2002) Lung nodules identification rules extraction with neural fuzzy network. IEEE Neural Inf Process 4(1):2049–2053
9. Linda G, Shapiro GC, Stockman G (2001) Computer vision, theory and applications
10. Magesh B, Vijaylakshmi P, Abhiram M (2011) Int J Comput Trends Technol
11. Magesh B, Vijaylakshmi P, Abhiram M (2011) Computer aided diagnosis system for identification and classification of lessions in lungs. Int J Comput Trends Technol 3(5):714–718
12. Hadavi N, Nordin J, Ali S (2014) Lung cancer diagnosis using CT-scan images based on cellular learning automata. IEEE, pp 154–159
13. Proceedings of 7th international workshop on enterprise networking and computing in healthcare industry, HEALTHCOM (2005), pp 378–383
14. Sammouda R, Hassan JA, Sammouda M, Al-Zuhairy A, El-Abbas H (2005) Computer aided diagnosis system for early detection of lung cancer using chest computer tomography images. In: GVIP 05 Conference, vol 3(5), pp 714–718
15. Suna T, Wanga J, Li X, Pingxin L, Liua F, Luoa Y, Qi G, Zhua H, Guo X (2013) Comparative evaluation of support vector machines for computer aided diagnosis of lung cancer in CT based on a multi-dimensional data set 16(3):519–524
16. Liu X, Ma L, Song L, Zhao Y, Zhao X, Zhou C (2015) IEEE J Biomed Health Inf 19(2)
17. Zhao B, Gamsu G, Ginsberg MS, Jiang L, Schwartz LH (2003) Automatic detection of small lung nodules on CT utilizing a local density maximum algorithm. J Appl Clinic Med Phys 4(3):248–260

Object and Obstacle Detection for Self-Driving Cars Using GoogLeNet and Deep Learning

G. Krishna Chaitanya and G. Maragatham

Abstract Self-driving cars are the latest innovation in which the car runs by itself. The self-driving cars can be called as autonomous cars. This involves many technologies like artificial intelligence, machine learning and deep learning. When coming to the self-driving cars, the main aspect which it needed to be taken care of is the obstacle detection and the object detection. The object detection by a car in more simpler words object recognition process done by a machine which involves the concepts of machine learning and deep learning. Deep learning helps in achieving the object and the obstacle detection. There are various algorithms which help in the object detection like artificial neural network, convolutional neural network, AlexNet, VGG Net, GoogleNet, etc. GoogleNet is the CNN architecture which makes the image recognition an easier task. For the self-driving cars, obstacle and object detection GoogLeNet are not much addressed in the recent works. So, it can be considered as a latest technology. In this paper, the recent works about the self-driving cars and object detection and obstacle detection and the future scope of it are discussed.

Keywords Self driving · Obstacle detection · Object detection · Artificial intelligence · Machine learning · Deep learning · GoogleNet

1 Introduction

Nowadays, self-driving cars are the trend in the market that is going to occupy the future market space. The key aspect of self-driving cars. Comes into picture that the whole concept lies on the automated driving based on the visual cues and inputs it receives through the sensors and analyse and understand and being able to predict decide and operate based upon the scenario or the situation.

G. Krishna Chaitanya (✉) · G. Maragatham
Department of Information Technology, SRM Institute of Science and Technology, Kattankulathur, India
e-mail: kg5070@srmist.edu.in

G. Maragatham
e-mail: maragatg@srmist.edu.in

D. J. Hemanth et al. (eds.), *Artificial Intelligence Techniques for Advanced Computing Applications*, Lecture Notes in Networks and Systems 130,
https://doi.org/10.1007/978-981-15-5329-5_30

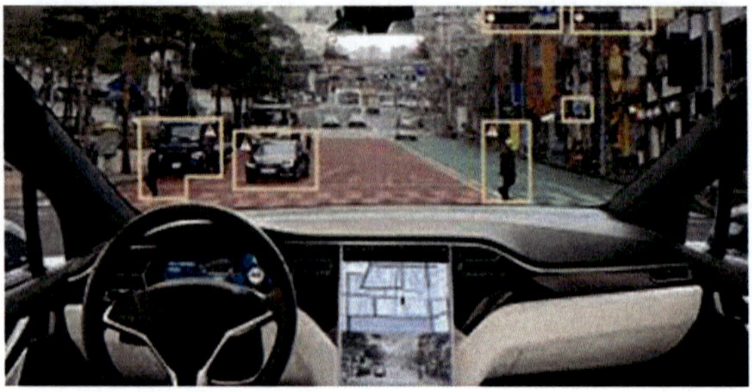

Fig. 1 Self-driving car detecting objects

Today, multiple international companies are trying to get in this marketspace due to its high viability and cost-effective solution it can provide in the future by eliminating the unnecessary manpower. For this, multiple companies/individuals/research scholars had opted multiple ways to implement the solutions by prototyping the multiple aspects involved in its final output product (Fig. 1).

So, first of all, a self-driving car is a vehicle which runs by itself without human effort or simply called as driverless cars. Some of the key aspects that are needed to be taken care of are obstacle and object detection, getting data from IoT devices and learnings from the past events to improve the model to predict and detect the objects.

In this paper, we deal with one of the major aspects that is the visual cue management which mainly includes object and obstacle detection. Now, let us see what is this object and obstacle detection, how do they differ and why they a prominent role and what are the different possible approaches to handle them.

2 Literature Survey

Obstacle is the objects that the cars generally need to avoid during the course of driving or in motion. Generally, this purpose of detection and identification can be done by the help of simple image processing if it happens in an structured environment where the 3D image of the objects present is already given as the reference to compare and act upon. This becomes the main reason why protos are easily implemented for this purpose. But when it comes to real time, it is completely useless because it is an unstructured environment where the accurate determination of size, shape, depth, range of the object/obstacle comes into the picture. Here, the constraints that need to be taken care also include position and orientation of the car, and also, the tilt (angle of view) should also taken into consideration [1].

Now, the key challenge happens at 3D reconstruction and matching the object/obstacle with the previous information present with servers. One of the approaches to do this is to use the approach of stereo matching which can be done in two ways one is with the help of bigdata and databases another is with the help of neural networks [1].

Using databases: In this method, we need to attain the datasets of images of multiple objects/obstacle from multiple scenarios and multiple environments and store them in a database. Whenever we came through the obstacle, we need to traverse the new dataset with all the datasets present in database linearly to find a approximate match which can be unreliable as it involves storing the dataset of the images in databases and getting a matching reference from them, and this method is prone to high chance of failure as it has risks of complete shutdown if the database is lost or corrupted or the scope of the input is limited due to environmental constraints [1].

With the help of neural networks: Neural networks have the ability to learn by themselves and produce the output that is not limited to the scope of the input provided to them, and the input is stored in its own networks instead of a database so that it can ensure the loss of data does not affect its working. The loss of a few chunks of data in one particular place does not prevent the network from functioning. Essentially, it works on a system of probability, and it can handle large amount of data sets; it has the ability to implicitly detect complex nonlinear relationships between dependent and independent variables; it has ability to detect all possible interactions between predictor variables—based on data fed to it, it is able to make statements, decisions or predictions with a degree of certainty [2].

But a simple neural network cannot satisfy our need of providing reliable output in real world due to timing constraint because where the decision needs to be taken in micro or Pico seconds where the speed also comes into the picture, so our focus needs to be shifted to much more complex neural networks which can satisfy our requirement that can be enhanced by the concept of deep learning [2].

Now, deep learning involves much more deeper complex networks than ANNs where system is self-teaching, learning as it goes by filtering information through multiple hidden layers, in a similar way to humans [2].

In deep learning, the complex neural networks store the different pieces of information in different layers that are arranged hierarchically. In this way, when an information comes, each level of neurons processes the information, provides insight and passes the information to the next, more senior layer. The best analogy to understand this is to refer how human brain works and stores information in terms of neurons [2].

Now, out of all the combinations and types of neural networks present, the best way to achieve required output is to use convolutional neural networks where in more than one convolutional layer is present where these are either completely interconnected or pooled before passing the result to the next layer [2].

The convolutional layer undergoes a convolutional operation on the input. Due to this which the network can be much deeper but with much fewer parameters. For this purpose, there are different multiple CNNs were already created and trained based on requirement of scenario such as GoogLeNet and AlexNet [3].

GoogleNet: GoogLeNet is a pretrained convolutional neural network that is 22 layers deep. The network trained can be loaded either on ImageNet or Places365 datasets. The network that is trained on ImageNet classifies all the pictures into one thousand object class, such as keyboard, mouse, pencil and many animals. The network trained on Places365 is comparable to the network trained on ImageNet, however classifies pictures into 365 totally different place classes like field, park, runway and lobby. These networks have learned totally different feature representations for a large vary of pictures. The networks both have an image input size of 224-by-224. Multiple pretrained deep neural networks can also be added for better functionality which can also be found in MATLAB [3].

The major advantage of GoogLENet is you can retrain a GoogLeNet network to perform a new task using transfer learning. When performing transfer learning, the foremost common approach is to use networks pretrained on the ImageNet dataset. If the new task is comparable to classifying scenes, then mistreatment the network trained on Places-365 will offer higher accuracies [3].

AlexNet: AlexNet contained eight layers; the initial five were convolutional layers, some are followed by max-pooling layers, and therefore, the last three were fully and absolutely connected layers. It used the non-saturating ReLU activation function that has shown improved training performance over tanh and sigmoid. Once identified potential field-based obstacle avoidance formulation is carried out by using the obstacle range, size information, car position and orientation. Finally, proportional derivative navigation control loop along with obstacle avoidance algorithm can be used [4].

3 Existing System

In the recent works, the self-driving cars implementation has been addressed only using artificial neural networks, machine learning and the deep learning concepts only. The idea of implementing it using the GoogleNet is not yet been addressed in the most of the works.

The obstacle and objects are recognised using deep neural network, convolutional neural network and deep learning concepts. Similarly the concepts that are used in order to detect the object and the obstacle, classification is done only by using the concepts of neural networks. Those are used because the CNNs give the best result in image recognition and object detection in the image and the video also.

Some of the algorithms like the autoencoder in the deep learning area have proven its capability in maintaining the accuracy in the image recognition (Fig. 2).

We can see that the objects are being detected by the car in the while on moving on road. Coming to the difference between obstacle and object, the object is anything that is lying on a roadside, but an obstacle is anything that interrupts the flow of any traffic, for example, in a moving car perspective, a vehicle parked on a roadside is called an object. If a pole suddenly falls due to some accident that is caused by the nature or caused by human error, then it can be taken as an obstacle (Figs. 3 and 4).

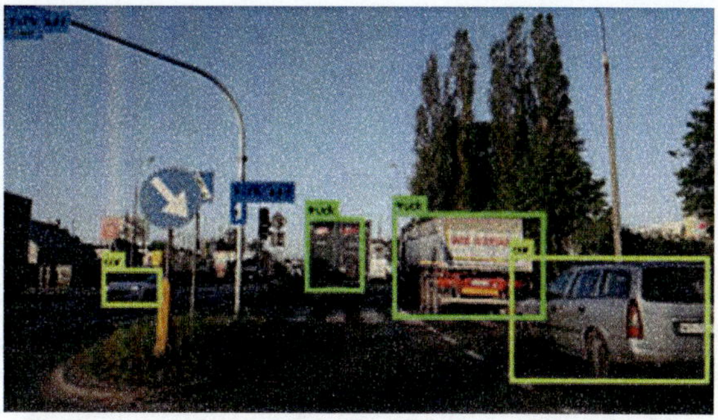

Fig. 2 Objects being detected by self-driving car while moving

Fig. 3 Car parked on roadside considered as an object

Fig. 4 Pole fell down due to a natural calamity considered as an obstacle

4 Proposed System

The literature survey and the existing system state that the deep learning concepts are used in detecting the objects. Now here, the GoogleNet and deep learning are the concept introduced in detecting the objects. The work flow of the proposed system goes as follows.

4.1 Input Data

The input data is in the form of a sequence of images and which consists of a driving route of a car. Each image is needed to be considered as an frame of a video, as video contains the sequence of frames. All of those images are needed to be combined into a single video so that it would become a video. The video would be more helpful in detecting the objects as a flow rather than taking each and every image and detecting the objects the images as the number of image is very high. The object detection or an obstacle detection would be more easy if it is a video because this process goes on a video flow, so it will be easy. Detection of an object on each and every image can be more time taking and confusing because each and every image is considered as a frame. The movements from the place to place might take large number of images. This might take it as the repetition of the work. Hence, the set of images are converted to the video.

4.2 Use Input Data to Detect the Objects Initially

When the video input is ready, then it is the flow of images as we know. Objects in the video of a moving cars perspective the external objects are only considered. Objects that are present outside the car can be objects initially. And those are needed to be detected initially as a foremost task. The objects that are present inside the car are stable and the consistent images which are common in all the scenarios. So, there is no valid point in detecting the objects inside the car. The objects that are needed to be detected are to be outside the car or in other words the objects that are present in the different environment which might become an obstacle in the different kind of scenarios. The detecting objects in the images are the primary part of the obstacle and object detection, depending on which the solution for all the scenarios and the situations is designed.

4.3 Segregate the Objects by Labelling

After detecting the objects, the objects needed to be segregated into the different types. In the perspective of the moving car, there are many things that may be present on the road. Then, there might be things like cars, trucks trees, poles, traffic signals, buildings nearby. All those objects are needed to be specified with the labels so that each object is segregated individually. Then, it is object needs to be differentiated from an obstacle using the GoogLeNet and deep learning algorithms. This process is necessary in order to make the car stop when an obstacle is in front of it rather than hitting it and causing an accident which is the thing to be avoided as a main idea. It must not stop recognising a parked car or any other vehicle or object which is lying by the side of a road, which is another unwanted thing and may lead to a huge disturbance in a traffic, and there is also a chance to cause an accident, which is again a dangerous incident that anyone would never wish to happen.

5 Future Scope

In future, the self-driving cars have the capability to change the world in such a way that there will not be any kind of news like the road accidents. The self-driving cars will be designed in such a way that at any cost, the car will obey all the traffic laws of the land to make sure the passenger reaches the destination safely. It is needed to start and reach from one place to another place nowadays. Most importantly, it will help the emergence service like ambulance, fire and police to react more quickly to avoid the loss or any disturbance in the society.

6 Conclusion

In conclusion, the most point to touch in the general aspect of self-driving cars is the strict obeying of traffic rules and easy reacting of emergence services. Coming to the technical aspect, the existing system has contributed a lot to the self-driving cars, and many improvements have been made in the recent works. The proposed system is a proposal of carrying the improvement in the self-driving cars using the latest technology called the GoogLeNet.

References

1. Nguyen VD, Van Nguyen H, Tran DT, Lee SJ, Jeon JW (2017) IEEE "learning framework for robust obstacle detection, recognition, and tracking. IEEE Trans Intell Transport Syst 18(6)
2. Yu H, Hong R, Huang XL, Wang Z (2013) Obstacle detection with deep convolutional neural network. In: 2013 Sixth international symposium on computational intelligence and design. Department of Information Science, Nanchang Teachers College, Nanchang, 330103, China, Department of Electronic and Information Engineering, Shijiazhuang Tiedao University, Shijiazhuang, 050043, China. Email: pdc1028@gmail.com, hongruxia2006@163.com
3. Salavati P, Mohammadi HM (2018) Obstacle detection using GoogleNet. In: 8th International conference on computer and knowledge engineering (ICCKE 2018). Faculty of Computer Engineering University of Isfahan Isfahan, Ferdowsi University of Mashhad. Email: Iran pouyan@eng.ui.ac.ir, Iran h.mahvash@eng.ui.ac.ir
4. Bui HM, Lech M, Cheng E, Neville K, Burnett IS, School of Engineering, RMIT University, GPO Box 2476, Melbourne VIC 3001 Australia. Center of Technology, RMIT University, 702 Nguyen Van Linh, District 7, Ho Chi Minh, Vietnam. Faculty of Engineering and Information Technology, University of Technology Sydney, PO Box 123, Broadway NSW 2007, Australia Corresponding author: H. M. Bui (s3372651@rmit.edu.vn). Object Recognition Using Deep Convolutional Features Transformed by a Recursive Network Structure.

Received October 26, 2016, accepted November 18, 2016, date of current version January 27, 2017. 10.1109/ACCESS.2016.2639543

5. Dairi A, Harrou F, Sun Y, Senouci M (2018) Obstacle Detection for intelligent transportation systems using deep stacked autoencoder and k-nearest neighbor scheme. IEEE Sens J 18(12)
6. Prabhakar G, Kailath B Sudha N, Kumar R (2017) Obstacle detection and classification using deep learning for tracking in high-speed autonomous driving. Electronic system design IIITDM Kancheepuram, Chennai, India. Autonomous Vehicle Program—R&D Tata Elxsi Limited Chennai, India. 978-1-5090-6255-3/17/$31.00 ©2017 IEEE. Email: eds14m006@iiitdm.ac.in, bkailath@iiitdm.ac.in, sudha.n@tataelxsi.co.in, rajesh@tataelxsi.co.in
7. Aswathy P (2018) Deep GoogLeNet features for visual object tracking. In: 2018 IEEE 13th International conference on industrial and information systems (ICIIS). Department of Avionics Indian Institute of Space Science and Technology Trivandrum, India, aswathyece2011@gmail.com. Siddhartha Department of Avionics Indian Institute of Space Science and Technology Trivandrum, India. siddhartha1994@yahoo.co.in. Deepak Mishra Department of Avionics Indian Institute of Space Science and Technology Trivandrum, India. vr.dkmishra@gmail.com
8. Zhang X, Song Y, Yang Y, Pan H (2017) Stereo vision based autonomous robot calibration. Robot Auto Syst 93:43–51
9. Sivaraman S, Trivedi MM (2013) Looking at vehicles on the road: a survey of vision-based vehicle detection, tracking, and behaviour analysis. IEEE Trans Intell Transp Syst 14(4):1773–1795
10. Nguyen VD, Nguyen TT, Nguyen DD, Lee SJ, Jeon JW (2013) A fast evolutionary algorithm for real-time vehicle detection. IEEE Trans Veh Tech 62(6):2453–2468
11. Kavukcuoglu K, Sermanet P, Boureau YL, Gregor K, Mathieu M, LeCun Y (2010) Learning convolutional feature hierarchies for visual recognition. Adv Neural Inf Process Syst l(23):14–23
12. Yu RCH, Zhang DP (2007) Obstacle detection based on a four-layer laser radar. In: Proceedings of the 2007 IEEE International conference on robotics and biomimetics, pp 218–221
13. Girshick R, Donahue J, Darrell T, Malik J (2016) Region-based convolutional networks for accurate object detection and segmentation. IEEE Trans Pattern Anal Mach Intell 38(1):142–158
14. Fakhfakh N, Gruyer D, Aubert D (2013) Weighted V-disparity approach for obstacles localization in highway environments. In: Proc. IEEE Intell. Vehicles Symp. (IV), pp 1271–1278
15. Benenson R, Mathias M, Timofte R, Van Gool L (2012) Pedestrian detection at 100 frames per second. In: Proc. IEEE Conf. Comput. Vis. Pattern Recognit. (CVPR), pp 2903–2910
16. Morales N, Morell A, Toledo J, Acosta L (2016) Fast object motion estimation based on dynamic stixels. Sensors 16(8):1182
17. Godha S (2017) On-road obstacle detection system for driver assistance. Asia Pacific J Eng Sci Technol, 16–21
18. Jia B, Feng W, Zhu M (2016) Obstacle detection in single images with deep neural networks. SIViP 10:1033–1040

Fast Bandwidth-Delay Routing Methods for Software-Defined Network (SDN)

V. Tejaswini, M. Sayeekumar, and G. M. Karthik

Abstract Multi-objective optimization methods for routing in SDN networks have come in existence in which QOS measurements is a challenging one. In order to improve the quality of service, new routing algorithm that quickly and efficiently calculates bandwidth-related restricted routes is proposed. The algorithm is determined for the SDN which is the backbone of networks that separate the control plane from the data plane and centralize it rationally in order to maximize the use of network resources. By considering the control framework that categorizes traffic flows into a limited number of division based on the delay-sensitive level, reduce computational complexity of that complex travel engineering task. The results of the research show that while fairly simple, the method proposed for quality of services supply leads to fewer rejected quality of services requests under a wide range of process performance than the difficult competitive ones.

Keywords Quality of services (QOS) · Software-defined network · Bandwidth · Traffic flows

1 Introduction

In software-defined networking (SDN), the network intelligence centralized information plane and brought together for programmable controller. The capability of this idea is reflected in the way that the nature of administration with QOS and traffic building (TB) can perform more expertly in all centralized systems. Control software dynamically handles SDN network device to provide network resources allocation and avoid blockage in network.

V. Tejaswini · M. Sayeekumar (✉) · G. M. Karthik
SRM Institute of Science and Technology, Kattankulathur, Chennai, India
e-mail: sayeekum@srmist.edu.in

V. Tejaswini
e-mail: tejaswinichowdaryv@gmail.com

G. M. Karthik
e-mail: karthikgl@srmist.edu.in

D. J. Hemanth et al. (eds.), *Artificial Intelligence Techniques for Advanced Computing Applications*, Lecture Notes in Networks and Systems 130,
https://doi.org/10.1007/978-981-15-5329-5_31

Traffic engineering calculations implemented till now in the writing overwhelmingly consider the issue of giving up transfer speed-ensured traffic burrows in the system. Nonetheless, with the large development of ongoing Internet development, delay-obliged calculations have gotten progressively required. In the event that we accept that the connection delays are known, discovering courses with postponement and transmission capacity ensures is attainable with shortest way first (SWF) calculations. In any case, SWF calculations rapidly make the bottlenecks for future traffic requests in light of the fact that no traffic engineering is working to adjust system data. On the opposite data, the blend of traffic engineering and data transmission postpone-obliged steering brings about issue unsolvable continuously. To be specific, traffic engineering strategies ordinarily improve the usage of the system assets via cautiously choosing the connection loads. Both the connection weight and postpone are the added substance imperatives, which imply that the directing issue comes down to the obliged way improvement, which is NP-finished. A typical way to deal with rearranging this issue is to discredit the connection delays. Be that as it may, the effectiveness of the calculations straightforwardly identifies with the size of blunders presented during discretization.

In software-defined network systems, the intricacy of course count is offloaded to software-defined network controllers. Confused bundles are cushioned or disposed of until the comparing stream sections are introduced by the controller. In this way, decreasing inactivity in correspondence between the system gadget and the related controller is of an incredible significance. At whatever point the heap of a controller arrives at a limit, the bundle handling dormancy increments considerably. In such cases, the controller handling idleness will end up being a non-immaterial factor altogether full circle inertness. Thus, setting up data transfer capacity-delay-ensured burrows with difficult task TE calculations in enormous data systems is unreasonable. That spurred us to think about a model of control plane which groups quality of service demands in a limited number of classes, in view of the affectability of traffic streams to add up to way time taken. For each quality of service method, we proposed another directing calculation that figures courses for delay-touchy (DT) data transmission when the arrange is introduced, and refigures that set just if network change happens.

1.1 Beginning Controller-Switch Connection

As appeared in Fig. 1, in the SDN-based remote work systems, just a couple of switches are straightforwardly associated with the controller. The main undertaking is to associate every one of the changes to (in any event one) controller by setting up introductory/fundamental steering. That implement an underlying (i.e., not-changeless) steering method where a software-defined networking controller will discover every one of the switches by connecting the system without thinking about whether the way it found is ideal or not. For accomplishing this, we utilize an OpenFlow-based steering calculation for beginning controller switch association with adjusting OLSR in two different ways.

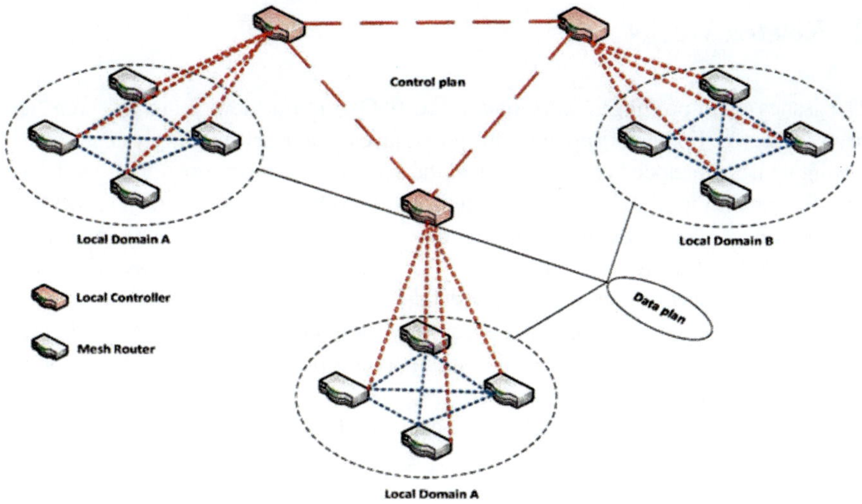

Fig. 1 SDN mesh topology reference model [1]

1.2 Control-Switch Data Optimization

During the underlying association through setup, in directing ways given as in this method will be utilized to put in new other options (i.e., briefest, ideal, or burden adjusted) ways. The control can choose to pick a way through these elective ways among inside and a connected since, at this point, the controller has a worldwide perspective on the system. From that point, it introduces the comparing rule to course parcels from the changes to itself in the switches by engendering the data utilizing the first not-optimized ways.

1.3 Steering Among Switches

After next step, the control would determine the briefest way directing through connection them and it sends at that point introduces these directing ways and relating sending rules utilizing the most limited ways setup in the past stage. As portrayed above, to accomplish the above steering procedure, we needed to change the OpenFlow customer.

2 Related Works

The issues in improving quality of service (QOS) is not new, main research has been subjected. Endeavors up until this point have been overwhelmingly centered on setting up transfer speed ensured ways in the system. The most broadly utilized data transfer capacity-obliged steering calculation is least bounce calculation [1] (MHA), which registers the way with the least number of possible connections (joins skilled to help the data transfer capacity request). In [2], WSP is also known as MHA, and that algorithm sets feasible short path with the highest bandwidth. The shortest widest path (SWP) calculation [3] pursues a contrary methodology and figures the largest possible way. On the off chance that there is a multiple, way with the most modest number of jumps is viewed as ideal.

The recently portrayed calculations accept that the connection remaining data transfer capacity is the main powerful data accessible. Thus, the choice course regularly comprises an impediment for future stream appearances. As an answer for this issue, calculations that consider places of entrance departure hub sets (IE matches) have been proposed. These calculations are particularly reasonable for spine systems, where the number of IE combines for a particular area is generally little [4]. Indeed at the point when all hubs are entrance departure hubs, almost certainly, a few subsets of the hubs are progressively significant and should be organized. Minimum interference routing algorithm (MIRA) [5] is used to select the least critical link in the network but does not interfere during routing which may lead to dissatisfaction of upcoming demand. Critical link concept was introduced by MIRA to determine which link should be avoided during routing between the ingress-egress pairing that will mainly focus setting up on the bandwidth path. By triggering MIRA profile-based routing, the deal with bandwidth-guaranteed flows for dynamic routing problem in multi-commodity network was proposed.

Data transfer capacity defer-obliged calculations have been a lot less researched in the writing. The basic arrangement was proposed in [6], which consistently picks the way with the littlest deferral in the wake of pruning every one of the connections with deficient data transfer capacity. Delay-weight capacity (DWC) algorithms calculate the weighted bandwidth paths which are inversely proportional to delay values of the path. Selection of the paths in network is done by considering IE (Ingress-Egress) pair, and end-to-end delay of the path is measured by seat request with delay constrains and bandwidth [5–8]; smaller value is the most powerful network since it satisfies request with little bandwidth requirement and the number of requests the network can take, called as capacity of the network which is measured by bandwidth of the path; it is easy to identify the most powerful network for IE pairing is the least bandwidth between nodes [9–11], and all the links that use most powerful network are removed which are basically below the least delay path. Consider maxima sum of network capacity of each IE pair and each network capacity is measured by the power of the network. Link which determined path bandwidth is taken as bottleneck link and is also known as critical link during routing the least bandwidth in network.

This is accomplished by utilizing weight work those increments with the degree of connection criticality [10]. Once the loads are resolved, joins with inadequate data transmission are reduced from mastermind diagram, and extended Dijkstra's most short path (EDSP) [12] could be used to enlist the route with the most negligible weight that satisfies the delay essential. A few endeavors have been made to use the advantages of SDN design for better steering execution [13–15]. Be that as it may, data transmission defers-obliged TE in SDN systems is still under-examined theme. An ongoing report [14] proposes versatile constant correspondence framework, adaptive routing, and priority for SDN (RT-SDN), which addresses steering and traffic wanting to offer through and through cutoff time guarantees on software-defined network stages. Adaptive routing and priority for SDN (RT-SDN) addresses consistent arranging while simultaneously expecting courses are given. That pursues substitute computations that select courses with information transmission and defer guarantees under the assumption that association delays are known early [9–11, 16–18], considering the way that through and through postponements are generally directed by causing delays in the considered circumstance of spine SDN.

QoS advancement system for SDN controller is proposed by Egilmez et al. [19], which intends to give relative confirmations in wording of bundle hardship and fulfill concede requirements of need traffic streams. The proposed structure is expected to compose quality of service traffic streams and secure best-effort streams in a comparable time by using dynamic stream development system. Be that as it may, it does not abuse asset reservation or potentially need lining systems to give start to finish transfer speed ensures. A programmed QoS power over OpenFlow convention was proposed by Kim et al. [20] that depends on rate forming [21]. Their controller configuration satisfies data transmission necessities of traffic streams just if interfaces on the briefest ways have adequate limit accessible. To meet defer necessities,

The proposed controller configuration maps traffic streams to statically arranged yield lines as indicated by their needs. Our work contrasts from the referenced recommendations as far as improvement objectives [17]. What is more, use case considered and proposed computationally undemanding calculation which will in general progressively improve the usage of spine arrange by utilizing SDN control plane, while giving supreme data transmission and defer ensures.

3 Methodology

By considering the issue of setting up traffic burrows in programming characterized spine systems, at present, wide-territory spine systems are frequently founded on Internet Protocol and multi-protocol label switching (MPLS) innovations. In the rest of the paper, that will assume Software Defined Networking consists of Multi-Protocol Label Switching like information plane and control plane. Such network will have many advantages, and all multi-protocol label switching administrations could be more helpful and dynamic in nature. More up-to-date forms of OpenFlow convention as of now enable to push, pop, and swap multi-protocol label switching names in software-defined networking systems. That considered model of Internet Service Supplier (ISS) spine is organized. Utilizing the phrasing obtained from IP/MPLS systems, the supplier arrange comprises of center routers and provider edge switches interconnected with point-to-point joins. MPLS innovations are used by center router switch so as to advance traffic among the provider edges (PEs). By the above situation provider edge (PE), center router switches are designated for software-defined networking/OpenFlow proficient switches that keeps up state-of-the-art perspective on the system state. All system control capacities, including steering, quality of service arrangement, asset reservation and traffic engineering, run as applications over the controller.

Virtual rented line services (VRLL) are given by controllers which are an essential piece for offering measuring of ISP suppliers. VRLL administrations can be utilized to give quality of service certifications to certain applications or then again to help VPN functionalities (e.g., interconnect extraordinary urban communities of an organization through the ISP arrange). Software-defined networking controller gets quality of service demands for such administrations, makes steering choices, saves assets in the system, creates MPLS names, and sets up the passages in the information plane by utilizing OpenFlow to circulate stream rules to connections along the courses. The controller looks after data with respect to reservations and each settled passage. In this way, at the point when another passage is mentioned, the controller can compute course which can satisfy quality of service prerequisites.

3.1 The Proposed Routing Algorithm

The calculation accepts that proliferation delays are known, just as the places of IE matches in the system. These suspicions are effectively supported and could prompt huge improvement in the system execution. Every entrance and each departure hub does not make fundamentally the IE pair. This is on the grounds that, for instance, virtual private network traffic will just start and exit at particular entrance and departure hubs. The data are accessible to the controller and changes not as often as possible. So as to perform quick course arrangement, the calculation runs in disconnected and online stage. In a disconnected stage, Yen's calculation is utilized to ascertain the main K-short path most limited ways (in terms of) for every IE pair. These ways ought to have the option to fulfill postpone prerequisites traffic in delay sensitive. Note that some IE combines will not have short pathways with delay. The last arrangement of delay-sensitive courses consistently incorporates every single disjoint way with postpone lower than delay. On the off chance that they are most certainly not remembered for the consequence of the Yen's calculation, they are unequivocally included toward the finish of a disconnected stage.

3.2 MDWCRA Algorithm

Maximum delay-weight capacity routing algorithm is used for finding the least delay path form all the links in the network. First calculate the weight of all the links from source to destination of IE pair (s, d) and compute the delay-weight capacity. Only bottleneck critical link is considered after eliminating other links while finding the next least delay path, for example, limited fields, where n is the prime whole number and p is the positive number. Binary bend calculations utilize littler key sizes contrasted with other cryptographic calculations, thus preparing rate has been expanded (Fig. 2).

I1E1 (1, 2, 3, 4) and I2E2 (1, 5, 6, 4) are identified by MDWCRA algorithm were (5, 6) and (2, 3) are the critical link. In the model above, if I1 sends a few littler solicitations of a similar total size (10 Mb/s), the accessible limit with respect to I2-E2 and I1-E1 will lessen considerably. The connection (2, 3) will be basic for all IE matches with limit tumbles to 3 Mb/s. In this way, the least basic connection is chosen for basic connections to be alluring. As a rule, that is not the situation, in light of the fact that the majority connections would be considered basic on the comparative weight. Likewise, non-delay sensitive (NDS) traffic will be steered extremely more long ways (so as to keep away from basic connections) and expend huge system assets. So as to give better adjusting of NDS load, the connection loads are made out of two sections line. The initial segment is an equal estimation of the remaining transfer speed. The subsequent part relies upon leftover data transfer capacity, however increments with the degree of the connection criticality. In this manner, the loads of non-basic connections are not really the equivalent. The overall significance of the interface criticality is constrained by α parameter ($0 \leq \alpha \leq 1$). On the off chance that traffic in delay-sensitive is required for predominant, α must be taken as high esteem.

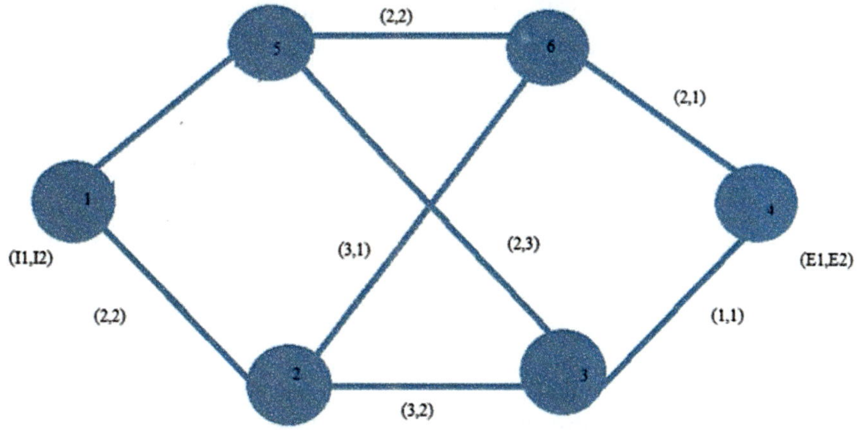

Fig. 2 Example of MDWCRA algorithm vector associated with each link (bandwidth, delay) reference example [16]

At the point when the loads are resolved, QoS demand is handled. Right off the bat, joins with lacking data transfer capacity are expelled from the organize diagram. At that point, if demand has a place with non-delay-sensitive class and Dijkstra's calculation find out the briefest way by using weighted diagram. On the off chance that DS demand is gotten, way with the least weight is picked from the arrangement of DS ways predetermined in a disconnected stage.

4 Implementation

4.1 Flow Migration Extensions for the Algorithm

The first flow migration extensions act as dynamic which is triggered by variation of traffic flow. The framework for delivering pair of traffic flow is done by primary algorithm, which runs until stoping criterion are given and re-trafficing will not be done without any conditions, inorder to avoid dynamic changes in traffic flow which causes unbalance network resource usage. As traffic streams show up and leave, it is conceivable that courses at first chose for seemingly perpetual associations after at some point cause a lopsided system stacking. In this way, propose two conceivable stream relocation expansions for the above calculation. The mainstream relocation augmentation acts proactively. It is activated by a flight of traffic stream. After some traffic burrow gets expelled from the system, various n top stacked connections are chosen, and for every one of them, beginning with the most noteworthy stacked one, stream relocation is endeavored. In this progression, all traffic streams utilizing the connection are arranged in dropping request regarding the mentioned defer bound, with the end goal that NDS streams are over the rundown, while streams with exacting defer bound are toward the finish of the rundown. At that point, the essential directing calculation (depicted in the past section) is run for every one of them, one by one, until the halting paradigm is met. In this way, the calculation does not attempt to recourse the traffic streams unequivocally.

On the off chance that one traffic stream is effectively moved to another course which does exclude that connection in any event one traffic stream for Ingress-Egress pair has been checked to be redirecting, and each fizzled from that point when the endeavor of stream movement comes up short, the Ingress-Egress pair is added to extraordinary rundown, meant as fizzled. The calculation checks substance of this rundown before every movement endeavor and does not attempt to recourse the traffic stream if its Ingress-Egress pair is in the rundown. This guarantees the quantity of the movement endeavors cannot surpass the quantity of Ingress-Egress combines in the system. As referenced above, n characterizes the quantity of joins which will be considered by stream movement system. That had picked n to be equivalent to the quantity of connections in the course of the withdrawing stream. This decision is sensible on the grounds that assets from n joins are liberated by the withdrawing stream, while simultaneously keeping away from the calculation costs that would result from much of the time endeavoring stream relocation on all connections. The second calculation for stream movement is responsive, activated just when burrow arrangement falls flat. The objective of the calculation is to discharge enough assets for QoS demand by redirecting existing traffic streams in the system. So as to figure out which connections are the most appropriate for stream relocation, the calculation characterizes the connection weight as a diminishing capacity of the leftover data transfer capacity. Dijkstra's calculation is used to compute the way with the least weight and for each connection of that way stream relocation is endeavored until one of the two halting criteria is met. The principal halting basis is met in the event that the connection has adequate assets to suit the showed up QoS demand. The second is equivalent to in the proactive stream movement calculation. Note that traffic streams are arranged in plummeting request as indicated by postpone prerequisites. Along these lines, the stream movement technique prior meets the halting conditions.

5 Results and Outputs

System lifetime is determined utilizing values which forms the follow record and delivers the outcome. The system lifetime is expanded a little contrasted and that of destination source distance vector routing, which is brought about by the capacity in course revelation. The cradle is utilized to store the Route Request so the way with high vitality can be taken, which brings about sparing vitality and all the more expanding in the system lifetime (Fig. 3).

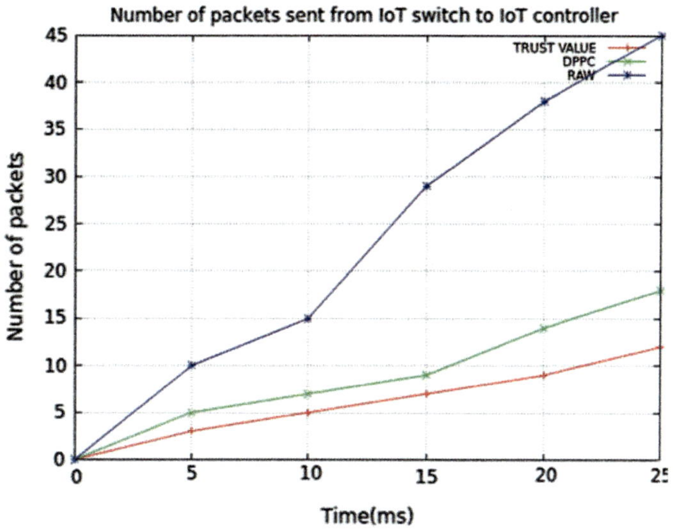

Fig. 3 Number of packets transmitted in SDN between switch and controller

Throughput is defined as the number of packets transmitted from source to destination in a particular time (m/sec). Delay requirement is set to 20 ms. During the packets transmitting, it would select the critical link with very low delay. This link has a weight that is inversely proportional total residual bandwidth and delay of the critical path. The packets are routed through selecting critical link with a least weighted link, and less delay request avoids unnecessary loading, which are achieved by using weight function with increasing critical link weights to avoid the insufficient bandwidth in the network, and bandwidth-delay requirement is satisfied by comparing the path with the least weight in network.

The collision of time on the bandwidth of IoT controller and switch channels in Internet of things is shown in Fig. 4. For trust-based algorithm in a DDoS attack, the bandwidth of the IoT controller and switch channels in Internet of things increases continuously from 0 to 10 s. Internet of things switch eliminates DDoS attack packets when trust value algorithm implemented in the IoT controller finds DDoS attacks in 10 s. Then, Internet of things switch need not have to transmit a greater number of packets-in messages to the IoT controller. The value of information packets obtained by the Internet of things controller simultaneously falls to the lower level till it is constant (Fig. 5).

Bandwidth controllers which avoid the insufficient bandwidth in the network while packet is transmitting from source to destination by setting the bandwidth capacity of each link, expected low queuing delay bandwidth with high-speed network. The value of the critical link of bandwidth delay can be calculated by dividing physical distance of each node in link by signal speed in the network. The information of link reservation is maintained by the controller and establishes new tunnel when it is required during packet transmitted from source to destination.

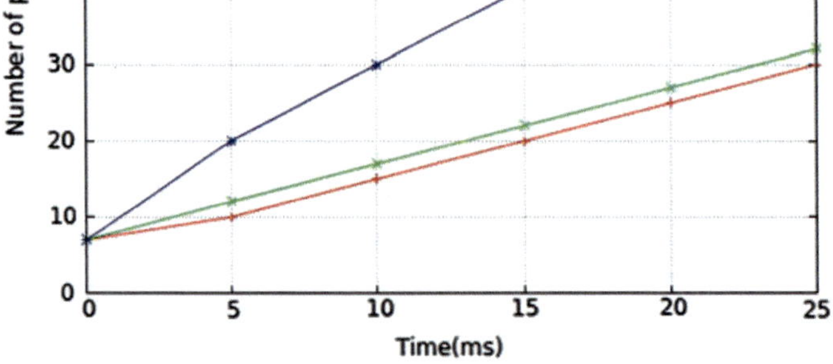

Fig. 4 Number of data received by controller

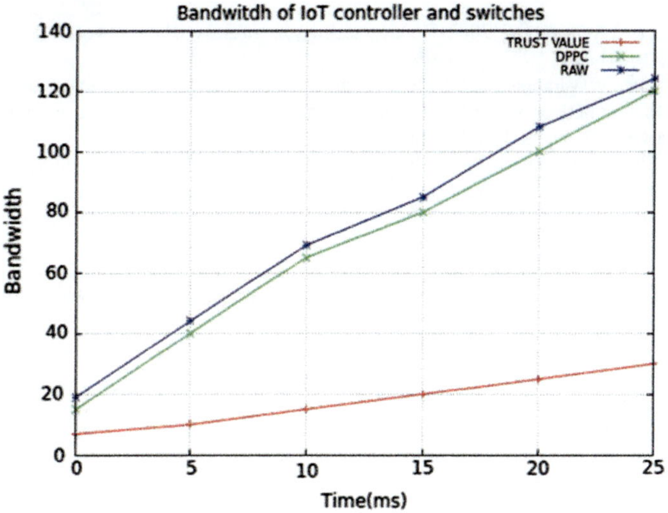

Fig. 5 Bandwidth of controller

6 Conclusion

In this paper, as investigated execution of transmission capacity delay-obliged steering calculation could be used for QoS provisioning in software defining network spine systems. In view of joined nature of SDN control plane, computational multifaceted design of transmission limit defer-obliged coordinating has been recognized as a critical confining part. Hence, straightforward QoS model which offers a limited set number of delay breaking points to the framework provider's customers. Further, two-stream relocation expansions for the calculation is progressed which abuse software-defined network controller's capacity to change courses "on the fly" to redesign execution as far as the demand blocking proportion. The research results indicated that the proposed calculations guarantee better usage of system assets than the aggressive arrangements. The upgrades in execution traverse a wide assortment of working conditions.

References

1. Elzain H, Wu Y (2019) Software defined wireless mesh network flat distribution control plane. Future Internet 11(8):166. https://doi.org/10.3390/fi11080166
2. Yousefi S, Mousavi MS, Fathy M (2006) Vehicular ad hoc networks (VANETs): challenges and perspectives. In: 2006 6th International conference on ITS telecommunications proceedings, pp 761–766
3. Haas ZJ, Deng J, Liang B, Papadimitratos P, Sajama S (2002) Wireless ad hoc networks. Encycl Telecommun

4. Deng H, Li W, Agrawal DP (2002) Routing security in wireless ad hoc networks. IEEE Commun Mag 40(10):70–75
5. Balakrishnan V, Varadharajan V (2005) Packet drop attack: a serious threat to operational mobile ad hoc networks. In: Proceedings of the international conference on networks and communication systems (NCS 2005), Krabi, pp 89–95
6. Peng M, Shi W, Corriveau J-P, Pazzi R, Wang Y (2016) Black hole search in computer networks: state-of-the-art, challenges and future directions. J Parallel Distrib Comput 88:1–15
7. Akyildiz IF, Wang X, Wang W (2005) Wireless mesh networks: a survey. Comput Netw 47(4):445–487
8. Chang J-M, Tsou P-C, Woungang I, Chao H-C, Lai C-F (2015) Defending against collaborative attacks by malicious nodes in MANETs: a cooperative bait detection approach. IEEE Syst J 9(1):65–75
9. Venkatesh K, Srinivas LNB, Mukesh Krishnan MB, Shanthini A (2018) QoS improvisation of delay sensitive communication using SDN based multipath routing for medical applications. Elsevier Future Gener Comput Syst 93:256–265
10. Reddy V, Venkatesh K (2020) Role of software-defined network in Industry 4.0. EAI/Springer Innovations in Communication and Computing-Internet of Things for Industry 4.0 Design, Challenges and Solutions, pp 197–218
11. Open Networking Foundation. Software defined networking: the new norm for networks. Web white paper, retrieved Apr. 2015
12. Aijaz A, Aghvami AH (2015) Cognitive machine-to-machine communications for internet-of-things: a protocol stack perspective. IEEE Internet Things J 2(2):103–112
13. Chen P, Cheng S, Chen K (2014) Information fusion to defend intentional attack in internet of things. IEEE Internet Things J 1(4):337–348
14. Tootoonchian A, Gorbunov S, Ganjali Y, Casado M, Sherwood R (2012) On controller performance in software-defined networks. HotICE 54:7–10
15. Yang Y, Muppala JK, Chanson ST (2001) Quality of service routing algorithms for bandwidth-delay constrained applications. In: Ninth international conference on network protocols, pp 62–70
16. Yang Y, Zhang L, Muppala JK, Samuel, Chanson T (2003) Bandwidth–delay constrained routing algorithms. Elsevier Comput. Netw 42:503–520
17. Naveen Chandar B, Arivazhagan N, Venkatesh K (2019) Improving the network performance using mp-olsr protocol for wireless ad hoc network (MANET). Int J Recent Technol Eng 8(3):5700–5707
18. Kim W, Sharma P, Lee J, Banerjee S, Tourrilhes J, Lee S, Yalagandula P (2010) Automated and scalable QoS control for network convergence. INM/WREN"10, San Jose, CA, pp 1–6
19. Tomovic S, Prasad N, Radusinovic I (2014) SDN control framework for QoS provisioning. TELFOR, Belgrade, pp 111–114
20. Ishimori A, Farias F, Cerqueira E, Abelem A (2013) Control of multiple packet schedulers for improving QoS on OpenFlow/SDN networking. EWSDN, pp 81–86
21. Jarschel M, Wamser F, Hohn T, Zinner T, Tran-Gia P (2013) SDN based application-aware networking on the example of youtube video streaming. EWSDN, pp 87–92

Detection of Atypical Activities—A Review

G. Malvika and S. Sindhu

Abstract Atypical activities are typically aberrations of scene entities (vehicles, human, animal activities, or environment) from normal behavior. Abnormal activities include eve-teasing, attacking, terrorist attack, harassment, robbery, unidentified objects, etc. The main aim of this paper is to detect abnormal activities in crowded places and to alert disabled people from these activities. Such "Atypical Activities" behavior typically translates to some kind of a problem like unidentified objects, harassments, robbery, etc. Atypical event discovery can be considered as coarse level video understanding, which sifts through peculiarities from typical examples. This paper gives a complete characterization of the detection of atypical activities writing into two stages.

Keywords Atypical activities · Abnormal activities · Unidentified objects

1 Introduction

Atypical activity detection is a subdomain of behavior understanding from surveillance scenes. Anomalies are typically aberrations of scene entities (vehicles, human, or the environment) from the normal behavior. Atypical activities include eve-teasing, attacking, terrorist attack, harassment, robbery, unidentified objects, etc. The proposed model will be trained to identify the abnormal activities in crowded scenes using some sequence of actions. It will indicate the person then and there about the attacker.

Atypical activity detection in jam-packed scenes is significant and testing some portion of the wise video reconnaissance framework. Be that as it may, abnormal event detection and restriction is as yet a difficult issue in clever video observation;

G. Malvika (✉) · S. Sindhu
Department of Information Technology, SRM Institute of Science and Technology, Chengalpattu, India
e-mail: gm9383@srmist.edu.in

S. Sindhu
e-mail: sindhus2@srmist.edu.in

D. J. Hemanth et al. (eds.), *Artificial Intelligence Techniques for Advanced Computing Applications*, Lecture Notes in Networks and Systems 130,
https://doi.org/10.1007/978-981-15-5329-5_32

however, some incredible advancement has been made in highlight extraction, conduct displaying, and peculiarity estimating. The most testing issue is that the meaning of the abnormal activity is inconclusive in the vast majority of these present reality observation recordings. By and large, occasions that are essentially unique in relation to basic occasions are characterized as irregularities, which implies that inconsistencies are characterized by ordinary occasions rather than orders or subtleties of themselves.

Disregarding how it is satisfactorily immediate to list a couple of sorts of assortments from the standard for a particular scene, for example, the closeness of a vehicle or a bicyclist in a person by strolling swarm, it is amazing to list all the bizarre occasions that could be conceivable inside that scene; along these lines, there are interminable helpful classes in a particular request task.

1.1 Video Surveillance

Video footage observance application has been around since the 1950s. In interminable evaluations, a broad combination of utilization areas has been tried especially in swarmed and security-sensitive spots, for instance, military area, railroad stations, open squares, and walkways/subway. The huge bit of video surveillance is not simply on the condition of the cameras for human eyes yet moreover for totally modernized acumen works out. In spite of the way that thing assertion and shape estimation issues have been explained by past appraisals in the past scarcely any years, with the exponential headway of front-line video observation gadgets and advancement, a gigantic measure of assignments can be improved the degree that the exactness and the faithfulness of target affirmation, following, plan, and lead assessment.

Video surveillance is utilized to assess picture assembling that adequately decreases the issue of perceiving a moving article. The issue of shortcoming and dependability to survey the size and position of the 3D object in a scene is moreover observed as it cements diverse awesome components that could influence the area, for instance, condition, lighting, sorts of article improvement, etc. In this way, the procedural kind structure of video perception should be outlined and gathered to get important data. Diverse value of video perception has also been settled, and this fuse acknowledgment figuring's, following a goal in the field, course of action unequivocal space and dismembering execution (for instance length/dimensions, concealing, and pace) for perceiving issues in swarm area. Additional analysis on the current innovative within the improvement of video police work which provides the history and progress of video perception structure from the primary up to the third time allotment.

2 Related Work

The author introduced a new unsupervised learning approach for the detection of atypical activities detection system based on convolution auto-encoder architectures. Existing frameworks revolve around the inconsistencies which appear outside scenes, and bearing in mind few troublesome uninhibitedly accessible UCSD variations from the norm recognizable proof datasets. The current system blocks the route toward cutting the commitment to fixes by the convolution standard of the convolutional neural system, which makes our structure major and clear. The present system shows the adequacy and nature of the proposed approach, displaying intense execution to existing methodologies. In like manner, the proposed situation can get from the noteworthy neural system structures for object divulgence and social affair assignments to configuration powerfully display day systems, so as to address the different models from the information video [1].

A basic however proficient system, AED-Net, is proposed given a self-administered learning technique. Crude information from observation video cuts is utilized to ascertain optical stream maps. The existing system is a critical level feature that is then evacuated by PCAnet, which is moreover used to choose the anomalism of close by strange events and worldwide atypical events. The exploratory results show that the structure performs well in perceiving both overall peculiar events and neighborhood abnormal events. Also, after an LRN layer was added to address the overfitting issue, the display of this structure improved. The structure accomplishes results that are superior to anything cutting-edge techniques, showing that it can viably remove movement designs from crude video and use them to distinguish peculiarities [2].

An unaided strategy is proposed for a worldwide strange group occasion discovery. To start with, the technique is figuring the STACOG descriptor of the info video arrangement. Second, the K-medoids bunching calculation is utilized to segment the STACOG descriptors of preparing outlines into a lot of groups and is then used to decide if an edge is typical or not utilizing a separation metric. The proposed irregularity location strategy was tried on the benchmark dataset: UMN dataset, PETS2009. Investigations show that the proposed technique accomplishes similar outcomes with the cutting-edge strategies as far as exactness while requiring low computational expense than elective methodologies [3].

A model affirmation procedure is proposed to choose irregular conditions in jam-stuffed circumstances. The existing model intended to adequately perceive both neighborhood and overall unpredictable models in transient and spatial circumstances. The created model learns an advancement heatmap of the district which is later utilized as a spread for seeing and following highlights over different edges, while in like way detaching the video plot into various non-covering areas and figuring improvement structures inside each phenomenal locale. Our model shows a massive improvement over the starting late proposed model's stumble. It is in like way littler importance it will, in general, be passed on in any circumstance with only a few days sooner intending to perceive occasions since it trains itself with time [4].

A procedure is proposed for isolating development features and as such has given a fruitful and quick system for development affirmation. The proposed feature extraction procedure relies upon close by auto-associations of the space–time tendencies and suitably gets the geometric characteristics, for instance, recurring patterns, of space–time development shape. The development is seen in the arrangement of pack of-plot features, which can enough expel the development characteristics in a computationally capable manner, not at all like a standard sack of features which depicts the development pitifully. In tests development affirmation using distinctive datasets, the proposed strategy indicated positive shows, when appeared differently in relation to various systems. In addition, the results were gotten with less computational weight and much snappier than steady [5].

Different inconsistency location strategies are investigated that can be applied for street organizing substances including vehicles, individuals, and their collaboration with the earth. Existing model treats irregularity location by accepting information as the essential unit enumerating the learning strategies, highlights utilized in learning, approaches utilized for inconsistency identification, and applied situations for abnormality discovery. Existing system expects to set a couple of future headings by investigating the holes in the present PC vision-put together procedures through dialogs with respect to different potential outcomes [6].

A profound learning way is proposed to deal with identifying genuine inconsistencies in observation recordings. Because of the multifaceted nature of these reasonable oddities, utilizing just typical information alone may not be ideal for peculiarity recognition. Existing system endeavors to abuse both ordinary and abnormal recordings. To maintain a strategic distance from work concentrated fleeting explanations of peculiar sections in preparing recordings, existing system gets familiar with a general model of abnormality location utilizing profound MIL system with feebly marked information. To approve the proposed methodology, another enormous scale oddity dataset comprising of an assortment of true inconsistencies is presented. The preliminary outcomes on this dataset show that our proposed variation from the norm acknowledgment approach performs generally better than check methodologies. In addition, existing model shows the estimation of our dataset for the task of peculiar development affirmation [7].

PC vision applications have come to depend continuously on superpixels starting late, yet it is not for each situation clear what involves a fair superpixel computation. With an ultimate objective to grasp the focal points and drawbacks of existing systems, there are around five top tier superpixel computations for their ability to adhere as far as possible, speed, memory adequacy, and their impact on division execution. Another superpixel estimation is displayed, simple linear iterative clustering (SLIC), which changes a k-recommends gathering way to deal with overseeing gainfully made superpixels. Despite its straightforwardness, SLIC sticks to limits correspondingly as or superior to past techniques. At the same time, it is quicker and more memory gainful, improves division execution, and is prompt to contact super voxel age [8].

The current system's fundamental point is to recognize atypical events through a small entertainment over a run of the mill base. Given a combination of customary planning models for instance, a picture gathering or an assortment of close by Spatio-brief patches proposed the sparse reconstruction Cost (SRC) over the common lexicon to check the ordinariness of the testing test. By indicating the earlier weight of each reason during inadequate changing, the proposed SRC is continuously strong stood apart from other extraordinary case affirmation criteria. To gather the over-finished ordinary bases into a more diminutive jargon, a novel word reference confirmation procedure with pack sparsity restriction is composed, which can be settled by standard curved improvement. By masterminding various sorts of Spatio-brief premises, our system can isolate both closes by and overall astonishing occasions. Meanwhile, as it does not depend upon object recognizing confirmation and following, it will, as a rule, be applied to swarmed video scenes. By resuscitating the word reference reliably, our system can be feasibly slackened up to the online occasion territory. Assessments on three benchmark datasets and the association with the top-level techniques bolster the upsides of our strategy [9].

The current model is another technique for the human activity blueprint by utilizing a persuading blend regarding another 3D incline descriptor with an optic stream descriptor, to address Spatio-regular intrigue focuses. These focuses are utilized to address video approaches utilizing a pack of Spatio-regular visual words, following the triumphs accomplished in article and scene depiction [10].

Another view–based way to deal with dealing with the delineation and insistence of human progression appears. The reason behind the depiction is a fleeting organization—a static vector-picture where the vector regard at each point is a segment of the progression properties at the relating a spatial zone in an imaging blueprint. The fundamental worth is a parallel worth exhibiting the closeness of headway and the ensuing worth is a bit of the recency of progress in a get-together. The procedure typically performs basic division, is invariant to arrange changes in speed and runs legitimately on standard stages [11].

The event zone subject to using features from a static structure can give poor results from the point of view of two principle edges: the state of the camera and the circumstance of the event that is going on in the scene. The past causes issues when organizing and test events are in different remarkable propensities from the camera to the attested circumstance of the event. The last can be a wellspring of issues when masterminding events to occur in any circumstance in the scene, and the test events happen in a position one of a kind about the status events. The two issues degenerate the precision of the static structure framework. Subsequently, this work proposes a system called a remarkable structure for event confirmation, which can deal with the two bits of the issue. In this work, the author has used the dynamic grid method to recognize four types of patterns: implosion, explosion, two-way, and one-way using Multimedia Analysis and Discovery (MAD) pedestrian dataset. The starter results show that the proposed framework can see the four sorts of event structures with high precision. Additionally, the exhibition of the proposed technique is better than the static cross-portion strategy and the proposed structure achieves higher precision than the past framework concerning the starting late referenced edges [12].

Programmed acknowledgment of human activities is a significant however troublesome issue in the zone of PC vision. In this paper, a novel methodology is acquainted with handle the issue. Right off the bat the human body is identified using form data and the body is followed by the cross breed technique. The most noticeable highlights are looked by utilizing the mean-move technique dependent on the body structure and the History Motion Image Information. At long last, a learning strategy depends on the different help vector machines which are utilized to learn activity types progressively. A strategy is proposed that coordinates the Bayesian structures with the SVM technique, which to a great extent improves the acknowledgment rate utilizing notable data. Investigations show that our framework can run progressively for the discovery of abnormal practices with constrained data and produces a vigorous outcome by utilizing memorable movement data [13].

The idea, the usage and the handy use of the calculation for the identifying of possibly hazardous circumstances in the group were exhibited. In light of trial results appeared in the paper and on a greater arrangement of all acquired outcomes, an end can be made that the existing calculation is adequately viable for recognizing the person on foot swarming close to section bottlenecks. Later on, a few improvements of the calculation should be possible, for instance, the consideration of the element of making insights of swarming almost a specific structure exit. Besides, an association of various cameras to the framework is arranged so as to empower the passerby's course forecast capacity related to the present circumstance in enormous structures [14].

This paper is a survey paper on crowd investigation dependent on PC vision. This work handled three significant issues in swarm investigation: individuals tallying/thickness estimation, following in jam-packed scenes, and group conduct understanding in a more elevated level examination, similar to the worldly advancement, fundamental bearings, speed estimations, and discovery of unordinary circumstances. With respect to combination, the survey was centered around swarm models that either utilizes PC vision calculations to remove genuine information data, meaning to improve the authenticity of the reenactment, or that are utilized to prepare/approve PC vision systems. A few contemplations about these issues are given straightaway [15].

The acknowledgment progressively of group elements openly puts are getting basic to evade crowd related debacles and guarantee the wellbeing of individuals. Investigation starts with a novel following strategy, in view of HOG descriptors, to at long last use pre-characterized models (for example swarm situations) to perceive swarm occasions. These situations are characterized utilizing measurable examination from the informational indexes utilized in the experimentation. The methodology is described by joining a nearby investigation with a worldwide examination for swarm conduct acknowledgment. The nearby investigation is empowered by a strong following strategy, and worldwide examination is finished by a situation displaying organizing [16].

Visual discernment in outstanding scenes, particularly for people and vehicles is beginning at now one of the most dominant research centers around PC vision. It has a wide extent of promising applications, melding access control in remarkable districts, human prominent proof a not too bad way off, swarm development bits of information and obstruct assessment, acknowledgment of odd practices, and keen surveillance using different cameras, etc. All around, the taking care of the arrangement of visual perception in ground-breaking scenes consolidates the accompanying stages: the showing of circumstances, acknowledgment of development, the portrayal of moving things, following, comprehension and delineation of practices, human distinctive verification, and blend of data from various cameras [17].

This paper audits and endeavors improvements and general techniques of stages associated with video reconnaissance and examines the difficulties and possibility for joining object following, movement examination, conduct investigation, and biometrics for standoff human subject distinguishing proof and conduct understanding. Conduct investigation utilizing visual reconnaissance includes the most progressed and complex investigates in picture preparing, PC vision, and man-made reasoning. There were numerous assorted techniques have been utilized while moving toward this challenge, and they shifted and relied upon the required speed, the extent of utilization, and asset accessibility, and so forth. The inspiration of composing and displaying a study paper on this theme rather than a how-to paper for an area explicit application is to audit and pick up understanding into visual reconnaissance frameworks from a major picture first. Exploring/studying existing accessible attempts to empower us to comprehend and answer the following inquiries better: Developments and procedures of stages engaged with a general visual reconnaissance framework; how to distinguish and examine conduct and purpose, and how to approach the test, on the off chance that you have openings [18].

It is a moving assignment to create viable and effective appearance models for hearty items following because of components, for example, present variety, enlightenment change, impediment, and movement obscure. Existing web-based following calculations regularly update models with tests from perceptions in late edges. Regardless of much achievement has been illustrated, various issues stay to be tended to. In any case, while these flexible appearance models are data subordinate, there doesn't exist a sufficient proportion of data for online estimations to learn toward the beginning. Second, the Web following estimations normally encounters coast issues. In light of self-instructed learning, slanted models are likely going to be incorporated and degenerate the appearance models. Proposed an essential yet reasonable and capable after count with an appearance model subject to features isolated from a multiscale picture incorporate space with a data free reason. The existing appearance model utilizes non-versatile arbitrary projections that protect the structure of the picture include space of articles. A sparse estimation organize is created to capably evacuate the features for the appearance model. Stuffed test photos of the front-line target and the establishment using the proportionate insufficient estimation system.

A coarse-to-fine glance through the procedure is grasped to moreover diminish the computational multifaceted nature in the acknowledgment framework. The existing compressive after figuring runs ceaselessly and performs well against top tier strategies on testing courses of action similar to capability, precision, and power [19].

Biometric conspicuous evidence systems have become hugely standard and huge by virtue of their high enduring quality and capability. Nevertheless, individual conspicuous confirmation a decent way off stays a troublesome matter. Walking will always be been seen as an essential biometric feature for human affirmation, recognizing verification. This might be viably secured from division and doesn't require any customer coordinated effort thusly mentioning it proper for objective fact. Nevertheless, the task of seeing an individual using step can be horribly impacted by fluctuating viewpoints making this task progressively testing. Existing technique handles this issue by perceiving Spatio-common features and performing expansive experimentation and getting ready instruments. A 3D convolution deep neural network is proposed for discrete unmistakable confirmation using step under various viewpoints. The present structure is generously increasingly capable to the degree reality and also performs the best way in every practical sense all core interests [20].

3 Comparison of Computer-Based Visions in Surveillance

See Table 1.

Table 1 Comparison of various existing methodologies used for anomaly detection

Ref.	Focus	Research areas
Achanta (2012) [8]	Simple linear iterative clustering Super-pixels	Comparing five state-of-the-art algorithms, concentrated on their boundary adherence, division speed, and performance
Zhang (2014) [19]	Fast compressive tracking	Robust tracking algorithm, compress features from foreground and background targets
Szczodrak (2016) [14]	Egress monitoring	Detecting potentially dangerous situations, detecting the pedestrian crowding near passage bottlenecks
Preechasuk (2015) [12]	Dynamic grid	Anticipated four sorts of occasion designs: implosion, blast, two-way, and single direction utilizing Multimedia Analysis and Discovery

(continued)

Table 1 (continued)

Ref.	Focus	Research areas
Thapar (2017) [20]	Invariant gait recognition network	Comparison of multi-view gait recognition with present State-of-the-art network, basic silhouette is used to decrease the run time
Mostafa (2017) [4]	Convolutional neural network	Detects abnormal motion and sudden changes, capable of identifying local as well as global abnormal models
Sultani (2018) [7]	Multi instance learning	Multiple Instance Learning and Automatically learn a deep anomaly ranking model
Nady (2018) [3]	Space–time auto-correlation of gradients	Processing STACOG descriptor of input video segment, K-medoids clustering algorithm used to parcel the STACOG descriptors of preparing outlines into a lot of groups
Wang (2019) [2]	Abnormal event detection network	Self-supervised framework AED-Net composed of principal Component Analysis Net (PCA-Net) and Kernel principal Component Analysis Net (kPCA)
Ming (2019) [1]	Variational autoencoder	For complex reconnaissance scenes variational auto-encoder with convolution part is utilized

4 Conclusion

In this paper, we have revised important computer vision-based survey papers on Anomaly Detection in crowded places. We investigated various abnormal activities and how we can detect them by using video surveillance. We treat Atypical Activity Detection by taking image frames from various videos. We tried to look into the best model for detection of Atypical Activities by comparing various Computer Vision-Based techniques.

References

1. Ming X, Xiasheng Y, Dongyue C, Chengdong W, Jiang Y (2019) An efficient anomaly detection system for crowded scenes using variational autoencoders. MDPI Open Access J
2. Wang T, Miao Z, Chen Y, Zhou Y, Shan X, Snoussi H (2019) AED-Net: an abnormal event detection network. Science Direct

3. Nady A, Atia A, Abutabl A (2018) Real-Time abnormal event detection in crowded scenes. J Theor Appl Inf Technol
4. Mostafa T, Uddin J, Md. Haider A (2017) Abnormal event detection in crowded scenarios. In: 3rd International conference on electrical information and communication technology
5. Kobayashi T, Otsu N (2012) Motion recognition using local auto-correlation of space-time gradients. Pattern Recogn Lett Science Direct
6. Kumaran S, Dogra D, Roy P (2019) Anomaly detection in road traffic using visual surveillance: a survey, IEEE
7. Sultani W, Mubarak S Chen C (2018) Real-world anomaly detection in surveillance videos. In: 2018 IEEE/CVF conference on computer vision and pattern recognition
8. Achanta R, Shaji A, Smith K, Lucchi A, Fua P, S¨usstrunk S et al (2012) Slic superpixels compared to state-of-the-art super-pixel methods. IEEE Trans Pattern Anal Mach Intell 34(11):2274–2282
9. Cong Y, Yuan J, Liu J (2013) Abnormal event detection in crowded scenes using sparse representation. Pattern Recogn 46(7):1851–1864
10. Ballan L, Bertini M, Del Bimbo A, Seidenari L, Serra G(2009) Recognizing human actions by fusing spatio-temporal appearance and motion descriptors. In: International conference on image processing, pp 3569–3572
11. Bobick AF, Davis J (2001) The recognition of human movement using temporal templates. IEEE Trans Pattern Anal Mach Intell 23(3):257–267
12. Preechasuk Jitdumrong, Piamsa-nga Punpiti (2015) Event detection on motion activities using a dynamic grid. J Inf Process Syst 11(4):538–555
13. Yufeng C et al. (2007) Abnormal behavior detection by multi-SVM–based bayesian network. In: International conference on information acquisition (ICIA). pp 298–303
14. Szczodrak Maciej, Czyzewski Andrzej (2016) video analytics-based algorithms for monitoring egress from buildings. Multimed Tools Appl 75:10733–10743. https://doi.org/10.1007/s11042-014-2143-7
15. Jacques JCS Jr, Mussef SR, Jung CR (2010) Crowd analysis using computer vision techniques. IEEE Signal Process Mag 27(5):66–77
16. Garate C, Bilinski P, Bremond F (2009) Crowd event recognition using HOG tracker. In: Twelfth IEEE international workshop on performance evaluation of tracking and surveillance (PETS Winter), Dec 2009, IEEE, Snowbird, UT, United States, pp 1–6
17. Weiming H et al (2004) A survey on visual surveillance of object motion and behaviours. Syst Man Cybern Part C Appl Rev IEEE Trans 34(3):334–352
18. Teddy Ko (2008) A survey on behaviour analysis in video surveillance for homeland security applications. In: Applied imagery pattern recognition workshop (AIPR), 37th IEEE
19. Zhang K, Zhang L, Yang MH (2014) Fast compressive tracking. IEEE Trans Pattern Anal Mach Intell 36(10):2002–15
20. Thapar D, Nigam A, Aggarwal D, Agarwal P (2018) VGR-net: a view invariant gait recognition network. In: Proceedings of the IEEE international conference on identity, security, and behavior analysis 2018 Jan 11–12; Singapore, IEEE, Singapore, New York

Probabilistic Optimization of Incorporating Security Ciphers and Encryption of Data Storage in Cloud

Haripriya Kaduvu, V. Lavanya, and M. Saravanan

Abstract Cloud computing is the next generation of IT organization. Cloud computing moves the software and databases to the large centers where the management of services and data may not be fully trusted. In this system, we focus on cloud data storage security, which has been an important aspect of quality of services. To ensure the correctness of user's data in the cloud, we propose an effective scheme with Advanced Encryption Standard and other cryptographic algorithms. Extensive security and performance analysis show that the proposed scheme is highly efficient. Development of efficient parallel data processing in clouds has come into existence and research for parallel security has been a challenge from the past few years. Parallel security is the data processing framework to explicitly exploit the dynamic storage along with data security. In the generation of IT organization, service level agreement on the basis of security is not an important one. But now it has become a mandatory matter to secure the sensitive data and for corruption detection.

Keywords Cloud computing · Resource allocation · Cryptography encryption · Decryption · Corruption detection

1 Introduction

A few patterns are aperture in broadcast computing, which is an Internet-based and appliance of PC addition to advancement. The beneath big-ticket and all the added ascendant processors with the artifact as an advice (SaaS) addition engineering are alteration server farms into pools of registering administering on a huge scale. The

H. Kaduvu · V. Lavanya (✉) · M. Saravanan
SRM Institute of Science and Technology, Kattankulathur, Chennai, India
e-mail: lavanyav@srmist.edu.in

H. Kaduvu
e-mail: kaduvuharipriya_na@srmuniv.edu.in

M. Saravanan
e-mail: Saran84gct@gmail.com

D. J. Hemanth et al. (eds.), *Artificial Intelligence Techniques for Advanced Computing Applications*, Lecture Notes in Networks and Systems 130,
https://doi.org/10.1007/978-981-15-5329-5_33

accretion arrangement alteration acceleration and dependable arrangement associations accomplish it even believable that Web audience can buy in accomplished administrations from advice and programming that lives on limited focuses. Moving advice into the broadcast accumulator offers absurd advantage to audience back they do not charge to anticipate about the challenges of absolute accessories the executives. The colonizer of broadcast accretion sellers, Amazon Simple Accumulator Service, (S3) and Amazon Elastic Compute Billow (EC2) are both accepted models. While these online administrations give a lot of added allowance and adjustable registering assets, this date move, be that as it may, is dispensing with the assignment of adjacent machines for advice stockpiling and abutment simultaneously. Therefore, audience are at the benevolence of their billow specialist co-ops for the accessibility and abidingness of their information. From the point of appearance of advice security, which has consistently been a cogent section of attributes of administration, broadcast accretion presents new testing aegis dangers for amount of reasons. For example, accepted cryptographic framework with the end ambition of advice aegis affirmation cannot be accurately accepted because of the client's accident ascendancy of advice beneath broadcast storage.

Consequently, acceptance of appropriate advice put abroad in the billow accept to be directed after acquirements of the absolute information. Considering altered sorts of advice for every applicant put abroad in the billow and the absorption of constant affirmation of their advice wellbeing, the affair of acknowledging rightness of advice stockpiling in the billow turns out to be abundant all the added testing. Broadcast accretion is not alone an alien advice stockroom. The advice put abroad in the billow ability be active by the clients. Notwithstanding, this activating aspect additionally makes accepted artlessness aegis procedures. The sending of broadcast accretion is fueled by server farms active in a synchronous, alternate, and appointed way. Singular client's advice is needlessly put abroad in abundant areas to added abatement the advice appropriateness dangers. In this way, broadcast conventions for accommodation accurateness affirmation will be of a lot of acceptation in accomplishing an affable and defended billow advice stockpiling framework in reality. Be that as it may, such cogent arena stays to be absolutely advised in the writing.

This procedure, while can be admired to agreement the accommodation accurateness after accepting audience accepting advice and they are on the accomplished apperception on individual server bearings and the all-inclusive majority of them do not anticipate about altered advice tasks. As an agnate methodology, scientists accept additionally proposed conveyed conventions for guaranteeing stockpiling rightness over altered servers. Once more, none of these broadcast affairs knows about altered advice activities. Therefore, their appliance in billow advice stockpiling can be absolutely constrained.

In this paper, we proposed an applicable and adjustable appointed artifice with absolute able advice abetment to agreement the rightness of client's advice in the certificate conveyance address to accord redundancies and affirmation the advice constancy. This development radically lessens the accord and accommodation aerial if assorted with the accepted replication-based record. To process/assign employments like applicant affidavit alteration to billow server, we will accord appointment of ambassador to animate addition to assumption server for archetype letters will be transferred by ambassador on primary server. Added our framework will by itself action advice in alongside augment to billow hubs for capacity. Clearly, these letters will be lumped in altered locations and with the encryption. Once administrator/client transfers capital archives, the axiological alive date will be able for 18-carat research. The document/assets stockpiling on billow is basal appliance of accumulated cloud. In any case, if any accident of advice happened because of infection advance or hacking, it will be harder to recover.

2 Related Works

Zuling Kang et al. in [1] "A Novel Approach to Apportion Cloud Asset with Diverse Execution Characteristics" 2013, the creators depict In a commonplace cloud computing environment, there will continuously be distinctive sorts of cloud assets and a number of cloud administrations making utilize of cloud assets to run on. As they can see, these cloud administrations more often than not have diverse execution characteristics [2, 3]. A few may be IO-intensive, like those information questioning administrations, whereas others might request more CPU cycles, like 3D picture handling administrations. In [4] the meantime, cloud assets moreover have diverse sorts of capabilities such as information handling, IO throughput, and 3D picture rendering. A straightforward truth is that apportioning a reasonable asset will significantly progress the execution of the cloud benefit and make the cloud asset itself more proficient as well. So, it is vital for the suppliers to apportion cloud assets based on the wellness of execution characteristics between assets and administrations. In this paper, they present a modern.

Chunguang Wang et al. in [5] "VCE-PSO: Virtual Cloud Embedding through a Meta-heuristic Approach" 2013, the creators depict resource portion, an essential and ceaselessly advancing piece of distributed computing, has been drawing in a great deal of specialists lately. Be that as it may, the greater part of current cloud frameworks [6]. Deliberate asset assignment just as arrangement of autonomous virtual machines, overlooking the presentation of a virtual machine is additionally relying upon other coordinating virtual machines and furthermore the net connections usage, which result in a poor proficient asset use. In this paper, they propose a novel model Virtual Cloud Embedding (VCE) to plan the cloud asset portion issue.

In [7], VCE views every asset demand as a basic unit instead of autonomous virtual machines including their connection limitations. To address the VCE issue, they build up a metaheuristic calculation VCE-PSO, which depends on molecule swarm streamlining calculation, to designate different assets as a unit thinking about the heterogeneity of cloud framework and assortment of asset necessities.

Rong Yu et al. in [8] "Toward cloud-based conveyance networks with economical resource management" 2013, the authors describe within the era of the net of things, all parts in intelligent transportation systems are connected to enhance transport safety, relieve traffic jam, cut back pollution, and enhance the comfort of driving

[9]. The vision of all vehicles connected poses a big challenge to the gathering and storage of huge amounts of traffic-related in this. During this article, they propose to integrate cloud computing into transport networks such the vehicles will share computation resources, storage resources, and information measure resources. The projected design includes a transport cloud, a wayside cloud, and a central cloud.

Parikh in [10] "A review on distributed computing asset assignment procedures" 2013, the creators portray cloud computing is a kind of registering which can be considered as another period of figuring. Cloud can be considered as a quickly rising new worldview for conveying processing as an utility. In [11, 12], distributed computing different cloud purchasers request assortment of administrations according to their powerfully evolving needs. So, it is the activity of distributed computing to profit all the requested administrations to the cloud purchasers. Be that as it may, because of the accessibility of limited assets, it is exceptionally hard for cloud suppliers to give all the requested administrations. From the cloud suppliers' point of view, cloud assets must be assigned in a reasonable way.

Sheng Di et al. in [13] "Versatile Calculation for Minimizing Cloud Errand Length with Expectation Blunders" 2014, the creators portray compared to conventional dispersed computing like network framework, it is non-trivial to optimize cloud task's execution due to its more limitations like client installment budget and separable asset request. In [14, 15], they analyze in-depth their proposed ideal calculation minimizing errand execution length with detachable assets and installment budget: (1) They infer the upper bound of cloud assignment length, by taking under consideration both workload forecast blunders and hostload forecast blunders.

In [16, 17], with such state-of-the-art bounds, the worst-case errand execution is unsurprising, which can progress the [18, 19] quality of benefit in turn. (2) They plan an energetic form for the calculation to adjust to the stack flow over errand execution advance, advance progressing the asset utilization. (3) They thoroughly construct a cloud model over a genuine cluster environment. Load balance scheduling [20, 21] in virtual machines has taken as a challenging to protect the sensitive data by using algorithms resources of cloud computing. Heterogeneous embedded computing optimization [22] services have come into existence to incorporate ciphers in the cloud data storage. Recently, there are many problems in the case of missing data. So, to find out the errors and solving that missing data is observed through nodes. It is very important to solve the missing data issues very accurately [23]. The way to provide the fastest communication without any delay using software-defined network has come into existence in recent time. It has mostly involved in the case of medical applications to secure the data without effecting to society [24]. Innovations in communications and computing of advanced technologies in the field of internet of things using software-defined networks have been executed for better performance of data [25]. Improvisation of the data performance in the network checking whether it is delay or fast using olsr network and ad hoc network for the advance technology [26].

3 Problems in Cloud

As there is no facility of data recovery accessible in cloud space, this is a significant application to demonstrate proficient strategy for missing information recuperation. The cloud's virtualized nature empowers promising new use cases for effective parallel information preparing. Be that as it may, it likewise forces new difficulties contrasted with exemplary group arrangements. The significant test we see is the cloud's darkness with prospect to abusing information region. For security reasons, mists frequently fuse arrange virtualization procedures which can hamper the derivation procedure, specifically when dependent on inactivity estimations.

To provide economical parallel processing for resource allocation we would like to contemplate bigger security whereas resources being allotted, information loss hindrance and economical allocation of cloud storage resources. several cloud infrastructures face issues in data processing of knowledge. Such issues could encounter because of hacking methodologies. Generally, cloud security and storages process the data by providing users in the combination of third-party users and providers. The two categories of security concerns in cloud computing are users and providers. Hosting an application and providing security to that application is a big matter in the security concerns of cloud computing. Securing infrastructure is another concern to provide in and outbound securities.

A. Need of Work

Main challenging topmost work to incorporate ciphers is to secure cloud storage for registered clients. Data hacking and server hacking are the main problems in the transfer of data. So to solve those problems, provide security against data hacking/server hacking. Administrator plays a main role to maintain user profiles and data profiles. Providing security for files through encryption concept is the main work to incorporate security issues. To make multi-parts of client files, multi-server facility for data storage is to be provided. Allocation of resources and files based upon storage capacity is to be done. Finally, recovery of infected or corrupted resources/files takes place to secure the data in cloud storage. Providing perfect quality of service and performance accuracy of data is a must. At the least regular observation of performance of data and solving the issues by optimization techniques should be provided.

4 System Architecture

To provide parallel information allocation and information security we have got information on multiple cloud nodes and just in case if any data/part of file is missing or virus-infected, then we will able to recover data/resources. For recovery purpose, we have got enforced self-derived formula that together works with encryption algorithms that are feasible, dynamic information allocation approach, and logical

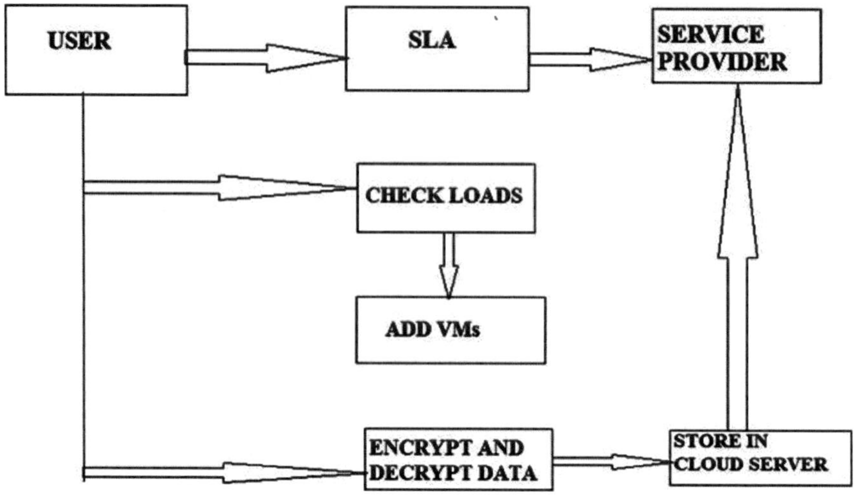

Fig. 1 System architecture referred model [27]

addresses of file. we have a tendency to aim to recover information 100% that is not doable with existing systems.

In our proposed framework, customer can have the option to enroll with cloud specialist co-op. Enlisted customer needs to login to transfer document which he/she need to store on cloud. Here, customer anticipates more elevated level of security for his/her assets. In this framework, server gets assets transferred by customer. Server audits got asset to channel prohibited information/documentation. The principle server gets assets transferred by customer. Consider client have transfer "test.txt" document proposed calculation encodes the record and parts into a number of information pieces. These splitted information pieces parallelly handled to store on helper cloud server-1 and assistant cloud server-2 appeared in Fig. 1. This framework gives greater security to information and there will not be any immediate access to client to assistant server's information.

i. **Client Registration Facility for New Clients and Login Facility for Existing Clients:** We have developed client registration and login window. This will enroll new client with framework to profit framework office. In this module, existing customer can sign into transfer his/her information record. In the event that client is enlisted, at that point, he/she gets ID and secret phrase. At the hour of login, client needs to give right ID and secret phrase. At that point, framework checks the client ID and secret phrase in the event that both are right, at that point client signed into the framework. Unapproved individual cannot have the option to login into the framework.
ii. **File Encryption and Upload:** In this module, when the client is enlisted, he/she transfers record on primary server and the most server scrambles the record and put away on server. The Progressed Encryption Standard (CPABE) calculation is executed for encryption of the information shown in Fig. 2. The most server

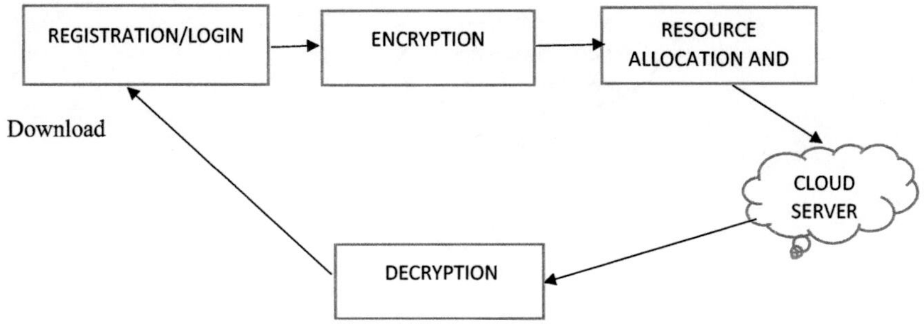

Fig. 2 Encryption and decryption referred model [28]

gets records transferred by client. At that point, record is splitted into a number of information chunks. The CPABE calculation will scramble the splitted record parts. This information will be parallelly handled to store on nodes.

iii. **File Decryption and Download:** In this module, client can download the records from server. The most server unscrambles the record utilizing CPABE calculation and gives to the user. The CPABE calculation is additionally executed for unscrambling of the information. This will give more security for information and there will not be any coordinate get to of client to assistant server's data.

iv. **Dynamic Resource Allocation**: In this module, we have made virtual hubs for allotting various servers for capacity. In our undertaking, we have made hubs for putting away the records appeared in Fig. 3. Assume record is scrambled and put away on various hubs. At the point, when we are attempting to download record at that point document parts are gathered from different hubs.

We have displayed a framework that employments virtualization innovation to store information assets (record) powerfully and back green computing by optimizing the number of servers in utilize. We have combined distinctive sorts of workloads (capacity) pleasantly and made strides the in general utilization of server assets. We have created a set of heuristics that avoid over-burden within the framework successfully whereas sparing vitality utilized.

v. **File Corruption Detection and Recovery:** In this module, client's document is splitted and scrambled utilizing CPABE calculation. That document parts are put away on various hubs. By utilizing heuristic algorithm, we have determined hash esteem and put away on database. At the point, when client needs to download his/her document, on the other hand, hash estimation of the current record is determined and checked with old hash esteem. In the event that both hash esteems are coordinated, and at that point, client gets his unique record. In the event that hash esteems are not same, at that point, we can say that record parts are undermined or tainted. At the point when we realize that document is ruined, we can recuperate this record.

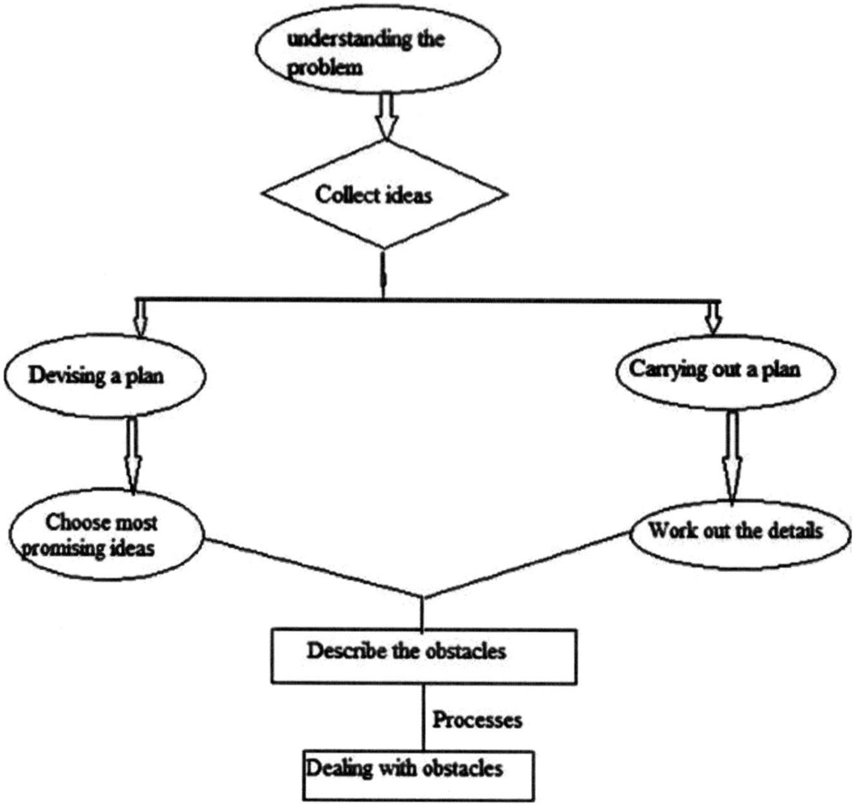

Fig. 3 Proposed algorithm referred model [29]

5 Experimental Section

Generally, heuristic algorithm is used in the most typical problems to embellish the issues in a practical way which is more sophisticated due to real constraints. When solving classy methods are very slow, heuristic algorithm came into existence, which solves the problems very quickly with more accuracy and compatibility.

Here, multi-objective function is proposed to incorporate ciphers that are nothing but securing the sensitive data without any breaches. With the use of cryptographic algorithms, data is secured that is stored and also encrypted in the cloud. Previously, this implementation took place in the domain of Amazon Web services (AWS). Now with the help of some optimized mathematical methods and objectives, implementation is done in Microsoft Azure. Encryption may be in java or any other programing language shows how data can be secured and how incorporation of ciphers is done.

6 Result Analysis

Comparison of encryption algorithms.

Comparison table of encryption algorithms referred model [30]

File	Average encryption time (ms)		
	CPABE	AES	DES
F1	0.00055	0.0008	0.00135
F2	0.0006	0.00095	0.00155
F3	0.0007	0.0012	0.0018
F4	0.00095	0.0014	0.0020

Table shows the file and time required for encryption. We have taken three different algorithms for comparison viz. CPABE, Advanced Encryption Standard(AES) and Data Encryption Standard.

Figure 4 Shows the consequence of examination of various encryption calculations, for example, CPABE, Advanced Encryption Standard (AES) and Data Encryption Standard (DES). The got outcomes from various encryption calculations are utilized to investigate the consequence of proposed framework. The acquired outcomes from encryption calculations are thought about and in the wake of contrasting the got outcomes it shows the CPABE is effective calculation for encryption and unscrambling.

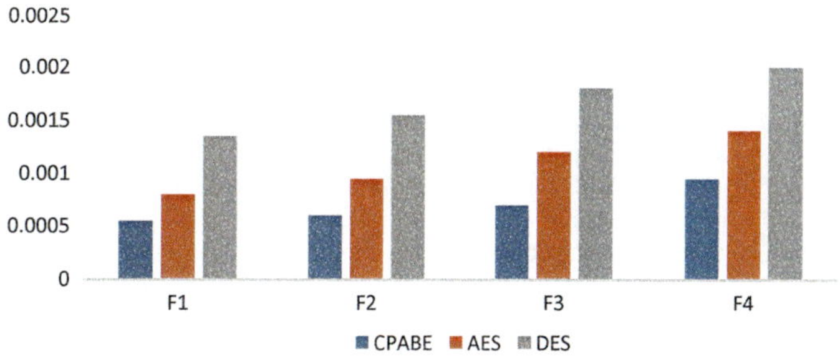

Fig. 4 Comparison of three encryption algorithms referred model [30]

7 Conclusion

We have implemented the framework to supply information security and ensure the rightness of user's information in cloud. Security is given through the encryption and decoding method. Moreover, framework is executed for parallel processing of information. Incorporation of security ciphers to address the issues regarding security by using some probabilistic methods is proposed. Comparison of encryption algorithms is taken place to detect the average time of performance for feasible requirements in future.

References

1. Kang Z, Wang H (2013) A novel approach to allocate cloud resource with different performance traits. In: 2013 IEEE international conference on IEEE, services computing (SCC)
2. Marston Sean, Li Zhi, Bandyopadhyay Subhajyoti, Zhang Juheng, Ghalsasi Anand (2011) Cloud computing—the business perspective. Decis Support Syst 51(1):176–189
3. Zhang S, Zhang SF, Chen XB, Huo XZ (2010) Cloud computing research and development trend. In: Proceedings of the 2010 second international conference on future networks (ICFN '10). IEEE Computer Society, Washington, DC, USA, pp 93–97. https://doi.org/10.1109/icfn.2010
4. Dillon T, Wu C, Chang E (2010) Cloud computing: issues and challenges. In: 24th IEEE international conference on advanced information networking and applications (AINA), pp 27–33
5. Wang C, Wu Q, Tan Y, Guo D, Wu Q (2013) VCE-PSO: virtual cloud embedding through a meta-heuristic approach, In: 2013 IEEE 10th international conference on IEEE, high performance computing and communications & 2013 IEEE international conference on embedded and ubiquitous computing (HPCC_EUC)
6. Wu L, Gag SK, Buyya R (2011) Sla-based resource allocation for software as a service provider (saas) in cloud computing environments. In: 2011 11th IEEE/ACM international symposium on Cluster, cloud and grid computing (CCGrid), IEEE, pp 195–204
7. Buyya R, CS Yeo, Venugopal S (2008) Market-oriented cloud computing: vision, hype, and reality for delivering it services as computing utilities. In: HPCC'08 10th IEEE international conference on High performance computing and communications, 2008, IEEE, pp 5–13
8. Yu R, Zhang Y, Gjessing S, Xia W, Yang K (2013) Toward cloud-based vehicular networks with efficient resource management, IEEE, Network, IEEE
9. Kumar A, Pilli ES, Joshi RC (2013) An efficient framework for resource allocation in cloud computing, In: 2013 fourth international conference on IEEE computing communications and networking technologies (ICCCNT)
10. Parikh SM (2013) A survey on cloud computing resource allocation techniques, In: 2013 Nirma university international conference on IEEE, Engineering (NUiCONE)
11. Bhardwaj K, Mahajan R, Surinder M (2016) Improved load management in cloud environment using MHT algorithm. Intl J Control Theory Appl 9(22):301–305
12. Bhardwaj AK, Mahajan R, Surender (2016) TTP based vivid protocol design for authentication and security for cloud. IEEE Xplore, pp 3275–3278
13. Di S, Wang C-L, Cappello F (2014) Adaptive algorithm for minimizing cloud task length with prediction errors, In: IEEE transactions on IEEE cloud Computing
14. Di S, Kondo D, Wan C (2014) Optimization of composite cloud service processing with virtual machines. IEEE transactions on computers, IEEE

15. Srinivasa KG, Kumar KS, Kaushik US, Srinidhi S, Shenvi V, Mishra K (2014) Game theoretic resource allocation in cloud computing, In: IEEE applications of digital information and web technologies (ICADIWT), 2014 fifth international conference
16. Surender K, Mahajan R (2015) A modified heuristic-block protocol model for privacy and concurrency in cloud. Int J Adv Comput Sci Appl (IJACSA) 6(9):179–184
17. Duy TVT, Sato Y, Inoguchi Y (2010) Performance evaluation of a green scheduling algorithm for energy savings in cloud computing. In: Proc IEEE International Symposium Parallel Distribution Process, Workshops Phd Forum, Chengdu, China, pp 1–8
18. Li J, Qiu M, Ming Z, Quan G, Qin X, Gu Z (2012) Online optimization for scheduling preemptable tasks on IaaS cloud systems. J Parallel Distribute Compute 72(5):666–677
19. Qiu M, Zhong Ming JL, Gai K, Zong Z (2015) Phase-change memory optimization for green cloud with genetic algorithm. IEEE Trans Compute 64(12):1–13
20. Luna JM, Abdallah CT (2011) Control in computing systems: Part II. In: Proc IEEE multi-conf syst control, Denver, CO, pp 32–36
21. Xu M, Cui L, Wang H, Bi Y (2009) A multiple QoS constrained scheduling strategy of multiple workflows for cloud computing. In: Proc IEEE International Symposium Parallel Distribution Process, Workshops Phd Forum, pp 629–634
22. Qiu M, Li H, Sha EH-M (2009) Heterogeneous real-time embedded software optimization considering hardware platform. In: Proc ACM Appl Compute, pp 1637–1641
23. Srinivas LNB, Ramasamy S (2017) An improvised missing data estimation algorithm for wireless sensor network applications. J Adv Res Dyn Control Syst 9(18):913–918
24. Venkatesh K, Srinivas LNB, Mukesh Krishnan MB, Shanthini A (2019) QoS improvisation of delay sensitive communication using SDN based multipath routing for medical applications. Future generation Computer Syst 93:256–265 2018
25. Reddy V, Venkatesh K (2020) Role of Software-Defined Network in Industry 4.0. In: EAI/Springer innovations in communication and computing-internet of things for industry 4.0 design, challenges and solutions, pp 197–218
26. Naveen Chandar B, Arivazhagan N, Venkatesh K (2019) Improving the network performance using mp-olsr protocol for wireless ad hoc network (MANET). Int J Recent Technol Eng 8(3):5700–5707
27. https://journals.plos.org/plosone/article?id=10.1371/journal.pone.0216067. Resource provisioning approach for cloud applications
28. https://www.geeksforgeeks.org/classical-cryptography-and-quantum-cryptography/
29. https://www.google.com/search?q=heuristic+algorithm&tbm=isch&ved=2ahUKEwibwNexwrXnAhWNh0sFHQUXB4YQ2cCegQIABAA&oq=heuristic +algorithm&gs_l=img.3..35i39j0j0i5i30l2j0i8i30l3j0i24l3.103560.110898..111215…1.0..4.213.5445.35j19j1……0….1..gws-wiz-img…..10..35i362i39j0i67.UObj8rjzuaw&ei =MCQ4XpvoIY2PrtoPha6csAg&bih = 657&biw=1366img…..10..35i362i39j0i67.UObj8rjzuaw&ei=MCQ4XpvoIY2PrtoPha6csAg&bih=657&biw=1366
30. https://symbiosisonlinepublishing.com/computer-science-technology/computerscience-information-technology32.php

Inventory Prediction Using Market Basket Analysis and Text Segmentation—A Review

B. V. R. Sai Teja and N. Arivazhagan

Abstract Recently, information mining has pulled in a lot of consideration in the data business and in a society where information keeps on developing consistently. The facts and knowledge acquired from huge data can be utilized for applications extending from advertising investigation, extortion identification, creation control, client maintenance, and science investigation. Due to new competition in the present market such as e-commerce sites, because of that so many offline markets are affecting. We will help to rectify this problem by helping the offline stores to maintain their customers. By recommending the present and upcoming patterns were trailed by the society. In our proposed system, we propose a strategy for the expectation of the client's buying behaviour so that we can manage the stocks in the shop. This paper gives a complete characterization of market basket analysis and text segmentation of the customer's data.

Keywords Market basket analysis · Text segmentation · Stocks · Behaviour · Customer · Shops

1 Introduction

Today business is advancing from the item focused to a client-focused condition. The inside and out comprehension of client behaviour is a basic achievement factor to fabricate long haul, productive associations with explicit clients in the worldwide aggressive commercial centre. In this manner, client conduct expectation is an urgent method for examination client social administration. A client behaviour examination gives knowledge into the various factors that impact a group of people. It gives you a thought of the intentions, needs, and basic leadership strategies being considered

B. V. R. Sai Teja (✉) · N. Arivazhagan
Department of Information Technology, SRM Institute of Science and Technology, Chengalpattu, India
e-mail: br2621@srmist.edu.in

N. Arivazhagan
e-mail: arivazhn@srmist.edu.in

D. J. Hemanth et al. (eds.), *Artificial Intelligence Techniques for Advanced Computing Applications*, Lecture Notes in Networks and Systems 130,
https://doi.org/10.1007/978-981-15-5329-5_34

during the client's adventure. This examination encourages you to see how clients feel about of your organization, just as if that observation lines up with their guiding principle. Another key business need is the capacity to foresee a client's general worth. A customer behaviour analysis adds effectiveness to this procedure by distinguishing perfect client qualities. By focusing on these personas, your business can pull in brand-steadfast clients before your rivals do.

Here we will discuss two methods:

1.1 Market Basket Analysis

Market basket analysis is one of the most broadly perceived and important sorts of data examination for promoting and retailing. The purpose behind promote bushel examination is to make sense of what things customers purchase together. It takes its name from the probability of customers hurling all of their gets tied up with a shopping bin (a "advertise container") during looking for nourishment. Acknowledging what things people purchase as a social event can be valuable to a retailer or to some other association. A store could use this information to put things as regularly as conceivable sold together into a comparative locale, while a rundown or World Wide Web dealer could use it to choose the plan of their stock and solicitation structure. Direct promoters could use the carton examination results to make sense of what new things to offer their prior customers.

Now and again, the way that things sell together is self-evident—each drive-through joint asks their clients "Would you like fries with that"? at whatever point they experience a drive-through window. Nonetheless, now and then the way that specific things would sell well together is a long way from self-evident. An outstanding model is that a grocery store playing out a bin investigation found that diapers and brew sell well together on Thursdays. Despite the fact that the outcome makes sense—youthful couples loading up on provisions for themselves and for their kids before the end of the week begins—it is not the kind of thing that somebody would ordinarily consider immediately. The quality of market container investigation is that by utilizing PC information mining devices, it is a bit much for an individual to consider what items shoppers would intelligently purchase together—rather, the clients' business information is permitted to represent itself with no issue. This is a genuine case of information-driven promotion.

At the point when it is understood that customers who get one thing are likely going to buy another, it is achievable for the association to publicize the things together, or to make the purchasers of one thing the target potential outcomes for another. If customers who purchase diapers are presently inclined to purchase ale, they will be impressively bound to if there happens to be a blended show essentially outside the diaper path. Also, if it is understood that customers who buy a sweater and accommodating pants from a particular mail-demand list have a proclivity towards buying a coat from a comparable stock, offers of coats can be extended by having the telephone delegates depict and offer the coat to any person who acquires to orchestrate the sweater and pants. Still better, the stock association can give an extra 5% discount on a group containing the sweater, pants, and coat at the same time and advance well the complete pack. The dollar proportion of offers is guaranteed to go up. By concentrating on customers who are starting at now known to be likely buyers, the feasibility of displaying is in a general sense extended—paying little personality to if the publicizing shows up as in-store appears, record position structure, or direct plans to customers. This is the explanation behind market bushel investigation—to improve the reasonability of publicizing and arrangements techniques using customer data successfully available to the association.

1.2 Text Segmentation

Text segmentation is the route towards parcelling created content into critical units, for instance, words, sentences, or focuses. The term applies both to mental techniques used by individuals when getting content and to counterfeit systems realized in PCs, which are the subject of ordinary language planning. The issue is non-piddling, in the light of the fact that while some created vernaculars have unequivocal word limit markers, for instance, the word spaces of formed English and the specific start, average and last letter conditions of Arabic, such banner are from time to time flawed and not present in each made language.

2 Related Work

In this project, we proposed market basket evaluation technique which fuses text division innovation and association rule mining innovation. Properties of matters can be delivered subsequently earlier than mining connection regulates with the aid of the use of text division advancement. This methodology has been utilized to a café geared up with digital mentioning structure to provide pointers to customers, where the tests had been done. The preliminary consequences show that the strategy is successful and true [1].

Market basket evaluation is a methodology for the one of a kind verification and assessment of the buying behaviours of the clients. A lot of lookup work has been beginning at now executed in exhibit bin investigation, and it has been determined that there are a couple of troubles associated with promote bushel examination. The first is that the prerequisites of the customers hold altering with recognize seasons and time. Consequently, the outcomes of market container examination are definitely in charge to seasons and time; subsequently, we need to perform it on and on. Another problem is associated with Apriori estimation which requires a repeated breadth of the complete database of patron trades to discover candidate sets and progressive element sets. Our work targeted on the utilization of pretending neural system methodology to vanquish these issues. We proposed a singular layer feed-forward in section-related neural framework strategy, which diminishes the time taken in the kept sifting of the database and moreover assembles the efficiency of the count number [2].

This paper centres on Apriori execution for association rule mining. Frequent item sets produced by Apriori calculation totally rely upon a base help limit. It was seen that Apriori burns through running time because of competitor item sets age; thus, the requirement for an increasingly vigorous mixture calculation for affiliation rule mining is unavoidable. In any case, for a value-based database where numerous exchange things are rehashed commonly as a superset in that kind of database, Apriori is appropriate for mining regular item sets. Along these lines, this calculation produces visit item sets completely. The calculation was executed utilizing PHP and MySQL database; the board framework was utilized for putting away the stock information. The calculation produces visit item sets totally and creates the precise solid guidelines. The proposed framework can be utilized to decide the purchasing behaviours of consumers with more prominent pace of exactness and in this way improve day-by-day deals [3].

In this proposed framework, which separates the significant points or problems from Skype client input origin and estimates the feeling related to those subjects utilizing Vibe metric, can be a genuine model around there. In contrast to different past research, which has concentrated on removing the client's opinions either internationally or in isolated subjects, our work centres around following the connected passionate directions over all the significant issues from Skype client input after some time. In addition, it likewise gives a stage to both contemplating client feelings and following how the feelings with respect to various significant themes correlatively change after some time by utilizing unstructured printed client criticism information and organized client action telemetry information [4].

In the client-focused commercial centre, the comprehension of client conduct is a basic achievement factor. The enormous databases in an association for the most part include multiplex information, for example, static, time arrangement, representative consecutive and literary information, which are independently put away in various databases of various areas. It represents a take a look at established targeted purchaser conduct forecast. In this investigation, a novel methodology called collaborative multiple kernel support vector machine (C-MK-SVM) is produced for dispersed consumer habits expectation using multiplex information. The changing heading technique for multipliers (ADMM) is utilized for the worldwide streamlining of the circulated sub-models in C-MK-SVM. Computational assessments on a practicable retail dataset are accounted for. Computational outcomes show that C-MK-SVM displays greater accurate client behaviour prediction performance and better computational speed than support vector machine and multiple kernel guide vector computing device [5].

In this proposed system, we advise a system for the gauge of the EC client's buy direct via joining unique classifiers reliant on hereditary calculation. The machine used to be tried and surveyed the use of web data from the most important EC association. We moreover tried the authenticity of our system all matters regarded game layout issues the use of composed via hand numerals. In the chief investigate, we show that the proposed remember can improve the conjecture accuracy of the purchase propensity. Moreover, in the consequent assessment, we in like manner exhibit that this machine has favoured execution over other merging strategies we endeavoured similarly as any person classifiers. In view of the attractive consequences acquired from these preliminaries, the wrapping up remarks can be sketched out as seeks after. At first, the proposed GA-based combining approach can be properly utilized to the want assignment of the customer's purchase propensity in all actuality case with extra precision than the widespread facts mining moves close. Second, this strategy is also an exquisite framework for several classifier mixes by using and large association problem areas. In the two cases, our technique suggests perfect introduction over man or woman classifiers and different recognized joining systems we endeavoured [6].

This examination builds up an item suggestion framework structured on apparatuses from the spatial insights writing. In our proposed AL technique, customers are believed to be organized on a joint-space (pick-any scaling) map in which persons that is near one any other offer similar thing tendencies. Instead of multidimensional scaling (which is used to end mine how combat brands are seen), pick-any scaling has been used to understand the association between manufacturer arranging and purchaser tendencies. This assessment abuses the houses of pick-any scaling to decide an extent of customer closeness reliant on previous buy lead. By mixing the AL spatial preference model with a pick-any guide, we can manufacture every other system for assessing aspect alternatives.

For clients in arm's database, we show observationally that the AL model gives outstanding guesses similar to benchmark methodologies [7].

In this paper, we propose a system called COREL for client's buy behaviour. This structure involves a two-organize method. Initially, the relationship between items is researched and abused to forecast client's inspirations, i.e. to construct an applicant item assortment. Second, clients' inclinations for item includes are figured out how to recognize which applicant items are destined to be acquired. These examination researches three classifications of item had include dynamic item, highlights that might be seen by the client yet not by the investigator and item includes that are static and perceptible by the examiner. We misuse the buy information from a web-based business site to create techniques to learn client inclinations for every one of these three classifications. At the point when an item obtained by a specific shopper is submitted to COREL, the program can restore the top n items well on the way to be bought by that client later on. Tests directed on a genuine data set show that client inclination for specific item includes and assumes a key job in basic leadership and that COREL incredibly beats the benchmark methods. The results demonstrate that our way to deal with figuring item notoriety is possible and that client inclination for item includes has a significant sway on buying choices [8].

In this paper, we bring the piece primarily-based approach into the buying list desire and a quick time later give an observational connection of the individual-level and the section primarily-based structures in this issue. Therefore, comprehended AI classifiers and clients' buy history are used, and the evaluation is carried out on a certifiable data set by using coordinating a motion of tests. The outcomes advocate that there is no simple champ in this relationship and the presentations of purchaser lead showing techniques rely upon the AI figuring used. The examination can bolster researchers and professionals to recognize exclusive portions of using client lead displaying tactics in the purchasing list gauge [9].

In a standard retail keep the habits of clients may additionally yield a ton to the store aide. In any case, with recognition to digital shopping it is stupid to hope to see and dismember client conduct, for instance, facial mirrors, matters they test or contact, etc. For this circumstance, clickstreams or the mouse improvements of e-customers may additionally give a couple of bits of know-how about their obtaining conduct. In this assessment, we have acquainted a model with separate clickstreams of e-customers and concentrate data and make assumptions related to their buying behaviour on a propelled commercial enterprise focus. In the wake of gathering facts from an Internet-based totally business promotion in Turkey, we played out a facts-mining application and extricated online clients' standards of conduct about purchasing or not. The model we current forecasts whether clients will or won't purchase their matters blanketed to buying bushels a propelled business focus. For the examination, decision tree and multi-layer neural framework conjecture records mining models have been used. Disclosures have been discussed at closing [10].

Reason for this exploration is, to discover client enthusiasm on items and base on examined yield, give best or closest to client premium items to promoting and deals division. With the goal that they can make offers and plans according to client intrigue. Venture asset arranging (ERP) contains numerous modules among which client relationship the executives manages advertising and pulling in clients. In this module for the most part salespeople contacts their clients and satisfies them with the best ideas to expand deals, yet reaching clients on irregular will bring about more endeavours with no ensured income. Introduced look into work is centred around this situation. The proposed calculation figures loan fee of clients on given item dependent on orders they had put in past. The proposed calculation does the investigation of items and clients have acquired previously and dependent on that creates result containing rate pace of client enthusiasm for items. This outcome can be passed to CRM office with the goal that they can contact clients dependent on their inclinations; along these lines ensured income. The association can likewise decide to actualize mechanized mail framework or IVR framework at that point to contact clients dependent on this outcome. This calculation depends on idea of information mining [11].

A huge difficulty that encounters associations mainly communicates interchanges business is "customer unsettles", and this takes place when a patron decides to go away an association's landline enterprise for every other connection contender. Thus, our focuses on the previous examination to manufacture a mannequin that will count on mix client through describing the customer's specific practices and characteristics. We will use statistics mining methods, for instance, clustering, portrayal and connection rule. The precision and exactness of the framework used are so fundamental to the accomplishment of any support trying. Everything considered if the association does not assume about a client who is going to go away their business, no suited pass can be made by means of that association towards that client [12].

This paper suggests that multiple kernel support vector machines (MK-SVMs)-based client beat expectation system is carried out to bind three information disclosure errands, which are highlight choice, class forecast and choice guideline extraction, into an entire structure. A two-organize cycle of different raised streamlining issues is intended for at the same time include determination and class forecast. In the light of the chose highlights, bolster vectors are utilized to extricate choice standards. An open CRM dataset is utilized to examine the presentation of this methodology. Exploratory consequences exhibit that MK-SVMs accomplish promising execution on the intensely slanted data set with the aid of techniques for a re-adjusted methodology, and the removed ideas accomplish excessive inclusion and low bogus caution with a modest variety of preconditions [13].

This paper is built an information distribution centre with automobile marketing subject by utilizing SQL server. Also, on this establishment, we receive the choice tree model and partner rule model to investigate the history information of the automobile marketing, get the automotive customer classification rules and clients' characteristics partner runs, and acknowledge forecast of clients' obtaining behaviour to raise the centre rivalries of automotive enterprises. This paper removes automotive marketing data, develops information distribution centre, receives a better ID3 choice tree model, and an affiliation rule model to do information extracting and afterwards acquires expectation data of automotive clients' behaviour. Trial and relative outcomes check the legitimacy and precision of the expectation results [14].

The purpose of this assessment paper is to take a gander at the association between buyer conduct variables and capacity to buy. First we investigate to find an association between client conduct to buy things on evolving variables, for instance, natural component, definitive component, solitary component and social component. Consequently, this paper proposes time-creating arbitrary backwoods classifier that the utilization novel part working to predict the conduct of purchaser that affects the choice of securing the component basically. Delayed consequences of arbitrary backwoods classifier are more precise than other AI figuring [15].

In this paper, I propose client behaviour investigation utilizing excitement examination. Particularly, we will likely disclosure drop-off clients from get to logs. It is imperative to discover drop-off clients prior and bolster them to remain in a help. Generally, forecast models are developed in AI and RFM investigation is utilized in showcasing field. In this paper, I gauge energy levels, which signify clients' actuation, from perceptions and apply them to forecast of revelation of drop-off clients. In evolutional tests, I utilize genuine online shop get to logs and talk about connection between energy levels and drop-off clients. I affirmed that many drop-off clients took lower eagerness levels in assessment point, and the energy level could be utilized to anticipate drop-off clients [16].

A scope of calculations was once utilized to crew on the net retail clients of a UK employer using verifiable exchange information. The prescient capacities of the classifiers were evaluated utilizing linear regression, Lasso and regression trees. In distinction to most-related examinations, groupings depend on explicit and promoting centred customer practices. Forecast precision on undeveloped clients used to be commonly ideal to 80%. The fashions completed (and appeared at) for grouping were logistic regression, quadratic discriminant analysis, linear SVM, RBF SVM, Gaussian process, decision tree, random forest and multi-layer perceptron (neural network). Postcode data was at that point used to characterize solely on socioeconomics received from the UK Land Library and comparative open records sources. Forecast precision stayed highest quality to 60%. In this proposed system, we have exhibited the plausibility of distinguishing the probabilities of purchaser buy emphasis, making it viable to goal huge patron social occasions. We have moreover exhibited how open information sources can be used to grow inward statistics and thus attain extended showing and efficiency. So a ways we have these days begun to uncover what is underneath [17].

The recency (R), repeat (F) and financial (M) values are comprehensively used in enterprise and the academic network to exhibit records of purchaser direct and graph for future advancing approaches. This paper proposes another mannequin for RFM desire for customer's situation to intermittent neural structures (RNNs) with remedied direct unit incitation work. The model uses an auto-encoder to address elements of data parameters (for instance consumer dependability number, R, F and M). The proposed mannequin is the first of its type in the composing what is more, which has quite a number open entryways for more improvement. The model can be elevated by means of the use of the entire all the extra getting ready data. It is fascinating to explore in addition buildings of the mannequin in auto-encoder and recursion levels. Clumpiness is another variable which can be viewed as an extra substance to R, F and M (for occasion RFMC) factors. Another pathway is thinking about a number of parameters of patron (for instance zone, age, etc.) for modified characteristic extraction and similarly enhancement of suggesting constructions [18].

We have proposed a statistics pushed and client-driven methodology for market basket analysis. Our dedication is twofold. In the first place, we have characterized Temporal Annotated Recurring Sequences (TARS). At that point, we have utilized TARS to manufacture a TARS-based indicator (TBP) for deciding clients' subsequent baskets. We have worked on probes true world data sets performing that TBP outflanks nice in class techniques and, conversely with them, it offers interpretable examples that can be utilized to accumulate bits of understanding on clients' buying behaviours [19].

We carried out a metamodel that altered arrangements the different components of data present in esteem based datasets. These estimations can be customer, thing, offer, target, business focus and trades. Our structure also has dynamic capacities with regard to finish rundown of abilities age and joins differing AI computations to learn the desire model. Our framework works through and through from feature working to uncovering repetitive probabilities of customers for things (or business focus, brand, site or store chain). Likewise, the foreseen intermittent direct of customers for different things close by their worth-based history is used by our offer improvement model I-Prescribe to propose things to be offered to customers with the goal of enhancing the appearance on the theory of given displaying spending plan. We exhibit that our dynamic features manage two one of a kind data challenge datasets, by sharing preliminary outcomes [20] (Table 1).

Table 1 Algorithms/techniques and drawback of existing system

SR.NO.	Paper Title	Algorithms/Techniques	Drawback
1	Market basket analysis based on text segmentation and association rule mining	Text segmentation and association rule mining	This algorithm is time-consuming
2	Market basket analysis using artificial neural network	Artificial neural network	Overall accuracy will be less
3	Frequent pattern and association rule mining from inventory database using Apriori algorithm	Apriori algorithm	The proposed structure can be used to choose the buying practices of buyers yet it does not give that much precision
4	Big data technology and ethics considerations in customer behaviour and customer feedback mining	Vibe metric	Printed client input are client produced content, which may contain grimy words and commotions. We ought to consider every one of these conditions by sifting these one-sided factors which may offer ascent to deceiving results
5	Distributed customer behaviour prediction using multiplex data: a collaborative MK-SVM approach	Support vector machine, Multiple kernel learning	Client behaviour forecast is a standard binary characterization issue. Accordingly, this examination just spotlights on the binary class and not to multiclass
6	Combination of multiple classifiers for the customer's purchase behaviour prediction	GA-based combining approach	The proposed system does not give much accuracy for classification
7	Predicting product purchase from inferred customer. Similarity, an autologistic model approach	Autologistic model approach	The prediction of the product purchase behaviour is different from person to person and the algorithm is not perfect for this operation

(continued)

Table 1 (continued)

SR.NO.	Paper Title	Algorithms/Techniques	Drawback
8	Predicting customer purchase behaviour in the e-commerce context	Customer purchase prediction model (COREL)	In this examination, be that as it may, brand inclination does not fundamentally improve the presentation of COREL when we isolate item brand level by the quantity of things under the brand
9	An empirical comparison of customer behaviour modelling approaches for shopping list prediction	Machine learning algorithm	In this paper, they have done the comparative test for customer behaviour
10	Analysis and prediction of e-customers' behaviour by mining clickstream data	Decision tree and multi-layer neural network	This investigation cannot be viewed as general guidelines for online clients or their practices, rather this examination shows the "practicability" of such a work
11	Efficient application of data mining for marketing and sales decision making in ERP	Data mining	In our calculation, we have considered deals orders and in that we have considered limitations that every item has been requested in amount one yet continuously it could be more, so the piece of calculation that is centred around item event include in deals request can be improved for more than one amount requested
12	Customer churn prediction model using data mining techniques	Clustering, classification, association rule	This exploration has applied the forecast model of the proposed client beat through utilizing various perspectives yet exactness is less
13	Building comprehensible customer churn prediction models: a multiple kernel support vector machines approach	Support vector machine, multiple kernel learning	In this paper, they did not include the financial prediction

(continued)

Table 1 (continued)

SR.NO.	Paper Title	Algorithms/Techniques	Drawback
14	Research on data mining algorithms for automotive customers' behaviour prediction problem	Data mining algorithms	This exploration is on data mining algorithms for the behaviour of customer prediction
15	Prediction of Consumer Behaviour using Random Forest Algorithm	Random Forest Algorithm	The primary burden of random forests is their multifaceted nature. They are a lot harder and tedious
16	Customer behaviour analysis with enthusiasm analysis	Machine learning	Exactness of the algorithm is not high and not clear also
17	Towards accurate predictions of customer purchasing patterns	Classifiers, regression, segmentation,	This paper isn't planned to direct increasingly point by point examination by embracing unaided and all the more critically exploratory methods to assist our comprehension of the variables that impact client conduct in a progressively nonexclusive setting
18	Customer shopping pattern prediction: a recurrent neural network approach	Recurrent neural networks	This paper isn't thinking about different parameters of client (e.g. area, age, and so forth.) for programmed include extraction and further improvement of recommender frameworks
19	Market basket prediction using user-centric temporal annotated recurring sequences	Temporal annotated recurring sequences (TARS)	The prediction is not accurate
20	Generic framework to predict repeat behaviour of customers using their transaction history	Logistic regression, machine learning algorithms	The exactness of the algorithm which is used in this paper is not much

3 Conclusion

Recently, the improvement of sharing of data is achievable through the system which has expanded massively, and because of that we can predict the customer's behaviour. Also, with this, we are going to actualize a model which can predict the customer's behaviour and the stocks of the product. This is a survey paper, where we have inspected various sorts of calculations and predictions of the data taken place until now. Thinking about the upsides and downsides in the list of reviewed papers, we are going to make a system that will forecast the client's behaviour and the stocks of the product with higher precision.

References

1. Wen-xiu X, Heng-nian Q, Mei-li H (2010) Market basket analysis based on text segmentation and association rule mining. In: First international conference on networking and distributed computing, IEEE
2. Bhargav A, Mathur RP, Bhargav M (2014) Market basket analysis using artificial neural network. In: International conference for convergence of technology
3. Adewole KS, Akintola AG, Ajiboye A.R. (2014) Frequent pattern and association rule mining from inventory database using apriori algorithm. Afri J Comput ICT 7(3)
4. Deng X (2017) Big data technology and ethics considerations in customer behavior and customer feedback mining. In: IEEE international conference on big data
5. Chen Z-Y, Fan Z-P (2012) Distributed customer behavior prediction using multiplex data: a collaborative MK-SVM approach. Knowl-Based Syst 35:111–119
6. Kim Eunju, Kim Wooju, Lee Yillbyung (2002) Combination of multiple classifiers for the customer's purchase behaviour prediction. Decis Support Syst 34:167–175
7. Moon Sangkil, Russell Gary J (2008) Predicting product purchase from inferred customer similarity: an autologistic model approach. Institute for Operations Research and the Management Sciences (INFORMS) 54(1):71–82
8. Qiu J, Lin Z, Li Y (2015) Predicting customer purchase behavior in the e-commerce context. Springer Science + Business Media, New York
9. Peker S, Kocyigit A, Erhan Eren P (2018) An empirical comparison of customer behavior modeling approaches for shopping list prediction. In: 41st international convention on information and communication technology, electronics and microelectronics (MIPRO)
10. Sılahtaroğlu G, Dönertaşli H (2015) Analysis and prediction of E-customers' behavior by mining clickstream data. In: 2015 IEEE international conference on big data (Big Data), Santa Clara, CA, USA
11. Bhadlawala S (2013) Efficient application of data mining for marketing and sales decision making in ERP. In: 2013 Nirma university international conference on engineering (NUiCONE), Ahmedabad, India
12. Mitkees IM, Badr SM, ElSeddawy AIB (2017) Customer churn prediction model using data mining techniques. In: 13th international computer engineering conference (ICENCO), Cairo, Egypt
13. Chen Z, Fan Z (2011) Building comprehensible customer churn prediction models: a multiple kernel support vector machines approach, In: ICSSSM11, Tianjin, China
14. Huang L, Zhou C, Zhou Y, Wang Z (2008) Research on data mining algorithms for automotive customer's behaviour prediction problem. In: seventh international conference on machine learning and applications, San Diego, CA, USA

15. Valecha H, Varma A, Khare I, Sachdeva A, Goyal M (2018) Prediction of consumer behaviour using random forest algorithm. In: 5th IEEE Uttar Pradesh section international conference on electrical, electronics and computer engineering (UPCON), Gorakhpur, India
16. Yanagimoto H (2016) Customer behaviour Analysis with Enthusiasm Analysis, In: 5th IIAI international congress on advanced applied informatics (IIAI-AAI), Kumamoto, Japan
17. Valero-Fernandez R, Collins DJ, Lam KP, Rigby C, Bailey J (2017) Towards accurate predictions of customer purchasing patterns. In: 2017 IEEE international conference on computer and information technology (CIT), Helsinki, Finland
18. Salehinejad H, Rahnamayan S (2016) Customer shopping pattern prediction: a recurrent neural network approach. In: 2016 IEEE symposium series on computational intelligence (SSCI), Athens, Greece
19. Guidotti R, Rossetti G, Pappalardo L, Giannotti F, Pedreschi D (2017) Market basket prediction using user-centric temporal annotated recurring sequences. In: 2017 IEEE international conference on data mining (ICDM), New Orleans, LA, USA
20. Kazmi AH, Shroff G, Agarwal P (2016) Generic framework to predict repeat behavior of customers using their transaction history. In: 2016 IEEE/WIC/ACM international conference on web intelligence (WI), Omaha, NE, USA

Analysis of Improvisation of Customer Satisfaction in Banking Sector Using Big Data Analytics

M. Aasritha, D. Hemavathi, and G. Sujatha

Abstract Customer satisfaction is the primary concern in the banking domain. Nowadays, the market is mostly customer centric and there is a huge competition among the business organizations. In order to overcome the huge competition, there is a need of customer loyalty which helps to retain the existing customers. Therefore, analysis is required to know about the customer spending patterns. The paper presents the comparative study of two different datasets, namely future loan status prediction and marketing targets from Kaggle, which helps to analyze the customer satisfaction in the banking sector. To determine the customer satisfaction, a relationship model was created between response variable and predictor variables using multiple regression method. The proper analysis of these data will provide a clear idea about customer aspects in particular services provided by bank.

Keywords Customer satisfaction · Big data analytics · Multiple regression

1 Introduction

At present, there is a need of big data analytics [1] because for every second the data will be generated. In order to process that huge amount of data, big data analytics is needed and it deals with 3 v's. First v is variety, which deals with the different data types. Generally, banks will deal with different types of data like customer ID, credit score, etc. Second v is velocity, it is defined at the rate of speed which at which new data is added to the database. Hundreds of transactions take place for every minute. Third v is volume, which is defined as the amount of space required to store the data.

M. Aasritha · D. Hemavathi (✉) · G. Sujatha
SRM Institute of Science and Technology, Chennai, India
e-mail: heeram007@gmail.com

M. Aasritha
e-mail: aasrithareddymodem@gmail.com

G. Sujatha
e-mail: sujathag@srmist.edu.in

D. J. Hemanth et al. (eds.), *Artificial Intelligence Techniques for Advanced Computing Applications*, Lecture Notes in Networks and Systems 130,
https://doi.org/10.1007/978-981-15-5329-5_35

There is a rapid increase in the number of banks day by day which results in the huge competition among the business organizations. So, there is a need of new strategies to satisfy the customers. Even research says that the customer satisfaction can be acquired by positive word of mouth and also by offering the suitable products or services [2]. Positive word of mouth plays an important role because customers are more likely to prefer particular banks, when they heard positive feedback from other people.

In this paper, a comparative analysis was done between two different datasets to analyze and improve the customer satisfaction. The first dataset consists of fields, namely credit score, current loan amount, annual income and current credit score, etc. In this dataset, credit score is considered as a target variable or response variable to predict the future loan status. The remaining variables like current loan amount, annual income and credit score are considered as independent variables or predictor variables. Multiple regression technique is applied for the data in order to establish a relationship between independent variables and dependent variables. The results of this dataset will provide a clear idea to the banks whether to provide loans or not to the customers. High credit score indicates that the customers were satisfied about particular product.

The second dataset comprises of data about marital status, education, loan and subscribed. In this dataset, subscribed is considered as the target variable or response variable to know whether the customers have subscribed to term deposit or not. The other variables like marital status, education and loan are identified as the independent variables or predictor variables. The dataset can be analyzed and able to identify whether the majority of the customers are subscribed to the term deposit or not.

The researchers also identified the attributes that are responsible for the customer dissatisfaction. The attributes are reliability, accuracy, security, assurance, handling and recovery. All these attributes play an important role in satisfying the customers. One of the important attributes is handling and recovery, which has the ability to overcome the conflicts and also provides the best way for handling the complaints. If the attribute fails to overcome the conflicts or handling the complaints, it will lead to the customer dissatisfaction. Therefore, service failure [3] should be avoided by providing quality services to the customers.

2 Literature Review

In this paper [4], the main focus is on the customer satisfaction in the retail banking. For analysis, they have considered 5000 customers of Italian bank which proved that there exists a nonlinear relationship between overall customer satisfaction and attribute performances. In this, multiple regression method was used and mathematical equation was formed in order to determine the overall customer satisfaction. Customer satisfactions will depend on many attributes like speed, accuracy, reliability, etc. One of the attributes is handling and recovery, which should have the ability to avoid the conflicts, efficiency to handle the complaint and should able to find and restore the errors as quickly as possible. It will lead to the customer dissatisfaction, if the complaint is not handled properly and the speed of restoring errors is less. In order to resolve these issues, three-factory theory of customer satisfaction methodology was implemented.

The paper [5] focuses on the importance of big data analytics considering the data which was stored from decades. The big data analytics help to overcome the serious disasters and also to understand the customer behavior. In this paper, the objective is to determine the reasons for the customer dissatisfaction. To improve the customer satisfaction, feedback analysis was used. The feedback analysis is necessary for banking sector [6] which helps to interpret the possible areas of improvement and also useful to recognize the gaps in services offered. Customers were also asked to give the rating for the bank based on the parameters like quality of service, speed of service and customer inquiries has reached successfully or not. For analysis, data was collected for a period of 42 months and dataset of 5000 records was considered. The study helps to capture the customer sentiment and also helps to evaluate the functioning of bank.

This paper [7] deals with the problems in retail banking with the help of statistics. The main problem in banking domain is heavy competition and frequent shift of customers. Therefore, the need of customer segmentation arises. In order to overcome the heavy competition and frequent shift of customers, there is need of effective tools, new products and services which helps to develop the customer loyalty. The research has focused on the two types of customers, first type refers to the selection of customers with less possibility of risk and the second type of customers refers to be more important because the customers were segmented in order to provide the most interesting and satisfactory product offers for them. The analysis was performed on 33,021 customers in 24 distinct campaigns which proved that the customer and home loan holders showed more interest, only when the credit cards were offered. And, it was concluded that Serbia is still a place for the retail banking and identification of specific segment of clients who responds positively to the new product offers.

In the past few years, banking domain has undergone numeric changes in terms of operating and providing effective services. This paper [8] presents the need of big data analytics, which helps to increase the revenue in the banking domain. The word big data means "bigger" in terms of size. Huge amount of data is being generated for every minute. The source of this data generation is by means of different services provided by banks. With the help of big data analytics, banks will develop their services [9] to the customers and it also helps them to analyze the customer's behavior and spending patterns, etc. In banking sector, big data is used for the fraud detection and prevention. It provides security as well as safety in the transactions and access. Big data also helps the customers to get the services precisely in the form of customer segmentation.

This paper [10] is about the analysis of customer behavior in the social media. In order to attain development in any domain, customer feedback is very important. At present, the market is mostly customer centric and there is a huge competition among the business organizations in order to achieve the customer satisfaction with the help of services. Nowadays, social media plays a vital role. Instead of giving direct feedback to the sellers, most of the customers disclose their views on certain brands in Facebook, Twitter, etc. So, it is inferred that the social media acts as a tool in order to analyze the customer behavior. By using customer's feedback in social media, the related data will be gathered and analyzed. The proper analysis of these data will help the companies to know what exactly customers think about particular brand or product.

3 Proposed Methodology

Regression analysis is a technique used for the prediction. The main objective of regression analysis is to establish the relationship between dependent variable and independent variables. It acts as an important tool for analyzing and modeling the data. There are different types of regression techniques available to make the predictions. These techniques are mainly characterized by three parameters, i.e., type of dependent variables, number of independent variables and shape of the regression line. The most commonly used regression techniques are linear regression, multiple regression and logistic regression.

Simple linear regression is a technique which consists of only one independent variable and one dependent variable. Multiple regression is a multivariate technique used for the prediction and forecasting. The objective of this model is to establish the relationship between independent variables and dependent variables. To predict the overall satisfaction of the customers, multiple regression technique was used. In *R*, regression model is created using lm() function. By using the input data, the model is used to determine the value of intercept and coefficients.

The multiple regression can be represented in the form of the mathematical equation:

$$y = a + b_1x_1 + b_2x_2 + b_3x_3 + \cdots + b_nx_n$$

where y is dependent variable or response variable. $x_1, x_2, x_3 \ldots x_n$ are independent or predictor variables. $a_1, b_1, b_2 \ldots b_n$ are coefficients.

Logistic regression is an underlying technique that is used to solve the classification problem. This technique is used to find the probability of a categorical dependent variable and works on binary data. In logistic regression, the dependent variable is a binary variable that contains information coded as 1 (yes/success) or 0 (no/failure). It is one of the popular techniques that fit models for the categorical data, particularly, for binary response data in the data modeling.

Every algorithm works best under the given set of conditions. Making sure that the algorithm fits the requirements ensures the better performance. In this paper, the analysis was performed on the overall satisfaction of the customers with the help of two datasets. Based on the results of two datasets, the problems faced by the customers and banks are identified. Along with the multiple regression, there are many techniques available in the multivariate analysis, but those techniques are not suitable for the requirements. In this paper, the main objective is to predict based on the results of response variable. Therefore, multiple regression technique fits the requirements that ensure the better performance.

3.1 Feature Selection

For the analysis, the attribute subset selection method is used. Attribute subset selection is one of the feature selection techniques [11] used to eliminate the unrelated attributes based on the *p*-value. The *p*-value helps to decide whether there is a significant relationship among variables at 0.05 significance level. From the results, *p*-value is much less than 0.05, it is obvious that there exists a significant relationship among variables (Fig. 1).

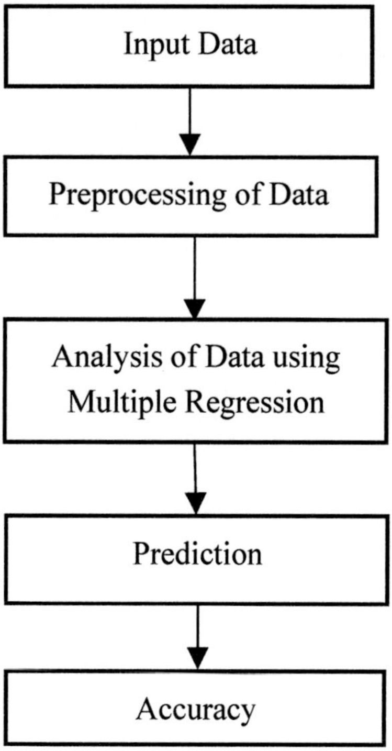

Fig. 1 Flowchart for analysis of customer satisfaction

4 Experiment and Results

In this paper, two different datasets, namely future loan status prediction and marketing targets, are used from Kaggle for the analysis of data. Comparison of two different datasets was done to analyze the customer satisfaction. For the analysis, multiple regression technique is used in order to determine the relationship between one response variable and multiple predictor variables. This technique is used, when there is a need to predict the output of one variable based on the other multiple variables. The variable that needs to be predicted is known as dependent variable or target variable.

4.1 Dataset 1: Future Loan Status Prediction

The dataset consists of 10,000 records. From the dataset, the credit score is considered as the target variable, which is used to predict the future loan status. The dataset was imported into the *R* environment. The objective of this model is to construct the relationship between credit score as the response variable with current loan amount, annual income and current credit balance as the predictor variables. The intercept (credit score) value will provide a clear idea whether to provide loan or not. Analysis was carried out on customer satisfaction using multiple regression technique. The results are in the form of intercept and coefficients which are used to predict the future loan status. Based on the results of two datasets, the problems faced by the customers as well as banks are identified.

In the input data, the minimum range of credit score is 585 and maximum range is 7510. From Table 1, it was clear that the credit score was more which implies customer loyalty. Therefore, financial services [12] help to increase the number of loyal customers and also to improve the financial performance of banks (Fig. 2).

4.2 Dataset 2: Marketing Targets

The dataset consists of 31,647 records. From the dataset, subscribed is considered as the target variable, which is used to predict whether the customer has subscribed to term deposit or not. After importing the dataset into the *R* environment, the aim of the model is to construct the relationship between subscribed as the response variable with marital, education and contact as the predictor variables. Based on

Table 1 Analysis to predict the future loan status

Intercept (credit score)	Current loan amount	Annual income
1306	$-4.402 * 10^{-6}$	$-9.97 * 10^{-5}$

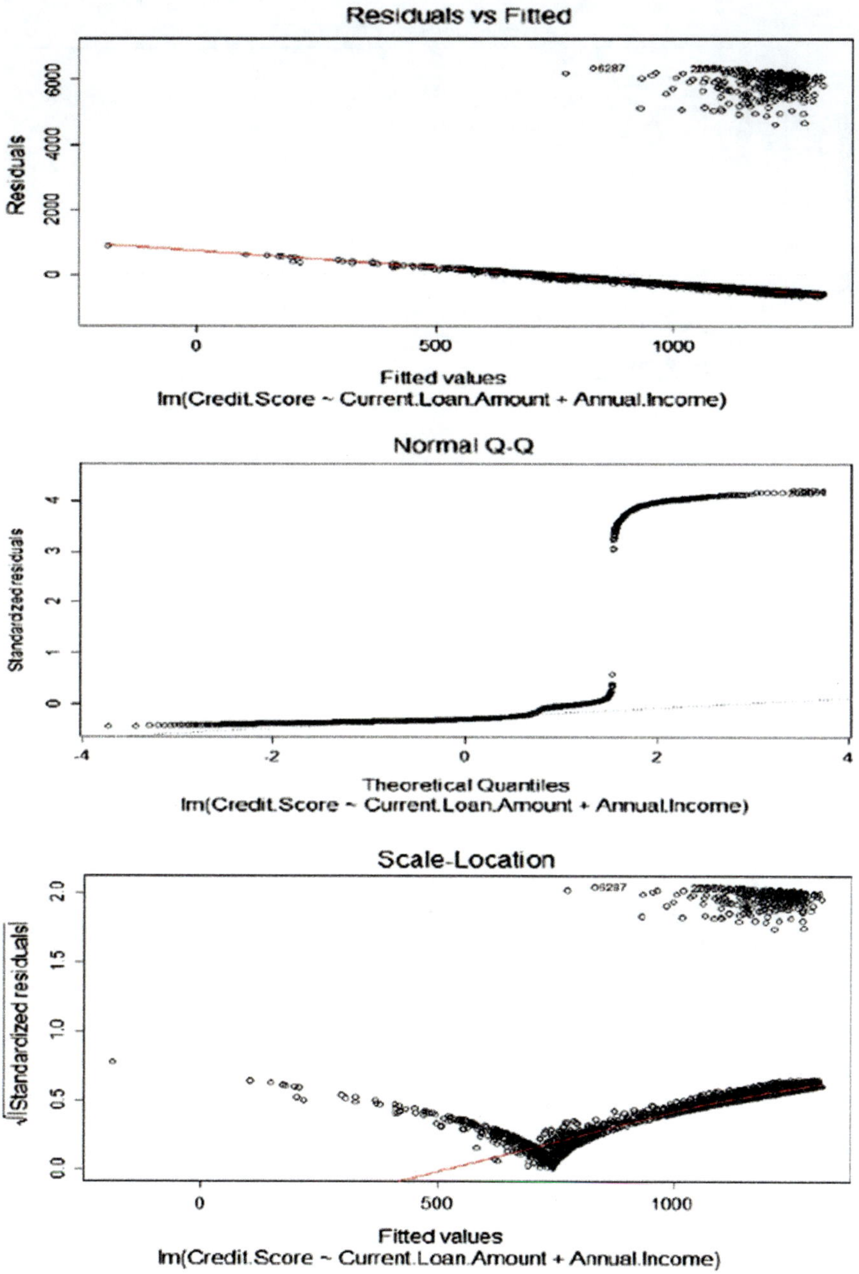

Fig. 2 Results of future loan status prediction dataset using multiple regression

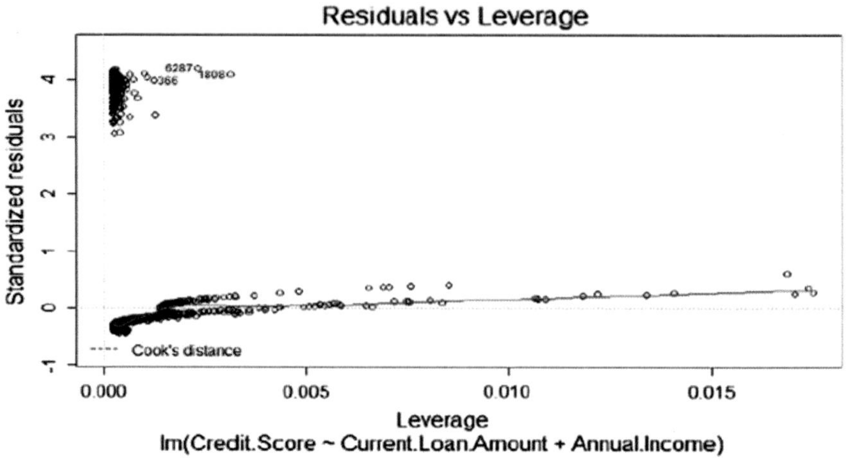

Fig. 2 (continued)

this result, prediction can be carried out whether the majority of the people have subscribed to term deposit or not. Analysis was performed on the overall satisfaction of customers using multiple regression technique. The results are expressed in the form of intercept and coefficients as numerical value. In order to determine whether the customer has subscribed to term deposit or not, the expected results should be in terms of categorical value. So, sqldf() function was applied on the data input to get the results in terms of categorical value. From the results, it is inferred that more number of customers has not subscribed to the term deposit. Therefore, it implies that the customer satisfaction was less.

From Table 2, it is very clear that only 3715 customers have chosen the term deposit subscription out of 31,647 customers. It is very obvious that the majority of the customers have not subscribed to the term deposit. It is inferred that there is a need for the segmentation of customers in order to resolve the issues because of the poor services provided by the banks. The customer segmentation helps the banks to divide the customers based on their requirements and helps them to provide suitable products to the customers based on their needs. The customer-based analysis plays an important role in the banking sector which helps to predict the future purchase of products based on the past purchase. Therefore, this analysis helps the banks to attain the success (Fig. 3).

Table 2 Results of marketing targets

	Subscribed	Yes	No
1	No	0	27,932
2	Yes	3715	0

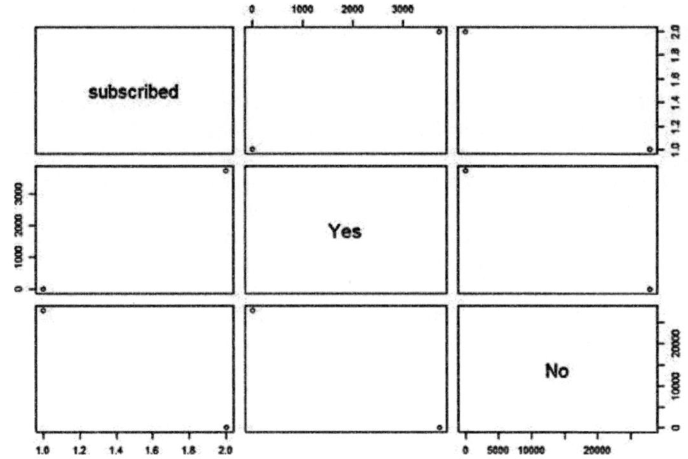

Fig. 3 Analysis of term deposit subscription using multiple regression

4.3 Comparison of Two Datasets

From the results of two datasets, it is very clear that the customer satisfaction was less in case of term deposit subscription. It can be due to various parameters like marital status, education, etc. In order to resolve this issue, customers have to be segmented based on their needs, which helps to increase the number of customers for the term deposit.

5 Conclusion

In this paper, two different datasets are considered to compare the overall satisfaction of customers. After the analysis, from the future loan status prediction dataset, it is inferred that the value of credit score was high. Therefore, banks can provide loan to the customers, which provide the customer satisfaction. The customer satisfaction also helps to provide the customer loyalty by improving the quality of services to the customers. Customer loyalty will play a crucial role in the banking sector which helps to maintain the long-lasting relationship between customers and banks that leads to the success of banks. From the results of term deposit subscription dataset, it is clear that 27,932 customers have not subscribed to the term deposit. Only 3715 customers have subscribed to the term deposit which implies that the customer satisfaction is very less. The subscription is very less due to many parameters like marital status, education, etc. So, there is a need to create proper awareness among the customers about particular product or service which helps to improve the customer satisfaction.

References

1. Wang S, Petrounias I (2017) Big data analysis on demographic characteristics of Chinese mobile banking users. In: 19th conference on business informatics. IEEE
2. Foroudi P, Gupta S, Sivarajah U, Broderick A (2018) Investigating the effects of smart technology on customer dynamics and customer experience. Comput Hum Behav 80:271–282
3. Haruna A, Rokonuzzamanb Md, Prybutokc G, Prybutoka VR (2018) How to influence consumer mindset: a perspective from service recovery. J Retail Consum Serv 42:65–77
4. Arbore A, Busacca B (2009) Customer satisfaction and dissatisfaction in retail banking: exploring the asymmetric impact of attribute performances. J Retail Consum Serv 16:271–280
5. Srivastavaa Utkarsh, Gopalkrishnanb Santosh (2015) Impact of big data analytics on banking sector: learning for Indian banks. Procedia Comput Sci 50:643–652
6. Mbaluka W (2013) Big data management and business value in the commercial banking sector in Kenya. University of Nairobi
7. Agaliotis K, Hadzic M (2015) Predicting retail banking consumer behavior using statistics. The Eur J Appl Econ
8. Srivastava A, Singh SK, Tanwar S, Tyag S (2017) Suitability of big data analytics in Indian banking sector to increase revenue and profitability. IEEE
9. Aggelis V, Christodoulakis D (2003) Association rules and predictive models for e-Banking Services. In: 1st Balkan conference on informatics
10. Kularathne SD, Dissanayake RB, Samarasinghe ND, Premalal LPG, Premaratne SC (2017) Customer behavior analysis for social media, 3(1)
11. Hemavathi D, Srimathi H (2019) An efficient approach to identify an optimal feature selection method using improved principle component analysis in supervised learning process. J Adv Res Dyn Control Syst 11(07-Special Issue)
12. Baptista G, Oliveira T (2015) Understanding mobile banking: the unified theory of acceptance and use of technology combined with cultural moderators. Comput Hum Behav 50:418–430

A Toolkit to Analyze the Task Allocation Strategies for Large Dataset in Geo-Distributed Servers

A. P. Aakash and P. Selvaraj

Abstract The evolving technological improvements on data handling with the geo-distributed datacenters pave the way for the cost-effective data maneuvres. The geo-distributed datacenter-based system planning poses an over whelming challenge in aggregating the appropriate data, distributing the server intensive computation, and setting up the seamless communication infrastructure. The datacenter providers have been striving to reduce the operational expenditure with different measures. The three segments, i.e., task assignment, data placement, and data movement essentially influence the operational cost of datacenters. We considered the cost minimization issue by streamlining of these three segments for advantageous benefits in geo-distributed servers. To delineate the endeavor of choosing the plausibility of the appropriate task allocation strategy, we suggested MATLAB-based toolkit that compares a 2D Markovian chain-based model and a MILP-based task allocation model. A mixed nonlinear programming constructs were linearized for the cost minimization. A MATLAB-based toolkit was proposed to explore the different task allocation strategies and their impact on the operational cost, communication cost, and the overall server cost. The existing datacenter simulation tools are having rigid and fixed functionality whereas the proposed toolkit offers different possibilities for the analysis and visualization of datacenter-related operations.

Keywords Datacenters · Cost minimization · Streamlining · 2D markov chain

A. P. Aakash (✉) · P. Selvaraj
Department of Information Technology, SRM Institute of Science and Technology, Chengalpattu, Tamil Nadu, India
e-mail: aa8641@srmist.edu.in

P. Selvaraj
e-mail: selvarap@srmist.edu.in

D. J. Hemanth et al. (eds.), *Artificial Intelligence Techniques for Advanced Computing Applications*, Lecture Notes in Networks and Systems 130,
https://doi.org/10.1007/978-981-15-5329-5_36

1 Introduction

Information explosion as of late prompts a rising enthusiasm for improved data manipulation from the remote servers those are distributed at distinct geographical places, e.g., Google's 13 servers cultivates in excess of 8 countries in four regions. Enormous information investigation has indicated its incredible potential in uncovering essential knowledge of information to improve cost-effective operations. Optimization of distributed Internet datacenters [1] in a massively parallel and distributed environment has just resulted into high cost because of its paucity for managing costlier server maneuvres with the corresponding assets. Gartner predicts that by 2015, 71% of datacenter spending will be required for the information handling and will reach up to $126.2 billion. Thus, it is essential to inspect the price reduction issue for gigantic information processing in geo-coursed server farms.

Various undertakings have been made to cut down the count or correspondence cost of datacenter. Datacenter re-resizing (DCR) has been used to decrease the OPEX by analyzing the various measures of datacenter. In perspective on DCR, a couple of works have explored the relationships among the geological scattering nature of servers and power-related factors in bringing down the overall cost. Huge data organization structures incorporate a complex record framework underneath, which courses data knots and their impressions over the server farms for fine-grained load balancing and big parallel information execution a challenging one. To lower the communication and processing cost, attempted to improve information domain by distributing the load on the datacenters in a manner to maintain a strategic distance from remote information stacking.

In spite of the fact that the above works have acquired some positive outcomes, they are a long way from accomplishing the cost efficient enormous information handling in view of the accompanying shortcomings. A general communication cost optimization framework for big data stream processing was proposed in paper. In geo-distributed datacenters, most calculation asset of a datacenters with low prominent information may remain inactive. Such low resource utilization further makes a considerable amount of load to be shared among the servers and hence results in greater cost. Hence, it is essential to explore and analyze the cost-effective operations involved with the distributed datacenters. This work has been focused on developing a toolkit to analyze the vital parameters of the datacenter operations.

1.1 Different Task Allocation Strategies

In a datacenter environment, the task is viewed as a basic entity that is responsible for the communication and processing of information. Some of the major task assignment frameworks are chief driven, group driven, individual driven, and director driven. Recognizing these strategies and their internal structures has become a vital need for the datacenter providers. This proposed work is viewed as a significant tool for the datacenter administrators to strategize and evenly distributing the load. Recognizing the pros and cons of various task allocation methodologies and comprehending their internal structures and its underpinnings in a geographically distributed environment are highly essential and beneficial.

1.2 Cost Estimation Models for Task Allocation Strategies

The task allocation is one central perspective in managing and arranging the computation loads all while satisfying the possible cost optimization ventures. There have been various task dispersion models proposed, including cost models and danger-based strategies. In any case, a productive blend of such models and a coordinating technique for task practices are comprehensively missing. In this article, we considered the existing models that reflect different points of view and consultation levels of undertaking segment decisions. In perspective on the datacenter planning, we sketch a system for exact evaluation and making decision on undertaking load by describing the model interface and the progressive solicitation of their usage.

1.3 Integer Linear Programming Models for Task Allocation Strategies

To overcome the shortcomings in analyzing various task allocation strategies, we studied the cost-reduction issue with the methods for joint optimization of various objectives. The information arrangement and steering in geo-spreaded server ranches involves various joint optimizations. The integer linear programming models were proposed for the various joint optimization problems in geo-distributed datacenter. To depict the task allocation and transmission in huge information processing pipe line, a two-dimensional Markov chain-based model was also proposed [2]. In integer linear programming (ILP), the various variables and the objectives of the task allocation strategies are trying to minimize the cost estimation time. The computational complexity of ILP-based cost estimation is NP-Hard. Hence, the ILP-based solutions are usually faster than the other meta-heuristic models for the minimal number of server nodes. If some decision making factors are not discrete, then the issue will be known as a mixed integer linear programming issue. The various logical cost-estimation models that have been proposed in the literatures are gathered and analyzed. The discrete-time and constant time models were also analyzed to understand its characteristics and constraints. In this work, we considered the cost reduction and optimization issue with the mixed integer nonlinear programming (MINLP) formulation. And a conducive user-friendly MATLAB-based tool is proposed to analyze the various task allocation strategies and visualize their characteristics with the required plots and analysis.

2 Related Work

2.1 Server Cost Minimization

Geo-distributed server farms are becoming a vital asset throughout the world to cater and manage a huge number of clients [3]. A server farm may comprise of huge quantities of servers and consume megawatts of power. A huge number of dollars on power cost have represented a substantial weight on the working expense to server farm providers. Subsequently, decreasing the power cost is a significant challenge for consideration from both scholarly community and industry. Among the techniques that [4] have been proposed so far for server farm energy, the strategies that draw in load balancing attracted a great deal of interest.

DCR and the underlying geographical positions are commonly optimized together to contemplate the cost optimization parameters. Fan et al. analyzed and compared the issues by expecting framework delay [5]. Leu et al. suggested managing provisioning approaches with respect to the given power spending plan [6]. Rao et al. proposed a model to manage costs by guiding customer requesting to geo-scattered server. Hence, there have been different techniques proposed for the server cost minimization issue.

2.2 Large Data Management

To manage the intricacies of large data management and the task calculation procedure, different models were proposed. The key challenge in massive data management involves the management of distinguished exchange among the various administrative partitions. Yazd et al. proposed a flexible data management model within the restricted datacenter scenario arrangement. They have also proposed an energy model in server by considering the regional properties. Hu et al. [7] proposed a device management model along with the existing massive-scale parallel computing infrastructures.

Ren and He [8] presented a coordinated and critical information assessment method for one of the world's most vital business enterprise frameworks at Fox Audience Network. Rao et al. [9] suggested a unique estimation model that minimizes the operational cost. Chen et al. analyzed the difficulty of organizing the three data management and cost optimization stages, i.e., guiding, setup, and optimization, of the MapReduce system and suggested a typical sense heuristic to battle expensive saving varied nature [10].

2.3 Data Placement

Shachnai et al. investigate a way to opt for the course of action of video-on-demand (VoD) file copies on the datacenters and therefore the proportion of weight limit consigned to every file copy therefore on confine the correspondence price whereas [11] guaranteeing the client expertise. Agarwal et al. proposed a knowledge set up instrument Volley for geo-appropriated cloud organizations with an intention to optimize WAN transfer speed price, server ranch limit limits, knowledge among conditions, etc. Cloud organizations use Volley by submitting logs of datacenter demands [12]. Donegal proposed a replication model that decouples data course and duplication to enhance the properties in scattered server ranches [13]. Oflate, Jin et al. suggested a linkage improvement plot that at an equivalent time, redesigns the virtual machine (VM) game setup and framework flow architecture that boosted the performance [14].

Existing work on server ranch cost upgrade, huge data the board or data course of action dominatingly bases on two or three major factors. To oversee gigantic [15] data dealing within geo-appropriated server farms coordinated energy cost management of distributed Internet datacenters in smart grid. Hence, we argue that it is essential to commonly think about information condition, task, and information coordination in a deliberated datacenter environment to optimize the operational cost.

3 System Model

In this section, we present the framework that acts as a tool for the datacenter administrators and researchers for the visualization and analysis of different task allocation strategies in datacenter environment.

3.1 Network Model

We proposed a geo-distributed datacenter model as showed up in Fig. 1; in that, all datacenters (DC) are linked with their switch. The I was considered as the set of datacenters, wherein each datacenter consists of Ji of servers that were linked with a switch, mi ∈ M with a cost of CL. The transference price CR for between server ranch traffic is more conspicuous than CL, i.e., CR > CL. It was assumed that all servers in the framework have a equal resource and limit. We presumed that J is the set of servers, i.e., $J = J_1\ S\ J_2 \ldots SJ|I|$. The entire structure can be shown as a graph $G = (N, E)$. The vertex set $N =$ MUJ fuses the set of M switches and J.

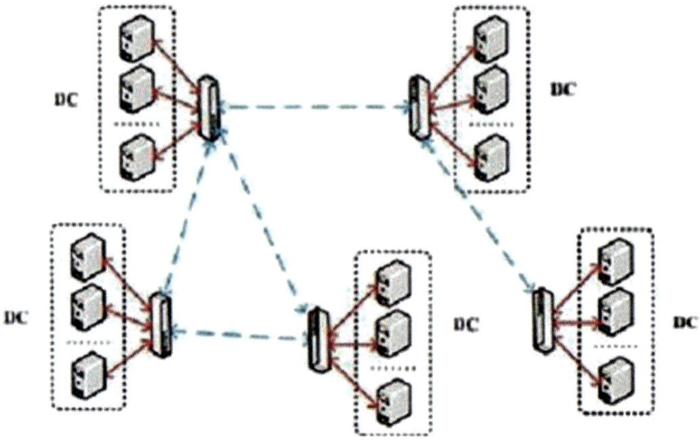

Fig. 1 Datacenter topology as a graph

3.2 *Task Model*

We assumed that the data were divided into K number of chunks, wherein each chunk 'k' has the size of $\varphi k(\varphi k \leq 1)$, which was attributed to the data-center growth limit. That is, for each data chunk, there are P copies stored in the distributed manner. The presence of the task at server ranches during a particular timespan can be represented as a Poisson process. In specific, let λk be the ordinary endeavor appearance rate referencing protuberance k. We implied the typical appearance pace of the job for irregularity k on datacenters j as $\lambda jk(\lambda jk \leq 1)$.

4 Existing System

There were different open source and commercial tools available for the simulation of datacenter operations. Some of the prevailing tools are Opendcim, NOC-PS, DCI manager, EasyDCIM, and Puppet Enterprise. These tools are used to simulate and explore the softwares and OS installed in the servers. These tools were used to monitor and analyze the power and temperature of the server nodes. The server provisioning, OS installation, sensor monitoring, and VLAN management were also can be monitored with these tools. But these tools have no provision to modify the constraints and other optimization parameters involved in the datacenter management. And moreover, these tools are not relying on the meta heuristic techniques like Markov model, etc., The ILP formulations can be modified to suit the specific need of the user. Hence, a user-friendly light-weight tool is highly essential for the datacenter administrators and the researchers. The following section states the proposed ILP-based light-weight tool for the datacenter management.

5 Proposed Approach

5.1 Proposed Mixed Integer Linear Programming (MILP)

We modulated the issue of datacenter task allocation as a mixed whole nonlinear programming problem and suggested a MATLAB-based tool that linearized it. The mixed integer linear programming (MILP) model that considered the cost minimization for distributed datacenters in smart microgrids-based power outages was proposed. In general, the MILP formulations were considered for the nonlinear issues with dependable and numbered components. The MILP has demonstrated to be an incredible solution for the sizable problem domains. At the same time, it consolidates algorithmic structure complexity from combinatorial to nonlinear. We have created a simple light weight MATLAB-based tool that executes the necessary MILP formulations to compute the desired output. We considered the following constants and variables while creating the application. We have used ***intlinprog*** tool box in MATLAB for the necessary computation (Table 1).

Table 1 Constants and variables used in the MILP formulation

Constants	
J_i	The set of servers in data centre i
m_i	The switch in data centre i
$W^{(u,\ v)}$	The weight of link $(u,\ v)$
φ_{k}	The size of chunks K
λ_{k}	The task arrival rate of data chunk k
P	The number of data chunk replicas
D	The maximum expected response time
P_j	The Power consumption of server j
$\gamma_{(u,\ v)}$	The transmission rate of link $\{u,\ v\}$
Variables	
X_j	A binary van able indicating if server j is activated or not
Y_{jk}	A binary variable indicating if chunk k is placed on server j or not
$Z_{jk}^{(\text{u},\ \text{v})}$	A binary variable indicating if link $(u,\ v)$ is used for flow for chunk k on server j
λ_{jk}	The request rate for chunk k on server j
ϑ_{jk}	The CPU usage of chunk k on server j
μ_{jk}	The CPU processing rate of chunk k on server j
$f_{jk}^{(u,\ v)}$	The flow for chunk k destined to server j through link $(u,\ v)$

5.2 *Computation and Visualization of Task Allocation Strategies with MATLAB*

MATLAB is a language that fuses estimation, recognition, and programming in hassle free manner. The Simulink offers various toolboxes that handle the assertion and underwriting of systems through displaying style checking, requirements conspicuousness, and system joining evaluation. The interface for the proposed tool is shown below. The task allocation strategies were simulated with respect to their task arrival rate and the number of server. The resultant graph shows the trend of operational cost for the various time series value of the task arrival rate and the server capacities. The results have proved that the MILP formulation yields least cost comparable to the Markov model.

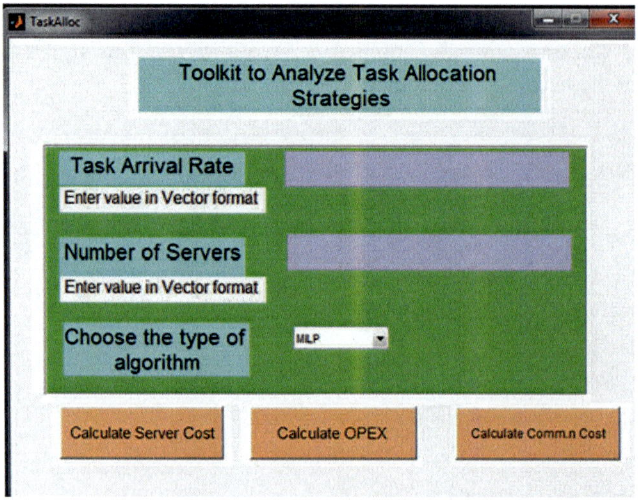

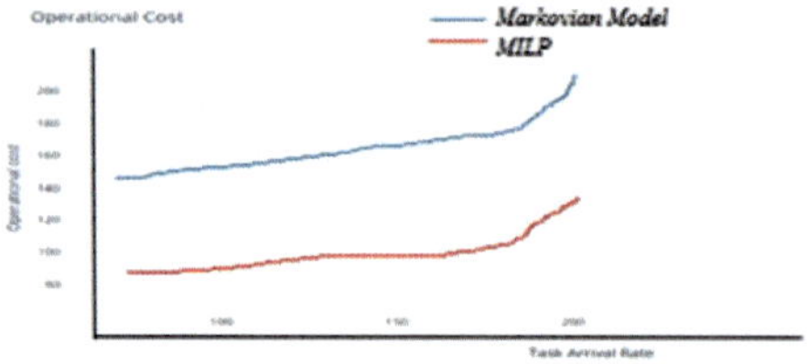

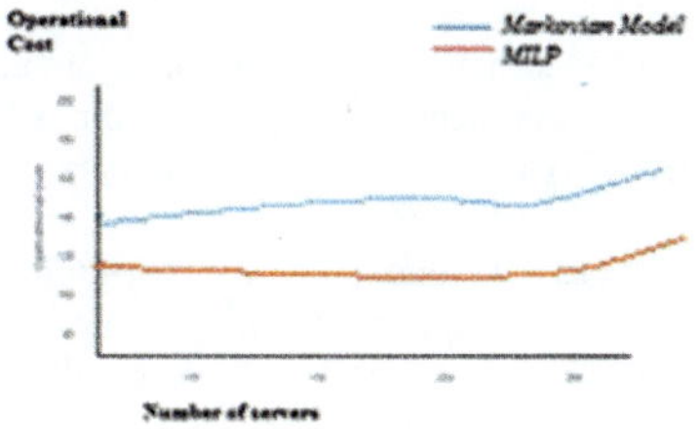

6 Conclusion

The technological improvements on data handling in the geo-distributed datacenters have paved the way for the cost-effective data maneuvres. The data positioning, task allocation, and server re-dimensioning have been considered as a very challenging multi-objective optimization problem. The huge geo-distributeddata server farm incurs high cost for the massive data applications. Hence, focusing on the task allocation strategies in datacenter is highly essential. For the analysis and visualization of datacenter task allocation, a MATLAB toolkit was proposed. The optimization problems such as data placement, task assignment, and data movement have been simulated with a easy to use intuitive GUI based tool. The appropriate MATLAB toolbox was used to simulate the different task allocation strategies. This proposed tool can be used as a guide for the datacenter administrators and the researchers for the visual exploration of the particular datacenter topology. The proposed tool provides an easy-to-use interface to modify the datacenter parameters and explore the various task allocation methods.

References

1. Gu L et al (2014) Cost minimization for big data processing in geo-distributed data centers. IEEE Trans Emerg Topics Comput 2(3):314–323
2. Rao L et al (2011) Coordinated energy cost management of distributed internet data centres in smart grid. IEEE Trans Smart Grid 3(1):50–58
3. Gandhi A, Harchol-Balter M, Adan I (2010) Server farms with setup costs. Perform Eval 67(11):1123–1138
4. Gao et al (2014) Carbon-aware energy cost minimization for distributed internet data centers in smart microgrids. IEEE Internet Things J 1(3):255–264
5. Yu L, Tao J, Yang C (2014) Energy cost minimization for distributed internet data centers in smart microgrids considering power outages. IEEE Trans Parallel Distrib Syst 120–130
6. Liu et al (2011) Coordinated energy cost management of distributed internet data centers in smart grid. Trans Smart Grid 3(1):50–58
7. Hu, et al. "A general communication cost optimization framework for big data stream processing distributed data centers." IEEE Transactions on Computers 65.1 (2015): 19–29
8. Ren S, He Y (2015). Coca: Online distributed resource management for cost minimization and geo-distributed data centers. IEEE Trans Comput 65(1):19–29

9. Rao L et al (2010) Minimizing electricity cost: optimization of distributed internet data centers in a multi electricity-market environment. In: 2010 Proceedings IEEE INFOCOM. IEEE
10. Guo Y, Fang Y (2012) Electricity cost saving strategy in data centers by using energy. IEEE Trans Parallel Distrib Syst 24(6):1149–1160
11. Zhang L et al Moving big data to the cloud: an online cost-minimizing approach
12. Zeng D, Lin G, Guo S (2015) Cost minimization for big data processing in geo-distributed data centers. Cloud networking for big data. Springer, Cham, pp 59–78
13. Yao Y et al (2012) Data centers power reduction: a two-time scale approach for delay tolerant workloads. In: 2012 Proceedings IEEE INFOCOM. IEEE
14. Guo Y et al (2011) Cutting down electricity cost in internet data centers by using energy storage. In: 2011 IEEE global telecommunications conference-GLOBECOM 2011. IEEE
15. Islam MA et al (2015) Online energy budgeting for cost minimization in virtualized data center. IEEE Trans Services Comput 9(3):421–432

High-End Video Data Transmission in Optimized Routing Framework for Wireless Networks

K. Navin, S. Murugaanandam, S. Nithiya, and S. Sivashankari

Abstract Protocols required to find the best path to reach the destination in wireless networks are specific for each application. Superior quality videos are in great demand in the present scenario. In this paper, the main aim is to minimize the distortion of high-end data traffic occurring over wireless networks. The proposed work is a new routing technique named minimum distortion routing (MDR). MDR concentrates to delineate a routing framework for minimizing distortion from source to destination effectively. The proposed work is to minimize the distortion experienced by the user and enhancing the video quality by catering to the application needs. Traditional link-based routing policies like expected transmission (ETX) count produce a large video distortion as they neglect the correlation among the links in a path. This results in complete communication path become congested, leading to high distortion. An analytical framework is an idle solution for minimizing the video distortion. Here, an optimized protocol for routing is built based on the framework's policy. The efficiency of the proposed protocol is verified using the NS2 simulator. The results indicate increased bandwidth utilization and reduced distortion within a short time interval when compared with the existing routing protocols. It also improved the efficiency of the network through increased throughput and decreased data loss when delivering packets from source to destination.

Keywords Distortion in wireless networks · Traffic congestion · ETX · Analytical model · Video distortion minimization · Wireless networks

K. Navin (✉) · S. Murugaanandam · S. Nithiya · S. Sivashankari
Department of Computer Science and Technology, SRMIST, Chennai, India
e-mail: navink@srmist.edu.in

S. Murugaanandam
e-mail: murugaas@srmist.edu.in

S. Nithiya
e-mail: nithiyas@srmist.edu.in

S. Sivashankari
e-mail: sivasans2@srmist.edu.in

D. J. Hemanth et al. (eds.), *Artificial Intelligence Techniques for Advanced Computing Applications*, Lecture Notes in Networks and Systems 130,
https://doi.org/10.1007/978-981-15-5329-5_37

1 Introduction

In recent days, usages of wireless systems are more widespread than earlier days [1]. Communications in cellular and wireless networks are unavoidable for humans in their daily routines. Data transmissions are detected and controlled by central base stations [2]. In multipath communication systems, there are many nodes involved between the source and destination. These nodes communicate among themselves along the entire path from the source to the destination. They are mainly used to route the data packets and identify the best route in the wireless networks. Multipath networks are dynamic in nature and exist on its own without any base station. These networks are capable of transferring high-end data like video traffic over it. Network phenomenon's like Internet Protocol Television (IPTV) and Voice Over Internet Protocol (VOIP) deliver a high-quality video by increasing bandwidth from time to time. Broadband plays a very important role in the above type of transmissions [3]. Effective usage of broadband services is not possible in tough terrains and rural areas because of the lack of technological facilities and their cost. This is the main reason why broadband services are not a hit in rural India. One path VLANs, mainly 3G and Wi-MAX are expensive and also require licensed access. So, it is expected that the multipath network can eliminate this issue and also provides high-quality service [2].

It is necessary to look at those VLANs offering multimedia services, as these are the biggest upcoming traffic sources. With the advent of smartphones, this traffic has gone to a further shoot up. Transmission of important and necessary information such as voice data, video data from vulnerable areas that are in immediate need for help is badly disrupted due to increasing network traffic. Usually, quality of service during video transmission is affected by the variety of compression mechanisms used at one end and also due to disruptions in the wireless channel. The losses that occur in the transit can be minimized by adopting video encoding standards, for example, I-type frames are a certain group of frame types as established by video encoding standards. The main consideration is I-type frames where data encoding happens independently. The end-level performance of video transmission can be detected using the Group of Pictures (GOP). GOP determines performance by taking into account the frame losses as a distortionmetric.

The most important functionality for video transmission is often looked upon which in turn leads to deduction in the quality of videos. The losses that happen along the path from source to destination are related to each other and are also specific to each application. In most cases, only a few links suffer high traffic resulting in distortion whereas other few are free and not utilized. The main error so far was concentrating on the network parameters rather than application in routing traffic

2 Related Work—Existing Systems and Their Limitations

Many different techniques from standardized organizations governing video encoding for the transmission of videos are available. The entire video data can be split up into numerous chunks and sent across different paths in the same network. Multiple description coding is the name given to the above technique [4, 5]. The process of reassembling the original video can be made successful by accounting the description in a network and also the quality of service (QoS) can be increased based on the sub-streams used. There are other techniques such as a layered technique for increasing the QoS. Different layers are added to the base layer for improving the performance but the main work takes place in the base layer. The layered encoding technique [6], [7] establishes video standards like MPEG-4 [8] and H.264/AVC [9] that provide the necessary orientation for the video transmission in a network [10, 11]. The splitting up of video clips into different frames facilitates better quality. All the frames area GOP. I-frame is the one, which carries information, and B and P-frames are used as sources. The video clip is split into different chunks. Temporary errors occur in the video streams due to compression and quantization [12]. An algorithm is proposed to estimate the difference between inter and intra-coding's. Peak signal-to-noise ratio becomes high because of the addition of the algorithm. The distortion minimization is taken as the base for determining the coding parameters [13]. The recurring frame loss rate is considered in simulation for observing the transmission on a network. Effects of wireless channel fading [14] and distortion are studied over a single hop already. Similarly, other studies have been performed on single links to study the effects [15]. Based on the length of errors among the frames, experimentations are done and the Markov chain system is established over multi-hop networks. A recursion model is used to show the distortion in a 2D scenario and works has been carried out based on that. Although all these experiments and studies have been carried out earlier, none have considered routing metrics as an important functionality, though the impact of routing plays a crucial role [13, 16].

In today's scenario, 4G networks are also used for determining the performance of the transmissions. H.264/SVC encoding is examined over mobile Wi-MAX [17]. The studies have concluded that the different encoding standards used on protocols have a direct impact on the performance. But none of these studies claim any report on the impact of routing protocols on video distortion. Researchers have been conducted upon various ad hoc and mesh networks [3, 18–20] regarding QoS and optimization among cross layers. Different evaluation metrics based on the network layers define QoS in several ways. Only throughput and delay constraints have been considered as performance metrics so far rather than application-specific metrics [5, 21, 22]. One more study based on QoS is done where disjoint paths from source to destination are used for transmission. Minimization of distortion is claimed to be achieved by selecting routes properly, but this selection of routes itself is complicated and is therefore considered heuristic [23].

3 Proposed System

The main aim of the proposed work is to minimize the distortion experienced by the user and to enhance the video quality by cantering to the application needs. A video clip can handle only a few packet losses using different schemes. The clip cannot be decoded if there are a large number of packets lost in a frame. The distortion is calculated by taking into account the unrecoverable video frames that are lost from source to destination. The approach used here is a multilayer design model. Rather than concentrating on one network quality metric, the aim here is to analyze the frame losses as an analytical model to delineate the dynamic behavior. The loss is calculated by associating the frame loss probability mapping with packet loss probability. It can thus be identified that routing is the optimized solution to reduce the source-to-destination distortion by finding the optimum path between them.

The proposed routing protocol is designed with the links being treated independently. The loss that happens during the transmission process can be found out using the dynamic approach. This becomes the base of the proposed routing protocol to minimize the distortion. I-type frames are the frames that carry the most important information and therefore if they are sent on a congested path, it affects the distortion metric considerably. So, if we find a path that is least congested, it will lead to the minimization of the distortion rate. The scope of this work is to eliminate the existing technical glitches in routing and to improve the performance metrics like quality of service, packet delivery ratio, throughput, bandwidth utilization, etc.

4 Proposed System

4.1 Usage of a Systematic Layered Approach

Proposed method is one of the important contributions of the system as the minimization in the distortion rate is achieved by finding the optimal route using a systematic approach. Also, the functionalities of physical and MAC layers are considered correlatively and put into best use.

4.2 *The Practicality of the Routing Framework*

The source keeps track of the routing information and the primary video is sent based on the practical optimal path.

4.3 *Severe Experimentations and Testing*

The minimization of the distortion rate is proven by carrying out various experimentations and testing and also by analyzing the different parameters that are considered to be important performance metrics. The factual figures prove the above. Peak signal-to-noise ratio (PSNR) rate is increased by 20%. The QoS is verified by considering different performance metrics and is proven to be at a higher rate than the existing protocols.

4.4 *Application Requirements Rather Than Network Requirements*

Taking into account that application-specific requirements rather than network requirements alone have given an edge over the traditional routing policies.

5 System and Protocol Design

Initially, the multi-hop routing setup is created using different nodes in a network. In the client and server, architecture is established. The client forwards the request to the server. Whenever the server is ready to process the request, it accepts the request from the client and processes it. In this process, the video frames are split into numerous packet chunks and the optimal path from the source to the destination is found out using the optimal path finding algorithm MDR. Then, the I-frames of the chunks are forwarded across this optimal least congested path. The video frame is thus sent successfully with minimal distortion.

The protocol design of MDR requires full knowledge of the network. ETX is used to determine the information about the network. The number of links and the nodes in the network as well as the neighbors of the nodes is updated in the routing table. This is taken as the estimate for processing the requests. After collecting the routing table information, the system estimates the route and sends the request to the destination. If the request reaches the appropriate node, the reply is obtained else the same procedure is repeated.

6 Algorithm

Procedure: Neighboring Node Discovery Input: Source node 'S', Destination node 'D' Output: Distortion Resistant Path.

Step 1: Begin

Step 2: Send Route request.

Step 3: For all nodes

If

The request message is for particular node D

Send Route reply to source S

Else

Forward request message to the next nodes.

Add the nearest node to the path.

End if End for

Step 4: return the path to the source.

Step 5: End.

7 Pseudocode

```
1: For the discovery of the neighboring node from
the network

/Neighboring node discovery Algorithm / Input:
source node S , destination node D Input: frame size
F

Output: route R from to S to D

Send route request

ReceiveACK(estimates,node---id's,messages) n□S

C ← F R ← [n/0] y ←(n,c)

Update y to R

Repeat

p* ← Next Optimum Node

C1 ← M[Cnew|Ccur=c]

n← p*

c←C1

y← (n,c)

Update y to R

Until y is the last node

2: To find the optimal path between the source and
destination

//Optimal Path Algorithm //

Input: Initial State y, last node l Input: set of
free nodes [Fn] Input: frame size F

Output: next node p* in path

A←n*c

 I←||A||

/*Optimal Calculation */

for j=I to 1 do
```

```
if j=I then
for all y belongs to A do Jj(y) ←   k(y)
end for
else
for all y=(n,c) belongs to A do
U(n) ←   {n' |n,n' 1 -hop neighbours } ¬
Ji(y,u)<---{g(y,u)+p(c,c'|u)J(y')
J(y)  ←  min(j(y,u))
P(y)  ←  arg min(j(y,u))
end for
end if
end for
n*P(y)
return n*
```

8 Implementation

The methodology, which we have used, comprises of the following four phases.

8.1 Phase 1: Multi-hop Routing Network

Existing protocols not considered multi-hop route settings, so in the proposed protocol design, optimize the routing in a multi-hop environment. In this module, the basic multi-hop setup is established along with the appropriate number of nodes in the network. The source and destination nodes are defined using NS2 as shown in Fig. 1. The environment is created using 30 nodes. Three different groups of nodes are in the environment. The interlinking connections between all the nodes along with the source and the destination are created. The node and server designs are concentrated so as to give a complete view of the network.

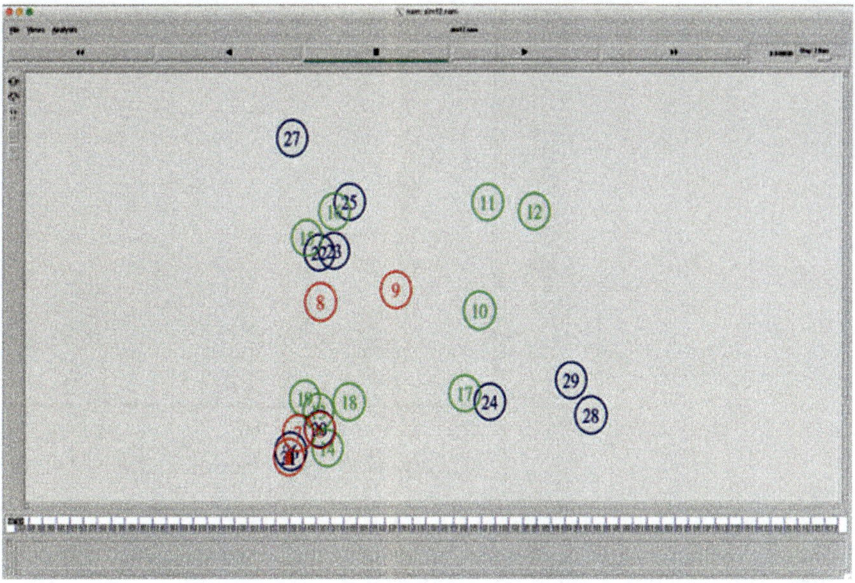

Fig. 1 Simulation environment is created using 30 nodes

8.2 Phase 2: Video Analytical Model

In the next phase, a request sent from the source to the server. This is done by constructing an analytical model in the intermediate device like a router. This model waits for the acknowledgment from the nodes after the request is being sent by the source. The server gives an acknowledgment whenever it is ready to process the request. The ETX estimates the path of the neighboring nodes in the network and updated in the routing table. This analytical model facilitates requests and information about the nodes in the network simultaneously.

8.3 Phase 3: Video Distortion Minimization

The main aim of this analysis is to minimize the distortion rate. The proposed methodology concentrates to implement an effective solution for distortion minimization. This results to finds an optimum path between source and destination. The optimum path is the path that is least congested for the I-frames to reach the destination without much distortion. The neighboring node discovery algorithm is used to calculate the nearest neighboring node or the next available hop for the delivery of data frames uses. After finding all the intermediate nodes from the source to destination, an optimal least congested and the shortest path is determined.

Fig. 2 Time versus average number of nodes throughput

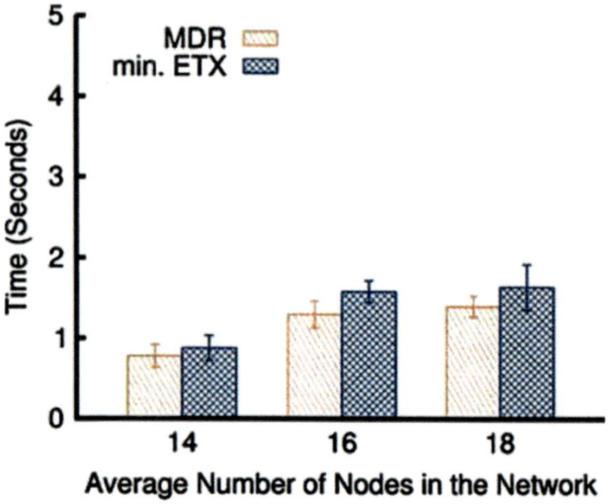

The proposed protocol design is used to compare the performance metrics of the existing protocols with the newly proposed protocol. The efficiency of the new protocol is proven using simulation results. Simulation results show that the improvement in the performance metrics has considerably increased the efficiency of the network compared to the already existing protocols' metrics.

In Fig. 2, the average number of nodes in the network is plotted against time and the result shows that in MDR, the number of nodes is covered in less time.

In Fig. 3, the bandwidth utilization, round trip time (RTT) and distortion are plotted against time and the efficiency of the MDR is highlighted. In Fig. 4, the throughputs of ETX and MDR are compared to highlight MDR's efficiency over ETX.

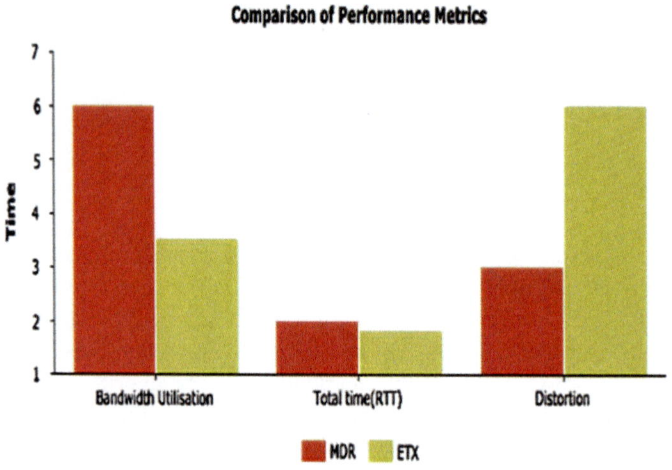

Fig. 3 Comparison of performance metrics

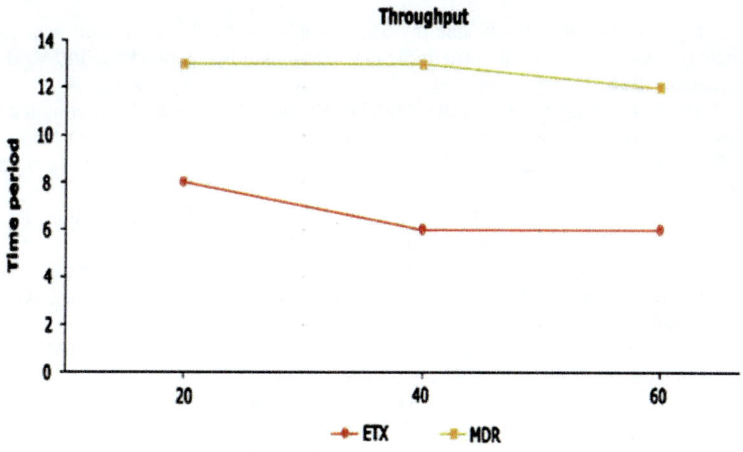

Fig. 4 Comparison of throughput

9 Conclusion

Existing wireless network systems are not efficient in terms of bandwidth utilization, RTT, and distortion. The proposed new routing protocol framework helps in effectively minimizing the total distortion during the video data transfer in the wireless network. The analytical framework designed with the help of new MDR protocol sends the frames in the least congested path to achieve minimum distortion. Using the NS2 simulator, the performances of the proposed protocol were evaluated and the results show increased bandwidth utilization and reduced distortion within a short time interval as compared to the performances of the existing routing protocols. It is also proved that the network's throughput has increased and data loss decreased when delivering packets from source to destination. So, the proposed routing protocol is proven to give much better results for video data transmission in wireless networks.

Future Work By reducing the distortion for high-end data, traffic in mobile wireless networks is most important in real-time. In the future work, it is possible to implement a protocol for low-end data in homogeneous and heterogeneous wireless sensor networks.

References

1. Braun T, Kassler A, Kihl M, Rakocevic V, Siris V, Heijenk G (2009) Traffic and QoS management in wireless multimedia networks. ser. In: Lecture notes in electrical engineering, vol 31. Springer, USA
2. Pei Y, Ambetkar VS, Modestino JW, Qi Q, Wang X (2005) Enabling real-time H.264 video services over wireless Ad Hoc networks using joint admission and transmission power control. Telecommun Syst 28(2):231–244

3. Papageorgiou G, Singh S, Krishnamurthy SV, Govindan R, La Porta T (2015) A distortion-resistant routing framework for video traffic in wireless multihop networks. IEEE/ACM Trans Netw 23(2):412–425
4. Mao S, Hou YT, Cheng X, Sherali HD, Midkiff SF, Zhang Y-Q (2006) On routing for multi-pledescription video overwirelessad hoc networks. IEEE Trans Multimedia 8(5):1063–1074
5. Mao S, Cheng X, Hou YT, Sherali HD (2006) Multiple description video multicast in wireless ad hoc networks. Mobile Netw Appl 11(1):63–73
6. Chakareski J, Han S, Girod B (2005) Layered coding versus multiple descriptions for video streaming over multiple paths. Multimedia Syst 10:275–285
7. ISO/IEC JTC1/SC29/WG11 (1999) ISO/IEC 14496—coding of audiovisual objects [Online]. Available: transmission of MPEG video over the internet. Signal Process Image Commun 15(1–2):7–24
8. Wiegand T, Sullivan GJ, Bjontegaard G, Luthra A (2003) Overview of the H.264/AVC video coding standard. IEEE Trans Circuits Syst Video Technol 13(7):560–576
9. Wang Y, Wenger S, Wen J, Katsaggelos AK (2000) Real-time communications over unreliable Networks. IEEE Signal Process Mag 17(4):61–82
10. Boyce JM, Zhang R, Regunathan SL, Rose K (2000) Video coding with optimal inter/intra-mode switching for Packet loss resilience. IEEE J Sel Areas Commun 18(6):966–976
11. Xiao J, Tillo T, Zhao Y (2011) Error-resilient video coding with end-to-end rate-distortion optimized at macroblocklevel. EURASIP J Adv Signal Process 1:80
12. Ivrlač MT, Choi LU, Steinbach E, Nossek JA (2009) Models and analysis of streaming video transmission over wireless fading channels. Signal Process Image Commun 24(8):651–665
13. Lee Y-C, Kim J, Altunbasak Y, Mersereau RM (2003) Layered coded versus multiple description coded video over error-prone networks. Signal Process Image Commun 18(5):337–356
14. Li D, Pan J (2010) Performance evaluation of video streaming over multi-hop wireless networks. IEEE Trans Wireless Commun 9(1):338–347
15. Wang Y, Wu Z, Boyce JM (2006) Modeling of transmission-loss induced Distortion in decoded video. IEEE Trans Circuits Syst Video Technol 16(6):716–732
16. Suroor S, Challa M, Syed ZA (2016) Enhancement to distortion-resistant routing framework for video traffic in wireless multihop networks. Sadia Suroor (IJCSIT) Int J Comput Sci Inf Technol 7(3)
17. Hanzo L, Tafazolli R (2007) A survey of QoS routing solutions for mobile ad hoc networks. IEEE Commun Surveys Tuts 9(2):5070
18. Migliorini D, Mingozzi E, Vallati C (2011) Performance evaluation of H.264/SVC video streaming over mobile WiMAX. Comput Netw 55(15):3578–3591
19. Alotaibi E, Mukherjee B (2012) A surveyon routing algorithms for wireless ad-hoc and mesh networks. Comput Netw 56(2):940–965
20. Wei W, Zakhor A (2004) Robust multipath source routing protocol (RMPSR) for video communication over wireless ad hoc networks. In: Proceedings if IEEE ICME, Taipei, Taiwan, pp 1379–1382
21. Chen J, Chan S-HG, Li VO (2010) Multipath routing for video delivery over bandwidth-limited networks. IEEE J Sel Areas Commun 22(10):1920–1932
22. Rong B, Qian Y, Lu K, Qingyang R, Kadoch M (2010) Multipath routing over wireless mesh networks for multiple description video transmission. IEEE J Sel Areas Commun 28(3):321–331
23. Liang YJ, Apostolopoulos JG, Girod B (2008) Analysis of packet loss for compressed video: effect of burst losses and correlation between error frames. IEEE Trans Circuits Syst Video Technol 18(7):861–874

Data Collection and Deep Learning for Traffic and Road Infrastructure Management

Surya Rajendran, Kayalvizhi Jayavel, and N. Bharathi

Abstract Managing the traffic of a city or a demographic is a deceptively significant task. The productivity of the people commuting through the region and the pollution caused by their vehicles can be critically impacted by the traffic management of the roads that they are in. Therefore, artificial intelligence systems can be a good alternative to traditional traffic management techniques. A framework for mining real-time data will help build the foundation for intelligent traffic management approaches that can be immediately tested instead of proposing a data collection subroutine every time. In addition to data collection, this paper proposes ways to derive insights, from which both short-term (real-time traffic routing and dynamic traffic signals) and long-term (building flyovers and installing new traffic signals) decisions can be made. Raw data is collected in the form of video feeds from the traffic cameras. Machine learning routines such as Deep Sort and Lucas–Kanade method are used to extract statistics (such as speed of vehicles and vehicle count) from the video feeds. Multi-layer perceptions are used to extract license plate information from the video feeds. This license plate information can be further extended for many security applications.

Keyword Traffic-management · Artificial intelligence · Internet of things · Computer vision

S. Rajendran (✉) · K. Jayavel
Department of Information Technology, School of Computing, SRMIST, Chennai, India
e-mail: suryarajendhran@outlook.com

K. Jayavel
e-mail: kayalvij@srmist.edu.in

N. Bharathi
Department of Computer Science and Engineering, SRMIST, Chennai, India
e-mail: bharathn2@srmist.edu.in

D. J. Hemanth et al. (eds.), *Artificial Intelligence Techniques for Advanced Computing Applications*, Lecture Notes in Networks and Systems 130,
https://doi.org/10.1007/978-981-15-5329-5_38

1 Introduction

The role of eyes with its visioning ability in bringing out the power of intelligence of the human brain is one of the natural phenomena in humans. The eyes are occupying the larger proportion than any other sense organs in studying the environment around us. The computer plays a major role in almost all domains with its virtue of processing more efficiently and faster. Many researchers are working to advance the computer to the new era of modeling as the human brain with the various artificial intelligence and machine learning techniques. To support the work, few researchers are focusing on realizing the power of vision to the computers. The computer's ability to view and study the surroundings is more challenging than expected. Even with four decades of research, many of the challenges in computer vision is unsolved, and the research is continuing in positive way in recent years.

The computer vision is seldom misinterpreted with image processing. The image processing is the capability of manipulating images from the existing images. Computer vision requires image processing for preprocessing the raw images. Specialized tasks such as medical imaging, optical character recognition, surveillance, fingerprint recognition and biometrics have already proven the success records of computer vision. Another gamut of applications is exploring the potentials of computer vision to evidence the productive inference. Image classification, segmentation, edge computing, etc., are relying on computer vision in order to respond better.

Road traffic images are classified and segmented using computer vision algorithms to detect the traffic violations, vehicle counts, amount of traffic, the alternate routes and traffic, speed of traffic, etc. In addition, the road condition where the amount of traffic is more is continuously monitored and broadcasted to other vehicles which are about to join in that road to not to proceed with this route. The roadside parking also keeps into account during the monitor and controlling of traffic and optimizing the running time of every vehicle.

The road traffic is categorized into two broad categories as regular traffic which is happened daily during the specific interval of time and the incidental traffic which happens rarely due to several reasons such as traffic signal violation, accident and traffic rerouting due to rallies. The later type is of less importance in optimizing the road traffic. The former type is always with greater interest especially to this work in order to study the nature of the traffic in day to day or peak traffic time (more than once in a day) basis. This study is used to optimize the solution which always tries to minimize the traffic density in the intended road.

The smartness is framed by dynamically rerouting the traffic in alternate routes is a solution if more than one route is available in that intended route, otherwise suggesting for new roads, expansion of roads and building of bridges and flyovers when the situation is worse as the time goes on. The major decisions are required well before keeping the construction time in reserve. These decision-making systems should be incorporated in smart cities system.

This work proposes the artificial intelligence and deep learning models for road traffic management to monitor and optimize using traffic cameras. The model is burdened with the task of decision making to approve and construct new roads, bridges and traffic signals to reduce the overall traffic congestion. The data required is image or video to be collected and manipulated with the group of interconnection cameras and dynamically forms a cluster to study the traffic in that location and contribute to the decision-making system of the overall smart cities system.

2 Literature Review

In [1], the simulation-based traffic model is generated using macroscopic, microscopic and mesoscopic flow models. The traffic models are evaluated and validated using both visual and statistical techniques. Besides, autonomous driving is also simulated and analyzed with nine different datasets such as KITTI, Cityscape, Udacity, BDDV and Oxford. Using motion planning algorithms and decision-making methods. Though this is focusing autonomous vehicles, the study of these algorithms contributes to the road traffic monitoring and management with some minor modifications.

In [2], convolution neural network-based object detection is studied and infers with classifying the vehicles and their counts. Real-time video frames are collected in roads of Delhi in India with camera installed to cover four different views. Objects are detected and indicated using rectangular bounding boxes with a label having any one vehicle value out of six categories. The observed results are used for getting the approximate value of air pollution and the transport status of that city. Besides, hardware cost and latency also evaluated for CNN-based implementations in this work.

The computer vision-based roadside occupation surveillance system (CVROSS) [3] was designed based on Internet of things to capture real-time images and processing. This work is proposed to estimate the parking gap smartly and support for decision making about vacancy and occupancy of roadside parking. AI-based fuzzy logic classification is applied to categorize the different kind of vehicles that are demanding roadside parking. This was tested in real time in the roads of Hong Kong based on smart transportation management.

Image processing-based computer vision is conquering the game of traffic monitoring [4] and optimization than any other traditional road traffic monitoring techniques such as inductive loop, magnetometer, active infrared, microwave radar, ultrasonic and acoustic. The positive side of video image processing is its wide coverage, continuity in vehicle monitoring, easy addition and removal of detection zones, abundant of data to process, etc. The traffic density, presence of ambulances, path for pedestrians, traffic rule violation vehicles, amount of congestion and alternate routes, etc., can be easily and more efficiently determined.

The vehicles are re-identified based on various sensor-based techniques [5] such as magnetic sensors, inductive loop detectors, GPS, RFID and cell phones and multi-sensors. The vision-based vehicle re-identification with handcrafted-based and deep feature-based methods which are also proved with better inference with datasets like compcars, vehicleID, boxcars, toycarReID, etc. The analysis is performed with two widely used performance metrics known as mean average precision (MAP) and cumulative matching curve (CMC).

The anomaly behavior in road traffic is analyzed and studied [6] based on computer vision (CV). The various CV-based methods are categorized with their learning strategy such as supervised, unsupervised and semi-supervised. Also, those are reviewed and categorized based on the approaches such as model-based, classification-based, prediction-based, reconstruction-based and clustering-based techniques. Besides, the various features that are extracted from images or videos serve as input to the specific technique, and applied scenarios and environments are also reviewed.

The road traffic density, vehicle condition and road condition [7] are all captured as video or image input to the system to optimize the running time of the vehicle and average waiting time in the traffic. The work is focusing on dynamically managed traffic system which changes the configuration of vehicle to match with the road condition with the available smartness in the vehicle.

Mobile telecommunication-based systems [8] with self-functions can be placed at road junctions to study busy traffic using image processing. Raspberry Pi is used to perform the processing functions for vehicle image segmentation, vehicle detection and finally come up with the vehicle count. The SVM classification in vehicle detection is used. In addition, vehicle tracking also implemented with continuous detection of vehicles with network of connected cameras.

The vehicle count is determined by computer vision [9] with the IoT setup of Raspberry Pi and its peripherals. The firebase cloud is used to store the data about the vehicle along the timestamp. Once the vehicle count crosses the threshold limit at one junction, alarm is raised in the remote control station which distributes the information to the required vehicles to alter their route.

Artificial intelligence-based convolution neural network [10] is developed to monitor the road condition and control the traffic. The unmanned aerial vehicle (UAV) is capturing the images and processing to analyze the road condition by identifying the multiple objects in the traffic. The different objects of interest are identified simultaneously with the AI-based CNN in order to detect the various possibilities of alternate routes in highly traffic locations.

3 Proposed Model

The first stage/phase of the data collection framework is the cameras. A grid of cameras strategically placed along the roads collects raw data in the form of videos. These videos are stored on a central server from which they are transferred to a processing pipeline that can process data to produce basic information such as (Fig. 1):

Speed of vehicles: The Lucas–Kanade [11] method for optical flow is used to determine the speed of an object by comparing its relative position in two images taken within a time interval. This information is useful to understand the overall flow of traffic in any given road.

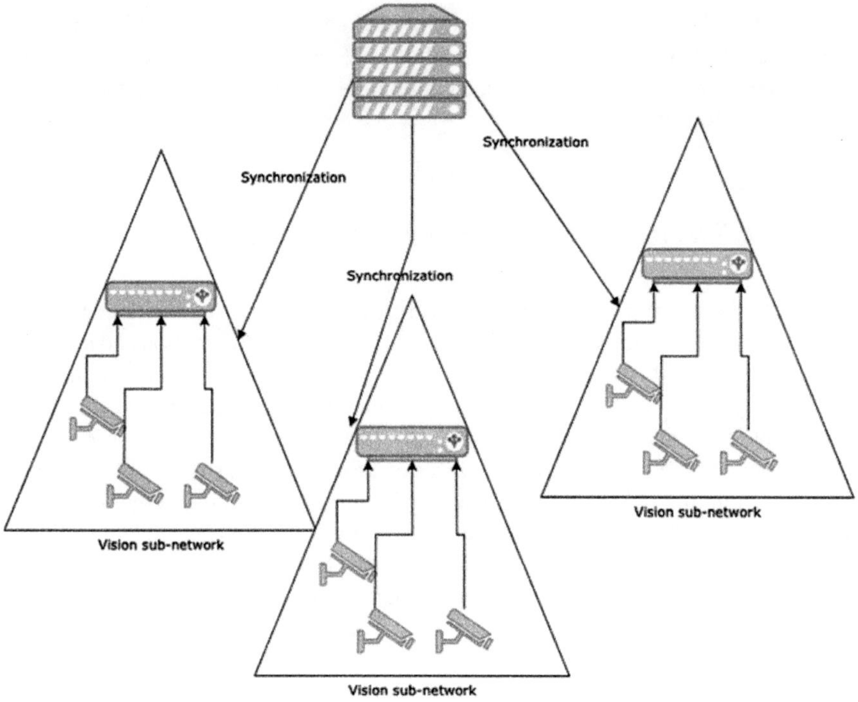

Fig. 1 A grid of vision subnetworks

Number of vehicles: Deep Sort [12] Algorithm is used to track the number of vehicles passing through the section of the road covered by a camera. Since it is a tracking algorithm, it can be used to track vehicles across cameras by stitching their feeds together.

Traffic jams: If all or majority of the vehicles are not moving along a section of the road, it can be recorded as a traffic jam.

Vehicle classes: Using YOLOv3, an object detection algorithm, the type of vehicle (e.g., car, bike, light truck, heavy duty truck and special purpose vehicles) that is traveling can be recorded.

3.1 Lucas–Kanade Algorithm for Optical Flow

The Lucas–Kanade method is a widely used differential method for optical flow estimation developed by Bruce D. Lucas and Takeo Kanade [11]. It uses least squares criterion with the assumption that the flow is essentially constant in a local neighborhood of the pixel under consideration and solves the basic optical flow equations for all the pixels in that neighborhood [13, 14]. This method can solve the hidden ambiguity by combining information from several nearby pixels. It is robust against image noise than the usual pointwise methods. As every method has a drawback, this too does not perform well in understanding the flow information in the interior of uniform regions of images.

3.2 Deep Sort for Vehicle Counting and Tracking

In recent times, there has been an increased interest in capturing overarching patterns in the image and not just details on the pixel-level. Object detection and tracking have traveled far in this regard. Detecting objects in an image is easier compared to detecting objects in a video. But this knowledge is extrapolated to extract objects from many static images and temporally analyzed to best predict its movement. Applications of being able to track the location of objects in a stream of images are enormous: counting objects/humans passing a region and tracking humans/animals movement. Both of which can be extended to variety of more advanced applications. Object tracking can be possible bucketed into two broad categories: Single object tracking—detect and track the location of a single object over multiple frames (e.g., tracking the trajectory of a cricket ball after a batsman stroke), multiple object tracking—detecting and tracking multiple objects present in a visual field and track till it exits.

Deep Sort uses three steps to solve the problems given above, namely:

1. Detect objects using Faster-RCNN.
2. Kalman filtering to model the objects as a dynamical system.
3. Association between new and old systems is computed using the Hungarian algorithm on bipartite graph matching.

In a nutshell, object detection is identifying objects independently in each frame. Object tracking in other hand is assigning IDs for every object identified and tracks them across frames. Deep Sort is a state-of-the-art method toward achieving these goals.

Challenges:

Real-world implementation is very challenging in comparison with testing environments. In a testing environment, the equipment used for capturing images/videos is high quality, and there are rare cases where noise interferes; however, in a real-world scenario, noise and low-quality cameras are the rule rather than the exception. Another technical issue that plagues real-world implementation is occlusion. Occlusion happens when in one frame the object is detected and goes undetected in few other frames. Here, the challenge is how the system would cope if the same object is detected after some time again. Often tracking of objects will be from various cameras, adding more complexity to the problem at hand. The problem is more pronounced when the camera is also in motion.

Another implementational issue is networking the various vision subnetworks and their respective nodes. Hence, networking them needs to be carried out in an efficient and simple manner.

4 Methodology and Analysis

Insights can be drawn from the information extracted from the visual feeds in the previous step which can help make both long-term and short-term decisions.

4.1 Long-Term Insights

Statistical information derived can be used to make long-term decisions. Since these insights will be based on a large volume of data, it is much more likely to assist in making sure that road infrastructure is effectively built.

Building flyovers: If vehicles along a long section of any road do not exit frequently and traffic is caused by vehicles that cross this section, then a flyover can be built to let the traffic flow more efficiently. The vehicles that travel the flyover can get to their destination without interruptions, and the vehicles that cross over can do so without having to wait for the road to clear up. This can also reduce much pedestrian accident that is common due to these situations.

Installing traffic signals: If an intersection frequently has traffic jams due to an equal or proportionate amount of traffic from all directions, then a traffic signal can be installed.

Building roads: Based on the type of vehicles that frequent the roads and the weather that the road has to endure, the rebuilding of the roads can be done more effectively. For example, roads that are used mostly by light vehicles and are in a dry region can be built with a lower thickness than a highway that has a lot of heavy vehicle traffic and frequent heavy showers. This will ensure longevity while conserving costs.

4.2 Short-Term Insights

While long-term insights can help make intelligent infrastructure decisions, short-term insights on the other hands can react to real-time changes in the order of minutes.

1. Redirection of traffic management force: Traffic management officials could be redirected to locations that suffer from heavy traffic at the moment. Locations that require extra attention could be identified by marking areas that have long and slow-moving traffic.

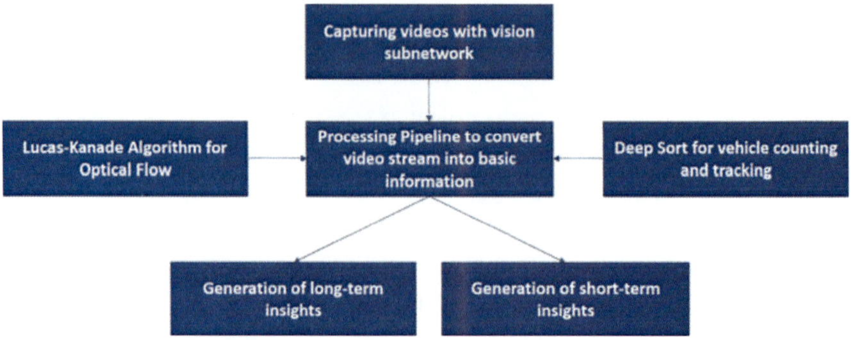

Fig. 2 Data flow of proposed model

2. Smart routing: Information about traffic jams, accidents, infrastructure breakdowns and construction can be transferred to mapping services such as Google Maps through an API.
3. Medical emergency support: When an ambulance is arriving through a heavily crowded area, traffic officers nearby can be alerted so that they may assist in quickly navigating the route. This could cut down casualties that are a result of delays in bringing a patient to the hospital.
4. Accident assistance: Typically, accidents that occur at night when the amount of people nearby is less are more likely to be fatal. However, with a smart vision network, medical assistance can be delivered to the necessary location immediately (Fig. 2).

5 Implementation

Implementation includes testing whether the vehicles in a video are accurately tracked and the speeds measured are reasonably within the margin of error. While one can use videos obtained from the Internet for testing the tracking system, it was more complicated to test the speed estimation system as the frame rate must be controlled and the actual speeds of the vehicle need to be known beforehand (Fig. 3).

For the test purposes, both systems were individually tested to troubleshoot and optimize accordingly. For the speed estimation system, real-time video was fed where the speed of the vehicles in the frame was measured by the drivers. A 720 p camera was used to record and stream the video to the computer using a CSI interface. The actual speed versus the speed as measured by the algorithm was compared to calculate accuracy (Fig. 4).

Fig. 3 Snapshot of tracking vehicles in a video

Fig. 4 Snapshot of detecting the vehicle count in a video

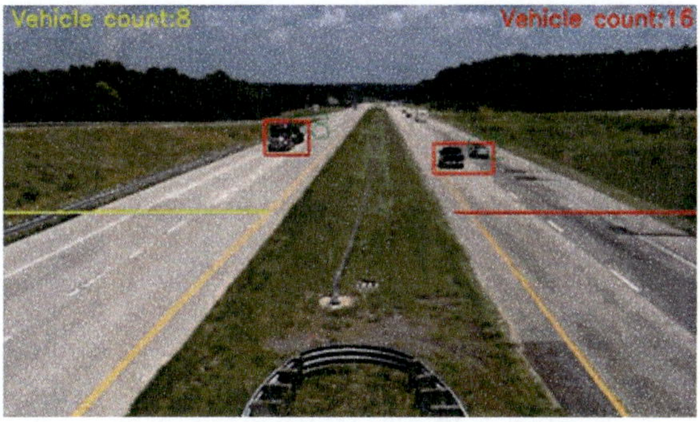

6 Results and Discussions

As observed above, preliminary testing with a few traffic cameras has displayed high accuracy in vehicle tracking and counting. Almost all vehicles were detected and processed. This is attributed to the Deep Sort algorithm.

On the other hand, since data that contains both the images and the vehicle speeds is required for testing the accuracy of the speed estimation system, a custom dataset with ten examples was obtained. With a margin of error (MoE) at 10%, the accuracy is 70% and at 15% is 90%. However, it is to be noted that speeds displayed by vehicles are almost always higher than actual speeds, so the accuracy could be much higher given that most of the predictions are on the lower side (Table 1; Fig. 5).

All other tasks outlined in the proposed model and methodology are built upon these fundamental predictions. In addition to that, the other tasks require data from an extended period of time that can be executed as future work.

Table 1 Ten sample data and its inference

Actual speed (in kmph)	Predicted speed (in kmph)	Result with 10% MoE	Result with 15% MoE
30	25.70	False	False
40	38.61	True	True
50	46.89	True	True
60	53.62	False	True
60	57.11	True	True
70	66.47	True	True
90	82.04	True	True
100	86.03	False	True
120	116.44	True	True

7 Conclusion

A framework is proposed to capture real-time video stream, governing traffic and providing insights based on video stream forms the core of smart traffic management. It is also implemented using machine learning routines such as Deep Sort and Lucas–Kanade method to identify individual vehicles and its count and to predict the speed of each vehicle in the stream, respectively. The inference from the implementation is 70% accuracy with 10% MoE and 90% accuracy with 15% MoE against real-time data. Besides proposing the framework and implementation, the proposed model is generating the short-term insights such as redirection of traffic management force, smart routing, medical emergency support and accident assistance and long-term insights such as building flyovers and laying roads and installing traffic signals for making decisions.

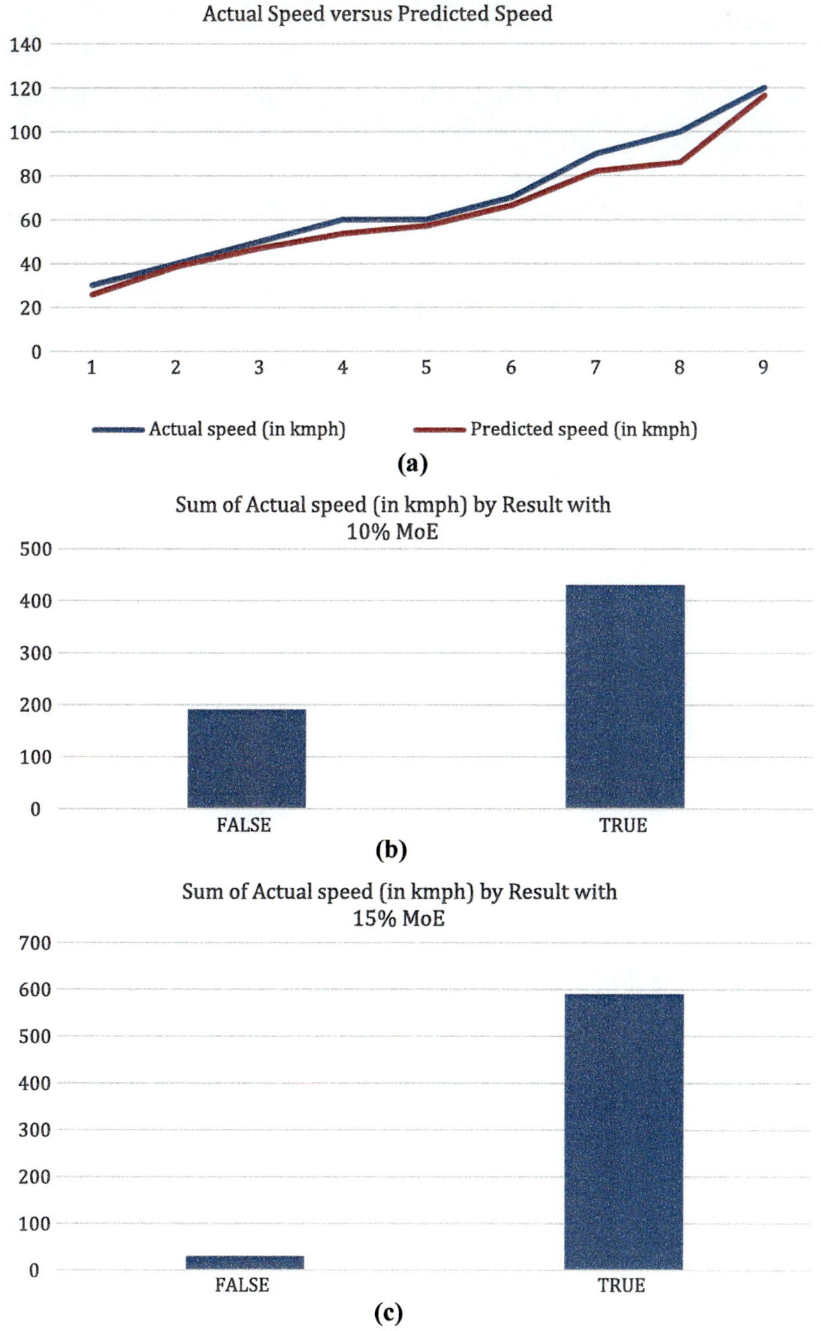

Fig. 5 **a** Actual versus predicted speed **b** and **c** 70% and 90% accuracy while measuring accuracy with 10% and 15% of MoE

References

1. Chao Q, Bi H, Li W, Mao T, Wang Z, Lin MC, Deng Z (2019) A survey on visual traffic simulation: models, evaluations, and applications in autonomous driving. computer graphics forum. https://doi.org/10.1111/cgf.13803
2. Chauhan MS, Singh A, Khemka M, Prateek A, Sen R (2019) Embedded CNN based vehicle classification and counting in non-laned road traffic. In: Proceedings of the tenth international conference on information and communication technologies and development—ICTDX '19. https://doi.org/10.1145/3287098.3287118
3. Ho GTS, Tsang YP, Wu CH, Wong WH, Choy KL (2019) A computer vision-based roadside occupation surveillance system for intelligent transport in smart cities. Sensors 19(8):1796
4. Jain NK, Saini RK, Mittal P (2018). A review on traffic monitoring system techniques. Soft Comput Theor Appl 569–577. https://doi.org/10.1007/978-981-13-0589-4_53
5. Khan SD, Ullah H (2019) A survey of advances in vision-based vehicle re-identification. Comput Vis Image Underst https://doi.org/10.1016/j.cviu.2019.03.001
6. Kumaran S, Dogra D, Roy P (2019) Anomaly detection in road traffic using visual surveillance: a survey. arXiv preprint arXiv:1901.08292
7. Osman T, Psyche SS, Ferdous JMS, Zaman HU (2017). Intelligent traffic management system for cross section of roads using computer vision. In: 2017 IEEE 7th annual computing and communication workshop and conference (CCWC). https://doi.org/10.1109/ccwc.2017.7868350
8. Rus C, Marcuş R, Stoicuţa O (2019) Road traffic monitoring system with self-learning function using the raspberry Pi platform. MATEC Web of Conferences, 290, 06009. https://doi.org/10.1051/matecconf/201929006009
9. Singh S, Singh B, Ramandeep Singh B, Das A (2019) Automatic vehicle counting for IoT based smart traffic management system for indian urban settings. In: 2019 4th international conference on internet of things: smart innovation and usages (IoT-SIU). https://doi.org/10.1109/iot-siu.2019.8777722
10. Yang J, Zhang J, Ye F, Cheng X (2019) A UAV based multi-object detection scheme to enhance road condition monitoring and control for future smart transportation. Artif Intell Commun Netw 270–282. https://doi.org/10.1007/978-3-030-22971-9_23
11. Lucas BD, Wolf T (1981) An iterative image registration technique with an application to stereo vision. In: From Proceedings of Imaging Understanding Workshop, pp 121–130
12. Murtagh F (1991) Multilayer perceptrons for classification and regression. Neurocomputing 2(1990/91):183–197
13. Lucas BD, Kanade T (1981) An iterative image registration technique with an application to stereo vision. In: Proceedings of imaging understanding workshop, pp 121–130
14. Bruce DL (1984) Generalized image matching by the method of differences (doctoral dissertation)

Sentiment Analysis of Food Reviews Using User Rating Score

Rutvi Patel and K. Sornalakshmi

Abstract This paper presents various techniques use in review of products to build a recommender system for a customer to help and identify which product is beneficial among all. The process of analyzing text form data and classify those customer reviews as negative, positive, and neutral is a process of sentiment analysis. Sentiment analysis is an increasing task nowadays because it includes many real-time applications. Rather than to identify just the opinion, system can extract the attributes of the expressions too. Binary classification on reviews based on their sentimental behavior is the most popular task of NLP nowadays. Which technique gives high accuracy and best prediction result by using different algorithm can be identified by this paper. This classification defines from the reviewer's score present in the topic.

Keywords Sentiment analysis · Product ranking · Feature extraction · NLP review analysis · Clustering

1 Introduction

What others think? Is a very important data during the basic decision-making process? Constantly each person defines different items via any Web-based e-commerce sites. Web page does not have any blueprint regarding intelligence system, so it is very hard for computer to understand the importance of semantics. So, organizations have to specially go deep into the sea and find that how their customers are thinking before purchasing a product and run their business according to that. Slant examination is the mechanical process of finding the conclusions and sentiment from the given reviews context by natural language processing (NLP). Conclusion investigation process identifies the sentiments form words like "positive," "negative," and "impartial."

R. Patel (✉) · K. Sornalakshmi
Department of Information Technology, SRM Institute of Science and Technology, Chengalpattu, India
e-mail: patelrutvi_bi@srmuniv.edu.in

K. Sornalakshmi
e-mail: sornalak@srmist.edu.in

D. J. Hemanth et al. (eds.), *Artificial Intelligence Techniques for Advanced Computing Applications*, Lecture Notes in Networks and Systems 130,
https://doi.org/10.1007/978-981-15-5329-5_39

Opinion examination is often prepared in three different levels like view point level, achieve level, and last is sentence level. Estimation of examination in report level always cares the full record as a solitary subject and orders as positive or negative conclusions. The sentiments conveyed in each sentence are grouped together in a sentence level [1].

The viability of varied AI strategies is inspected for grouping of online surveys utilizing models contrived from an audit corpus utilizing regulated learning techniques. Likewise, strategies for removing item highlight recognition are inspected and present a way for concluding descriptive word extremity when the extremity is obscure [2].

Conclusion investigation tends to those issues with new strategies for order. Its objective is often to play out an easy twofold arrangement (positive class and negative class), or to play out a more than one class characterization. Audits contain individuals' conclusions on an assortment of subjects along these lines' assessment mining must be adjusted to the concentrate subject explicit surveys. Supposition examination is often performed on a worldwide level theme or on an increasingly explicit point. Assumption examination is performed on nourishment formula by using an application. The target of the appliance is used to rank different plans consisting of center fixing hooked into surveys. This spares time of the clients scanning for the simplest formula for a fixing [3].

Double orders of slant on surveys are an inexorably mainstream task in natural language processing. In place of grouping positive audits and negative surveys, system arranges audits into extremely negative, negative, impartial, positive, and amazingly positive classes legitimately from the analyst's score on a subject. A basic RNN classifier, a modified RNN classification, and a GRU classification are trained. This investigation might be a valuable instrument to help cafés with bettering comprehend analysts' supposition about nourishment and may be utilized for various undertakings, for instance, recommender systems [4].

This examination depends on clients' audit for fine nourishment on Amazon. Each audit is often factorized as content component and non-content highlights. For the non-content component, traditional straight regressor (Flexible net, Edge, Tether) and regressor by groups (Irregular woodland regressor, KNN regressor) are utilized to organize on them while for the content highlights, unigram, bigram, and blended model are utilized. Lucidness of the audit content is likewise introduced [5].

From notion order, examination conducted for the abstract information inside the content thus mines the conclusion. Assessment investigation is that the system by that information is separated from the emotions, evaluations, and feelings of individuals concerning elements, occasions, and their traits. In choosing, the assessments of others majorly affect clients ease, making choices with reference to online looking, choosing occasions, items, and substances. The methodologies of content assessment examination and enormous work a specific level like expression sentence or archive level. This paper targets dissecting a response for the estimation order at a fine-grained level, explicitly the sentence level during which extremity of the sentence could be given by different common classes as positive, negative, and neutral [6].

The impact of the repetition studies within the Amazon audits datasets on a couple of proposal strategies. The rationale for this investigation is to specialize in this issue (might go overlooked on different datasets also), to survey the potential effect of data repetition on execution assessments, and by and enormous to boost consideration regarding the importance of dataset arrangement and cleaning even in absolutely scholastic undertakings. For instance, the one discharged by McAuley and Lekovic, in spite of blemishes, may be a significant and acknowledged support of the community [7].

Many models for predicting a task in data mining are learned; however, using the proper model is extremely important for an honest result and reducing the time period. During this assignment, accomplishing a predictive task employing a supervised learning model is attempted; however, it is said that goal can still be achieved with a suitable error. Latent factor model is satisfied the wants even using limited features. During this report, the predictive process from the start to the ultimate results using this model and comparing with rectilinear regression model and ridge regression which two had used most of the time in previous work. This determined ideally how this type of special dataset will help in building the model correctly using necessary features and eventually obtain a satisfactory result [8].

Customer reviews on e-commerce Web site contain customer feedback on various products. A research based on china Internet network information center is done where 82.1% of online customers read reviews of products and 41.1% of customers read online reviews before making the final purchase of online products. The analysis from customer reviews gives insight information about products and helpful in making business decision. Sentiment analysis of customer reviews on product in which the text is an efficient auxiliary method to analyze quantum of product reviews given by customer on online sites. Traditional research on customer reviews is based on text lever. To get the satisfaction level of the customer can be obtained from data procession techniques like lexicon-based and machine learning techniques. Both techniques are competent in online review sentiment analysis task. However, feedback of reviewer's, as user generated contain (UGC), which always contain some level of sensitive polarity content towards, product, expressed within the sort of emotional vocabulary around target words. The thing is that an equivalent expression may represent different polarity in meaning when changing different evaluation targets as shown in the following two comments. The processor speed for the mobile is fast enough. The power consumption for mobile is just too fast within the two online reviews above, the appraisal relationship (speed, fast) and (power consumption, fast) are often extracted different. An equivalent suggested word "fast" gives different sentiment when appraising different targets. The previous one review is positive one and later one is negative one. Generally, sentiment analysis methods fail to check spot context emotional sentiment polarity. In the previous research, the related issues are already faced and attributed the matter further to domain diversity by making specific lexicons for every subject book, movies, hottest, electronics for sentiment analysis of online customer reviews. During the study, work will be done into the matter of sentiment analysis of fine-grained level using fuzzy.

Logic: The aim is to identify their polarity of feedback of a particular product which might not be explored by eWOM of products from food online reviews. The important point of this study is summarized as below:

- Design a completely unique fuzzy product ontology mining technique, to support social analysis.
- Develop a fine-grained sentiment analysis technique which is supervised by semantic analysis at different level.
- eWOM exploration system with fine-grained sentiment analysis technique. This system can perform at different level product features [9].

Customer reviews mining is among the foremost important research areas where the field like text mining and knowledge extraction. It retrieves both subjective as well as objective information from different types of text like structured, semi-structured, and unstructured text formats. The subjective information provides opinions like love, joy, sadness, etc., whereas the target information gives only information. Customer reviews mining performs the task of segregating the customer reviews on the based on positive and negative for ranking products on a specific area of interest. As the online platforms reached to millions of users, the context increases exponentially. It makes difficult for a seller to know more about the customer [10].

The exponential increasing of Internet user base led to increase in the online shopping users. This change in user behavior may be useful information for e-commerce sites especially the Web agent. One of the services of Web agent is an online hotel reservation where the service is merged with facilities to provide reviews of the hotels that are booked online by different users. The feedbacks commonly used as a benchmark of customer's satisfaction and based on the feedback of their services and things are required to improve for a service provider. An outsized number of reviews cause difficulty in manual analyzing and conclude the whole reviews. Therefore, there is a requirement to study the prevailing reviews automatically [11].

2 Literature Review and Related Work

In this paper [1], survey of various technologies is done. Sentiment analysis is the process of detecting the sentiment shown within the text review. Here, a survey of various techy has produced a good result using the review score ratings. The limitation of this technique is it only works better just the open sentiment like scores. The result was not good enough for hidden values. In further work, predication-based method is going to be implemented with old techniques. More features are going to be extracted to handle the hidden sentiment analysis [1].

The quantity of training data was gathered before the study started, and the incontrovertible fact is that all test data were reviews contributed enough to the positive result. The output from the insight of product features subtask did not have enough test dataset. It may be contended that it is a more complex subtask than text-level sentiment analysis, but more care should be given to give the provable result. The results from the polarity reasoning subtask were something of a reflection compared with the other subtasks in this paper but show well-meaning promise for future studies even though the result described in this paper were poor [2].

Here, sentiment analysis generates the new method to classify the data. The application used by a paper [3] ranks different recipes which are available on their dataset and dataset is taken from the online Web services. This type of analysis saves the user's time and gives the best result according to human mindset for the particular surfed product or an ingredient. The analysis is done based on lexicon-based analysis approach. The words used as a positive like "testy," "bad," etc., are used to calculate the score of each review present in the dataset and store those words in a bag-of-words.

The review whose score is higher comes in the first place and the ranking goes the same. So, this approach will always be useful for the customer to choose the "which product is best as per to the requirement [3]?"

Different neural network techniques are defined here. This technique involves two different versions which are RNN and GRU. This technique is used to perform the sentiment classification of Amazon Company's dataset. After performing that the accuracy for the classification of 4 classes is 68.75% and the accuracy of the 5 class is 51.74% for the test dataset. In this paper, the technique used is padding zero which helps to improve the accuracy and by using bidirectional RNN with increasing the number of hidden layer size the accuracy can be improved further.

Here, the approaches are used for the sentiment classification are bag-of-words and network-based approaches. By using the first approach, the bag-of-words is used for the unordered data and then to form it into the sequence and also perform the word embedding technique. For the baseline approach, the SVM model is used to perform the Multinomial Naïve Bayes' classification technique. Based on the study of this paper, LSTM model is used to predict the GloVe embeddings and Word2Vec embeddings based on a specific domain. By the research, the GloVe embeddings are more helpful compared to the Word2Vec embeddings [10].

Comparison of two different dataset based on balanced distribution and primary distribution is done but the accuracy was achieved best while primary distribution using Word2Vec embedding. Because there was less difference among AUC score of balanced distributed dataset by using baseline model. The sampling techniques are used to reduce the overfitting problem. By comparing two techniques, baseline approach and LSTM approach, the LSTM approach gets higher AUC score for the NLP-based tasks. The LSTM model is basically used to identify the relationship which cannot be done by the baseline models [12].

Through a group of experiments on different sorts of regressions, it is found that the text feature contributes significantly to the performance of rating predictions. It is consisted of the intuition that the emotion of review reflects the rating in stars. Users tend to offer a better rating with review consisted almost of positive words. In the near future, combining the text feature and non-text feature to perform prediction can be attempted. It would have better performance [5].

Reporting various techniques for detection and rendering of knowledge is tried while working on this paper. Various methods for tongue processing and machine learning algorithms could be understood. The way to do data preprocessing and data cleaning could be understood. The way to apply machine learning algorithms is learned [6].

The method is efficient enough for the test case of iPhone 5 reviews on Amazon. For sentiment analysis, the approach integrates prior sentiment analysis methods. Classification of reviews along with sentimental analysis augmented the accuracy of the system which in turn provides accurate feedback to the user [13].

RNNMC works efficient than the baseline. The inadequate classification of trees, RNNMC, is also able to surpass in every metrics, and then Naïve Bayes classifier is used as a baseline algorithm and in that average word, vectors are used as an input features, which proves that phase-level structure always helps sentiment analysis task. For the recursive neural network algorithm, every node's classification is very important. By this model, the highest accuracy achieved is 80% using dataset of Stanford sentiment tree bank and also result represented with improper labeling. This model is not sufficient enough to achieve accuracy of more than 80%. So, concluding RNN algorithm needs strong supervision. However, the reviews and documents from the online sites are conducting very less labels, so it constrains in making them efficient model but these results are still better because without proper labeling it still works better. Moreover, it has many techniques to increase supervision and get good accuracy of the test model [14].

The method is precise enough for the test case of iPhone 5 user feedback on Amazon. For sentiment analysis, techs have designed their own method that integrates prior sentiment analysis approach. Classification of reviews along with sentimental analysis increased the accuracy of the system which in turn provides accurate feedback to the user [15].

Sentiment analysis is the process of identifying the emotion expressed in the document or feedback given by the user. The methodology for extraction of the food reviews based on score combined with prior text analyzing methods. The approach has produced a very efficient result using the score ratings. The restriction of this approach is it works superior only for the open sentiments like rating or scores. The results were not promising for unseen sentiments [11].

Through a set of experiments on different kinds of regressions, it is found that the text feature contributes significantly to the performance of rating predictions. It consists of the intuition that the emotion of review reflects the rating in stars. Users tend to give a higher rating with review consisted almost of positive words [16].

The significance of this method has a clear logic and a simple computation process. It is also important for early and elevating theories and methods for ranking products through online reviews by user. In terms of future research, to support consumers to use the proposed method to make acquire decision more easily, the support based on the proposed method needs to be efficient. Besides, the circumstances that one or more substitute for new products with few online reviews needs to be measured [17].

Extensive study has been conducted in the related academic areas. In this section, the literature of polarity detection, reviews, and feedback obtained from online customer reviews are reviewed. The correlated works of ontology-based sentiment analysis are also reviewed [9].

A new structure is planned based on their customer given review and the aspect of the product. The main aim of the algorithm is to rank a product based on their aspects by using the correlation technique (rank correlation). This can be done by using below four different ways.

1. The aspects were identified for the relevant products.
2. Based on their aspects, the spearman's correlation technique can be applied for the ranking.
3. Supervised learning techniques can be used by using different classification algorithms.

Using this way, outcomes define that there is a strong relationship between positive and negative product reviews. By comparing different technologies, SVM gives better results than NB & MaxEnt method, and NM & MaxEnt methods give better result compared to LDA [10].

3 Methodology

Sentiment classification of the customer review is used to pick, elaborate, and extract the feature from the text data. Feature selection from defined data is used to collect the knowledge from the text reviews and perform understanding steps according to that. The information of step preparation of data defines the following tasks:

- By removing the unnecessary information of the data like: name of the customer and given review date.
- Performance review analysis: discovering the part of speech or adjectives and count the number of frequencies of that particular word.
- Perform sentiment analysis task.

Sentiment analysis now collects all preprocessed data and performs classification task on that which classifies data into two different classes like positive and negative. Efficient machine learning technique is used to group subjective words within a sentence itself. This subject detection can be implemented by using part of speech (POS) technique. This POS technique is used to identify the sentiments from that subjective group and filter them out. POS is also used to remove the words which does not contain any emotion or sentiment, and the words which are used or a verb, adjectives, and nouns used in a sentence are used to define the class of the review. Frequently used words are stored as a vector of a text. These features are going to further process and convert into small letter alphabets and remove all punctuations and stop words. Then system prepares the word matrix in the next step. Further, the scores will be assigned to every class. These conclusion words are summarized and grouped together in word cloud. As a final result, the reviews containing 5 start and 4 start are considered as positive review, then the reviews containing 3 start are considered as neutral, and reviews containing 1 start and 2 start are considered as negative reviews for the different product [1].

Sentiment analysis with consideration of online review is done by a machine learning technique to elaborate and extract the polarity of sentiments by applying word processing theory to a text corpus and the result will be matched with their manual result which is calculated by the humans. As per the survey, the result calculated by the human is accurate up to 80% of the text data, and based on some theory, it defines that model is precise up to 80% not more than that by the human assessment, which is a golden standard consideration [2].

The statistical approach of lexicon-based approach (LA) is based on the customized algorithm. Here, the customer has to fill some detail about the particular product or the ingredient which is like: the name of product or the ingredient of the particular product. Now, this all values will be passed as a parameter to respective programming class and from there it will be stored in a particular column of a database. Now, the system will define the reviews regarding this product from the specified Web site and perform lexicon-based analysis on that text reviews. To analyze, bag-of-words technique is used to identify the positive and negative words based on their review score. To identify the positiveness or the negativeness of the word, the review should be more broken into sentences and sentences will be more broken into words and the corpus are generated now and those words are checked with prior stored words and based on their review score. The outcome is representing the product which is containing more positiveness by the user review [3].

Fuzzy product ontology is used to identify the taxonomy relationship between different but similar product. As an example, resolution is defined as a subclass of the parent class "screen" for the mobile or a laptop or a device like that. It also defines the non-taxonomy relationship between the different features of the product because of the expression like "screen size" is related to "big" or "small" or any noun. Additionally, the sentiment of the subject is "big" or a "small" or any noun. Additionally, the sentiment of the subject "big" is put into positive class and is decided by the fuzzy product ontology. The taxonomy association between the different product feature and the non-taxonomy relationship between the different products define the sentiment ultimately, which cannot be defined directly from the Web pages, so as to identify the relationship of the product features from the labeled data will be available in the corpus [9].

In this paper [10], an aspect-based review ranking for the product is proposed, which defines several key components like:

- The process to collect the review or information.
- Preprocess that reviews or the information.
- Feature extraction of the particular product.
- Classification of that reviews into different classes like positive, negative, and neutral
- Then ranking the product based on their class and the score of the product [10].

- **Mathematical Formulations**

In the defined paper [4], RNN algorithm is slightly customized here in this paper. Rather providing the classification of sentiments, this paper focuses on the prediction of each slice. The customized algorithm modifies to scale back on the prediction of the reviews and backpropagation.

$$h^{(t=k)} = \sigma\left(W^{(hh)} h^{(t+k-1)} + W^{(hx)} x^{(t=k)} = b_1\right) \tag{1}$$

$$\hat{y}^{(t+T-1)/T} = \text{softmax}(W^{(s)} h^{(t+T-1)} + b_2) \tag{2}$$

Here, T defines the number of steps in the following forward propagation.

In this paper [6], GRU algorithm is used which defines each step as below formula:

$$\begin{aligned} Z^{(t)} &= \sigma(W^{(s)} x^{(t)}) + U^{(s)} h^{(t-1)}) \\ r^{(t)} &= \sigma(W^{(r)} x^{(t)}) + U^{(r)} h^{(t-1)}) \\ h^{(t)} &= \tanh(r^{(t)} U h^{(t-1)} + W x^{(t)}) \\ h^{(t)} &= (1 - z^{(t)})\ \ h^{(t)} + z^{(t)} h^{(t-1)} \end{aligned}$$

- **Model Creation**

Several sorts of regressions are implemented with the feature involving text content and therefore the regressions only specialize in non-text feature. The best one among the regressions that does not considers text feature, random forest regression, has rock bottom MSE about 1.0. But the computation of random forest regressor is very expensive. It requires far more time when the number of decision tree exceeds 100, for instance, even more than half hours for 1000 trees. It is not efficient. However, in general, the regressions with text feature beat the regressions who do not consider the text content. They are more efficient. They only need several minutes for the ridge regression to compute even for the large dataset. The most powerful regression with the text feature is the mixture of unigrams and bigrams representation [5].

- **Sample Datasets in product reviews**

The dataset of Amazon food fine product reviews was collected from the Web pages of Amazon by McAuley and Lekovic. This dataset is a public dataset which is available for all. Approximately 35 million reviews are available which contain data of 18 years, up to March 2013. These reviews are containing different categories like books, movies, clothing, etc. These categories are treated as a separate dataset within the evaluation methods. In this dataset, review's helpfulness is represented by the numerical rating. This helpfulness is used to characterize that how many other users valued this review. This dataset also provides the information about which products were bought together, which products are bought after viewing advertisement, what products a user bought and what the user viewed, and therefore the sales ranks of products. Now, the further more category of this dataset is represented in a provided Table [7].

The dataset is used of Amazon e-commerce site, which is publicly available to analyze the reviews of the Amazon selling product. The total reviews present in the dataset are around 35 million. This data contains the record of 18 financial years, which are up to March 2013. This dataset reviews contain too many different products like books, movies, clothing, etc., which all are considered as a different dataset while applying the algorithm. This dataset contains column like review text, a numerical rating of the helpfulness, score of the review, etc. This paper [14] mentions that "Which products are brought together by the number of customers," "what products are bought by customer by multiple time viewing some advertisement," "which product are related to currently considered product," and also provides the ranking of the bestselling product. Here, the below table defines the dataset sample from the paper [14] (Table 1).

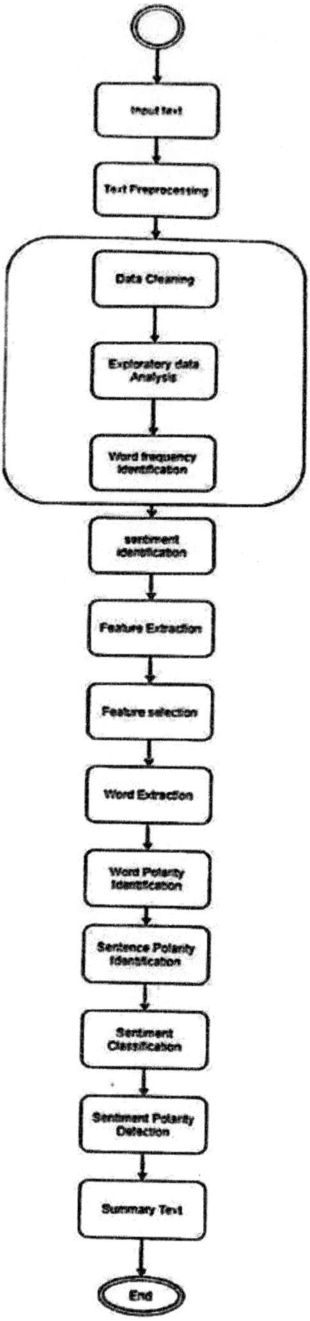
Text Preprocessing
Data Cleaning
Exploratory data Analysis
Word frequency identification
sentiment identification
Feature Extraction
Feature selection
Word Extraction
Word Polarity Identification
Sentence Polarity Identification
Sentiment Classification
Sentiment Polarity Detection
Summary Text
End

Table 1 Amazon reviews datasets [18]

Dataset	#Users	#Items	#Reviews
Amazon instant video	312,930	22,204	717,651
Arts	24,071	4211	27,980
Automotive	133,250	47,577	188,728
Baby	123,837	6962	184,887
Beauty	167,725	29,004	252,056
Books	2,588,091	929,264	12,884,488
Cell phones and accessories	68,041	7438	78,930
Clothing and accessories	128,794	66,370	581,933
Electronics	811,034	82,067	1,241,778
Gourmet Foods	112,544	23,476	154,635
Health	311,636	39,539	428,781
Home and Kitchen	644,509	79,006	991,794
Industrial and scientific	29,590	22,622	137,042
Jewelry	40,594	18,794	58,621
Kindle Store	116,191	4372	160,793
Movies and TV	1,224,267	212,836	7,850,072
Music	1,134,684	556,814	6,396,350
Musical instruments	67,007	14,182	85,405
Office products	110,472	14,224	138,084
Patio	160, 832	19,531	206,250
Pet supplies	160,490	17,523	217,170
Shoes	73,590	48,410	389,877
Software	68,464	11,234	95,084
Sports and out doors	329,232	68,293	510,991
Tools and Home Improvem	283,514	48,059	409,499
Tools and games	290,713	53,6000	435,996
Video games	228,570	21,025	463,669
Watches	62,041	10,318	68,356

In the paper [12], the number of record size is 568,454. This dataset includes 46 million words and contains 2.8 million sentences so the average of the sentence per review is 5 [12].

- **Data Processing**

Here, the used dataset is the public dataset of Amazon Company which is containing the column called as reviews which test column of the present paper. The reviews present in the Amazon dataset can be classified into two different classes like positive and negative where negative ratings come under 1 and 2 stars and positive reviews come under 4 and 5 stars and the rating which contains 3 stars comes under neutral. So, like that data is classified into different categories or classes. So, the data which contains 3 stars are not useful for the classification so that are not considerable for the processing of data. So, by noticing the skewness of 5-star data, paper used distribution technique to equally distribute data. For that distribution of data, two different sampling techniques can be used which are defined below:

- Split dataset into equal number of positive and negative reviews which are around 82,000 of the same datasets.
- To randomly choose 164,000 reviews from the same dataset which technique is known as "Primary distribution."

Here, in second sampling technique, the positive reviews are higher than the negative based on ratio. Here, to analyze data and split whole dataset into 80:20 ratios were 80% of data are used as training data and 20% data are used as testing data. So, to avoid data suffering, primary distribution of data can be used.

- **New Memory Generation**: Here *xt* defines input, *h*(*t*-1) defines hidden states, and ct defines possibility of the new word.
- **Input Gate**: The input gate function is used to check the memory until the new word is coming. Input gate function is also storing the input word so that past hidden state can define whether the word is useful of not, if it is useful then only it stores it in memory.
- **Forget Gate**: The forget gate is used as an analogical function to the input gate instead of identifying the helpfulness of the input word, and it checks that whether the word which is stored in a past memory is beneficial for the processing of present state. In this state, forget fate function focuses on input word first than focuses on past hidden state and then generate a future (ft).
- **Final Memory Generation**: In fourth stage, the model takes possible output from the forget gate ft and based on that removes the past memory ct-1. Here, it also takes possible output from the input gate and generates the new memory ~ ct. Here, it adds two results to get the final output as ct and store in a memory.
- **Output/Exposure Gate**: The use of this gate is to split the memory of hidden states which are present in every state in a LSTM. This gate assigns the memory ct which is present in a hidden state ht. The final output signal is generated by it [8].

Physical world data are partial, noisy, and inconsistent, which is known as row data or an unstructured data. This type of data generates the missing attribute values, error consumption, etc. In decision-making process, the understanding of knowledge is more important. So, to improve the standard of knowledge, the data should be preprocessed to improve the accuracy. As per the presented paper [10], the preprocessing task includes the following procedures:

- Remove punctuations, tags, and stop words.
- Stemming and lemmatization are reduced.

- **Approach**

Sr No.	Approaches & Algorithms	Summary
1	K-mean algorithm	The limitations of K-mean algorithm and improve the speed and efficiency of K-mean algorithm and result in optimal number of cluster.
2	Bag of Words (BoW)	Very recently, the model of the bag of words has become so popular in order to produce accurate predictions out of unstructured text data.
3	TFIDF Vectorizer	The need is to classify the set of documents according to the type.
4	Hierarchical Clustering	Hierarchical clustering is a method of cluster analysis which seeks to build a hierarchy of clusters.
5	Naive Bayes and SVM	It is generally used to determine the emotions, sentiments and summarization from large data and that information can be used to make some predictions.

If compared with the performance of RNN and the benchmark, they are very similar. However, the model has supported the review input variable and therefore the benchmark was supported the summary variable. Even if the reviews were very detailed with vocabulary specific to the products, the RNN is able to identify relevant information to determine if these are positive or negative reviews with an F1 score.

4 Summary

Ref. paper no.	Contribution	Algorithm	Accuracy	Findings
1	Reduced overfitting	Linear regression, ridge regression, latent factor model	N/A	Latent factor model is better compared to others
2	N/A	Linear SVC, Naïve Bayes, logistic regression, TFID vectorizer, bag-of-words, linear SVM, K-means clustering	N/A	N/A
3	N/A	RNN, GRU, hyper-parameter tuning	RNN (train acc.–80.38%, test acc.–51.74%), GRU (train acc.–80.14%, test acc.–50.09%)	Improved by adding bidirectional RNN, tuning of hidden layer size and number of steps
4	1.52% (Amazon fine food reviews) in the F1-score as compared to using simple tree-LSTMs	Sentiment analysis with RST, discourse-based sentiment analysis with deep Learning	1.52% increased	Suffer from vanishing or exploding gradients so LSTM is better
6	N/A	SVM, multinomial Naïve Bayes, long-term, short memory RNN	From 87.98 to increased 95.75	Word2vec domain is better with LSTM, LSTM with RNN is better than SVM & mul. NB
7	N/A	ML approach, lexicon-based approach, latent aspect rating analysis, bag-of-words, Web crawling	N/A	N/A

(continued)

(continued)

Ref. paper no.	Contribution	Algorithm	Accuracy	Findings
8	Calculated the trust scores with reduced time and space requirements and limited loss in accuracy	Approximation algorithm	N/A	N/A
9	N/A	Random forest, SVM (linear kernel & RBF kernel), linear regression	N/A	N/A
11	N/A	Count vectorizer, TF-IDF vectorizer, Latent semantic analysis, Word2Vector, TF-IDF weighted word2Vec, multinomial Naïve Bayes, logistic regression, svm, extremely randomizedtrees	Unigram TF-IDF vectorizer–70.38 & unigram + bigram TF_IDF vectorizer–71.28, unigramTF-IDF (with LSA)–69.98, multinomial NB–53.88, logistic regression–55.51, SVM–55.81, extremely randomized tree–70.70	We can get comparison between multiple classifying algo.
12	N/A	Rectangular label distribution, NLP techniques (unigram & bigram)	N/A	N/A
13	RNNMS, Hyperparameter tuning, different activation functions	N/A	N/A	RNNMS is better than baseline algo.
14	N/A	Feed-forward neural network, LSTM, baseline algo, collaborative filtering, matrix factorization	N/A	Feed-forward neural network & LSTM combination beats baseline algo.

References

1. Sasikala P, Sheela LMI (2018) Sentiment analysis of online food reviews using customer ratings. Int J Pure Appl Mathe 119(15):3509–3514
2. Wallin A Sentiment analysis of analysis of amazon reviews and perception of product features
3. Rao S, Kakkar M (2017) A rating approach based on sentiment analysis
4. Feng H, Lin R (2016) Sentiment classification of food reviews. Stanford University
5. Zheng C, Zhang Y, Huang Y (2016) Rating prediction on Amazon fine foods review. University of California, San Diego
6. Bhati V, Kher J (2019) Survey for Amazon fine food reviews. IRJET
7. Basaran D, Ntoutsi E, Zimek A Redundancies in data and their effect on the evaluation recommendation systems: a case study on the amazon reviews datasets.University of Southern Denmark
8. Ashok kumar J, Abirami S (2018) Aspect-based opinion ranking framework for product reviews using a sperman's rank correlation coefficient method
9. Analysis towards hotel reviews: an evaluation study. In: 4th international conference on computer science abd computational intelligence 2019, 12–13 September 2019
10. Liu Y, Bi J-W, Fan Z-P (2016) Ranking products through online reviews: a method based on sentiment analysis technique and intuitionist fuzzy set theory
11. Chu Y, Pham Y, Duong N, Wu Y Predicting Rating of amazon fine food from reviews CSE 190
12. Mikolov T, Corrado GS, Chen K, Dean J (2013) Efficient estimation of word representations in vector space. In: Proceedings of the international conference on learning representations. Arizona, USA. pp 1301–3781
13. Bhatt A, Patel A, Chheda H, Gawande K (2015) Amazon review classification and sentiment analysis. Int J Comput Sci Inf Technol 6(6):5107–5110
14. Wu J, Ji T (2017) Deep learning for amazon food review sentiment analysis. Stanford J Sci Technol
15. Barry J (2016) Sentiment analysis of online reviews using bag-of-words and LSTM approaches. School of Computing, Dublin City University, Ireland
16. Mikolov T, Sutskever I, Chen K, Corrado G, Dean J (2013) In: NIPS'13 Proceedings of the 26th international conference on neural information processing systems, vol 2. Nevada, pp 3111–3119
17. Rizka Putri Nawangsari, RetnoKusumaningrum, Adi Wibowo Word2Vec for Indonesian Sentiment
18. He R, McAuley J (2016) Modeling the visual evolution of fashion trends with one-class collaborative filtering. WWW

A Review on Big IoT Data Analytics for Improving QoS-Based Performance in System: Design, Opportunities, and Challenges

M. Safa and A. Pandian

Abstract The Internet of things (IoT) is a term generally utilized couple with genuine correspondence. The Internet of things (IoT) hypothesis anticipates the enhancement of our present condition toward novel improved spaces, for example, keen urban areas, smart home, matrix, computerized well-being, and programmed environmental contamination control. Numerous ventures are confronting parcel of difficulties and weaknesses regarding QoS. The maximum focus toward the perceivable and exactness is encouraged by IoT on QoS. Internet of things (IoT) is used for a decade for efficient and reliable communication. Nevertheless, such intermittent information's are not profitable without logical power. Diverse huge information, IoT, and investigation plans have been able to get profitable learning into immense information delivered by IoT devices. This paper examines the cutting edge look into endeavors focused toward big IoT data analytics and also, an incentive by proposing significance which attempts to assess the effect of IoT on inventory and QoS.

Keywords Internet of things · Data analytics · Quality of service · Challenges · Solution · End-to-end visibility · Big data

1 Introduction

A dynamic improvement of technologies in various fields has been observed due to the integration of two major systems: Internet of things and big data analysis. These combinations bring out tremendous changes, which fulfill the demands of the organizations and individuals. The data collection process is important factor, distributed

M. Safa (✉)
Department of Information Technology, SRM Institute of Science and Technology, Kattankulathur, Kancheepuram, India
e-mail: safam@srmist.edu.in

A. Pandian (✉)
Department of Computer Science Engineering, SRM Institute of Science and Technology, Kattankulathur, Kancheepuram, India
e-mail: apandiansrm@gmail.com

D. J. Hemanth et al. (eds.), *Artificial Intelligence Techniques for Advanced Computing Applications*, Lecture Notes in Networks and Systems 130,
https://doi.org/10.1007/978-981-15-5329-5_40

computing together with the rise of another age of less expensive, and littler remote gadgets with various correspondence conventions have laid the way to the arrangement of the IoT which has assumed a noteworthy job on the huge information scene [1]. Big data is basically partitioned into three fundamental amounts: volume, assortment, and velocity. More prominent open doors were made with potential capacity to break down and use immense measures of IoT information with QoS, incorporating applications in keen urban communities, transport and matrix frameworks, vitality meters, and patient wellbeing observing framework, home apparatuses, observation cameras, actuators, shows, vehicles, and industrial machines [2].

These applications have a scope of QoS necessities in IoT that has made a change in big data analytics testing in view of the handling and accumulation of data through various types of sensors. This set of data's can be normally classified based on effort and administrations [3]. To provide QoS-based service in the IoT, it is important to guarantee appropriate systems at each layers of the IoT engineering. For example, there will be a delay in each layer from top to bottom. Blend to IOT(QoS), big data analytics, means to help in business advancement to accomplish better outcomes and understanding of data, and subsequently, can settle on the best choices as a result [3].

Additionally, enormous information to support investigation plans which rapidly removes learned information using information mining systems that help with making future desires, recognizing late examples and choosing [4]. A deferral in any layer can prompt unsatisfactory QoS for security basic applications, for example, computerized driving frameworks which require steady input to look after control. So as guarantee that we can give ensured administration-based security for implementing the basic requirements, and the involvement of QoS in all aspects of IoT engineering is very essential to understand [5].

Exhibiting information investigation, IoT, and QoS into huge information require colossal assets, and IoT can offer a staggering blueprint. Authentic assets and real jobs of the stages are given by IoT associations to astounding correspondence among different passed on applications. Such a framework is sensible for the basics of IoT applications and can decrease two or three difficulties later on of huge information examination [5]. This creative affiliation extends the probability of realizing IoT in a prevalent heading. Issues on analysis of data and gadget recognition can be easily solved, once we acquire the sufficient knowledge on conjunction of IoT and big data analytical techniques. Hence, the challenges and upgradation in the field of data analysis applied in the brilliant city can be accomplished.

2 Outline of IoT and Big Data

The general outline on IoT and big data is focused for analyzing.

2.1 Internet of Things (IoT)

A Required platform having contrasting open doors in IoT at different quantities like sensors and gadgets to impart flawlessly inside a keen situation and empowers data sharing crosswise over stages in an advantageous way [4]. The combination and upgradation of IoT with big data strengthens the Internet advances as an opening to the various remote headways. It has been noted as a continuous determination in splendid urban zones with eagerness for making savvy systems, for instance, keen office and retail, brilliant horticulture, shrewd transportation, and shrewd human services [1].

In recent years, IoT had raised as a upgrading pattern level on all- encompassing electronic hardware to encourage day by day life tasks, including, wrist-watches, candy machines, crisis alerts, and carport entryways, and home apparatuses, for example, iceboxes, microwaves, forced air systems, and water warmers are associated with an IoT arrange and can be controlled closely based on the requirements [6].

The band of gadgets and items is interconnected through an assortment of correspondence arrangements, such as ZigBee, WiFi, and GSM-based systems. These specialized gadgets broadcast their data and get instructions from remotely forced gadgets to enhance living standards [7].

2.2 Big Data Analytics

The analytical procedures of big data of dedicated diverse sets of data which contain a complete pack of various data to uncover inconspicuous models which includes relationships, feature patterns, customer tendency data, and other significant business data. The capacity to dismember huge proportions of data can enable a relationship to oversee huge data that can impact the business [5]. Data excavators and authorities to isolate an enormous volume of data that may not be bridled utilizing standard contraptions but rather overwhelmed analysis of big data [8].

Analysis of big data requires advancement that can change a great deal of sorted out, randomly structured, and semi-composed data into progressively realistic data and metadata to organize legitimate techniques [5]. Diverse computations are used in informative mechanical assemblies for separating the data, discover examples, patterns, and relationships over a respectable assortment of data. Big data examination is an authentic test for a few applications because of data thickness and the versatility of significant figuring's that assistance such strategies [2].

2.3 Existing Analytics Systems for Big Data

A different systematic sort utilized by the IoT applications and the various analytics system based on their assessment has been discussed.

Real-time analytics implies that big data is prepared as a volume of data when the business client gets consumable bits of knowledge without surpassing a time period apportioned for basic leadership or a deliberate framework triggers an activity or a warning. For sensor inputs, real-time data analytics is proficient [5].

Off-line analytics is utilized when a fast reaction is not requisite. It helps in data obtaining effectiveness. To lessen the expense of data organize transformation, Off-line analytical design based on Hadoop technique is widely used in most of the Internet operations [5].

Memory-level analytics is associated when given data value is abundant than the memory level; analytics is reasonable for directing the real scenario. The two vital specialized segments that expand the handling execution in memory-level analytics are as follows: One is columnar data stockpiling, where the one-dimensional linear data structure is utilized. Another is massively parallel preparing, where multi-center, multi-string processor works on data to decreased access latencies [5].

BI analytics is when the memory level and volume of data are proportional, also, BI analytics permits simple investigation of data volumes. To recognize new chances and actualize a viable methodology in the focused market to handle the huge volume of data and to enhance productivity and stability, BI is utilized [5].

Massive analytics is connected when the volume of data is better than the total limit of the BI investigation item and conventional databases. The Hadoop distributed framework is utilized for data stockpiling and guide/diminishes for data examination. Massive analytics makes the business storm cellar and builds showcase intensity by separating significant qualities from data [5].

3 E2E QoS in IoT

QoS at each level of the IoT construction is considered as an essential component in security-based applications. Establishment of human needs in the field of automated driving vehicles is severely influenced by the disturbance in layers of the physical sensor system. To have a proper balance and to avoid delayed responses, it is very much essential to implement the QoS-based approaches. This analysis is very much important to point out the needy regions that require more attention [3].

At the point when there is an appropriate number of perceived methodology in the analysis of several divisions of the IoT layers in determining the transaction of data. In a healthcare application, [3] the data to be transferred from lower to higher layers. To achieve service-level agreement, effectively, we need to minimize the delay or raise an alert [9].

There are various different QoS factors, be that as it may, other than postponement, for example, security and dependability which clients might need to consider while asking for an administration. For instance, a client may ask for a wind sensor in a specific area no sweat of utilization however will acknowledge an all-encompassing deferral.

3.1 Data Extraction

The heterogeneity of gadget limits the QoS need of various applications that incited two or three specialist examining the execution of conventional layered responsible for this condition and rather have proposed the utilization of cross-layer strategies [10]. For this framework, we depict of the articles which use to investigate the several layers which encase the system. This reveals us to design which consists of one layer, for instance, the strategy of connecting the system layer with cross-layer for analyzing the system layers and other physical connections.

3.2 Layers in IOT Architecture

For highly critical applications, particularly in domains, for example, health care, we would hope to see an E2E focus on the QoS approaches, from the device to the user to guarantee well-being, the same number of these applications are life critical and a difference in any layer could cause delays and errors [6].

Physical Sensor: This layer leads as a door from which the real sensors. The values are estimated dependent on their prerequisite [11].

Deployment: How the nodes are really deployed in the zone? For example, in healthcare monitoring system, different sensors are deployed in the human body for continuous monitoring [12].

Physical Layer: This layer go about as a starting session, where the information can be send and get by defining the links and physical angles [13].

Data-Link Layer: The link layer helps in transferring the information from the physical layer toward network layer. The data packets are transmitted into bits which additionally figures out how to transmit mistake-free information by controlling stream control and error control.

Network layer: The network layer is vital for switching and routing and in addition, configuring the IP address too. Congestion control and packet sequencing are done in network layers [11].

Application Layer: The application layer gives access to the network benefits and in addition, mistake handling and information stream over the network.

Middleware: A technique to give access to diverse possessions and sustain interoperability within different applications [14].

Cloud: To store, oversee, and process information dependent on demand, *N* number of networks are connected in remote servers [11].

3.3 Enabling QoS in the Internet of Things

The service models can be easily characterized and designed with the application of the Internet of things (IoT); the fundamental quality of service (QoS) and its implementation in various IoT applications can improve the needs in various fields of services. Nevertheless, the QoS terms can be also be applied in the wireless sensor networks (WSN) as it is the important component in the Internet of things (IoT) [15]. Hence, the integrations of WSNs and applications in the IoT give a clear idea to fill the blank spaces in the arrangement of QoS. The requirement of the QoS also can be accomplished by servicing the various models of IoT to disclose the various categories of the IoT applications [16].

IoT layer	Cross-layer QoS parameters	QoS parameters and QoS metrics
Application layer	IoT exposure	Service time relies upon accessibility
		Administration delay
		Service accuracy dependent on load and need
		Data exactness
		Cost dependent on system arrangement and administration utilization
		Maximum resources possible per unit cost and punishments with respect to benefit corruption
		Based on fault tolerance
	Responding time	Service performance cost, load, and reliability
	Energy level	Data transmission, delay, packet loss rate, jitter
Network layer	Utilization/effectiveness	Deployment of network resources
		Sending networks(life time)
		Unwavering quality, throughput, real-time
		Examining parameters
Physical layer	Channel condition, power	Time management and synchronization
		Area/portability
	Transmission and modulation	Detecting and in-citation coverage

3.4 QoS Implementation in IoT

Most by far of the QoS usage use distinctive calculations and utilitarian modules for accomplishing QoS in IoT. The order of the practical modules, for instance, versatile and non-versatile modules in various layers for accomplishing QoS upgrade. To characterize the request of IoT using modules in constructing a crosslayer in the QoS System with adaptability for the basic parameters of QoS in IoT [8].

1. QoS-oriented sensor selection in IoT System concentrated on middle ware, benefit service-oriented computing by utilizing WuKong. The level of programming abstraction for the clients just needs to concentrate on their sensible plan of an application [17]. They defined the administration matchmaking problem as a maximum weighted bipartite matching problem, and analyzed the execution between greedy matching and integer linear programming (ILP) solution. They find that ILP solution is ideal yet tedious and may not be adaptable in expansive-scale IoT frameworks.
2. Trust-based QoS Routing Algorithm for Wireless Sensor Networks: In view of three variables: Energy capable, packer power, and QoS constraints are considered to alter the LEACH protocol. The calculation attempts to discover the course with greatest extraordinary vitality and trustworthiness so as to boost arrange misuse and enhance its execution. Initially, controlling the vitality and inclusion scale is adjusted to place the balance in load over the period of group head determination. Besides, trust assessment instrument is planned to expand the unwavering quality of the system in the phase of node clustering. At long last, in the time of data transmission, check and ACK system approve the information transmission [17].
3. **Quantitative Evaluation of QoS Prediction in IoT**: This Paper Proposes

   (i) A traffic flow management core value, which sort out Machine Type Communication (MTC) traffic flows network assets sharing inside Evolved Packet System (EPS) [4].
   (ii) An entrance component as a Wireless Sensor Network (WSN) entryway for giving an overlay get to channel between the Machine Type Devices (MTDs) and EPS.
   (iii) It tends to the impact and connection in the heterogeneity of uses, administrations of terminal devices and QoS issues among them. This work conquers the issues of network asset starvation by controlling drop of network execution. The plan is approved through reproduction, which shows the proposed traffic flow management arrangement beats the present traffic strategy [18].

4. **QoS by Priority Routing in IOT**: It is a framework which utilizes another encrusted network architecture approach for sudden and capricious extensive-scale changes, normally pervasive in Internet of things (IoT). The pictures or video information caught must be steered immediately. This is a tempestuous

assignment in extensive scale, by utilizing a specialists in different wireless networks are steered relies upon the need. The extent of this investigation is to stay away from deferral, obstruction amid exchange of information from different wireless networks [8].

This examination manages (I) impediment free information exchange under many limited networks (ii) low preparing overhead (iii) taking care of practical disappointment situations, organized traffic in an adaptable way (iv) robust to both topology disappointments and traffic varieties.

5. **The** IoT service selection scheme is a model to select the suitable service that satisfies a client's requirements. This scheme considers device, resource, and service. To dynamically aggregate individual QoS ratings and select physical services, the IoTSS scheme designs a physical service selection method that considers a user preference and an absolute supremacy correlation among the physical services [8].
6. **The Approximate Dynamic Programming-based Prediction (ADPP) scheme** is an assessment approach utilizing prediction systems to acquire precise QoS. Not at all like the traditional QoS prediction approaches, the ADPP scheme is acknowledged by fusing an approximate dynamic programming-based online parameter tuning technique into the QoS prediction approach [19].
7. The SQoSS scheme is a layered QoS scheduling scheme for service-oriented IoT. The SQoSS scheme investigates ideal QoS-aware service arrangement utilizing the information of every part service.
8. **The Intelligent Decision-Making Service (IDMS) scheme** builds a setting-oriented QoS demonstrate as indicated by the analytical hierarchy process (AHP). Utilizing this hierarchical clustering algorithm, the IDMS scheme can impact intelligent decisions by thinking about the clients' input. The prior investigation has pulled in noteworthy thoughtfulness regarding effectively take care of the QoS control issue.
9. In view of the Markov amusement demonstrate, the proposed plan can reasonably allot IoT resources while boosting system execution. In multi-operator situations, a diversion theory approach can give a viable decision-making system for asset allotment issues. To check the outcomes, we play out a reenactment and affirm that the proposed scheme can accomplish enhanced framework execution contrasted with existing schemes. [19]

4 Applications and Service Models in the IoT

Various applications have different prerequisites in necessities of QoS. For every area of use, in other words, transportation, logistics, healthcare checking, smart condition, personal and social and inventive, sort every explicit application in one of the three administrations dependent on their characteristics.

4.1 Transportation and Logistic

IoT is relied upon to assume a key job as a developing innovation in the territory of transportation and logistics. In coordination's, RFID monitors containers, beds, and crates. However, IoT gadgets produce huge volume of information on a standard premise. Hence, controlling information examination empowers undertakings to build bits of knowledge from the enormous measures of information created through IoT advances. Checking ecological parameters occasionally gathered information and this manner is not intuitive. Assisted driving and increased maps likewise done on powerful way [9].

4.2 Health Care

Tracking the patients in reality is critical. Different sensor data's are created through well-being checking electronic gadgets. In this manner, applying information investigation to information assembled from lethal screens, electrocardiograms, temperature screens, or blood glucose level screens can help human services stars to successfully assess the physical conditions of patients. Furthermore, information examination enables social protection specialists to dissect veritable ailments in their at a beginning time stages to encourage save lives. In addition, security of patients is ensured based on the clinical idea and efficient fast examination. What's more, we can enhance consumer loyalty, securing, and maintenance [1].

The author describes about the health care system based on IoT. A framework is proposed to make a continuous brilliant remote-observing and ready framework for patients, utilizing IoT and cloud-based capacity and investigation of patient's well-being data. The framework is productive in working up a social insurance checking model with a fast cautioning framework utilizing GSM alarms and GPS-based area following. Different kinds of information investigation instruments can be conveyed to acquire wide learning on the information that have been gotten which can help in research and in giving moment care to the patient. Further, the framework can be refreshed considering different parameters, for example, effectiveness, enhancement, quality of service (QoS) [7].

4.3 Smart Environments

The quick growth in the populace thickness in urban communities demands that administrations and an infrastructure be given to assemble the requirements of city populace. In this way, there has been an extend in the demand for embedded devices, such as sensors, actuators, and smartphones, most critical to impressive business potential for the new time of the Internet of things (IoT), in which all devices are fit for interconnects and communicate with one another over the Internet. Accordingly, Internet advances gives a method for incorporate and shares a typical correspondence medium [9].

5 Conclusion

The improvement in the fields of smart and sensor devices has been noticed over decades. The combination of IoT and analytical big data is right now at a trend where data to be prepared, modification in data, and investigating vast volume of data at a high event are fundamental. Moreover, we investigated the space by examining different open doors achieved by data analytics in the IoT worldview. For viable data transmission, IoT is implemented for the administration of QoS by enhancing the asset usage. At long last, the feasibility of expansion in with respect to QoS-IoT towards ordered distinctive IoT applications utilization. QoS models plans are inferred and characterized by different research networks. Scholarly associations are in light of the cautious examination and comprehension of administration segments, empowering innovations, message/information characterization, application space territories, and collaborations between every one of these modules/segments.

References

1. Yang P, Xu L (2018) The Internet of Things (IoT): informatics methods for IoT-enabled health care. J Biomed Inf 87:154–156
2. Gubbi J, Buyya R, Marusic S, Palaniswami M (2013) Internet of Things (IoT): a vision, architectural elements, and future directions. Future Gener Comput Syst 29:1645–1660
3. Wang J, Wang H (2014) Trust-based QoS routing algorithm for wireless sensor networks. IEEE Xplore
4. Bauer H, Patel M, Veira J The internet of things: sizing up the opportunity. McKinsey
5. Marjani M, Nasaruddin F, Gani A, Karim A, Hashem IAT, Siddiqa A, Yaqoob I (2017) Big IoT data analytics: architecture, opportunities, and open research challenges. IEEE Access
6. Yang P, Amft O, Cao Y et al (2016) Special issue on The Internet of Things (IoT): informatics methods for IoT-enabled health care. J Biomed Inform 63:404–405
7. Safa M, Pandian A, Ramalingam VV Geetha G (2018) Smart tracking e-health system: bringing precocity to internet-of-things based pervasive healthcare systems. Int J Pure Appl Mathe 119(15):1485–1493
8. White G, Nallur V, Clarke S (2017) Quality of service approaches in IoT: a systematic mapping. J Syst Softw
9. Tiainen P (2016) New opportunities in electrical engineering as a result of the emergence of the Internet of Things. Tech. Rep., AaltoDoc, Aalto University
10. Al Nuaimi E et al (2015) Applications of big data to smart cities. J Internet Serv Appl 6:25
11. Atzori L, Iera A, Morabito G (2010) The Internet of Things: a survey. Comput Netw 54(15):27872805
12. Nef MA, Perlepes L, Karagiorgou S, Stamoulis GI, Kikiras PK (20121). Enabling qos in the internet of things. In: CTRQ 2012: the fifth international conference on communication theory, reliability, and quality of service

13. Tsiatsis HV, Mulligan C, Karnouskos S, Avesand S, Boyle D (2014) From machine-to-machine to the internet of things: introduction to a new age of intelligence. Elsevier, Amsterdam, The Netherlands
14. Duan R, Chen X, Xing T (2011) A QoS architecture for IOT. In: proceedings of international conference cyber, physical society and computing (iThings/CPSCom), pp 717–720
15. Gorodov EY Gubarev VV (2013) Analytical review of data visualization methods in application to big data. J Elect Comput Eng Art. no. 969458
16. Tsai C-W (2015) Big data analytics: a survey. J Big Data 2(1):1–32
17. Kambatla K (2014) Trends in big data analytics. J Parallel Distrib Comput 7:2561–2573
18. Evans D (2011) The internet of things: how the next evolution of the internet is changing everything. CISCO white paper 1:14
19. Bellavista P, Cardone G, Corradi A, Foschini L (2013) Convergence of MANET and WSN in IoT urban scenarios. Sens JIEEE 13(10):3558–3567
20. Vithya G, Vinayagasundaram B (2014) QOS by priority routing in internet of things. Res J Appl Sci Eng Technol 8(21):2154–2160

Personality Prediction in Candidates using a Picture Based Test

Aditya Sattiraju, Saumya Roy, and D. Viji

Abstract The HR department always has a tough role to play when it comes to selecting candidates by shortlisting resumes. Personal interviews and group discussions are conducted with an aim to analyze a person's communication skills, reaction time, problem-solving skills, general knowledge, etc. Each applicant has a unique personality while every role in a company demands a certain set of qualities in a candidate to be able to maintain efficiency and productivity in a team-based environment. Different models that are commonly made use of to characterize the personality of an individual are the Big Five model, the Myers–Briggs Type Indicator (MBTI), DiSC assessment, etc. Most of the existing systems make use of a standard questionnaire in order to collect responses from candidates. These questions are meant to reveal a person's strengths and weaknesses along with his or her comfort zones but do not always provide accurate results due to the candidates' choice of responses provided irrespective of their true nature. The assessment of a candidate's traits in the proposed system is done by using the responses recorded for a picture-based test instead of a questionnaire which is then subjected to the OCEAN model to obtain results.

Keywords OCEAN · MBTI · Picture-based test

1 Introduction

In today's age, the sheer population of the number of candidates applying for a certain job opening through both in-campus and lateral entry routes makes it extremely difficult for the interviewers and the HR team to identify a suitable candidate for

A. Sattiraju · S. Roy · D. Viji (✉)
Department of Computer Science, SRM Institute of Science and Technology, Kattankulathur, India
e-mail: vijid@srmist.edu.in

A. Sattiraju
e-mail: sattiraju.aditya@gmail.com

S. Roy
e-mail: roy.saumya21011999@gmail.com

D. J. Hemanth et al. (eds.), *Artificial Intelligence Techniques for Advanced Computing Applications*, Lecture Notes in Networks and Systems 130,
https://doi.org/10.1007/978-981-15-5329-5_41

the post among the lot. Each domain in the company requires a certain type of mental from the employee in order for his or her tasks to be completed in an efficient and productive manner while working in a team-based environment. In a lot of cases, the candidate is hired based on the technical strengths in his resume and is assigned to a domain which he may not be suitable for or comfortable with mentally. This disrupts the balance of the company as there is a steep decrease in efficiency and work satisfaction. It is therefore crucial to identify a candidate's personality traits during the screening process of the recruitment phase. Some of these traits include openness, conscientiousness, extraversion, agreeableness, neuroticism, etc. (OCEAN) [1]. Based on these traits, one can assess the candidate's mental portfolio in an efficient manner in order to hire him for a certain position in the company.

Each company has multiple domains for which recruitment must be done and the employees must be screened beforehand. Each domain requires a certain type of mental and temperament from the employees for their tasks to be completed in an efficient and productive manner in a team-based environment. During the recruitment process, it is quite hard for the HR interviewers to make out the exact personality traits of each candidate due to time constraints. The most common approach employed is to pose a psychological questionnaire whose answers are recorded and assessed. There is however a good chance of the candidates trying to game the test and predict the ideally required answers to certain questions. This causes candidates to be assessed in an inaccurate manner which leads to them being assigned to domains that they are not suitable for which hinders the efficiency of the company. This may lead to a lack of satisfaction in the employees and their respective teams in the workspace which causes them to burn out mentally, further decreasing productivity. It is therefore necessary to introduce and improve psychological screening processes during the recruitment phase.

Identifying a person's traits will enable us to predict his emotional, behavioral, and cognitive patterns as well as their psychological health. From a business perspective, understanding one's psychographics (their personalities, attitudes, and values) can provide us with a significant advantage over basic demographic and persona-based approaches. This system can be used for recruitment processes in numerous industries. It can be used as a screening method for students before they select a stream to pursue post their secondary and senior secondary classes (tenth and twelfth grades, respectively).

2 Literature Survey

In the last decade, due to a large amount of information being circulated on social media platforms, like Facebook, Instagram, and Twitter, several works have started taking advantage of it. The various posts and comments made by users have been used as datasets in order to predict the personality of individuals as people of a similar personality tend to post and discuss similar topics. The analysis is seen to be done using LIWC and SPLICE [2] both of which involve making use of the words used in different posts to help the machine learning algorithm to predict different personalities. LIWC [3] is a text analysis technique which helps us segregate groups of words so that we can categorize them according to their usage. By categorizing them, it can compare their usage with various groups of people who have a varied personality. This technique has shown to have an accuracy of 74.2%.

Several standard algorithms like K-nearest neighbors (KNN), multinomial naïve Bayes [3], support vector machine (SVM), regression tree, multi-tasking regression [4] can be used in order to train the required model for the purpose of prediction. The usage of KNN and naive Bayes [3] gives a range of values which can be plotted and clustered in order to identify various personality groups. The online posts made by users are mostly in words which can be translated into vectors. An automatic personality assessment can be developed by using the word count and the vector space model [5].

Another method used is the implementation of a standard questionnaire and collecting the answers to it. The responses are compared with the existing dataset, and by using the id3 algorithm [6], prediction trees are generated at the end.

A unique approach has been used to predict the personality of the audience using a picture-based test (Rorschach inkblot test) [7]. In this test, the audience is shown a number of pictures (inkblots) in black and white and is asked to answer what they make out of it. The tester notes their response and compares it with similar responses, thus providing the personality using clustering of the data acquired. Rorschach test involves an extensive study of the pictures. The test not only evaluates its results on the answers provided by the test takers but also on their behavior, actions performed, and their way of expression during the answering process. All these factors are combined as people having similar personality traits will behave or react in a similar way when presented with the same picture although the responses may differ in the overall process.

By using methodologies such as CART and classification algorithms, [8] we can simplify the results produced by the datasets. Traits such as extrovertism and introvertism can be inferred from this. Classification algorithms are very useful in determining personalities as every individual has a different combination of personality traits and thus judging them based on a fixed scale will prove to be inefficient.

Experiments have been designed to be executed in a random order, with each of the conditions appearing more than once. The subjects of the experiment are then required to fill in a NEOFFI questionnaire (paper version), [1] consisting of 60 different statements. The statements use the Likert 5-point scale to accept the response of the subject ranging from a degree of agreeableness to disagreeableness. All the 5 OCEAN model [9] dimensions have 12 statements associated with them. After the implementation of multiple models, the accuracy of the method was found to be 65%.

Another work proposed a system which has been divided into multiple modules with the explanation for each module in detail as follows: The initial module will contain the image of a sample handwriting which is taken from the user from an online source. This module is followed by the second which performs preprocessing of the image that is done by using various techniques to smoothen the image and remove any leftover noise for better results. A value is decided which will serve as a threshold for the system. The threshold value of the pixels greater than the selected value is set to 1 in the binary image, and the latter which proves to be lesser than the selected value is set to 0. By performing this activity, the image is converted into the necessary binary image for the implementation of the neural network.

Similar experiments on the collection of data from the PAN CLEF 2015 conference 1 have been carried out [10]. The dataset comprised of 14,166 tweets from 152 different users, with an average of 100 tweets collected per user. The tweets ranged across various topics of discussion containing a vocabulary of 17,369 distinct words in the English language. The grouping in the dataset was done by the user, and values of the five personality traits were labeled for each user under the following labels: openness, extraversion, agreeableness, conscientiousness, and neuroticism (OCEAN model) [1]. The range of each personality trait lies between 0.5 and −0.5, with a 0.5 indicating the presence of the trait and a −0.5 indicating the absence of it.

Another approach had been using the BFI tested score and scaling them to predict personality. The BFI scores used standard deviation and mean value of the Sina Weibo users [11]. The OCEAN model's positive correlation with each other (among the 5 traits) [9] was indicated by the usage of the Pearson correlation coefficient. Neuroticism was the only trait negatively correlated.

Another research was conducted to analyze various personalities from twitter feeds. In this case, the naive Bayes [3] method was made use of to figure out the candidate's personality. The model was created with four factors: choleric, sanguine, phlegmatic, and melancholic. The tweets were analyzed using Linguistic Inquiry and Word Count (LIWC) [2] program, and Waikato Environment for Knowledge Analysis (WEKA) was executed on it.

Similar modeling in order to determine personality has been developed using the Five-Factor Model (FFM) [12] and the LIWC [2] which were used to extract features and annotations from publicly obtainable benchmarks. The model used a semi-supervised regression in order to improve the results of personality prediction. The multivariate models often outperformed the univariate ones, but the differences between the two models were not significant. It was found out that no common features were identified, which performed well on all the social media platforms and datasets.

In one of the papers, machine learning models were examined in order to predict the personality traits of gamers based on the character design chosen prior to the start of the game. The data required to perform this study was obtained by collection using basic survey and research techniques. The total number of participants in the research who responded to this survey was 94, among which most of the players were older than 15 years. Experimental results showed that the proposed model was able to predict the degree of openness, agreeableness, and conscientiousness of the selected game (OCEAN model) and the character's pre-design, with an accuracy of 66.7%.

A unique approach by making use of the multimodal framework was proposed for the prediction of impressions of extrovertism, neuroticism, agreeableness, conscientiousness, and openness (OCEAN model) [1] along with likeability and attractiveness across varying contexts. The research differed from the others as the annotations obtained over time were visual-only and audio-only. These were then compared to their audiovisual annotations. The proposed model was a time-continuous prediction approach [13] which factored in the temporal flow while analyzing relationships instead of treating time as a constant or an individual component. The research showed that the best results were obtained when the audiovisual annotations at a decision level were used to learn regression models. Continuously, it generated data provided better insights and the predictions were dynamic in nature.

A research stated that the signature of a person could be analyzed to obtain crucial information about the thought process and emotional state of the person. This is another form of text analysis [4]. Traits such as honesty and fear could be deduced from the signature analysis. The patterns observed during the process were the appearance of a dot on certain letters, an underscore or a line below the signature, disconnected strokes or curved beginnings, etc. The database consisted of 60 signatures collected from 10 different people. A 500 dpi resolution scanner was used for the process. Each signature was split into 5 parts, left, right, top, bottom, and middle. The personality was predicted by using an artificial neural network [14] along with a structural identification algorithm. The accuracies obtained were in the range of 92–100%. The model was to be used to predict traits in counseling, criminology, and medical science.

2.1 *Inference from the Literature Survey*

The survey papers depict various methods of research conducted on different classification models that are used for predicting one's personality. They contain techniques used for the identification of different traits within a person that are normally hard to detect and stay hidden. The Big Five model, also known as the OCEAN model [1] (Openness, Conscientiousness, Extraversion, Agreeableness, and Neuroticism), has been used the most with respect to an individual's personality prediction over the past few years. The result of employing this test leads to the assignment of one of the five major traits mentioned earlier to a candidate. We have therefore chosen to apply this model in our project due to its reliability, efficiency, and the fact that it has been scientifically approved. The DiSC model on the other hand categorizes one's personality into four traits instead of five, namely Dominance, Influence, Steadiness, and Conscientiousness. The model is easy to understand and accurate when implemented. It is most helpful in situations where utility, application, and interpersonal behavioral change are important, like sales, marketing, and leadership. Optimum results were obtained in both cases with minimal computational effort which reflects on the efficiency of the algorithms in recognition and classification of personality traits in an individual.

2.2 Algorithms Used

MBTI: The MBTI (Myers–Briggs) [6] is an indicating test which has been designed to identify and analyze an individual's personality, his preferences, and strengths. The questionnaire was originally developed by Katherine Briggs and Isabel Myers.

Depending on the answers provided to the questionnaire, people have been identified to have one out of the 16 different personality types. The aim of the questionnaire is to let candidates explore further in order to understand their own personalities which include their individuality, likes and dislikes, strengths and weaknesses, career choices and preferences, etc.

The questionnaire consists of four different scales:

Extraversion–Introversion (E-I)

Extraverts are known to be "outward-turning" and action-oriented. They normally prefer more frequent social interaction and feel energetic after having interacted with others. Introverts are on the other hand known to be "inward-turning" and quite thought-oriented. They normally prefer deep and meaningful conversations and social interactions and feel energetic after spending some time alone with themselves. Everyone exhibits introversion and extraversion to some extent, but most individuals have a certain preference for one or another.

Sensing–Intuition (S-N)

This involves looking at how people collect data and information from their environment. Everyone spends some time sensing and intuiting about various things. According to the Myers–Briggs Personality Type Indicator [6], individuals usually have a tendency to be dominant in a certain area. People who prefer sensing tend to pay a great deal of attention to reality. They focus on their attention on facts and minute details while enjoying being exposed to hands-on experience. The people who prefer intuition generally pay more attention to repetitive things such as patterns. They prefer thinking about the future, and other abstract and diverse theories.

Thinking–Feeling (T-F)

This involves focusing on people's tendencies to make certain decisions based on the date and information that they gather through their intuition. People who prefer pondering and thinking stress more on objective data available along with facts. They are found to be consistent and logical when deciding. Those who prefer sensing and feeling are known to more likely to consider emotions and people when concluding something.

Judging–Perceiving (J-P)

This revolves around how people normally tend to deal with their environment. Those who prefer judging tend to follow structure and make firm and solid decisions. People who tend to perceive things are found to be more open and flexible. These two tendencies are known to interact with the rest of the scales.

Each type is listed at the end after the analysis process through its four-letter code such as ISTP and ISTJ.

Text-based Semantics [4]: Semantics is the linguistic study of meaning in language. The language can be formal, programming, etc. Text mining techniques have become quite common and essential for knowledge discovery with an increase in the volume of digital text. This includes social media, the Web in general, or other internal organizations. Text mining employs certain methods to analyze data (unstructured) and identify patterns that were initially unknown. The data used for mining can vary from simple phrases and sentences to remarks and signatures [9]. There are normally five steps in this process as follows:

Problem identification: This involves the identification and specification of the application objective and scope.

Preprocessing: This step involves preparing the data for pattern extraction. Raw data is generally converted into some organized format which can be used as input for various algorithms used for knowledge extraction.

Post-processing: The extracted knowledge is evaluated in this step. If the knowledge obtained successfully meets the objectives of the process, it can be made available for other users. If the objectives are not reached, then another process cycle must be repeated.

Knowledge usage: This is the last step in the process where the knowledge extracted from the text is used in some application or domain. The efficiency of the knowledge is proportional to the number of people who have made use of it and have obtained satisfactory results in their work.

OCEAN: The OCEAN [1] model, also known as the Big Five personality test, is used to measure the five key dimensions of an individual's personalities. These are Openness, Extraversion/Introversion, Agreeableness, Conscientiousness, and Neuroticism.

Openness: Insight and imagination are some of the major characteristics of this trait. This domain is used to measure one's level of creativity and hunger/desire for knowledge. People normally tend to be creative and bold if they possess this trait. On the other hand, people have the tendency to be conventional or down to earth if they do not possess this trait.

Conscientiousness: This involves looking at the level of care that a person takes in his work and life. This includes organized or disorganized levels, the ability to make, and stick to plans and decisions. People are known to be disciplined and careful when they fall under this category. On the other hand, people who do not possess this trait tend to procrastinate quite a lot and make mistakes in the most basic tasks.

Extraversion/Introversion: This domain is used to measure one's level of sociability. This includes the amount of outgoing and quiet levels, other factors such as drawing energy from crowds, problems in communicating, and working with other people. An extrovert draws his energy from meeting and interacting with people in a social sitting. Such people are active in nature. Introverts are on the other hand quieter in nature. They find their refreshment in spending time alone with themselves rather than attending social events.

Agreeableness: This dimension is used to measure how well an individual gets on with other people. For example, Do you tend to put your needs before others? An agreeable person is often affectionate and compassionate while being trustworthy. Such people often take a keen interest in helping others out or volunteering in pro-social activities. People who are low in agreeableness tend to focus on their own

Table 1 Algorithm accuracy comparison

S. No.	Algorithm/model	Accuracy (%)
1.	Multi structural algorithm	87
2.	Artificial neural network	63
3.	Multimodel prediction	59
4.	Bayesian prediction	84
5.	K-nearest neighbor	60
6.	Support vector machine	61
7.	Multinomial Naïve Bayes	63
8.	OCEAN model	66.7
9.	Sequential minimal optimization (SMO)	74
10.	Bayesian network	71

interests over everything else. The concern toward other proceedings is relatively low.

Neuroticism: This is a domain used to measure the depth of an individual's emotional reactions. For example, Do you react in a neutral manner or negatively when you face bad news, do you obsess and worry about tiny details, etc. Some of the important characteristics are emotional instability, tendency to sulk often, unhappiness, etc. People with high levels of neuroticism are easily irritable and experience anxiety on a frequent basis.

Table 1 shows the different algorithm accuracy percentage. We chose to apply this OCEAN model to the responses of the picture-based test in order to accurately assess the candidate's personality traits.

3 Proposed Work

A candidate's personality is normally assessed by using psychometric questionnaires containing multiple questions/phrases to which the responses are in the form of degrees of relatableness. For example, Do you feel comfortable while sharing your thoughts and opinions in front of others? The options vary from Strongly Agree to Strongly Disagree. In a lot of cases, the required answers to certain questions can be predicted by the candidates taking the test. In order to avoid this, we have chosen to implement a picture-based test. The candidates are shown 5–10 different pictures, each of which can be interpreted in numerous ways. They are then required to note down the first three adjectives that they can come up with to describe the picture when they look at it. The responses are recorded and subjected to the OCEAN [4] model test in order to analyze their personality traits. The dataset that we are using to crosscheck and compare the user input consists of responses recorded for each of the pictures from people of the age group between 15 and 25 years. Our current target audience is set at high school and undergraduate students appearing for various interviews. The responses may differ for people belonging to other age groups. The final result will be the assignment of one of the 5 major personality traits from the OCEAN model [9] to the candidate. The reason behind choosing to use this model is the reliability and efficiency of said model. Various studies and researches that have been carried out on the Big Five personality traits have determined that the results obtained are universal. The studies have been focused on people hailing from over 50 diverse cultures. Scientists after further refining and making progress in the above study have arrived at a conclusion that these traits may also have biological roots. The five traits are known to normally affect or influence how individuals behave or react in different environments and situations that they find themselves in. Their behavior is also dependent on unpredictable, situational variables, but the major traits that underlie are mostly responsible for the output/response. The biggest application of the OCEAN model has been seen in professional workplaces. Most studies have focused on predicting a person's performance and social behaviors through the Big Five personality traits. Managers can work on developing and improving workplace cultures in order to improve the relationships between the workers and build trust by having a deeper understanding of all five traits.

To make the assessment foolproof, another module will be added which will ask a few words from the test taker to describe the picture. It has been found [2] that people with similar personalities tend to use a common group of words which can then be clustered into groups in order to properly identify their personality. The vocabulary of individuals tends to shape their personality as the phrases and objects used in speech are molded in order to provide a mental image to the people listening to them. The framework of the proposed work consists of:

Data Extraction: This includes the identification and collection of the dataset deemed suitable for application in the model. The dataset for the required adjectives for each picture is collected from people between the age group of 15–25 through the help of online Google forms. The dataset of pictures has been collected online under the guidance of Dr. Harini Atturu (Consultant Psychiatrist at Care Hospitals, Gachibowli, Hyderabad).

Preprocessing of data: Preprocessing is done in order to filter out the outliers. Data preprocessing helps in data cleaning and transformation. By doing this, we improve the efficiency of the system. The steps involved in data preprocessing are data cleaning, case-conversion, POS tagging, removal of stop-words and punctuations, etc. The adjectives stored in the database need to be error free.

Feature Extraction and Selection: After the data has been preprocessed, annotations, and features which are found to be relevant for the prediction are selected and extracted. The words used by the individuals are grouped according to their meaning, intensity, and frequency of usage. Adjectives with a lower frequency are not discarded, and the words are still used and updated in the order of priority based on the change in frequency of usage.

Classification: Classification involves the testing, training, and prediction of the model. After these processes have been performed, a dataset which is not seen or known before is handed over for prediction. Classification is done on the new dataset and the personality is predicted based on the inputs of the dataset followed by its results.

4 Implementation

The implementation of the project is required to be done in multiple stages. The user interface provided to the candidates consists of a form which collects and stores his personal details. Post-completion of the candidates is directed to the picture-based test. The users are provided with 5–10 different pictures in a sequential manner which can be interpreted in multiple ways. They are required to enter the first three adjectives that they come up with when they look at any given picture on the screen. The candidates are advised not to dwell too much on their responses and proceed further once they finish describing each picture. The picture dataset has been collected online under the guidance of a consultant psychiatrist, Dr. Harini Atturu. The training dataset for the adjectives used to describe each picture has been collected through online Google forms. The participants who contributed to this dataset belong to the age group of 15–25 years. The current target audience is limited to this age group. The user input provided by the candidates is cross-referenced with the existing training dataset to find similarities or patterns. Common adjectives used to describe a given picture are clustered together to make it easier to determine the user's traits. The OCEAN model [1] is then applied to the candidate responses stored in order to analyze his or her personality traits accurately. The five main traits that come under the OCEAN model [9] are Openness, Conscientiousness, Extraversion/Introversion, Agreeableness, and Neuroticism. Each of us tends to possess at least one of these five traits based on both our nature and nurture. The knowledge of knowing which category a candidate falls under can be used in a diverse manner. The candidate's mentality and temperament may lean toward a certain domain in a company which makes it easier for the interviewers to make a decision during the hiring process. The training dataset of adjectives is meant to be updated in a continuous manner based on the number of people who take the test. Newer words may be used to describe a given picture which showcases the creativity and thought process of a candidate. Similarly, the picture dataset needs to be updated periodically in order to avoid repetition in the test.

5 Conclusion

The innovation aspect involved in this process is the introduction of a picture-based test instead of a regular questionnaire in order to assess a candidate's personality. The initial target audience has been limited to the age group of 15–25 years. This includes college and high school students who are appearing for various interviews.

The results obtained from the OCEAN model is used to determine certain personality traits that the candidate possesses. Using these results to try and understand, one's psychographics (their personalities, attitudes, and values) can provide us with a significant advantage over basic demographic and persona-based approaches. The dataset consisting of the pictures used for the initial test can be increased over a period of time in order to avoid familiarity and repetition. The dataset consisting of the adjectives used to describe each of the pictures is bound to be expanded in time. This is a progressive step toward perceiving any given picture in highly creative and different ways. With the increase in the number of candidates taking the test, there is a proportional increase in the dataset used for reference which leads to the refinement of the accuracy that the test is able to provide. The target audience can be further diversified and increased as and when other age groups are involved.

References

1. Küster L, Trahms C, Voigt-Antons JN (2018) Predicting personality traits from touchscreen based interactions. 13 September 2018 IEEE Conference Paper
2. Tadesse MM, Lin H, Xu B, Yang L (2018) Personality predictions based on user behaviour on the Facebook social media platform. IEEE Access 17 Oct 2018
3. Pratama BY, Sarno R (2015) Personality classification based on Twitter. In: 2015 international conference on data and software engineering
4. Dandannavar PS, Mangalwede SR, Kulkarni PM (2018) Social media text—a source for personality. 2018 IEEE Conference Paper
5. Hassanein M, Hussein W, Rady S, Gharib TF (2019) Prediction of personality traits from social media using text semantics. 15 February 2019 IEEE Conference Paper
6. Abdulrahman R, Alsaedi R, AlSobeihy M (2018) Automated student to major allocation based on personality. 23 August 2018 IEEE Conference Paper
7. Dewangan RL (2017) A study of social desirable biasness in rorschach inkblot test. 05 October 2017 Research Gate Article
8. Ge L, Tang H, Zhou Q, Tang Y, Lang J (2016) Classification algorithm to predict students' extraversion-introversion. 24 November 2016 IEEE Conference Paper
9. Suryapranata LK, Kusuma GP, Heryadi Y, Abbas BS, Ahmad AS (2017) Personality trait prediction based on game character design using machine learning approach. In: International conference on innovative and creative information technology (ICITech)
10. Moreno DR, Gomez JC, Almanza-Ojeda DL (2019) Prediction of personality traits in Twitter users with latent features. 25 March 2019 IEEE Conference Paper
11. Xue D, Hong Z, Guo S, Gao L, Wu L, Zheng J, Zhao N (2017) Personality recognition on social media with label distribution learning. IEEE Journal 5:13478–13488
12. Tareaf RB, Berger P, Hennig P, Meinel C (2018) Personality exploration system for online social networks: Facebook brands as a use case. In: 2018 IEEE/WIC/ACM international conference on web intelligence (WI)
13. Celiktutan O, Gunes H (2017) Automatic prediction of impressions in time and across varying context: personality, attractiveness and likeability. In: IEEE transactions on affective computing, vol 8(1), January-March 2017
14. Lokhande VR, Gawali BW (2017) Analysis of signature for the prediction of personality traits. In: 2017 1st international conference on intelligent systems and information management (ICISIM)

Security Enhancement and Deduplication Using Zeus Algorithm Cloud

Abhishek Kumar, S. Ravishankar, and D. Viji

Abstract Data deduplication method allows the cloud users to manipulate their cloud storage efficiently by avoiding the storage of certain repeated data having the same bandwidth. To ensure that the data is secure and confidential, and the data is stored by encrypting it. We can use various encryption types like the RSA, DES, and AES. Convergent encryption algorithm is commonly used to retrieve storage capacity and lower data upload throughput and yet poses two challenges. The first problem is that they might be unsafe for statistical attack. The other thing is that they have convergent key management problems. The current paper assists in the usage of key-sharing approach which is verified by the proof of ownership to rectify these problems. As in only the initial data uploader, the holder encrypts the data with a randomly selected convergent key and then disperses the key in the cloud, and the key can only be accessed by users who have the reported data. In the existing framework, we found out that in the convergent encryption, the key of any given data can be generated by anyone in a deterministic way. An attacker can generate a key from plaintext encrypt the plain text and check if the resulting cipher text is already existing or not. Thus to overcome the issues, we propose the much more secure Advanced Encryption Standard (AES) without the third-party administrator (TPA) and to reduce the burden of computational overhead on the admin. We have used the Zeus algorithm to stop the duplicate data string in the database. Zeus algorithm offers better deduplication process and work to reduce the computational overhead of the admin. Analysis shows that this design is more effective, and the proposed security model remains stable.

A. Kumar · S. Ravishankar · D. Viji (✉)
Department of Computer Science, SRM Institute of Science and Technology, Kattankulathur, India
e-mail: vijid@srmist.edu.in

A. Kumar
e-mail: abhishek.k1308@gmail.com

S. Ravishankar
e-mail: ravi1998edge@gmail.com

D. J. Hemanth et al. (eds.), *Artificial Intelligence Techniques for Advanced Computing Applications*, Lecture Notes in Networks and Systems 130,
https://doi.org/10.1007/978-981-15-5329-5_42

Keywords Data deduplication · Cloud security · Cloud storage · Advance encryption standard · Zeus algorithm

1 Introduction

Cloud storage facilitates for storing and monitoring data and is also viewed as an easy way to manage volumes of data. With the proliferation of cloud storage, a huge volume of data is stored repeatedly. To save storage space and data management costs, the technique of deduplication is used to remove unnecessary information through physically providing just one instance of the very same content. However, this involves a high risk of loss of data confidentiality, data theft, integrity, and confidentiality on cloud servers, secure data transmission, integrity verification without much overhead or computing costs, access rights management, and security while sharing files with other users. We use cryptographic techniques to enhance and provide data protection in cloud computing: Cryptography is a way of using mathematics to encrypt and decrypt data. Once the information gets encrypted, it can be stored on an unprotected network (such as the Internet) or on the cloud storage system, so that none other than the intended receiver can read it.

There are several symmetric key algorithms that can be used such as AES, DES, 3DES, and asymmetric key algorithms such as RSA, Diffie–Hellman data encryption, and decryption algorithms. Many other encryption algorithms and deduplication techniques have been used earlier to improve security of data on the cloud like message-locked encryption (MLE) and interactive message-locked encryption [1], homomorphism encryption and recently used convergent encryption (CE) [2]. All the techniques are implemented with the help of a third party which basically acts as an admin and manages all data being uploaded or shared by the users. But being too dependent on third-party auditor (TPA) has issues which result in failure of the whole system [1–3]. Some unusual and unwanted behavior in TPA may cause the entire cloud system to decrease and further decrease system performance. TPA also uses additional hardware and cryptographic coprocessors that are expensive and require maintenance.

Currently, as cloud storage usage is being used vastly worldwide many companies and organizations have started operating their storage to a cloud environment; thus, removal of duplicate data plays a vital role than reducing storage expenses [4]. Coherent data deduplication in combination with cloud-based storage structure is introducing new possibilities for doing more with stored data. In existing framework, if a user uploads a file, it will be stored n number of times as a duplicate data. Due to this duplicate data entry, the normal storage capacity won't be enough to store the data and we need to buy the extra storage for storing the data. It won't allow the data which is being uploaded, and it won't occupy more space for storing the data.

A number of security frameworks which deal with secure deduplication and secure key management are provided in this survey. Security frameworks are evaluated based on their performance and security features. Overhead deduction includes overhead computation, overhead storage, and overhead communication involving the system. Scalability is the ability of the deduplication system to work with the increased file size and increased demand. Reliability is the ability of the system to be consistent with the repeated deduplication operation and helps to avoid data loss.

Cloud Architecture

Deduplication can be performed on a range of server systems, such as multi-cloud single site and hybrid cloud. Convergent keys and proof of ownership (PoW) [5] and [6] processes can be used in one cloud infrastructure to secure data from data theft and data breaches [2]. This is the most frequent approach followed by many of the commercial CSPs. Multi-cloud architecture splits the file into several shares and is spread through several cloud servers to avoid failure in a single storage architecture. To avoid disaster recovery in multi-cloud, optimized scheduling strategies can be applied to achieve data reliability and short recovery time. In hybrid cloud architecture, authorized data deduplication is performed. In this method, outsourced data is stored in public cloud and all data management operations are handled in private cloud. The user can perform duplicate check, if the user meets the specific privileges. Secure deduplication can be done by encrypting the user file with various privilege keys.

Threats:

There are endless problems in the cloud like trust, security, and privacy of the outsourced info. The data controller outsources the data to the cloud service company, which in turn leads to security risks over the safety of the outsourced data. Deduplication strategies in cloud storage include security risks and the likelihood of disclosing information about the file material contained in cloud storage. The security concerns cover the privacy of the data stored in the cloud and threats to security from active and passive foes. According to the adversaries creating the secret link, a secure communication mechanism is required between the cloud and the customer. This can be done by securing certain routing protocols to the communication channel.

2 Related Work

Convergent encryption (CE) was first introduced by Douceur [7] used in [2] deduplication that will save cloud space and lower data upload and throughput retrieval still it has two major issues. This paper overcame the problems in [8] but still two main issues are convergent encryption is easily attacked by an offline brute force when selecting data from the data sets. Other problem is the convergent key handling issues. CE ensures that each user has an individual master key which will encrypt their convergent key and then store them in the cloud storage that is different for all users that store the exact key for the same files. If we increase the number of customers and files stored in cloud, then number of convergent key also increases. Therefore, many keys are stored repeatedly and key management issues arise. In this current structure, only the data's initial uploader encodes the file with random key and after that, it uploads the key on the hosting cloud; therefore, the key can be retrieved only by users who own data.

Another paper shows the problem of secure information storage, which offers deduplication [1]. This paper offers anonymity for relevant communications, depending on the parameters of the program. They expanded the previous algorithm for message-locked encryption (MLE) to interactive message-locked encryption (iMLE) where various protocols are available for upload and download. This scheme provides security for data that is not only connected to each other but also allows the device parameters to be relied on. It shows that communication is not a realistic assumption, because current systems of deduplication are already interactive.

Deduplication is efficacious if many customers store exactly a certain data in the cloud; however, it causes security and ownership problems. Proof of ownership method allows all owners of the same data on cloud to prove to the cloud storage provider that they securely and effectively control the data. Nevertheless, several clients encrypt their data to provide protection before distributing it to the cloud storage, but this halter deduplication due to the variance property of numerous cryptographic algorithms. Many deduplication strategies have been suggested to resolve the issue by granting the exact same key to each data owner on the cloud for the similar data. But most solutions suffer from security flaws because they do not search for alterations in the initial data that is changed in the cloud storage service. This gives the cloud provider control of all data including when data access changes continuously through the use of secure randomized convergent encryption algorithm and collective key distribution. It avoids data loss not only to the user suspended, but to the cloud server too. This scheme safeguards data protection against any cloud incoherence breach [6]. Their results indicate that the current proposal is reliable from the past proposed models, while the overhead is negligible in computational terms.

With many uses of cloud services, which use artificial intelligence analysis, other applications such as image recognition systems have been used. Classification service, a classifier owner who acts as a service provider, lays down a rule that allows a user to request the analysis of their data. But, such an owner has to be online all the time and be equipped with bandwidth and resources for computing. The owner may outsource the service to another provider but there is an issue that protects data privacy and classifier privacy. In this paper, it proposes a new scheme for a classifier proprietor to delegate a remote server to provide users with a privacy classification service [9]. They also developed efficient classification protocols for two different classifiers in the proposed scheme. They have implemented the demo of the plan and conducted practicals. The tests show essentially the scheme is feasible analysis of their data.

Data confidentiality and storage capacity are two requirements for data cloud storage. Techniques used in this paper are proof of data possession (PDP) and proof of retrievability (POR) that ensure data integrity on the cloud [4]. Proof of ownership (POW) increases storage capacity as unnecessary duplicate data is safely removed on the cloud [5, 6]. Nevertheless, the two methods used together provide data integrity, storage capacity, contribute to non-trivial metadata duplication which contradicts POW's goals. Recent attempts at this issue show that the cost of computing and communication has risen and has also been shown to be unsecured. This provides a new approach for enabling effective and safe auditing of data integrity with cloud storage deduplication, and this problem was solved in this paper which uses techniques such as polynomial based computing tags and homomorphic linear computations [4]. These methods allow all the files and their respective authentication tags to be deduplicated from the files. Both are achieved with data integrity and storage deduplication. The proposed scheme also features the user uses constant real-time communication and computational cost. The new scheme therefore outperforms current POR and PDP schemes, thus offering deduplication.

As users transfer data to cloud service providers, several incidents involving data breaches rendering both sides encryption a popular requirement. It presents a great idea which differentiates data by popularity. They developed an encryption model based on their research, which guarantees the protection of uncommon data and gives vulnerable security and increased storage and throughput benefits for open files [10]. Therefore, deduplication can be beneficial for open files, thus ensuring that encryption protects controversial data. This scheme shows that it is secure in random oracle mode under the symmetric external decisional Diffie–Hellman assumption.

The change has been made from the paper [10], and the importance of data can be subdivided into many parts of levels of security. Use of symbolic logic uses safe encoding to protect all unwanted personal data in order to recognize protection and use convergent encoding to encrypt big data to perform deduplication of cipher text. If data popularity shifts, there is a big storage issue. There are many strategies for accessible data and uncommon data, and when the popularity of the data changes, the cipher text also changes. They are following symmetric cryptosystem thresholds in this scheme to provide cipher text protection for information and data deduplication [11]. Cloud users do not have to create new authentication by themselves and be active when the popularity of the data shifts.

Use of ABSC providing privacy and unverified access control of sensitive information and is much more successful than conventional "encrypt-then-sign" or "sign-then-encrypt" methods [12]. This uses encryption based on attributes and signature logic based on attributes. In this, various individuals can work in isolation and provide them with secret key without knowing the users credentials. The suggested scheme acknowledges the protection in the current system and supports other functional features such as authentication, covertness, inviolable, and secure verification.

3 Algorithm Used

Advanced Encryption Standard: It is widely used symmetric block cipher algorithm. It handles key sizes such as AES 128,192,256, and each of these ciphers has 128-bit block size. AES is capable of protecting sensitive data from attackers and does not enable them to hack through encrypted data relative to other algorithms. It is based on two common techniques also known as the substitution and permutation network (SPN) for encrypting and decrypting data. SPN is a sequence of mathematical operations performed in block cipher algorithms.

Zeus algorithm: Zeus possesses two important features. At first, it is able to identify and sort the requests common to multiple applications and only execute them once required, sharing the results among resource-saving applications. Second, its architectural solution using edge computing makes Zeus scalable in relation to the number of compatible VNs and applications, and suitable for processing delay-sensitive applications, only executing them if needed, sharing the results between them.

4 Proposed System

In our proposed system, we have developed a user interface, where different users can registers and upload files which are encrypted using Advance Encryption Standard algorithm on to the server. We use this algorithm as the main aim is to improve the security issues that existed in CE algorithm [2]. Other registered users can see the uploaded file name and can request for a particular file by sending the request to the file owner. Once the request is accepted by the owner, the requested user can download the file by decryption using secret key onto their system. During the file upload, the file is checked for deduplication using Zeus algorithm which also works to reduce the computational overhead of the admin.

4.1 Deduplication Check Using Zeus Algorithm

When the user uploads a file, UI checks for deduplication using Zeus algorithm. Figure 1 shows Zeus algorithm, and the user tries to upload to a cloud a file. At first, the file is subdivided into chunks. Due to the multiple upload of chunks in Zeus, the user verifies if the amount of parts is equal or not and if the amount of the last part is exactly similar to the already given part size. If the amount is not the same then we create a bit sequence of suitable range and add this bit sequence to the file at the end. Then, the similarity is checked on $h(ci)$, $h(ci + 1)i$, (where i belongs to $1, 2, 3\ldots, n$), on the two of chunks as shown is step 8. After the cloud receives $h(ci)$, $h(ci + 1)i$, this checks if the parts involved are correct or not as in step 9. In other scenarios, the output is returned as 2 to the client. Users access $c1/c2$ or specifically upload $c1$ and $c2$ to the cloud steps 13–20 based on the response given

Using Zeus algorithm, we can easily check for duplicate data on the server. The Zeus algorithm is already in-built in our proposed system so that the user workload is reduced as once the users upload the file he/she will directly get to know if the file uploaded is duplicate or not. Also, the communication cost is also minimal.

```
Input: file f with chunk size φ, and dirty chunk list L
user partitions f into chunks c1, . . . , cn
user sets n* = n
if bit length |cn| 6= φ
user performs padding to cn
if n is odd
user picks random chunk cn+1 and n* = n + 1
for i ∈ {1, 3, . . . , n* − 1}
user performs duplicate check on hh(ci), h(ci+1)i
if h(ci) ∈ L/ and h(ci+1) ∈ L/
cloud replies 1 or 2 according to Table 6
else
cloud replies dc response 2
if user receives dc response 1
user uploads ci ⊕ ci+1 to the cloud
if cloud does not receive ci ⊕ ci+1
L = L ∪ {ci , ci+1}
else
user uploads ci and ci+1 to the cloud
if cloud does not receive ci and ci+1
L = L ∪ {ci , ci+1}
```

Fig. 1 Zeus process

4.2 *File Encryption and Decryption Using Advance Encryption Standard Algorithm*

Operations in AES:

AES is an algorithm that is iterative. It has a set of multiple processes involving substitutions and permutations. AES does all of its computations on bytes instead of bits. Hence, this algorithm handles a plaintext of 128 bits as 16 bytes. Such 16 bytes are structured as a matrix, in four by four.

Encryption Process:

AES encryption involves many series of rounds linked to each other and each round has four sub-processes: Byte Substitution (SubBytes), Shiftrows, Mix-Columns, and Addroundkey. When the file is uploaded onto the cloud, using a random generated key, using this key the file is encrypted and using the AES algorithm which includes many rounds each having four sub-processes each. When the file is encrypted, it is then uploaded onto the server.

Decryption Process:

The decryption of cipher text in AES is the same as process of encryption but in the opposite order. There, too, each round has four reverse order processes—Add round key, Mix columns, Shift rows, and Byte substitution. Since each sub-process is in reverse order in each round, it is necessary to implement the encryption and decryption process separately, although they are very close to each other. When the owner accepts the request of user to download a file, then the user will get secret key from which the user can decrypt the file and download the file onto their system.

5 Result and Discussion

We analyzed and evaluated encryption and decryption process of different cryptography algorithms based on different sizes of files as input.

From Tables 1 and 2, we can evaluate that throughput is better for AES than other three algorithms, and the average time is also better for AES than DES, 3DES, and RC2 algorithms.

Zeus algorithm reduces the overhead of user and overall computation cost as it requires minimal hardware and software requirements. Zeus delivers a better understanding of privacy because deduplication is performed on the basis of chunks of a single file, so verification or presence of duplicate files is more reliable and quicker than other deduplication techniques. Therefore, the combination of AES encryption

Table 1 Performance of different encryption algorithms

Input size (KB)	DES	3DES	RC2	AES
50	29	54	57	56
100	83	81	91	90
250	116	112	122	113
900	240	300	268	258
5345	1296	1466	1570	1237
Average time (s)	352.8	402.6	421.6	350.8
Throughput (MB/s)	4.01	3.45	3.246	4.174

Table 2 Performance of different decryption algorithms

Input size (KB)	DES	3DES	RC2	AES
50	50	53	65	63
100	57	57	90	60
250	73	78	96	77
900	152	171	183	171
5345	783	835	904	655
Average time (s)	223	238.5	267.6	205.2
Throughput (MB/s)	6.347	5.665	4.985	6.452

and decryption algorithm and Zeus algorithm deduplication check enhance our system by reducing computation overhead, improving security, and also, our system is not dependent on any kind third-party auditor.

Therefore, the combination of AES encryption and decryption algorithms and Zeus algorithm deduplication check enhances our system by reducing computation overhead, improving security, and also, our system is not dependent on any kind third-party auditor.

6 Conclusion

In the whole paper, we introduced an efficient system of deduplication focused on Advanced Encryption Standard algorithm which is one of the powerful algorithms which is adopted on hardware and software worldwide. Data is encrypted only by the initial uploader. The user that wants to download the data has to request the admin. Only after the admin accepts the request, the user gets a key to decrypt the data. We have got rid of the trusted entity to keep the data safe from internal attacks. Key issues with the cloud storage are data protection and effective storage, and our system works to improve the security of the data by using the AES algorithm which is the most efficient compared to other algorithms. And for storage issues, we have used the Zeus algorithm, which improves the efficiency of the deduplication and reduces the computational overhead of the admin.

References

1. Bellare M, Keelveedhi S (2015) Interactive message-locked encryption and secure deduplication. In International Workshop on Public Key Cryptography-PKC
2. Wanga L, Wanga B, Songa W, Zhili Z (2019) A key-sharing based secure deduplication scheme in cloud storage. Elsevier, Boca Raton
3. Boneh D, Franklin M (2003) Identity-based encryption from the weil pairing, SIAM
4. Yuan J, Yu S (2013) Secure and constant cost public cloud storage auditing with deduplication, IEEE
5. Liu M, Yang C, Jiang Q, Chen X, Ma J, Ren J (2018) Updatable block-level deduplication with dynamic ownership management on encrypted data, IEEE
6. Hur J, Koo D, Shin Y, Kang K (2016) Secure data deduplication with dynamic ownership management in cloud storage, IEEE
7. Douceur JR, Adya A, Bolosky WJ, Simon P, Theimer M (2002) Reclaiming space from duplicate files in a serverless distributed file system, IEEE
8. Kwon H, Hahn C, Koo D, Hur J (2017) Scalable and reliable key management for secure deduplication in cloud storage. In: 2017 IEEE 10th international conference on cloud computing (CLOUD)
9. Li T, Huang Z, Li P (2017) Outsourced privacy-preserving classification service over encrypted data, IEEE
10. Stanek J, Sorniotti A (2016) Enhanced secure thresholded data deduplication scheme for cloud storage, IEEE

11. Hou H, Yu J, Hao R (2019) Cloud storage auditing with deduplication supporting different security levels according to data popularity, IEEE
12. Zheng H, Qin J, Hu J, Wu Q (2015) Threshold attribute-based signcryption. In 2015 IEEE 2nd international conference on cyber security and cloud computing

Subjectivity Detection for Sentiment Analysis on Twitter Data

C. Sindhu, Binoy Sasmal, Rahul Gupta, and J. Prathipa

Abstract With the quick increment in the quantity of web clients, the Internet has an enormous measure of data produced by the clients. Many people share their views regarding a topic on social media platforms such as Facebook and Twitter and give their feedback or review about a product on e-commerce web sites such as Amazon and Flipkart which leads to a huge amount of data. The identification of subjective statements from the data is known as subjectivity detection. To automate the analysis of such data, sentiment analysis is used. The aim is to find the opinionative data and classify it according to its polarity, i.e. positive, negative or neutral feedback, known as sentiment classification and then analysing it which is known as sentiment analysis. However, before performing sentiment examination, the information is exposed to different pre-processing procedures which finally give the desired optimized output. This allows us to get to know about the public's mood or opinion about a particular topic. This summarization helps the concerned organization or public to improve their product or service based on the feedback received.

Keywords Twitter · Sentiment analysis · Subjectivity detection · Opinion · Corpus

1 Introduction

With the enormous development of number of web users, the quantity of tweets every day on Twitter has likewise expanded definitely. Mining of sentiment from these tweets is helpful for the organizations and associations. For instance, it very well may be utilized as a sub-module in suggestion motors and so forth.

Sentiment analysis [1] is relevant mining of content that plans to group content into positive, negative and impartial. Sentiment analysis is an issue which incorporates different NLP sub-problems which are to be settled which incorporate mockery identification, element acknowledgement and subjectivity recognition and so forth.

C. Sindhu (✉) · B. Sasmal · R. Gupta · J. Prathipa
Department of Computer Science and Engineering, SRM Institute of Science and Technology, Chennai, India
e-mail: sindhucmaa@gmail.com

D. J. Hemanth et al. (eds.), *Artificial Intelligence Techniques for Advanced Computing Applications*, Lecture Notes in Networks and Systems 130,
https://doi.org/10.1007/978-981-15-5329-5_43

Subjectivity detection [2] has picked up significance with the quick development of data created via web-based networking media which requires the identification of subjective data (opinion) and objective data (fact). Subjectivity identification can be substantially more testing than polarity recognition; however, it has been under-explored because of the supposition that most of the information via web-based networking media is objective. For example, "My favourite pair of shoes is sold out" is an objective statement because it is a fact, and "This pair of shoes is very good" is a subjective statement because it tells about the opinion of the person.

Subjectivity detection helps to get to know about the opinion of the users about a particular product and topic which indeed helps the concerned organization or public to improve their service or product dependent on the feedback received.

2 Related Work

Systematic literature audit process is used in this overview. First we scanned for some related papers, research reports that are comprehensively worried about subjectivity identification or opinion mining from the content.

Detection of user's opinion and classifying its polarity, i.e. positive, negative and neutral, is known as polarity detection (PD) [3]. Previously done work in sentiment analysis was either knowledge-based or sentiment-based. But recently there have been various studies that utilize various machine learning techniques to classify the text. Supervised machine learning techniques are comparatively better than unsupervised machine learning techniques in performance but it is expensive to acquire the huge amount of labelled data required for supervised learning, whereas it is comparatively easy and less expensive to acquire unlabelled data for unsupervised learning.

Numerous specialists are putting their endeavours to identify the best technique for subjectivity identification. Albeit, a portion of the algorithms give great outcomes such as support vector machine (SVM), maximum entropy, Naïve Bayes [4] and so forth; however, no technique can resolve every one of the difficulties. The vast majority of the researchers detailed that SVM has high precision [5] than different algorithms. The different algorithms and the data sets used in different papers have been mentioned in Table 1.

3 Subjectivity Detection Approach

A framework was implemented in which the first step is to classify messages as subjective and objective tweets (subjectivity detection). The second step is to classify the subjective tweets into positive and negative (polarity detection).

Usually, a purely objective sentence does not carry any sentiment, and a purely subjective sentence usually tends to lean towards a positive or a negative sentiment.

Table 1 Subjectivity detection tasks in sentiment analysis

Paper	Data set	Algorithms	Features
[6]	Own Twitter data	SVM	Meta-features (part of speech (POS), polarity-MPQA)
[7]	Manually annotated tweets	corpus-based, dictionary-based, log-liner regression	WordNet, POS
[8]	Own Twitter data	SVM	unigrams, emoticons, hashtags, lexicon [30]
[9]	Own Twitter data	FCA	WordNet, OpenDover
[10]	SemEval-2013	clustering-based word sense disambiguation (WSD), lexicon-based classifier	WordNet, SentiWordNet
[11]	SS-Tweet	Senti strength	polarity, negations, emphatic lengthening
[12]	Twitter data	Unigram, bigram, uni-bigram	POS, SentiWordNet
[13]	Movie reviews	Statistical, maximum entropy	Emoticons, negations
[14]	Starbucks twitter data set	Dynamic architectural artificial neural networks	Polarity, hashtags
[15]	Twitter data	SVM, Naïve Bayes, maximum entropy, hybrid approach	Unigram feature
[16]	Spatiotemporal social (STS), healthcare reform (HCR) data set	LexRatio, maximum entropy, LProp, N-gram, hashtags, emoticons, lexicon [3]	Emotion detection, Twitter follower graph
[17]	Customer review Twitter sata set	Naive Bayes, maximum entropy, SVM	WordNet, sentiment classification

Though there are a few exceptions, for example, "The food made me sick" is an objective sentence with sentiment, and "I believe he came to the college yesterday" is a subjective sentence with no sentiment. Classifying a sentence as objective or subjective is done by using libraries such as TextBlob created by Steven Loria, and tools such as Opinion Finder (http://mpqa.cs.pitt.edu/opinionfinder/). A filtering mechanism is also implemented to have a control on the level of subjectivity in the training set by using a subjectivity threshold.

Another approach that was previously implemented, which was later scraped due to inconsistencies in the results and lack of accuracy, was from a given tweet, we map its POS using a POS dictionary (http://wordlist.sourceforge.net/pos-readme). POS tags are used to indicate sentiment tagging in a tweet. Objective messages usually consist of adjectives or interjections. We get the prior subjectivity and polarity

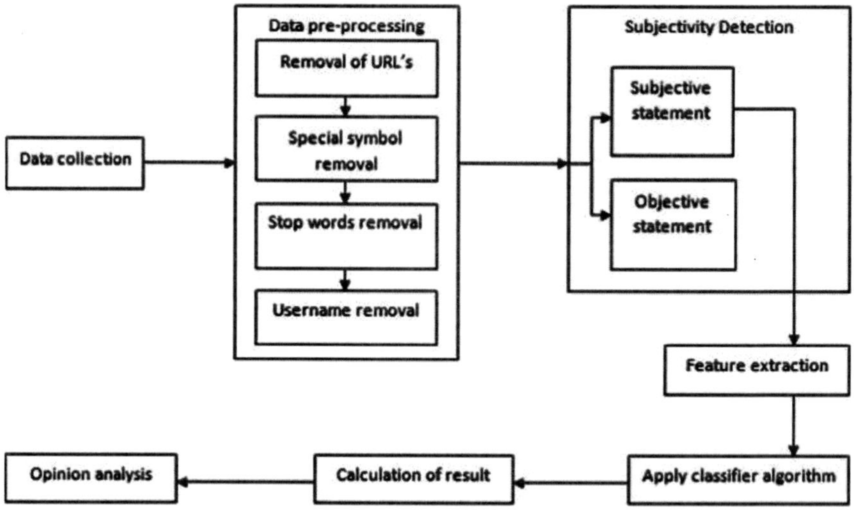

Fig. 1 Subjectivity detection for sentiment analysis on Twitter data

Table 2 Data set annotation scheme

Sentiment	Annotation
Positive	4
Neutral	2
Negative	0

information from the subjectivity lexicon used in [18]. The following methodology in Fig. 1 is adopted for the subjectivity detection.

3.1 Data Collection

The data set that is used consists of around 1.6 million tweets for training and 5000 tweets for testing [19]. The tweets in the data set are categorized into positive, negative and neutral. The data set is very versatile and consists of various categories such as company, movie, location, person, product, event and misc. The emoticons were removed for the training and the test data (Table 2).

3.2 Data Pre-processing

The data extracted from Twitter contains various contents which do not contribute to the sentiment of the user; therefore, it has to be first pre-processed. Pre-processing

[20] includes four basic steps—removal of URL, removal of special symbols, removal of stop words and removal of username. In removal of URL, any kind of link which is tweeted by the user and does not contribute to the sentiment analysis is removed. Removal of special symbol step deals with removing various symbols which do not have any actual sentiment, e.g. full stop (.), punctuation mark (!) and so forth. Stop words [21] removal step removes the stop words, words such as a, the which do have no effect on sentiment analysis should be removed and the conversion of emoticons to its equivalent word. Finally, in the username removal step every user's username starts with @ which has no effect on the sentiment analysis is removed, e.g. @username.

3.3 Subjectivity Detection

As previously mentioned, the first step is to classify the tweet into subjective and objective and remove the tweets based on their subjectivity scorekeeping only the tweets having score higher than the specified threshold.

This step is introduced to achieve higher accuracy. The pre-processed data is taken and is classified into subjective or objective statement using a subjectivity classifier. All the tweets having a subjectivity score lesser than the specified threshold are filtered out, and the classifier is trained with only the remaining tweets. It is observed that as the subjectivity threshold is increased, significant amount of tweets gets filtered out.

3.4 Feature Extraction

A data set contains numerous ascribes that add clamour to the data and influence exactness. The commotion likewise bit by bit expands the time required to assemble the model. Feature extraction basically combines ascribes into a reduced feature set. The selected features and their blend assume a significant job for identifying the sentiment of the text.

Selection of features [22, 23] from the extracted features can possibly improve the arrangement exactness, restricted in on a key feature subset of opinion discriminators and give more prominent understanding into habitually happening ascribes and qualities.

The extracted features focus on a document vector whereupon machine learning strategies are applied to group the extremity of the content utilizing the got document vector.

3.5 Apply Classifier Algorithm

There are generally three approaches which include: **Supervised learning** [24, 25] is a sort of learning in which we train the machine with the information which is well labelled. The machine learning approach pertinent to sentiment examination, for the most part, belongs to supervised classification. In machine learning-based methods, two sets of records are required: training set and a test set. Machine learning techniques such as naïve Bayes, SVM, maximum entropy and so forth are used. **Unsupervised learning** [26] is a sort of learning wherein we train the machine with the information which is not labelled. Classification is performed by comparing the features of a given text with sentiment lexicons whose sentiment values are determined prior to their use. Clustering methods such as k-means, mean shift clustering and so forth are used. **Reinforcement learning** (RL) [27] is the field that reviews the problems and procedures that attempt to retro-feed its model to improve. To achieve this, RL needs to be able to "sense" signals, consequently choose an activity and afterwards look at the result against a "reward" definition. RL attempts to make sense of what to do to boost these prizes, yet it does this without any support.

Supervised learning methods for classification by using machine learning [28] algorithms such as Naive Bayes, SVM and maximum entropy have been found to give good accuracy. SVM was used as vast majority of researchers claimed it to be more accurate than the other algorithms, so we decided to use SVM to build the classifier.

3.6 Calculation of Result

Calculating the polarity of the user's statement using the approach described. The most generally used assessment measurements are accuracy, recall, precision and F-score. The confusion matrix is shown in Table 3.

Table 3 Confusion matrix showing the performance of a sentiment analysis method

	Is positive	Is negative
Positive prediction	TP	FP
Negative prediction	FN	TN

4 Evaluation

First we see the effects of the subjectivity threshold parameter. From the results obtained, it can be clearly observed that the tweets get filtered out to an ever-increasing extent with an increase in subjectivity threshold parameter as shown in Fig. 2.

Here we see the tweets remaining after the filtering process from TextBlob and Opinion Finder tool. TextBlob utilizes a function that finds a tweet's subjectivity level, whereas Opinion Finder tool denotes which segments of the message are subjective. This helps us to find the level of subjectivity of a tweet. SentiOutlook is used, which we created, to find the best-suited filtering for our experiment.

$$\text{Subjectivity Level} = \frac{\text{Length of subjective parts}}{\text{Total length of the tweet}} \tag{1}$$

For the experiment, we pick an optimal threshold value of 0.5, factoring that the model should be trained on a progressively conventional data set and the subjectivity level can be calculated using (1). The relation of the accuracy with subjectivity threshold can be seen in Fig. 3.

Using the SVM classifier, we obtained satisfactory accuracy, precision, recall and *F*-score as shown in Table 4.

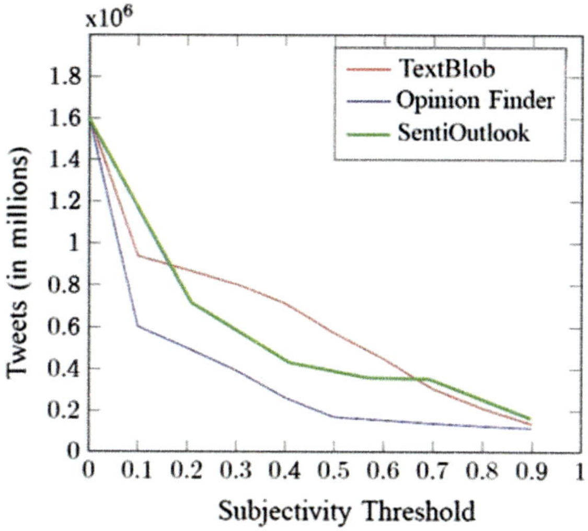

Fig. 2 Remaining tweets with subjectivity threshold

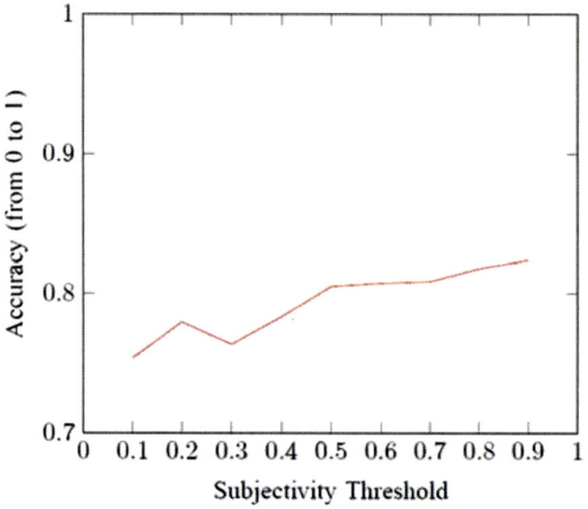

Fig. 3 Accuracy against subjectivity threshold

Table 4 Performance matrices using SVM

Matrices	Values	Formula used
Accuracy	82.66	$\frac{TN+TP}{TN+FN+TP+FP}$
Precision	78	$\frac{TP}{TP+FP}$
Recall	86	$\frac{TP}{TP+FN}$
F-score	84	$\frac{2\times precision\times recall}{precision+recall}$

5 Discussion

Twitter has a large amount of data in the form of tweets which includes the comments, opinions and reviews of the public regarding a particular product or service. Therefore, sentiment analysis comes into play to mine the opinion of the users. Many researches have been done on this but there is still a lot of scope in increasing the accuracy of the system. We came across various techniques which can be used to improve the accuracy but hardly any work is accomplished [29] on them such as oxymoron words, misspelled words, etc., and these problems should be considered in any future work done on this topic. Also, the example we took in the introduction "This pair of shoes is very good" is actually a subjective statement; however, our system detects it as an objective statement. It will also be quite interesting to go beyond just the positive and negative and extract more information and patterns from these data.

6 Conclusion

The growth of social data is exponential, which has given rise to new aspects, such as the subjectivity detection. Subjectivity detection is a natural language processing task that consists of differentiating subjective data (opinions) from objective data (facts). By using subjectivity detection, we can filter out the tweets that are objective and find the tweets that are subjective and carry out sentiment analysis only on the subjective data. The accuracy of the sentiment analysis can further be increased by implementing methods to fix the misspelled words and correcting the use of any abbreviated or shortened words. The use of oxymoron words is another factor that can affect the accuracy of sentiment analysis which can be corrected by using a new model to detect and replace the oxymoron words with equivalent words that are effectively analysed by the sentiment analyzer.

References

1. Neri F, Aliprandi C, Capeci F, Cuadros M, By T (2012) Sentiment analysis on social media. In: IEEE/ACM international conference on advances in social networks analysis and mining (ASONAM), pp 919–926
2. Satapathy R, Chaturvedi I, Cambri E, Ho SS, Cheon Na J (2017) Subjectivity detection in nuclear energy tweets. Computacion y Sisttemas 21(4):657–664
3. Wilson T, Wiebe J, Hoffmann P (2005) Recognizing contextual polarity in phrase-level sentiment analysis. In: Proceedings of the human language technology conference and the conference on empirical methods in natural language processing (HLT/EMNLP), pp 347–354
4. Parveen H, Pandey S (2016) Sentiment analysis on Twitter dataset using Naive Bayes Algorithm. In: IEEE trans 2nd international conference on applied and theoretical computing and communication technology (iCATccT), pp 416–419
5. Pang B, Lee L, Vaithyanathan S (2002) Thumbs up? Sentiment classification using machine learning techniques. In: Proceedings of the ACL-02 conference on empirical methods in natural language processing, vol 10, pp 79–86
6. Barbosa L, Feng J (2010) Robust sentiment detection on twitter from biased and noisy data. In: Proceedings of the 23rd international conference on computational linguistics: posters (COLING'10). Association for Computational Linguistics, Stroudsburg, pp 36–44
7. Kumar A, Sebastian TM (2012) Sentiment analysis on Twitter. Int J Comput Sci 9(4):372–378
8. Zhang L, Ghosh R, Dekhil M, Hsu M, Liu B (2011) Combining Lexicon-based and Learning-based Methods for Twitter sentiment analysis. Technical report, HP Laboratories
9. Kontopoulos E, Berberidis C, Dergiades T, Bassiliades N (2013) Ontologybasedsentiment analysis of Twitter posts. Expert Syst Appl 40(10):4065–4074
10. Ortega R, Fonseca A, Montoyo A (2013) SSA-UO: unsupervised twitter sentiment analysis. In: Proceedings of the 7th international workshop on semantic evaluation—2nd joint conference on lexical and computational semantics (SemEval'13). Association for Computational Linguistics, pp 501–507
11. Thelwall M, Buckley K, Paltoglou G (2012) Sentiment strength detection for the socialweb. J Am Soc Inform Sci Technol 63(1):163–173
12. Gurkhe D, Rishit B (2014) Effective sentiment analysis of social media datasets using Naive Bayesian classification
13. Duric A, Song F (2012) Feature selection for sentiment analysis based on content and syntax models. Decision Support Syst. 53:704–711

14. Saif H et al (2016) Contextual semantics for sentiment analysis of Twitter. Inf Process Manag 52:5–19
15. Bahrainian SA, Dengel A (2013) Sentiment analysis and summarization of twitter data. 2013 IEEE 16th international conference on computational science and engineering (CSE) IEEE
16. Speriosu M, Sudan N, Upadhyay S, Baldridge J (2011) Twitter polarity classification with label propagation over lexical links and the follower graph. In Proceedings of the first workshop on unsupervised learning in NLP (EMNLP'11). Association for Computational Linguistics, Stroudsburg, PA, pp 53–63
17. Gautam G, Yadav D (2014) Sentiment analysis of Twitter data using machine learning approaches and semantic analysis. In: 2014 seventh international conference on contemporary computing (IC3), IEEE
18. Riloff E, Wiebe J (2003) Learning extraction patterns for subjective expressions. In: Proceedings of the 2003 conference on empirical methods in natural language processing (EMNLP-03), pp 105–112
19. Go A, Bhayani R, Huang L (2009) Twitter sentiment classification using distant supervision. CS224N Project Report, Stanford, p 12
20. Jianqiang Z, Xiaolin G (2017) Comparision research on text preprocessing methods on twitter sentiment analysis. IEEE Trans 5:2870–2879
21. Saif H, Fernandez M, He Y, Alani H (2015) On stopwords, filtering and data sparsity for sentiment analysis of Twitter. In: Proceedings of 9th Language Resources Evaluation Conference (LREC), Reykjavik, Iceland, 2014, pp 80–81
22. Mansour R, Hady MFA, Hosam E, Amr H, Ashour A (2015) Feature selection for twitter sentiment analysis: an experimental study. Computational linguistics and intelligent text processing: 16th international conference, CICLing 2015, Cairo, Egypt, April 14–20, 2015, Proceedings, Part II, Springer International Publishing, pp 92–103
23. Chandrasekhar G, Sahin F (2014) A survey on feature selection methods. Comput. Elect. Eng. 40:16–28
24. Dhanalakshmi V, Dhivya B, Saravanan A (2016) Opinion mining from student feedback data using supervised learning algorithms. 1–5. https://doi.org/10.1109/icbdsc.2016.7460390
25. Read J (2005) Using emoticons to reduce dependency in machine learning techniques for sentiment classification. In: Proceedings of ACL-05, 43nd meeting of the association for computational linguistics. Association for Computational Linguistics
26. Ko Y, Seo J (2000) Automatic text categorization by unsupervised learning. In: Proceedings of the 18th conference on computational linguistics, vol 1. Associations for computational Linguistics, pp 453–459
27. Frenay B, Verleysen M (2016) Reinforced extreme learning machines for fast robust regression in the presence of outliers. IEEE Trans Cybern. 46(12):3351–3363
28. Chaudhari M, Govilkar S (2015) A survey of machine learning techniques for sentiment classification. IJCSA 5(3):13–23
29. Patil H, Atique M (2015) Sentiment analysis for social media: a survey, pp 1–4. https://doi.org/10.1109/icissec.2015.7371033
30. Ding X, Liu B, Yu PS (2008) A holistic lexicon-based approach to opinion mining. In: Proceedings of the conference on web search and web data mining (WSDM)

Review on Hybrid Recommender System for Mobile Devices

R. Lavanya, Tanmay Khokle, and Abhideep Maity

Abstract Movie recommendation as a task is complex as there is a large variety in the preference of users across the board, and there are numerous genres of movies. A hybrid approach is more conducive for movie recommendations on mobile devices as it combines the best of the most widely employed methods. This allows the results to be scalable when handling big data tasks as is very common in movie recommendations given the large amount of reviews. In a number of reviews, every user also voices their opinion and analyzing this sentiment is crucial in order to understand the emotional investment a person has in the movie. Therefore, after using a hybrid approach, this study uses sentiment analysis to filter the results further.

Keywords Collaborative filtering · Content-Based filtering · Hybrid system · Sentiment analysis

1 Introduction

Nowadays, almost all people carry a mobile device with them. The rising popularity of streaming services and better apps has added to the comfort of using a mobile device. Coupled with the portability and ease of use of a mobile device, this has led to a mobile device being one of the primary ways people access media; where movies are arguably the most popular form of media accessed by users. Along with simply consuming the content, people look for relevant, useful, and quick recommendations for choosing which media to access. These recommendations not only help the users decide what to watch next, but also keep their interest high by showing them content

R. Lavanya (✉) · T. Khokle · A. Maity
Department of Computer Science, SRM Institute of Science and Technology, Kattankulathur, India
e-mail: lavanyar@srmist.edu.in

T. Khokle
e-mail: tkhokle11@gmail.com

A. Maity
e-mail: abhideepm@gmail.com

D. J. Hemanth et al. (eds.), *Artificial Intelligence Techniques for Advanced Computing Applications*, Lecture Notes in Networks and Systems 130,
https://doi.org/10.1007/978-981-15-5329-5_44

they are likely to enjoy. For choosing a better system for movie recommendations, it is important to compare the existing frameworks in use, as done by Patel et al. [1]. Recommendations for movies is a challenging task because it is difficult to provide results in near real-time, results that are relevant, accurate and useful to the user.

Movie recommendation is a complex task as it has to take into account a huge base of users and make the predictions accordingly. This is because people have vastly different tastes and preferences for movies, and certain choices may prove to be outliers to popular works. Therefore, it is important to consider a large number of them, if not all, to encompass the preferences of a large enough population. In order to solve this issue of having a huge user base, many techniques for recommenders have been proposed. Based on information available online, it is apparent that the Collaborative Filtering technique of recommendation is probably the most widely used and thus most widely implemented technique in this domain. A summary of some published articles relevant to these issues is given in Table 1.

In Collaborative Filtering, items are recommended by measuring similarity between users' preferences, which can be measured by calculating the correlation between their watched movies, the reaction of the viewers to them and the rating given to these movies thereafter. The main drawback of Collaborative Filtering is that the data available for users is often sparse. Collaborative Filtering makes predictions based on the history of interactions done by the user and in case there is little or no history of interaction, the Collaborative Filtering algorithm fails or is not able to provide useful recommendations.

In general, people tend to prefer watching movies which are rated higher, and have been given mainly positive reviews by other movie watchers and avoid movies that are rated poor. However, user reviews of movies also contain other important information such as their likes, dislikes and preferences. Users express their preferences and feelings about movies to supplement the ratings that they give. The feelings contained in these reviews affect the choice of other potential viewers on the media delivery platform in choosing whether or not to watch a particular movie. A more effective and efficient method of recommendation needs to be used in order to overcome the limitations of Collaborative Filtering. Thus, sentiment analysis of the reviews is very useful as it helps us recommend movies by understanding the human thought and emotions attached to the experiences of the user.

Spark seems to be an effective tool to implement a good recommender system as it takes advantage of the concept of storing the data in Resilient Distributed Datasets [12]. Resilient Distributed Datasets store the data in logical partitions and stores the intermediate results as well, making it more effective in calculations than a simple Data Frame and enables it to give out nearly real-time results.

In this paper, a number of approaches are analyzed to assess which approach would be the most effective for the purpose of movie recommendations and provide the best scalability on a mobile device such as a mobile phone. Judging these works based on their advantages, shortcomings and results, a preferred direction for the techniques can be decided. Although the one of the most common methods for movie recommender systems is Collaborative Filtering, it may be beneficial to take into consideration certain approaches that use a different approach, as they may

Table 1 Comparison of recommendation system techniques

S. No.	Techniques Used	Outcome	Pros	Cons
1	• Knowledge- based recommender systems • System of context aware semantics [2]	• For cold-start situations, the recommetz has a fair prediction accuracy	• Accurate Predication of movies based on the location and timing	• This mechanism is information is not available online • Complimentary mechanism to increase the accuracy is still being worked on
2	• Content-based and collaborative filtering recommendation [3]	• Substitute hybrid recommendation system improves performance of CF • Switching hybrid recommendation method significantly reduces the execution time	• Hybrid recommendation system increased the performance of collaborative filtering • The MoRe system can be used for any type of content	• Since algorithms implemented were on a specific dataset, it limits the conclusion to the movie domain
3	• Collaborative filtering • Genetic algorithms (K-means) [4]	• Compared to the existing clustering-based CFs, the accuracy and timely performance is better	• Provides high prediction accuracy • Capable of producing effective ratings based on traditional systems for movies	• Cannot deal with high dimensional data practically • Variation of clusters to be studied
4	• Cloud computing • Automata • Sentiment analysis [5]	• Sentiment analysis with automata achieves personalized recommendations	• The timeliness is good in comparison to most sentiment systems in common use	• The score does not vary for a low number of users
5	• Recommendations based on social network • Mobile devices • Sentiment analysis [6]	• Sentiment metric used performs well for music recommendation	• The eSM improves the music recommendation system • The system is not taxing on the hardware	• The study is limited to music recommendation systems
6	• Collaborative filtering, sentiment analysis [7]	• Opinion words give a more accurate outlook toward the sentiment	• The model makes it easier to integrate sentiment analysis into collaborative filtering	• Scope of data is limited

(continued)

Table 1 (continued)

S. No.	Techniques Used	Outcome	Pros	Cons
7	• User segmentation, clustering, collaborative filtering [8]	• Computational time and accuracy is better in comparison with traditionally employed CF systems	• The model performs better than collaborative filtering on the feature segregation by age • Performs better than traditional filtering	• Improvement in filtering is not constant when filtered by gender of person • Only improves collaborative filtering without solving the problem of content information
8	• Collaborative filtering • Content-based filtering • Boltzmann machines [9]	• Good performance over cold-start data	• Can learn by utilizing the higher order of interactions • Can predict binary user actions very well	• Non binary actions may not be used with this system
9	• Collaborative and content-based filtering • Diversity [10]	• Better accuracy and coverage than the usual system employed for recommendation in a cold-start environment	• It works great on the movie Lens dataset and shows a good alternative to hybrid system	• Dataset not tested with different types • Very restricted data used in limited range
10	• Semantic web • Knowledge-based system [11]	• Performs well for movie recommendation task	• Yields better results than state of the art solutions in cinema domain • Semantic differences allow for addition of features that improve user satisfaction	• Not very accurate for domain other than movie recommendation • Not scalable

provide benefits that outweigh a straightforward implementation of Collaborative Filtering (such as the system used by Rosa et al. [13]). By reviewing these previous works in the field of recommender systems, the overall effectiveness of using a hybrid system should be compared to other methods.

Based on the comparison, the best approach to movie recommendations can be chosen and tested. The benefits or implications of using sentiment analysis along with the chosen system, for reasons already mentioned earlier, also need to be tested.

2 Related Work

Collaborative Filtering is one of the most preferred approaches for solving the problem of movie recommendations for users. A range of solutions have been suggested in order to further improve Collaborative Filtering systems, such as employing a combination of K-means clustering, and genetic algorithms [4]. This method performs better than a traditional clustering-based approach. However, similar to other clustering methods, it shows a sharp increase in the mean error as the amount of data increases, which is not desirable. Collaborative Filtering struggles to scale well, which is a big problem because if a recommender system is to be deployed on a platform where there are a huge number of users and movies, then it would take a long time to compute the predictions, which is not preferable or efficient. To combat this, a combination of Collaborative Filtering and Self-Organizing Map Neural Networks could be employed as explored by Lee et al. [8]. However, they tested this on data comprising of only fifty-four movies and 174 users. It needs to be tested for data with more depth and a greater number of users and movies. This is necessary because the field in which the recommender system will be employed in, could potentially have thousands of movies and an even greater number of users. Therefore, the system needs to be tested for how it would scale in such an environment. An advantage of self-organizing maps mentioned is that they do not need as much computation as Collaborative Filtering, but the method was tested on a very restricted range of only well-known movies, so it puts into question whether it will maintain this advantage over Collaborative Filtering when the data used is not curated and has higher dimensionality. Additionally, it completely ignores similarities between the movies themselves, which is also a drawback of utilizing only Collaborative Filtering.

Thus, approaches for movie recommendations have moved away from Collaborative Filtering and prefer other techniques to solve the problem more effectively. One such popular technique is sentiment analysis. Sentiment analysis utilizes language processing to take into account the opinions of the views posted as written reviews. This is important as most people use good reviews as an important metric when deciding whether to watch a movie or not. Krishna et al. [5] suggest that using learning automata for sentiment analysis will help in improving the results. The learning automata approach is beneficial as it boosts the speed of reaction when faced with a request for a new recommendation. Similarly, [6] conclude that a using an enhanced sentiment metric will yield a better recommendation in turn. Their result was limited by the fact that it was tested only for music recommendations which has overlapping features and is a different task from movie recommendations. Movies, on the other hand, have more depth to them as they have content with deeper features such as a plot, theme and music as well.

A sentiment metric was also used by Leung et al. [7], in combination with Collaborative Filtering and produced more conducive results for the recommendations. However, it was again limited by the scope of the data. Here, reviews of users with less amount of reviews were ignored completely. Additionally, no data related to the actual movies besides from the ID, name and rating was used. This underlines the general issue with using only a sentiment analysis method or using sentiment analysis along with Collaborative Filtering, which is that it doesn't take into account the similarity between the movies.

Some other works on movie recommendations used data that included information on the movie theaters as well as their timings to create context aware semantic nets [2]. The method including theater data may not have a very positive impact because a lot of people watch movies on their devices directly. Additionally, there may be a bias for movie watchers at a location for a particular movie theater, so it is not the best feature to consider for reliable recommendations. One approach may be to take advantage of the capabilities of an Unified Boltzmann Machine to improve performance and tackle the cold-start problem of recommendation systems [9]. A cold-start problem arises when there is no history available for users, specifically new users. It becomes incredibly difficult to recommend movies to a person if there is no information on what kind of movies they tend to watch or the ratings they would give to these movies. The issue with using Unified Boltzmann Machines is that they cannot be extended for recommendation purpose beyond binary actions. Additionally, this method needs to be tested on data that combines the individual features it was tested on to see whether it continues to work well when faced with practical and realistic data.

To combat these issues, it may be better to explore a hybrid model in order to mitigate the drawbacks of the individual methods. The system proposed by Wang et al. [4], is not practical in an environment of high dimensionality. Another suggested method deals with employing a Diverse Collaborative Prediction [10]. Their hybrid between a Collaborative Filtering and Content-Based approach suggested quite evidently gives better results than traditional techniques., However, the dimensionality of data used for testing was limited as only three features were considered in the evaluation, and it needs to be corroborated by using different kinds of data. Further, this method completely disregards review data that is so crucial to gauge user opinions and their objective views on the movie. A survey on the effective hybrid systems has been performed by Burke [14].

Use of a hybrid with a knowledge network in a socially aware environment has shown promise [11]. However, it is not very scalable. Lekakos and Caravelas [3] presented a pragmatic comparison of Collaborative Filtering, Content-Based system and a hybrid method (combining the two) for a movie recommendation system. It supports the idea that a hybrid system combining a Content-Based recommender and Collaborative Filtering method is more effective. Taking into consideration the results and implications of these previous works, the approach of using a hybrid system seems to be more effective. However, a combination of a hybrid system and a sentiment analysis model can improve these results even further, and make the system scalable at the same time.

3 Comparison of Hybrid Recommender Systems

In the related work a number of algorithms have been utilized for the implementation of a hybrid recommender system. A comparison of algorithms must be made to determine which one is most conducive to the task.

3.1 Collaborative Filtering with Self-organizing Map

Lee et al. [8] have used a combination of Collaborative Filtering and Self-Organizing Map to reduce the Mean Absolute Error. This method worked particularly well on a prediction based on neighbors. This method gave better results than simple Collaborative Filtering in most clusters. However, in some cases it was outperformed by the Collaborative Filtering model, which calls for further improvement. This model can be further extrapolated to being used on other datasets to improve the confidence. This method uses demographic data to cluster similar users. The demographic clusters used in this method take into consideration a very narrow view of the features, namely age and sex of the person.

3.2 Collaborative Filtering with Content-Based Method

An Item-Based Collaborative Filtering model has been employed by Ghazanfar and Adam [15], in combination with a content-based recommender model. The Collaborative Filtering is achieved through predicting the neighbors, through an item matrix, whereas the content-based recommendations are brought forth using TF-IDF method of approach. Movielens and FilmTrust datasets were used in this paper to check the accuracy of the predictions. Mean Absolute Error was used here as a standard for the accuracy of the system. On comparison with [8], the Mean Absolute Error is more. However, the depth of features is also more in the method by Ghazanfar and Adam [15]. It may be beneficial to use the latter as it reflects user data generated in a more realistic sense. Moreover, the utilization of Content-Based method helps it to be more scalable.

A clustering method for judging similarity using matrices was used in [16]. Both the group as well as item data was used for judgment. Adjusted cosine similarity was used along with correlation based similarity to generate the similarity matrices.

Table 2 Error prediction of recommendation algorithms

S. No.	Algorithm	Best average mean absolute error
1	Collaborative filtering with self-organising map neural network [8]	0.72 (based on demography—sex)
		0.61(based on demography—age)
2	Collaborative filtering with content-based filtering [15]	0.79
3	Clustering with hybrid system of Content-based and collaborative filtering [16]	0.75
4	Naïve hybrid [15]	0.82

Grouping by items in this paper helped the system tackle cold-start issue more effectively than a traditional hybrid method. The data was derived from MovieLens for this approach as well. The Mean Absolute Error for this system is lesser in comparison to the ones used by Lee et al. [8] and Ghazanfar and Adam [15], in most cases. However, its main drawback is that like the others, it does not consider the review data which may be crucial in judging user outlook toward movies. Furthermore, the complexity of calculations is also more in the method employed by Ghazanfar and Adam [15] as compared to [8] and in cases of smaller clusters, more than [16]. A contrast between accuracy of the hybrid systems discussed is given in Table 2.

Although the error on MovieLens is lesser in the method applied by Li and Kim [16], it also has more time complexity than [15]. The features however, are also lesser in [8], which may explain the lower error in prediction. The data used by Lee et al. [8] approach is not very large, which also contributes to the high accuracy in comparison to the Movielens 100 k used by the other two approaches. Therefore, the approach used by Ghazanfar and Adam [15] seems to be the most promising out of the three systems. A common drawback of all of these systems is that they do not take into consideration the review data of the users. In [15] the comparison between the performance of the proposed system and other systems is given. The proposed method performs significantly better than the simple hybrid between Collaborative Filtering and a Content-Based one.

A performance on a cold-start scenario is also necessary to judge the capability of a recommender. Table 3 gives a comparison between the performance of the method proposed by Ghazanfar and Adam [15] and Li and Kim [16] on the cold-start data.

Table 3 Mean Error rate for recommendation Algorithm

S. No.	Algorithm	MAE—cold start	MAE—no cold start
1	Collaborative filtering with content-based filtering [15]	1.06	0.79
2	Clustering with hybrid system of content-based and collaborative filtering [16]	0.76	0.75

As we can judge by the comparison the method by Li and Kim [16] performs better on the cold-start problem and may provide an advantage over the method used by Ghazanfar and Adam [15]. However, it must be noted that it also takes a longer time to compute and may be a critical factor. Realistically, the system needs to be able to deal with cold-start data if faced with such. In this case clearly one method performs better, but at the cost of complexity.

The approach used by Leung et al. [7] to utilize sentiment analysis in tandem with Collaborative Filtering helped incorporate crucial review information. The frequency of opinion words formed the basis for the positivity of the reviews. This method, while taking reviews into consideration, did not include the similarity between the movies themselves. There is a possibility to improve on this by using a method of hybrid recommendation inspired from the work of [15, 16] to combine with the sentiment analysis similar to the one used by Leung et al. [7].

4 Conclusion

In comparing the related work on hybrid systems and recommenders, a hybrid system along with sentiment analysis shows the most promise for movie recommendations. By combining the most widely employed methods in recommendation, it allows the results to be scalable when handling big data tasks as is very common in movie recommendations given the large amount of reviews. In a number of reviews, every user also voices their opinion and analyzing this sentiment is crucial in order to understand the emotional investment a person has in the movie.

References

1. Patel B, Desai P, Panchal U (2017) Methods of recommender system: a review. In 2017 international conference on innovations in information, embedded and communication systems (ICIIECS), pp 1–4. IEEE
2. Colombo-Mendoza LO, Valencia-García R, Rodríguez-González A, Alor-Hernández G, Samper-Zapater JJ (2015) RecomMetz: a context-aware knowledge-based mobile recommender system for movie showtimes. Expert Syst Appl 42(3):1202–1222
3. Lekakos G, Caravelas P (2008) A hybrid approach for movie recommendation. Multim Tools Appl 36(1–2):55–70
4. Wang Z, Xue Y, Feng N, Wang Z (2014) An improved collaborative movie recommendation system using computational intelligence. J Visual Lang Comput 25(6):667–675
5. Krishna PV, Misra S, Joshi D, Obaidat MS (2013) Learning automata based sentiment analysis for recommender system on cloud. In: 2013 international conference on computer, information and telecommunication systems (CITS), pp 1–5. IEEE
6. Rosa RL, Rodriguez DZ, Bressan G (2015) Music recommendation system based on user's sentiments extracted from social networks. In 2015 IEEE international conference on consumer electronics (ICCE). IEEE, pp 383–384

7. Leung CW, Chan SC, Chung FL (2006) Integrating collaborative filtering and sentiment analysis: a rating inference approach. In: Proceedings of the ECAI 2006 workshop on recommender systems, pp 62–66. 2006
8. Lee M, Choi P, Woo Y (2002) A hybrid recommender system combining collaborative filtering with neural network. In: International conference on adaptive hypermedia and adaptive web-based systems. Springer, Berlin, Heidelberg, pp 531–534
9. Gunawardana Asela, Meek Christopher (2009) A unified approach to building hybrid recommender systems. RecSys 9:117–124
10. Shrestha J, Jo GS (2009) Enhanced content-based filtering using diverse collaborative prediction for movie recommendation. In: 2009 first Asian conference on intelligent information and database systems. IEEE, pp 132–137
11. Carrer-Neto W, Hernández-Alcaraz ML, Valencia-García R, García-Sánchez F (2012) Social knowledge-based recommender system: application to the movies domain. Expert Syst Appl 39(12):10990–11000
12. He Z, Zhou X (2016) A fast and better hybrid recommender system based on spark. Netw Parall Comput 147
13. Ozcan A, Oguducu SG (2010) A recommendation framework for mobile phones based on social network data. In: Software engineering, artificial intelligence, networking and parallel/distributed computing. Springer, Berlin, Heidelberg, pp 139–149
14. Burke R (2002) Hybrid recommender systems: survey and experiments. User Model User-Adap Inter 12(4):331–370
15. Ghazanfar MA, Adam PB (2010) A scalable, accurate hybrid recommender system. In: 2010 third international conference on knowledge discovery and data mining. IEEE, pp 94–98
16. Li Q, Kim BM (2003) Clustering approach for hybrid recommender system. In: Proceedings IEEE/WIC international conference on web intelligence (WI 2003). IEEE, pp 33–38

Efficient Machine Unlearning Using General Adversarial Network

S. Deepanjali, S. Dhivya, and S. Monica Catherine

Abstract According to a recent study conducted by Forbes, the collective world data is expected to raise by 175 ZB in 2025 accounting to a 61% increase in data generation. As human, we are favoured with the nature of forgetting irrelevant data. The neurons in our brain are designed to filtrate the noise in the information in order to create more space to store relevant data. This neurophysical principle is witnessing its significance in the field of artificial intelligence due to data intensification. There is a great need for developing systems that have the capability to forget the data on the user's request without compromising its system accuracy as data are the backbone of any deep learning system. Our approach combines the process of batching training data and general adversarial networks for data augmentation networks to achieve a complete and faster data forgetting model.

Keywords Machine unlearning · Data forgetting · General adversarial network · Data augmentation

1 Introduction

1.1 Need for Machine Unlearning

The advancement and popularity of social media have resulted in a high level of information invasion. These platforms use our information for tracking user behaviour like cookies and tamper higher sensitive user information. As a result, The EU

S. Deepanjali (✉) · S. Dhivya · S. Monica Catherine
Department of Information Technology, SRM Institute of Science and Technology, Kattankulathur, Chennai, India
e-mail: deepanjs@srmist.edu.in

D. J. Hemanth et al. (eds.), *Artificial Intelligence Techniques for Advanced Computing Applications*, Lecture Notes in Networks and Systems 130,
https://doi.org/10.1007/978-981-15-5329-5_45

government has reinforced the 'right to erasure' law by which the citizens have the power to request the system to remove their data permanently [1]. The accuracy of the system is not scaled down when a restricted amount of user requests data removal but when the extensive amount of user requests the same the training data gets downsized thereby reducing the system's decision-making capability.

1.2 Space and Time Complexity

The training data are trivial and expensive both in terms of system space and computational time. They serve as the knowledge base to decision system. In real life these decision plays an influential role. Increased inappropriate and garbage data can lead to a high false-positive rate which thereby degrades the accuracy of the system. These data are usually removed as a process of data cleaning which can also be considered as unlearning process. Our approach focuses in scoring both timeliness and complete data forgetting.

1.3 Security Perspective

Polluting training data is one major issue that concerns system security. Detecting the intrusion data and cleaning plays a major role in building a model. In the above paper [2] GAN is used to identify the anomalies in image data. The essence of machine unlearning is to remove data and its complete lineage in order to defend the system from data tampering.

2 Related Work

The field of intentional forgetting is flourishing among the researchers in recent years due to the requisite for improving the cognitive capability of the model. A recent survey by [3] categorizes forgetting into two types. Type 1 formulates oblivation of individual data items whereas type 2 focuses in disremembering of data lineage or relation among the individual data. However, multiple research work discussed below focuses on type- 1 forgetting. The main challenge of machine unlearning is to develop asymptotically time-balanced system. The work proposed by [4] focuses on faster unlearning process using statistical query learning. Here the data is computed as the summation of a set of data and in order to scale down the process of retraining model from search, the data to be forgotten are subtracted from the summation.

Jilek et al. [5] abstracts human nature in context to organizational memory for decentralizing and self-organizing data. They figured a threshold value called memory buoyancy for data relevancy. Adaptive synchronization was used to amplify their model where forgotten data were moved to cloud, for indispensable data retrieval. Intentional data forgetting can also be enforced in distributed model [6] where the prime focus is given to multiagent system. The main challenge of teamwork in distributed system is information overload. AdaptPRO the model developed has conquered this challenged by belief-desired- intention architecture. As discussed earlier, profound work has been carried only for type-1 machine learning, creating a huge gap for exploration in type-II data forgetting.

3 Background

As discussed in the introduction, there are multiple reasons which can result in machine Unlearning. Our work focuses on User's dismay and request to remove their data from the platform. We can standardize the unlearning process as individual users U_i has reserved their information as data points in the training data $T\{I_1 \ldots I_n\}$, where I_i denotes unitary data point in the T_d. Any prediction model learns to make a decision from these data points. When a user finds that his information has been tampered then he requests the system to intentionally forget i.e. Unlearn his data represented as I_{noise} from T_d resulting in retraining the model from the scratch. The data to be unlearned can be requested from single user to multiple users (Fig. 1).

Data abundance is a fundamental demand for deep neural networks. As the amplitude of labelled data increases the decision-making capability also increases. In data forgetting the critical situation may arise when multiple users say thousand user requests to remove the data concurrently. This data removal will affect the model accuracy to a greater extent. Data augmentation plays eminent assistance in meeting the aforementioned situation. Data augmentation enlarges the size of training data using the already available data. Hence, DA can be considered as in-sync approach with intentional data forgetting. This mapping can be represented as

$$\theta : T_d \to T_d' \tag{1}$$

In the above expression, T_d represents the training data $\{I_1 \ldots I_n\}$ where $I_{\text{noise}} \subset T_d$

I $_{\text{noise}}$ generally denote the data to be removed as per user requisition. T_d' represents the new augmented training data set where I_{noise} is replaced with the augmented data along with original data points which contains both original data points I_i and Inflated data I_{aug}

$$T_d' : I_i \cup I_{\text{aug}} \tag{2}$$

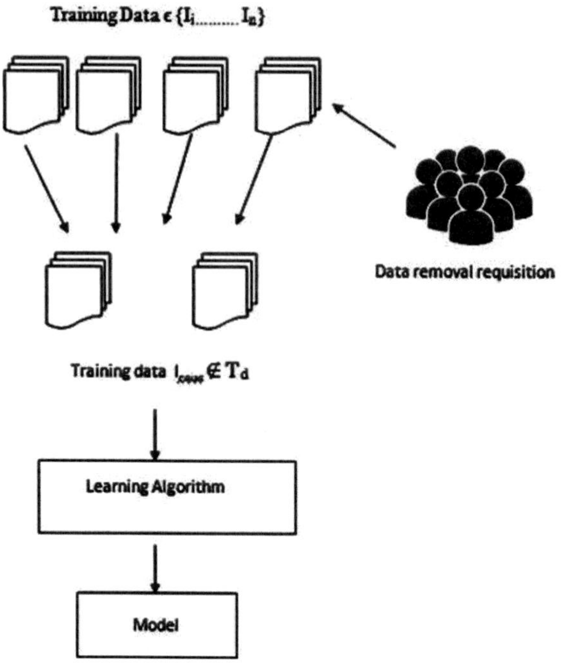

Fig. 1 Machine unlearning environment

There are multiple approaches for augmenting data with respect to Computer Vision [7]. In our approach, we apply the general adversarial network for data augmentation. GAN has seen notorious growth since its development by good fellow [8]. GAN has been applied in a variety of areas where computer vision being top. GAN as their name indicates comprises two artificial networks as shown in Fig. 2. Generator Network fakes the data sample by taking random noise as input, where data points can be audio-video or numerical data as in our case. Discriminator network functionality is identical to lawman where the output is labelled as [0, 1] based on originality of the generated data from the generator network. This result is repeatedly given as feedback to the generator to improve its performance. The word adversarial symbolizes that two t networks act as a rivalry or competitive network in order to increase their efficiency. The objective function of the entire process can be represented as (Fig. 2).

$$\min_{G} \max_{D} V(D, G) = \xi_{x \sim p\text{data}(x)}\left[\log D(x)\right] + \xi_{z \sim p(z)}\left[\log(1 - D(G(z)))\right] \quad (3)$$

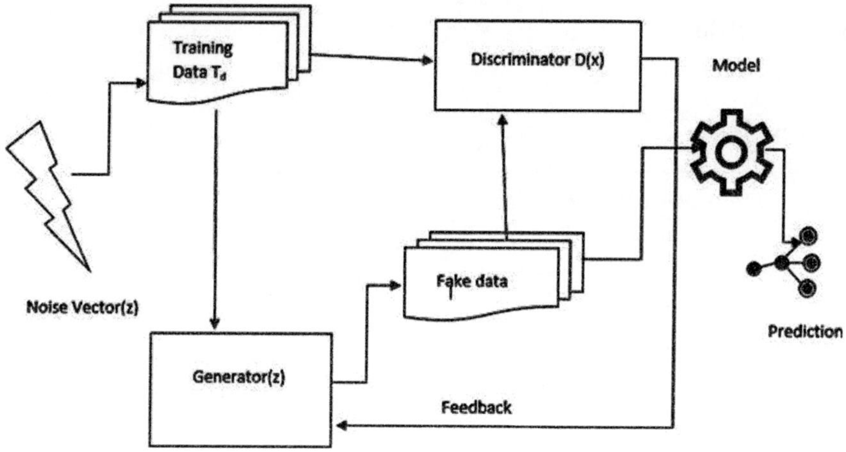

Fig. 2 General adversarial network model

4 Proposed Methodology

The proposed work applies a methodology to combine Batching and Data augmentation using GAN architecture. The data forgetting process focuses on two main aspects completeness and asymptotically faster algorithm design. The later mentioned factor can be resolved by incorporating by batching the training data.

The training data T_d is branched into a set of smaller batches $\{G_1 \ldots G_n\}$, The process of splitting the batches is based on the correlation threshold value of feature in T_d calculated using Pearson correlation method [9]. The two major motive of grouping the data is to avoid the overhead of retraining the system from scratch and also the batch T_d^i from which data was removed can alone be augmented using GAN. Note, that each G_i embodies unitary labelled data points I_i. The I_{noise} is a part of training data which user wishes to remove it for multiole reason like security breech or if the system no longer need its data. As a result the data has to be removed from system's database. It is searched among the batches and removed, the group from which data was removed is represented as G_{noise}, this batch can be utilized as knowledge base to the discriminator network $D(z)$ for classifying fake and real data. The generator network randomizes the noise vector and produces replacement data which can be used to train the model. Thus making system time accelerated. For experimental purposes, we make use of leakyRelu as activation function binary cross-entropy as loss function and Adam optimizer. The following parameter was selected because of their standard nature.

5 Algorithm

INFOGAN

Input : User $U_i \rightarrow$ *Request*; *TrainingData* T_d
Output Training Data T'_d, Where $I_{noise} \not\subset T'_d$

---

for each Feature f_i in T_d **do**
$r = \frac{n\sum f_i f_{i+1} - \sum f_i f_{i+1}}{\sqrt{n\sum f_i^2 - (\sum f_i)^2}}$
end for
$SelectedFeature \leftarrow Max(r)$
$Group(T_d) \rightarrow \{G_i \ldots G_1\}$
$UserU_i \rightarrow Request(T_d)$
$Initializerequest(I_d) \rightarrow I_{noise}$
Locate $G_i \subset I_{noise}$
Delete I_{noise} from G_i
$G_i \rightarrow GeneratorNetwork$
$G(z) \rightarrow GeneratedData$
Append G(Z) into G_i

$$Discriminator_{Feedback} = \begin{cases} 1, & \text{if } G_d \rightarrow real. \\ 0, & \text{if } G_d \rightarrow fake. \end{cases} \tag{4}$$

Generator network $\leftarrow Discriminator_{Feedback}$

---

6 Data and Result

The objective of our methodology is best applicable for database containing personal information. The befitting database is available as open-source patient Information dataset [10]. This data is collected from 130 hospitals ranging from the year 1998 to 2008 is utilized for our experiment. This dataset is used to model a system with decision tree for predicting the readmission of patients. The structure of the data contains approximately 100,000 patient data with 50 features including personal information like race, age, and lab test results. The reason for forgetting can happen in two cases

Case 1 The individual patient can request to remove their information when they speculate that their data is tampered.
Case 2 When hospital management considers the patient information is irrelevant.

The impact of data forgetting can be visualized when multiple user requests for data forgetting. We have studied the progress of our methodology by varying the number of user's requests. The grouping of data into batches is based on Pearson correlation which resulted in ten batches. Each individual batch contains approximately 9000 individual patient data. The procedure is followed as a comparison for traditional machine unlearning process and for the proposed steps. In the following graph, the number of user's request is represented in the x-axis and y-axis represents the accuracy obtained by training decision tree. The graph clearly depicts that the accuracy of decision tree decreases by 20% for traditional method as the number of data forgetting request increases, whereas the accuracy is maintained consistently when GAN was Incorporated.

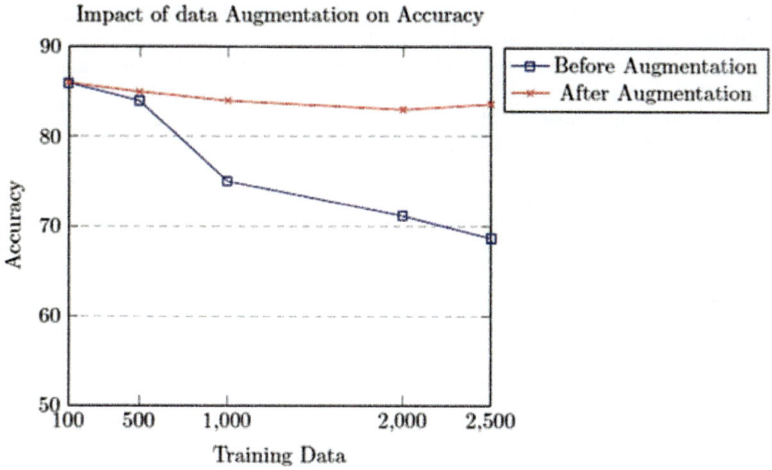

7 Conclusion

In the previous sections, we have clearly discussed the surging demands and challenges of machine unlearning in various fields. Although the forgetting process comprises of two types, we have focused only on Type-1 data forgetting. The results discussed in the previous section clearly depict that data augmentation using GAN can clearly increase the accuracy with a reduced training time of the system model. Future research in this area converges on fixating Type-II unlearning process, where the complete lineage of data will be removed by figuring all the relation among the individual data points.

References

1. BBC News-Technology. https://www.bbc.com/news/technology-49808208
2. Schlegl T, Seeböck P, Waldstein SM, Langs G, Schmidt-Erfurthb U (2019) f-AnoGAN: fast unsupervised anomaly detection with generative adversarial networks. Med Image Anal 54: 30–44, 99–110

3. Eiter T, Kern-Isberne G (2019) A brief survey on forgetting from a knowledge representation and reasoning perspective. Künstl Intell: 9–33. https://doi.org/10.1007/s13218-018-0564-6
4. Cao Y, Yang J (2015) Towards making systems forget with machine unlearning. In: 2015 IEEE symposium on security and privacy, San Jose, CA, pp 463–480. https://doi.org/10.1109/sp.2015.35
5. Jilek C, Runge Y, Niedeŕee C, Maus H, Tempel T, Dengel A, Frings C (2019) Managed Forgetting to support information management and knowledge work. https://doi.org/10.1007/s13218-018-00568-9
6. Reuter L, Berndt JO, Ulfert A-S, Antoni CH, Ellwart T, Timm IJ (2019) Intentional forgetting in distributed artificial intelligence. Künstl Intell 33(1):69–77
7. Cao Y et al (2019) Recent advances of generative adversarial networks in computer vision. IEEE Access 7:14985–15006. https://doi.org/10.1109/ACCESS.2018.2886814
8. Goodfellow IJ, Pouget-Abadie J, Mirza M, Xu B, Warde-Farley D, Ozair S, Courville A, Bengio Y (2014) General adversial nets. In: NIPS' 14: proceedings of the 27th international conference on neural information processing systems, vol 2, pp 2672–2680, Dec 2014
9. Weston J, Mukherjee S, Chapelle O, Pontil M, Poggio T, Vapnik V (2000) Feature selection for SVMs. Adv Neural Inf Process Syst 13(7):668–674
10. Dataset. https://www.kaggle.com/brandao/diabetes

IoT-Based Traffic Congestion and Safety Management with Street Light Control System

Utkarsh Maheria, C. Fancy, and M. Anand

Abstract Right now, where vitality is a significant concern around the world, it is our duty to spare vitality adequately. With the headway in the innovative segment, mechanization framework has had a significant impact in everyday existence with computerization being favored over the manual framework, offering ascend to the indispensable need of making things increasingly robotized. One of the primary motivations behind this task is to improve the efficiency and concoct a shrewd framework which can control the road lights and settle on choices by thinking about the light power. Here the day and night mode can be distinguished by fixing a specific power an incentive on photodiode sensor. Another exceptional piece of this task is to keep up the traffic signal consequently with no assistance of traffic police by estimating the blockage on a specific street. In the current innovatively affected world, automobile overloads during times of heavy traffic are one of the prime concerns. During times of heavy traffic, crisis vehicles like rescue vehicle stall out in jams. Because of this, crisis vehicles don't arrive at their goals in time, now and again which prompts loss of human lives. In this way a framework has been formulated which is utilized to give leeway to any crisis vehicle by turning all the red lights to green on the way of the crisis vehicle, henceforth giving a total green wave to the ideal vehicle. Utilizing the comparative innovation, Theft discovery has been executed giving one more advantage to the client of the framework in recognizing the taken vehicle using the RFID reader.

Keywords IoT system · Congestion · Safety management · Street light control system

U. Maheria · C. Fancy (✉) · M. Anand
SRM Institute of Science and Technology, Kattankulathur, Kancheepuram, Tamil Nadu, India
e-mail: fancyc@srmist.edu.in

D. J. Hemanth et al. (eds.), *Artificial Intelligence Techniques for Advanced Computing Applications*, Lecture Notes in Networks and Systems 130,
https://doi.org/10.1007/978-981-15-5329-5_46

1 Introduction

In this era, the vitality is a basic segment without which we can't envision living or playing out any of the undertakings. Vitality is a need. With the assistance of fuel vitality, we run our vehicles and move from here to there. With coal, power is given. In this way, it appears glaringly evident that how much valuable the vitality is for our lives. With such significance, it is exceptionally essential that we spare vitality also for our forthcoming future needs.

One of the primary motivations behind this undertaking is to design an astute framework that will spare vitality by controlling the road light dependent on the light force falling upon the sensor just as the movement of individuals around them. The working of road lights can be effectively observed through the use of IoT. The framework will likewise assist with keeping up the traffic signal naturally with no assistance of traffic police by estimating the clog on a specific street. During times of heavy traffic, crisis vehicles like rescue vehicle, stall out in jams. Because of this, crisis vehicles are not able to arrive at the accident location in time, coming about into loss of human lives. Along these lines a framework should be executed which can be utilized to give leeway to any crisis vehicle by turning the red light to green on the way of the crisis vehicle, henceforth giving a green wave to the ideal vehicle [1, 2]. Another execution is of the burglary discovery framework which is another vital piece of the proposed framework where the taken vehicle will be distinguished by the RFID system introduced adjacent to the traffic light and the framework will naturally send the message to the administrator just as all the police headquarters about the area of that specific sign. Consequently, this undertaking endeavor to coordinate answers for traffic-related issues just as vehicle robbery into one framework alongside the use of IoT for checking the road lights [3].

1.1 Purpose

The aim of this project is to coalesce a set of different systems proposed for solving different problems involving traffic congestion, vehicle theft, and energy wastage and to develop a prototype with improved efficiency and addition of a few more functionalities along with providing IoT connectivity for monitoring the status of streetlights as well ensuring respite to the technicians who usually wandering around the streets looking for a faulty light to repair [4]. The system will not only help in ensuring safety to the vehicles but also help in saving lives of people involved in an accident or suffering from any major injury by allowing the ambulance to pass through the traffic and reach its destination well within time.

1.2 Scope

This project proposes an embedded system integrated with different functions focusing on traffic and vehicle theft-related problems and solving them along with the application of IoT for monitoring the streetlights. The traffic congestion is dealt with installing sensors along the road which would measure the density and accordingly change the traffic lights [5]. The RFID tags installed with the emergency vehicles would also help in switching the red light to green for easy passage during emergency situations. A similar concept is used with vehicle theft detection where the tags help in identifying the vehicle and generating an alert message to be sent to the police as well as the traffic administrator. This will be the main advantage of the system which will help the police to track down stolen vehicles easily.

1.3 Need for Traffic and Safety Management

Although there have been many proposals in the same scope, nobody has yet devised an integrated system that could implement all these separate functions into one. We wish to create a coalesce of such systems which can

- Help solve the congestion problems that occur on a day to day basis at the traffic signal.
- Ensure that the ambulance gets an easy passage through all traffic signals without having to wait for the light to turn green or other vehicles to grant passage.
- Make it easier to identify faulty streetlights.
- Help the police detect stolen vehicles.
- Tackle economic losses that occur on a daily basis due to delayed work and save lives of people involved in an accident.

2 Related Works

2.1 Works Related to Traffic Congestion Systems, Street Light Monitoring, and Vehicle Theft Detection

Various blockage checking and the executive's frameworks have been proposed in writing over the previous years however, they are restricted to the design part and lack experimental results or only the prototype has been proposed. Sadhukhan and Gazi [6] is one such paper where the author has displayed traffic clog control. Here the proposed work comprises of traffic density checking module (TDMM) and traffic the management module (TMM). Where TDMM measures the length of traffic line present before signal traverse point so as to decide the density of traffic clog.

Then again, TMM endeavors to powerfully change the activity time of traffic signals dependent on the assessed density of traffic blockage on various streets interfacing with the intersections area so as to check the clog appropriately. Researchers implement a traffic system with implementation of IoT using networking-based idea with several software applications which takes traffic updates from the application users, based on their location [7]. Swathi et al. [8] utilizes shortest route identifications, IR sensors to gauge traffic thickness. But, IR sensor is affected by temperature and stickiness. Thus, the outcome which was created by the IR sensor was not precise. Researchers have mentioned the traffic system scenario inside Pakistan where traffic density is estimated utilizing camera and a few sensors [9]. In view of the sensors' information, Pakistan controls the traffic. They likewise utilized a smoke sensor to identify the crisis circumstance, for example, fire mishap. Camera sensor might be influenced by downpour and mist. In addition, it's not savvy.

This paper presents a method to detect the stolen vehicle using the RFID technology where it sends the location using GPS and allows the ambulance to pass through the signal using the RFID tags itself [10]. It only makes it more complex as there is no such requirement of GPS and IoT instead the location can just be sent from the signal itself using gsm module.

Mentions the use of IoT in monitoring automated street lights which switch on and off based upon ldr sensor readings [11]. But it's just a proposed idea and no implementation as a prototype have been shown. It also uses Zigbee module to combine the output of 3–4 led lamps and operate as one unit.

Implements the IoT-based street light switching from dim to on to off and monitoring the status of street light using IoT along with detecting the temperature around the street light [12].

3 Proposed Method

The designed system consists of several sensors which perform various functions helping in providing multi-functionalities to the system. The system is divided into the following modules.

3.1 Traffic Congestion Clearance

This part consists of the clearance of traffic based upon the density as calculated by the ultrasonic sensor and accordingly the traffic lights are controlled. If the density increases at a particular signal then the traffic lights are turned to green thereby eliminating the chances of congestion occurring at a particular intersection.

3.2 Street Lighting and Monitoring

This part consists of the PIR and LDR sensor employed for the footpath lights & the lights on roads, respectively. The PIR sensor detects the motion of the people and automatically increases the intensity of the footpath light while the LDR sensor is responsible for turning the road lights ON and OFF in the absence and presence of light, respectively. The status of light is monitored through the application of IoT where the data is sent to the cloud and analysis of the output is done through graphs.

3.3 Clearance and Alarm for Ambulance

As soon as the ambulance's RFID tag comes under the radar of RFID reader installed beside traffic signal, the signal automatically turns green thereby allowing the ambulance to pass through the traffic signal smoothly and to reach its destination within the time frame. Alongside this, we have added an alarm that triggers as soon as the rescue vehicle is detected and informs the public to clear the way for an ambulance. The LCD screen put up at the traffic signal also displays a message to clear the way for ambulance.

3.4 Vehicle Theft Detection with Alarm

This part employs the use of RFID tags and the same RFID reader installed on the traffic lights. As soon as the RFID reader identifies the tag associated with the stolen vehicle it turns the signal red and makes the system send messages to all the police stations about the area of traffic signal where the stolen vehicle is present. Another implementation is of the alarm present at the signal which goes off informing the public about presence of theft vehicle and displays the license no. of theft vehicle on to the LCD screen installed at the junction.

The Fig. 1 shows the analytics of the street light monitoring system through which we track if the street light has been working or not while Fig. 2 has been designed to explain the proposed integrated system.

4 Conclusion

This project is proposed for providing an integrated system ready to use to solve some of the major problems that our society currently faces. This project aims to help solve the traffic congestion problem, save time, and as many lives as possible. It

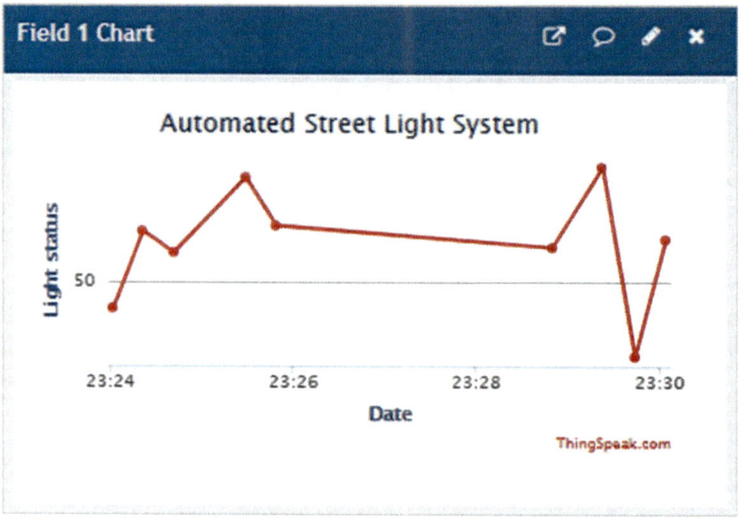

Fig. 1 Analytics of the street light monitoring system

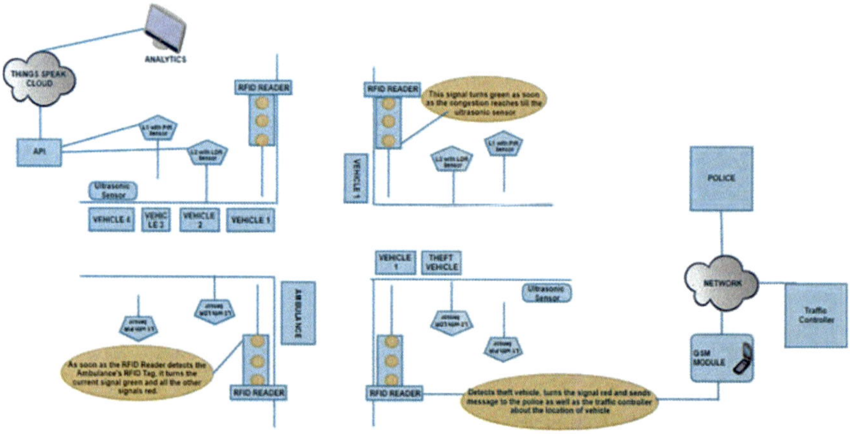

Fig. 2 Proposed integrated system

also leads to subsequent saving of energy which is the prime need in today's world. It also gives an edge to the police in finding the stolen vehicle and gives respite to the technicians in saving their time to manually identify the faulty street lights.

5 Future Work

Currently, many of the signals are being operated with the DC current while Solar energy lacks implementation in the real world traffic signals which can ensure a lot more energy saving. Also, better sensors with good range can be used to detect the density of the vehicle and different algorithms including those of Machine Learning can be employed to switch the traffic signal.

References

1. Shinde SM (2017) Adaptive traffic light control system. In: 2017 1st international conference on intelligent systems and information management (ICISIM)
2. Masum AKM, Chy MKM, Rahman I, Uddin MN, Azam KI (2018). An Internet of Things (IoT) based smart traffic management system: a context of Bangladesh. In: 2018 international conference on innovations in science, engineering and technology (ICISET). https://doi.org/10.1109/iciset.2018.8745611
3. Bhate SV, Kulkarni PV, Lagad SD, Shinde MD, Patil S (2018) IoT based intelligent traffic signal system for emergency vehicles. In: 2018 second international conference on inventive communication and computational technologies (ICICCT)
4. Saifuzzaman M, Moon NN, Nur FN (2017) IoT based street lighting and traffic management system. In: 2017 IEEE region 10 humanitarian technology conference (R10-HTC)
5. Savla DV, Savla HR, Kansara KB (2018) Brainy streets an automatic lighting system. In: 2018 2nd international conference on inventive systems and control (ICISC)
6. Sadhukhan P, Gazi F (2018) An IoT based intelligent traffic congestion control system for road crossings. In: 2018 International conference on communication, computing and internet of things (IC3IoT). https://doi.org/10.1109/ic3iot.2018.8668131
7. Mahalank SN, Malagund KB, Banakar RM (2016) Device to device interaction analysis in IoT based smart traffic management system: an experimental approach. In: 2016 Symposium on colossal data analysis and networking (CDAN). https://doi.org/10.1109/cdan.2016.7570909
8. Swathi K, Sivanagaraju V, Manikanta A, Kumar SD (2016) Traffic density control and accident indicator using WSN. Int J Mod Trends Sci Technol IJMTST Traffic 2
9. Javaid S, Sufian A, Pervaiz S, Tanveer M (2018) Smart traffic management system using Internet of Things. In: 2018 20th international conference on advanced communication technology (ICACT), pp 393–398
10. IOT based stolen vehicle detection and ambulance clearance system. Int J Eng Res Technol IJERT. In: ICONNECT—2017 conference proceedings, ISSN: 2278-0181. Published by www.ijert.org
11. Mary MCVS, Devaraj GP, Theepak TA, Pushparaj DJ, Esther JM (2018) Intelligent energy efficient street light controlling system based on IoT for smart city. In: 2018 international conference on smart systems and inventive technology (ICSSIT). https://doi.org/10.1109/icssit.2018.8748324
12. Dheena PPF, Raj GS, Dutt G, Jinny SV (2017) IOT based smart street light management system. In: 2017 IEEE international conference on circuits and systems (ICCS). https://doi.org/10.1109/iccs1.2017.8326023

Effective Networking on Social Media Platforms for Building Connections and Expanding E-commerce Business by Analyzing Social Networks and User's Nature and Reliability

R. Lavanya, Anushka Saksena, and Aparnika Singh

Abstract "One's network is their net worth" is the common saying most successful people follow. Building a good network can help in both personal growth and that of one's business. However, a network must be reliable and involves people with a positive attitude, a likeable nature, and an affinity to attract other people through their power of influence on social media. These are a few characteristics that help in establishing a network that is trustworthy and in targeting the right audience who can help expand an e-commerce business through their strong impact of reviews, posts, likes, and friend circle. Through this paper, we aim to target people who are more socially likeable and open to building new social connections on social media platforms (Facebook is our platform of focus), which can be used for building individual connections or advertising and promoting products through effective networking.

Keywords Social networks · Nature prediction · Networking

1 Introduction

Data mining may be defined as the procedure of detecting discrepancies, similarities, and patterns in humongous volumes of datasets to anticipate results. Data mining is vital since it is capable of identifying and clearing all noise in data, and then making good use of relevant data to determine likely outcomes and increase the rate of making informed decisions. The best insights can be obtained when large and complex datasets are used. Advances in processing speed have facilitated the

R. Lavanya (✉) · A. Saksena · A. Singh
Department of Computer Science, SRM Institute of Science and Technology, Kattankulathur, Chennai, India
e-mail: lavanyar@srmist.edu.in

A. Saksena
e-mail: anushkasaksena_amit@srmuniv.edu.in

A. Singh
e-mail: aparnikasingh_pratap@srmuniv.edu.in

D. J. Hemanth et al. (eds.), *Artificial Intelligence Techniques for Advanced Computing Applications*, Lecture Notes in Networks and Systems 130,
https://doi.org/10.1007/978-981-15-5329-5_47

shift to easy and automated data analysis as opposed to tedious and time-consuming practices used over the past few years [1].

Over the past few years, media has witnessed a huge transformation. People tend to build friends online, connect with people, and sometimes refer to them for advice when they cannot confide in people they know personally. Consumers are looking for information regarding products and services using social networking sites. Not only does a social media platform benefit a plethora of e-commerce and emerging businesses but also helps with networking. Adding a reliability factor to this eases the process and makes it more trustworthy [2, 3].

Understanding a user's nature by analyzing the number of likes, friendships initiated, and likes received eases the process of communication with the person. For example, if a user X has initiated a 100 friendships and user Y has initiated only 10 friendships, it can be said that user X posses a higher tendency to be approachable and active on social media as opposed to user Y. User X may also be more aware and have more knowledge about current trends and events. The high friend count implies a greater social circle which indeed implies higher outreach. This can be beneficial for connecting with new people or selling more products, as per one's requirements. A person is more likely to accept requests from someone having mutual friends and more willing to buy products tried and tested by someone in their social circle subject to good reviews being received.

1.1 Social Networks and Social Media

Social media, when introduced, was primarily used for socializing with people online and forming groups with like-minded people who share common interests. Gradually, this gap was bridged when people started using social media to expand their e-commerce business and network efficiently in order to connect with other people worldwide [2].

1.2 About Facebook

Facebook is one of the most widely used and popularly known social media platforms. It can be verified from Internet sources that Facebook leads in number of users when compared to any other platforms. This gives us more amount of data and parameters to analyze. That was our motivation behind choosing Facebook over any other platform. Facebook also has groups and pages for people with similar interests to connect [2, 4].

2 Scope of Paper

(i) Compare accuracy, precision, and recall values for different algorithms and identify which one performs best.
(ii) Display the social network graphically and identify which users are most likeable and approachable based on parameters like likes received, friendships initiated, items liked.
(iii) This inference can be used to expand one's personal network and build more effective connections or promote products and services by targeting the most popular and influential users based on their likeability of their social circle.

3 Related Work

Yukuo C., Jing Z. et al. worked on predicting trust in Alibaba e-commerce platform. They tested the model built by them called e-trust and e-trust s on five datasets—Ciao, Epinions, Advogato, Alibaba, and Ali-large. Epinions and Alibaba suited their purpose best. Accuracy and efficiency performance were measured for e-trust and e-trust s models. E-trust takes hours for single iteration while e-trust s outperforms CF by 30.09% (return rate), 45.45% (behavior rate), and 42.08% (rating rate) [5].

Ida M., Francesco et al. aimed to identify early adopters and their applications in recommendation systems. Early adopters are people who are very well versed with new upcoming trends and people who follow them can also be targeted for networking and advertising purpose. They collected data from Yahoo's toolbar and prepared an early adopter graph. Values were compared to previously used Jaccard and Bernoulli values. Precision-at-k and log-likelihood were used to evaluate parameters like similarity and percentage of recommended pages visited by the user. An improvement of 20% for precision-at-15 was obtained. One limitation is that some amount of noise and sparsity may be present in data [6].

Lars B. et al. combined supervised random walks (SRW) with SRW with logistic regression to predict and recommend links in social networks. They worked on Facebook data and evaluated AUC and precision values. The major benefit of this work was that it can be expanded to missing links and detect anomalies in data. The model outperforms unsupervised approaches and feature extraction approaches. The drawback is that real networks are extremely sparse [7].

Mohsen J. et al. used a matrix factorization technique with trust propagation for recommendations in social networks. They applied CF and matrix factorization on Epinions and Flixster dataset. The model captures the transitivity of trust and is evaluated for precision and recall values. The problem of cold start may be an issue [8].

Yi-Fen Chen analyzed herd behavior in purchasing books online. He defines five hypothesis and analyzes each one of them using statistical models and parameters like standard mean deviation and ANOVA. The study of each hypothesis results in a different inference. The conclusion vital for our project is that "people pay attention to reviews from other users" [9].

Liu S., Zhang L., Yan Z. predicted pair-wise trust in social media networks based on ML. They used LR, naïve Bayes, and decision tree algorithms on Epinions, Flixster, and Ciao dataset to evaluate precision and recall values using CM. They analyzed the feature extracting process and recognized its crucial role in social media networks. According to their research, a lot of issues based on trust labeling exist in datasets currently [10].

Amit G., Francesco B., and Laks L. strived to learn influence probabilities in social networks. They worked on Flickr dataset using Bernoulli distribution and Jaccard index to evaluate precision and recall values. The proposed model can predict time in which users perform action but all assumptions are centered around viral marketing which ignores the effect of time and assume edges have constant influence probabilities [11].

Yoon-Jo-Park aimed to predict the usefulness of online reviews using LR and random forest using Amazon dataset. The shortfall of this project was that it did not verify the reliability of the view characteristics [12].

Early prediction of market success can greatly improve any e-commerce business by predicting repeated ratios and sales. One can know if user is a good target to expand an e-commerce business and be a reliable and loyal customer over time [13].

Table 1 compares the numerous algorithms, datasets, and performance measures used in order to come up with the most efficient system. Our work is based around the detailed study of this comparison of methodologies.

4 System Methodology

4.1 Framework

Figure 1 describes the following procedure. We begin by comparing the different datasets and cleaning and preprocessing data using interpolation. It is vital to extract features that are essential for reliability prediction and understanding the nature of the user. We then train algorithms with extracted features. The accuracy, precision, and recall values are compared, and the csv file is passed as input to R studio where graphs and plots are visualized.

Table 1 Comparison of different trust prediction techniques

	Objective	Algorithm/strategy used	Dataset	Relevance	Performance measure	Limitation
1.	Trust relationship prediction in Alibaba E-Commerce [5]	e-Trust	Alibaba Epinions	Picks up attribute features while ignoring network structures because of labeled relationships	Precision of top *k%*	Data sparsity is an issue that has an effect on the results
2.	A trust-based collaborative filtering algorithm for e-commerce recommendation system [14]	Slope one, weighted slope one, bipolar slope one	Amazon dataset	Combination of trusted data and user similarity has enhanced the prediction accuracy	Mean absolute method (MAE), RMSE	Cold start issues
3.	Early adopters role in web page recommendations [6]	Early adopter model	Yahoo	Identifying early adopters and recommending users who follow them	Precision	Exploit user browsing and behavior data to build steadfast n/w of influence
4.	Predicting and recommending links in social networks [7]	Supervised random walks (SRW), LR	Facebook data	Can be expanded to missing links, anomaly detection	AUC, Precision	Real networks are extremely sparse
5.	Predict pairwise trust in social media using ML [10]	LR, naïve Bayes, decision tree	Epinions, FilmTrust, Flixster, and Ciao	Analyzed the feature extracting process and realized its importance in social media networks	CM, precision, recall, accuracy	Research issues based on trust labeling
6.	Predicting trust relations among users in a social network: the role of influence, cohesion, and valence [15]	Degree discounted out-link similarity	Product dataset	It develops a trust-based ranking of users that is in very good agreement with the ground truth	Chi-squared test and a subsequent Cramer's V metric	Need to improve the verification method to test widespread effects of trust inference process
7.	Learning influence probabilities in social networks [11]	Bernoulli distribution, Jaccard index	Flickr dataset	Predicts time in which users perform action	Precision, recall	All assumptions are centered around viral marketing which ignores the impact of timing and assumes edges have consistent influence probability

(continued)

Table 1 (continued)

	Objective	Algorithm/strategy used	Dataset	Relevance	Performance measure	Limitation
8.	Predicting the helpfulness of online customer reviews across different product types [12]	SVR, LR, R and F, and M5P	Amazon dataset	the determinant factors for each product category were explored which further helps us realize what parts are more important when considering review rating	MAE	Did not verify the reliability and validity of their view characteristics extracted by LIWC
9.	Review rating prediction based on user context and product context [16]	RRP model based on user and product context	English dataset	Aiming at the problem of user context dependency of sentiment words, a review rating the prediction method based on review content and user context is proposed	Recall, precision	Method does not regard common affection of user and product on review texts in capturing the sentiment of review texts. Regardless of existing methods or our method, only the review content information is modeled in RRP
10.	Recommendations in social network using matrix factorization [8]	CF, matrix factorization	Epinions, Flixster	Transitivity of trust are captured in the model	Precision, recall	Cold start

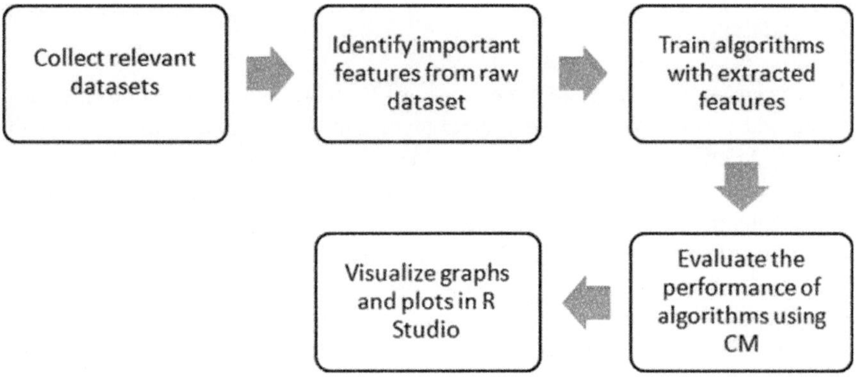

Fig. 1 Pipeline of workflow

4.2 Modules

4.2.1 Confusion Matrix

This is defined using python libraries in Jupyter notebook.

A CM represents the performance of a classification model on a set of test data whose real values are familiar. We store values of accuracy, precision, and recall for the three models—LR, naïve Bayes, and classification algorithm.

$$\text{Accuracy} = (\text{TP} + \text{TN})/(\text{TP} + \text{TN} + \text{FN} + \text{FP}) \tag{1}$$

Equation (1) represents how frequently the classifier is right in predicting values [10].

$$\text{Precision} = \text{TP}/(\text{TP} + \text{FP}) \tag{2}$$

Equation (2) represents how frequently a value that is predicted as true correct [10].

$$\text{Recall} = \text{TP}/(\text{TP} + \text{FN}) \tag{3}$$

Equation (3) represents the true positive rate [10].

It should be noted that all three Eqs. (1), (2), and (3) must be calculated and not just one to determine best performance. Equations (2) and (3) normally have an inverse relationship.

4.2.2 Classifiers

Logistic regression. There are three types of LR algorithms, namely binary, multinomial, and ordinal. We use the ordinal algorithm as it can categorize data with ordering. Ordering levels make it easier to identify users with stronger influence and social likeability. A threshold value is used to classify data into appropriate classes. For example, predicted value ≥ 0.5 then classify user as friendly. One must aim to maintain a balance between specificity and sensitivity while selecting a threshold [10].

Naïve Bayes. With relation to our dataset, we assume that no pair of features is dependent. For instance, we assume number of likes has nothing to do with the number of friendships initiated by a user. Also, each feature is assigned equal weight or importance. Knowing only the likes and friend count alone cannot predict the outcome accurately [10].

$$P(A|B) = [P(B|A)P(A)]/[P(B)] \tag{4}$$

Equation (4) represents the basic formula for naïve Bayes classifier where *P(A|B)* is the posterior probability and *P(B|A)* is the likelihood of the events occurring [10].

Classification Tree. CTs sort tree based on feature value and have no parameters to tune. However, the tree needs to be built each time new parameters are introduced. Also, while pruning the latter minority nodes may be pruned, this is a limiting factor in case of imbalanced datasets. Combining pruning with sampling is a good way to mitigate this issue [10].

4.2.3 Algorithm Comparisons

On comparing the accuracy, precision, and recall values, we observe that LR performs best on the dataset used followed by naïve Bayes algorithm followed by classification tree.

Figure 2 represents the format the output is obtained on comparison of three

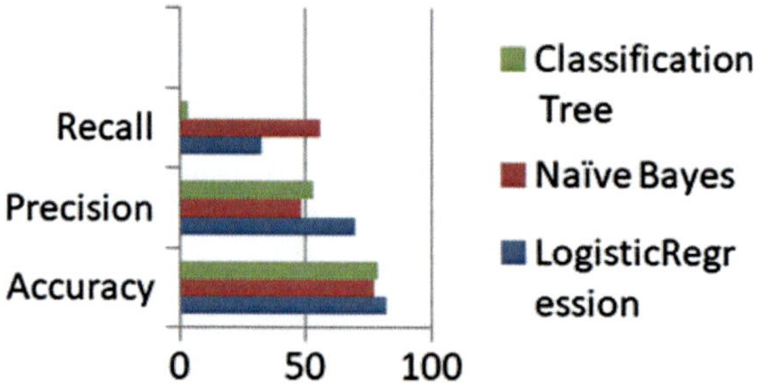

Fig. 2 Parameters evaluated and compared on different algorithms

algorithms using confusion matrix. It is evident that logistic regression performs best as opposed to classification tree which performs the poorest.

4.2.4 R Studio for Visualization of Graphs

To further enhance the appeal of our project and make it easily understandable by any user, we use R studio to visualize the social media network graphically and plot functions like number of likes against user id. The analyzed dataset is imported as a csv file into R for further analysis and visualization.

Figure 3 shows the range of user id against the number of likes on posts. This gives us a fair idea of which user is more active and has a tendency to like content.

4.3 Pseudocode

Step 1 Preprocess and interpolate data
Step 2 Train training data: X_train, Y_train
Step 3 Define confusion matrix (CM)
Step 4 Import estimator object (model)
Step 5 Create instance of estimator
Step 6 Use training data to train estimator
Step 7 Evaluate model for accuracy, precision, and recall using CM defined in step 3
Step 8 Repeat for other methodologies—naïve Bayes and classification tree
Step 9 Compare outcomes of step 7.

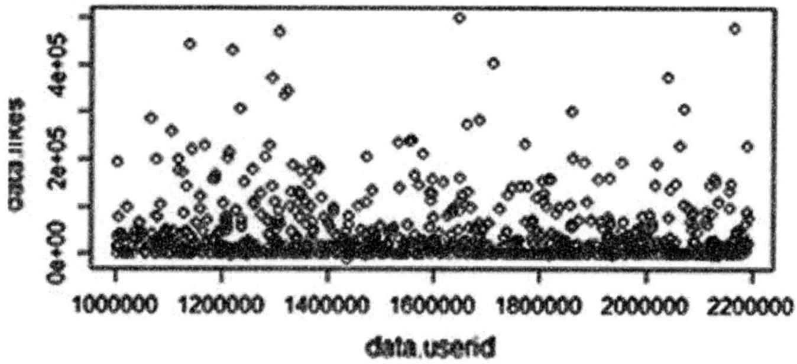

Fig. 3 Scatter plot representation of number of likes against user id

4.4 Datasets

Kaggle. Kaggle has one of the widest collections of datasets online to suit every purpose. While the dataset contained over 1500 rows, it had a lot of discrepancies that we mitigated using interpolation. It was the dataset that suited our requirements best. Also, user id is in numeric form and therefore can be used as input to algorithms that require data sorted in ascending or descending order. We were able to plot graphs in R studio using various parameters and analyze social network. The rows of the dataset were as follows: user identification number, age, date of birth-day, month, year, gender, tenure, number of friends, number of friendships initiated, likes, likes received along with sources (desktop or application). It was relatively easy to identify which user was more friendly in nature by analyzing the amount of friendships_initiated and likes. This also gave an insight into which user would be more willing to accept new proposals and more prone to building new connections [10].

Network data repository. This site specializes in datasets pertaining to social media analysis, and hence, is best suited for graphical analysis. Parameters of this dataset are as follows: number of degree, number of nodes and edges, average degree, global clustering coefficient, etc. One of the biggest drawbacks of this site is that its dataset is only available in.mtx format. Converting this into a csv file for analysis is quite challenging without proper technical tools and knowledge [10].

5 Result and Discussions

On comparing the algorithms, we found that LR gives the best accuracy, precision, and recall values.

A precision–recall curve can also be obtained for the same. It is evident that LR has the best outcome on the dataset used.

Table 2 demonstrates the output from different classifiers obtained. We obtain the best accuracy and highest precision values using logistic regression and lowest recall value as compared to other algorithms.

Table 2 Evaluating and comparing accuracy, precision, recall values for different algorithms

No.	Parameter evaluated	Logistic regression	Classification tree	Naïve Bayes
1	Accuracy	81.9778	78.8000	77.0667
2	Precision	69.8276	53.0103	48.3899
3	Recall	32.5628	36.2814	55.8794

6 Conclusion

We observe that LR provides the best output for the dataset considered. A high precision and low recall value are obtained. This implies that a lot of positive instances are missed (high false negative) but those predicted to be positive are undeniably positive (low false positive). We obtain an accuracy, precision, and recall value of 81.9%, 69.82%, and 32.5%, respectively, using LR. The same dataset is fed into R studio for a visual representation of social networks and plotting various parameters like likes against user id and friendships initiated versus user id which accurately represents the nature of the user. The proposed project can be improved by changing the threshold value of LR. It should also be kept in mind that naïve Bayes' assumptions are not fully practical in real life as all features are not independent of each other [10]. In order to improve nature prediction and trust prediction a credibility score must be assigned to each user. This will ensure 100% trustworthiness.

References

1. Bharati M, Ramageri B (2010) Data Mining Techniques And Applications. Indian J Comput Sci Eng 1
2. Singh M, Singh G (2018) Impact of social media on e-commerce. Int J Eng Technol UAE 7: 21–26. https://doi.org/10.14419/ijet.v7i2.30.13457
3. Crandall D, Cosley D, Huttenlocher D, Kleinberg J, Suri S (2008) Feedback effects between similarity and social influence in online communities. In: Proceedings of the ACM SIGKDD international conference on knowledge discovery and data mining, pp 160–168. https://doi.org/10.1145/1401890.1401914
4. Falch M, Henten A, Tadayoni R, Windekilde I (2009) Business models in social networking
5. Cen Y, Zhang J, Wang G, Qian Y, Meng C, Dai Z, Yang H, Tang J (2019) Trust relationship prediction in Alibaba e-commerce platform. IEEE Trans Knowl Data Eng: 1–1. https://doi.org/10.1109/tkde.2019.2893939
6. Mele I (2012) Early-adopter graph and its applications to web-page recommendation. In: CIKM'12: Proceedings of the 21st ACM international conference on information and knowledge management. Sapienza, University of Rome, Italy, Aristides Gionis and Francesco Bonchi, Yahoo!ResearcherBarcelona, Spain), pp 1682–1686, Oct 2012
7. Backstrom L, Leskovec J (2011) Supervised random walks: predicting and recommending links in social networks. In WSDM'11, pp 635–644
8. A matrix factorization technique with trust propagation for recommendation in social networks-RecSys'10. In: Proceedings of the fourth ACM conference on recommender systems, pp 135–142, Sept 2010
9. Chen Y-F (2008) Herd behavior in purchasing books online. Comput Human Behav
10. Liu S, Zhang L, Yan Z (2018) Predict pairwise trust based on machine learning in online social networks: a survey. IEEE Access 1–1. https://doi.org/10.1109/access.2018.2869699
11. Amit G, Lakshmanan LVS Learning influence probabilities in social networks. University of British Columbia, Vancoucer, BC, Canada, and Francesco Bonchi, Yahoo!ResearcherBarcelona, Spain
12. Singh J, Irani S, Rana N, Dwivedi Y, Saumya S, Roy P (2016). Predicting the "helpfulness" of online consumer reviews. J Bus Res 70. https://doi.org/10.1016/j.jbusres.2016.08.008
13. Fourt LA, Joseph WW Early prediction of market success for new grocery products. American Marketing Association

14. Jiang L, Cheng Y, Yang L, Li J, Yan H, Wang X (2017) A trust based collaborative filtering algorithm for e-commerce recommendation system. J Ambient Intell Human Comput
15. Nikhita V, Srinivasan P Predicting trust relations among users in a social network: the role of influence, cohesion and valence. Department of Computer Science and Engineering Ohio State University, vedula@cse.ohio-state.edu, srini@cse.ohio-state.edu. Valerie L. Shalin Department of Psychology and Kno.e.sis Wright State University
16. Wang B, Xiong S, Huang Y, Li X (2018) Review rating prediction based on user context and product context. Appl Sci 8:1849. https://doi.org/10.3390/app8101849

Context-Based Sentiment Analysis on Amazon Product Customer Feedback Data

C. Sindhu, Dewang Rajkakati, and Chinmay Shelukar

Abstract A sentiment is a mindset, opinion, or decision that a sensation triggers. Sentiment Analysis is the procedure of using text analytics to mine different sources of data for opinions. It is also referred to as opinion mining and explores the emotions of a person in relation to certain products. The internet is indeed rich with resources relating to sentiment analysis such as feedback of a product on an e-commerce website and tweets sent by an individual on twitter social networking site. Context-based sentiment analysis is an exciting tool that continues to evolve as Artificial Intelligence and Machine Learning become more sophisticated. It is a precise science that can be used to aid new product development as well as for marketing and public relations. Analysis will be able to process product data at scale in an efficient way. Customer behavior can be analyzed and it can be found out what products are in demand and which products are not doing well. This model plans to perceive the properties of products and stores which play a pivotal role in growing sales. Word Sense Disambiguation is used to distinguish two similar sounding words from each other. Customer behavior can be analyzed and it can be found out what products are in demand and which products are not doing well. Through this, it is proposed to find out sentiments involved in each product whether they be positive or negative, based on which sentiment analysis will be carried out. Naïve Bayes and Support Vector Machine Algorithms have been used for classification. The model will be robust with a high precision and straightforward addition of attributes from various sources to the set.

Keywords Context-Based · Sentiment analysis · Support vector machine · Naïve Bayes · Customer feedback · Word sense disambiguation

C. Sindhu (✉) · D. Rajkakati · C. Shelukar
Department of Computer Science and Engineering, SRM Institute of Science and Technology, Kattankulathur, Chennai, India
e-mail: sindhucmaa@gmail.com

D. Rajkakati
e-mail: dewangraj7@gmail.com

C. Shelukar
e-mail: chinmayshelukar3@gmail.com

D. J. Hemanth et al. (eds.), *Artificial Intelligence Techniques for Advanced Computing Applications*, Lecture Notes in Networks and Systems 130,
https://doi.org/10.1007/978-981-15-5329-5_48

1 Introduction

Recently, sentiment analysis of product customer feedback, an application topic in text mining, and computer linguistics research has become very popular. We would like here to analyze the association of product feedback from Amazon with customers' ratings. We use conventional algorithms, including analysis by Naïve Bayes Algorithm (NBA) as well as Support Vector Machine (SVM) algorithm. The core of sentiment analysis is the function of the classification of texts, and different words contribute differently. For current sentiment analysis research, distributed word representation is most often used. Moreover, distributed word representation takes only semantic word information into consideration, but does not take the meaning of the language into account. One is able to improve one's interpretation of these algorithms by analyzing the results. These may also be an alternative to other forms of rating fraud methods. The fixed aspects of each product and store will be described [1]. Analysis will be able to process product data at scale in an efficient way. It builds on strategies for the recovery of knowledge and defines key terms that represent feelings [2]. The contexts will be used to generate supervised learning features. Project proposes to build a predictive model for acquired sales data and predict sales of each product at a specific store. Understand attributes of stores and products which will play a crucial role in intensifying sales. Most social networking sites publish their Application Programming Interface (API) [3] based on an analyst's viewpoints, encouraging researchers and developers to collect and analyze data. Users can publish their own content through various blogs, social networking channels, and websites, based on their point of view. Nonetheless, there are certain limitations in such online data that may impede the sentiment analysis process. Some flaws include that people may post feedbacks whose quality may not be guaranteed. Like online Spammers will post spam on the forum. Some spam is obsolete while some spam can have fake opinions.

Word Sense Disambiguation (WSD) is a challenge in Natural Language Processing (NLP) to decide the "sense" of a word by using the said word in a specific context, an often unconscious mechanism in people. WSD is a problem of conventional classification: If a word and its possible senses are specified in a dictionary, classify the word into one or more sensory classes in context. Description characteristics (adjacent words in this case) justify the classification. WSD is a Resolution process whose word meaning in a given context is triggered by the use of the term, a mechanism that does not seem to be usually present in individuals. Some challenges we try to address include a sentiment analytical system that relies upon sentiment-based tokens for its analysis. For comments that absolutely have underlying emotions, the program may not work well. These underlying emotions are known as inherent sentiments. An inherent sentiment is generally expressed by neutral terms, making it quite difficult to judge its sentiment polarity [4]. Like an instance, a phrase like "This Item is described as" has a challenge, this positive feedback consists of only neutral terms, which occur intermittently. With these limitations in mind, this model aims to solve these problems. In order to increase the categorization of study levels, more characteristics will be extracted and grouped into vectors. To perceive the existence of such an opinion within the reach of a certain product the subsequent action is for the emergence of such a sensation in the sense of a substance to be recognized.

This paper contains data from a variety of customer feedbacks obtained by Amazon.com from February to April 2014. Every product analysis must first be tested and put under scrutiny before it can be made available. Subsequently, each analysis should eventually have a ranking, which can be used as the basic truth. To explain the ground truth, in case of positive, negative, or neutral thought, it is like a sticker of certain opinions [5]. The rating is in fact based on the system is positive, negative, or neutral. Customer behavior can be analyzed and it can be found out which products tend to do well and which do not. Algorithms such as NBA and SVM will be implemented to detect pivotal terms in text and analyze its context. Use detected terms to generate supervised learning features. It will improve sales based on sentiments portrayed for each product. This approach aims to distinguish customers' positive and negative opinions about various products and create and design supervised research to polarize large quantities of feedback. Our repository contains customer feedbacks and assessments, which we obtained from Amazon Consumer Feedback. On the basis of this, we extracted the repository features and developed several supervised models. Not only do conventional algorithms like NBA, SVM, and K-nearest neighbor (KNN) come in these models [6] but a few other algorithms are used which will be talked about in Related Work. The precise nature of these models is contrasted and the biased behaviors towards attitudes are better understood.

2 Related Work

Further research has been done in previous years to understand the significance of text resources [7]. The studies published in previous years on sentiment analysis have increased, as can be seen from the Knowledge Web statistics. Sentimental analyzing or mining of opinions is one of the themes of this study [8], based on a bunch of texts, we can research people's opinions, perceptions, behaviors, feelings, issues, incidents, topics, and their characteristics. This approach is used in different ways. Like an instance, corporations always want to gain feedback and perceptions about their products and services for the public or customers. Prospective clients would also like to learn current users' thoughts and feelings before using a service or purchasing a product. Eventually, analysts use this knowledge to examine market trends and customer perceptions in-depth that may lead to better bond market prediction [9]. Nevertheless, to look up, track webpages, and spread information about what they contain continues to be a daunting challenge thanks to various sites augmentation. Generally speaking, each site contains a lot of thought, not always easy to decode in lengthy posts. In general, it is arduous for the common human reader to locate the related webpages and to summarize the information and opinions [10]. In fact, it actually is complicated and difficult to teach a sarcasm recognition system, since machines are currently not yet able to think like people. In sentimental analytics, profound neural networks are also common. Some researchers have utilized a convolutional network to mark the semantic position so that task-specific design engineering cannot be overtaken [11]. Instead, the authors suggested that Recurrent Neural Networks (RNN) be used to improve composition in tasks such as sensing feelings.

We want to use all conventional algorithms in this model including NBA, KNN, SVM, and tricks for profound learning. As we compare the precision of these models, we want to better know how these calculations function in activities such as the study of emotions [12]. A considerable amount of papers were surveyed for this project. The scope of this survey includes all there is in an online supermarket which is an abundance of items, it is a matter of what to choose in the end. Though not all items belong to the same crux some items may not be selling so much. Due to which it might be a waste in buying them [13]. The applications of sentiment analysis include product feedback, customer support, and reputation management. Through our specific model, we can identify which items in an online supermarket are the ones that sell the most and which are the ones that are not doing well in sales. A rather important usage of our project includes word sense disambiguation. Work hereafter includes testing the categorization scheme using other datasets. Most of the datasets used in the survey papers are either from Amazon or Twitter. The commonly used algorithms are Naïve Bayes, Support Vector Machine, Latent Dirichlet Allocation (LDA) [14], Rule-based Sentiment Analysis (RBSA) [15], bidirectional long short term memory (BiLSTM) [16] and Random Forest Algorithm (RFA) [17]. In Table 1, we have discussed the algorithms, methodology, measures, and demerits of various papers used for the survey.

3 Word Sense Disambiguation

According to a word and its potential meanings, as defined in a dictionary, a word instance must be categorized into a class or a number of its meaning classes is the definition of WSD. The context attributes (like the adjacent terms) produce the evidence for data. Such as a mouse can be both the rodent and the electronic mouse for computers [18].

3.1 Dataset

The data used for this study is a compilation of Amazon product feedbacks. The four main categories include fashion, books, electronics, and home decor. Each feedback is divided into (a) ID of Consumer, (b) ID of product (c) Ratings (d) time schedule of feedback (e) whether the feedback was helpful or not. There were over 5 million feedbacks for 18,000 products [19]. Dataset chosen is quite large and contains many

Table 1 Survey table

Paper No.	Algorithm	Sources	Methodology	Measures	Demerits
[29]	SVM	Jingdong.com eLong.com	Extended sentiment dictionary used for meaningful sentiment recognition of comment texts	Precision—75.6% Recall—76.3% F1—76%	Weight of sentiment words need to be further refined
[9]	SVM	Amazon.com	Study level as well as sentence-level research categorization carried out	Recall—82.5% Precision—87.2%	Generalization of sentiment classifier limited
[16]	NBA, BiLSTM	Ctrip.com	Long short term memory used is bidirectional where probability of word based on full left and right contexts	Precision—86.02% Recall—84.13% F1—85.06%	BiLSTM absorbs a lengthy time in training model
[28]	SVM	Amazon.com	Uses case based reasoning strategy in order to learn from past polarizations	Accuracy—84.5% Precision—82% Recall—86.6% F1—91%	User Interaction dynamics are particularly complex
[10]	SVM	Twitter.com	Flexible approach and does not require manually coded resources	Precision—68% Recall—77% F1—72.3%	Learning of optimal kernel combination
[14]	LDA	Citysearch.com	Simple method which takes into account influence of aspect on sentiment polarity	Precision—70.4% Recall—80.4% F1—75.1%	Insufficient to rely only on lexicons
[4]	SVM	tencent.com	Provides know-how alternate of sensible transportation techniques	Accuracy—78.6% Precision—78.4% Recall—78.8% F1—79%	The venture to position into effect TSA procedure is not efficient
[20]	NBA	Amazon.com imdb.com	Provides survey covering techniques, methods and challenges	Precision—61.9% Recall—93.4% F1—74.2	Need to deal better with negation expressions

(continued)

Table 1 (continued)

Paper No.	Algorithm	Sources	Methodology	Measures	Demerits
[8]	NBA	Facebook.com	Paper targets setting up a language resource to overwhelm language specific problems	Accuracy—84.5% Precision—85.4% Recall—84.1% F1—84%	The lexicon does not contain informal opinion words
[7]	NBA	Github.com	Employs wide range of feature sets	Precision—74.8% Recall—59.4% F1—66.2%	Unable to deal with ASR outputs
[15]	RBSA	Facebook.com	Proposed avenues for integration of sentiment analysis to human-agent interaction	Precision—59.4% Recall—55.1% F1—57.1%	Not an adequate psycho-linguistic model
[22]	SVM	Twitter.com	Provides efficient and fast processing of data. Can be used for both small and large data	Accuracy—85.5%	Short informal texts tend to have many misspellings
[27]	SVM	Amazon.com	Uses rule-based approach to deal with real problems	Accuracy—78.7% Precision—76.6% Recall—82.5%	No implementation of TSA into existing ITS
[23]	SVM	Facebook.com	Uses ontology-based sentiment analysis, for potential guidance, the correct sentimental values in the ontology are stored	Precision—66% Recall—56% F1—59%	Subjectivity classification is not automated
[26]	RBSA	Facebook.com	Constructs and validates a gold standard list of lexical features	Precision—78% Recall—55% F1—63%	Suffers from lack of coverage of sentiment-relevant lexical features
[17]	RFA	Twitter.com	Uses multi-class classification	Accuracy—60.2% Precision—60.8% Recall—60.2% *F*-measure—59.7%	Many tweets present more than one sentiment

(continued)

Table 1 (continued)

Paper No.	Algorithm	Sources	Methodology	Measures	Demerits
[25]	SVM, NBA	Amazon.com Twitter.com	Uses a variety of algorithms and compares them	Accuracy—70.1% Precision—69.7% Recall—70.1% *F*-measure—69.9%	Documents assigned to contain opinions may include factual sentences as a part
[1]	SVM, NBA	Wordnet.princeton.edu	Does aspect classification followed by polarity classification	Accuracy—70.73%	Doesn't address word sense disambiguation
[2]	NBA	Twitter.com	Uses Stemming and Temporal Mining	Precision- 81.3% Recall- 78.9%	Proposed system cost is a bit expensive
[18]	NBA	Twitter.com	Has real-time implementation as it has taken tweets from the 2016 US presidential election	Accuracy—90.21% *F*-measure—89.98%	Only uses Twitter data and does not characterize influence extent of contrasting metrics in order to emphasize a feeling

Table 2 Datasets frequently used in various papers

Paper No.	Dataset	Source
[29]	Amazon.com	http://www.cs.jhu.edu/~mdredze/datasets/sentiment/
[9]	Twitter.com	https://www.kaggle.com/kazanova/sentiment140
[16]	Weibo.com	https://github.com/GYXie/weibo_dataset

reviews as given above. Having such a large dataset has its advantages as there is a large range between the worst review and the best review. In these cases having more reviews is a boon as we can look at the larger picture. In Table 2, we have talked about datasets that were frequently used in previous papers.

3.2 *Sentiment Sentence Extraction*

In this data, all idiosyncratic content was derived for future study. All feelings are contained in the subjective material. At least one positive or negative word will be in a sentiment phrase. Sentences are initially first tokenized into English words which were isolated [20]. The full sentence is used as sometimes taking lone words doesn't tell us the full picture. Taking the full sentence makes us understand the context of the sentence or paragraph [21].

3.3 Sentiment Phrase Identification

Adjectives and verbs are words that through negative prefixes, can express opposing feelings. Like instance, the following statement contained in the analysis of an electronic device "The built-in camera on this phone has its uses but so far nothing great." The word, "great" is a positive word [22]. Nonetheless, a sentence like "nothing great" depicts unpleasant experiences mostly. Those sentences, therefore, must be classified. Another example we can use is when one says that the doll looks pretty ugly. The word is pretty is a positive word but ugly is a negative word. Though when one uses both of them at once the overall result of the word is negative.

3.4 Word Sense Disambiguation

This decides what meaning of a word in a particular context is triggered by using the word. In one or more of its sense types, an incident will be classified. Features like adjacent terms help in distinguishing. Such as a mouse can mean a computer mouse as well as a mammal. Though when someone types mouse cage we know it means the animal. While if someone types a keyboard with a mouse we know they mean the computer mouse. Neighboring words help in classifying in this case [23]. Figure 1 has talked about the workflow for the model and steps involved in the WSD Methodology.

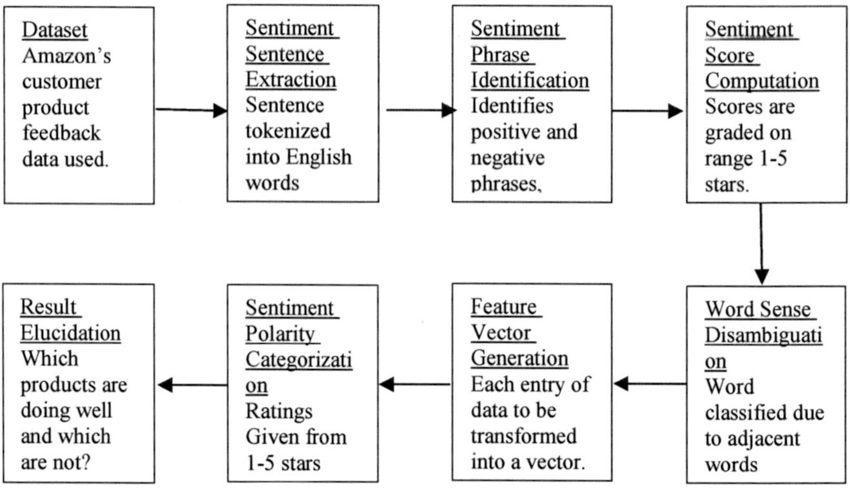

Fig. 1 Word sense disambiguation methodology

3.5 Sentiment Score Computation

Scores are graded on range 1–5 stars. A rating of 1–2 stars is taken as negative, 3 stars is neutral and 4–5 stars is taken as positive [5]. (1) star is generally known as poor reviews which means that the product itself was abysmal. (2) stars mean average nothing special. (3) stars is good. (4) is for a very good product while. (5) stars are for exceptional products.

3.6 Feature Vector Generation

Features are usually taken from sentiment tokens whose information is drawn out from the training dataset. A vector shall not have a copious load of features or else it will suffer from curse of dimensionality which makes available data become sparse while the volume of the space increases at high speed. The vector function consists of four components: 2 hashes centered on a binary array, an average sentiment rating, and a basic truth label [24].

3.7 Sentiment Polarity Categorization

When one goes through the procedure of polarity categorization of sentiments it can be found to be bipartite: given that categorization happens at both sentence and review level. When a sentence is given, the target of categorization at sentence-level is to segregate into negative or positive with regard to the sentiments conveyed. In Fig. 2

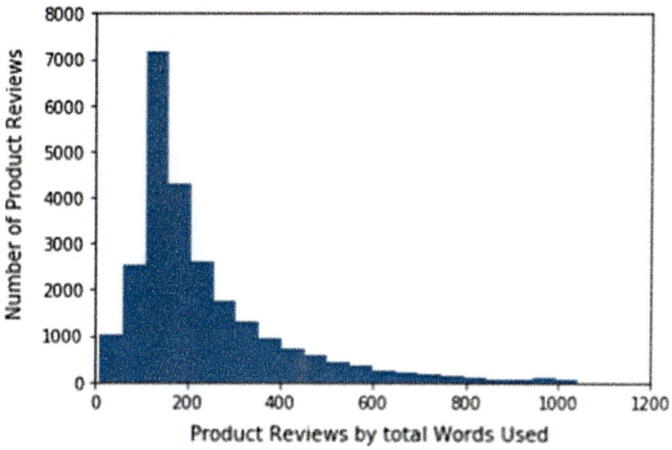

Fig. 2 Analysis of the product reviews

the number of product reviews has been compared with product reviews with total words used. For this method, training data require fundamental real-world tags that specify the positive or negative value of the given sentence [14]. Nonetheless, ground truth tagging can become a genuinely difficult issue, owing to the huge number of data that is present. Considering manual tagging of each of the sentences is quixotic, a machine tagging method is then chosen for an elucidation. If more number of negative tokens are present than positive ones, sentences will get tagged as negative, and conversely having more number of positive tokens will render the sentence as positive [9].

3.8 Result Elucidation

From the results obtained we can identify which product is selling more by the amount of positive reviews for them and the ones which are not doing well albeit the ones with the negative reviews can be looked into and we can find out what are the problems with these products [25] and try to improve upon them. So that sales of the product gaining negative reviews increases. In our model we have used two algorithms; NBA and SVM. The NBA works like suppose a set of training details occurs, such as *T*, where the n-dimensional function variable is each different. Thus, a variety of training data are available. $Y = ya, yb \ldots yz$, implies *z* measurements contrived *z* characteristics tuple [10]. Then we infer m classes where *ca*, *cb*, …, *cn*. According to tuple *m*, the classifier forecasts *m* resides in *Ci* providing: $P(cj|m) < P(ci|m)$, where $j, i \in [1, m]$ and j != i. $P(ci|y)$ will get calculated as:

$$P(Cj|Y) = \Pi P(y|Cj) \tag{1}$$

For the linear and nonlinear data set, SVM conducts data classification. SVM is a conventional algorithm for the classification of linear and nonlinear data in the Machine Learning (ML). Given training data with binary outputs, SVM tries to find a hyperplane as the decision-making field, in order to maximize the differentiation of positive and negative samples. SVM converts the data into a bigger dimension and solves the conundrum by locating a linear hyperplane if data is not linear [26]. The kernel used for such nonlinear data is called Gaussian Radial Basis function (RBF). SVM was used to classify each feedback to a label of "positive", "neutral", or "negative". Once data can be differentiated linearly, the optional boundary separates data from one class to another from a vector kernel via SVM. A linear kernel can be entered as a logical one:

$$V * X + d = 0 \tag{2}$$

X works as a training tuple, *V* is a function of weight and $V = va, vb \ldots vn$, while *d* works as scalar [27]. In Fig. 3 the Relation between the two classifiers mainly SVM and NBA has been compared with the three evaluation criteria mainly precision,

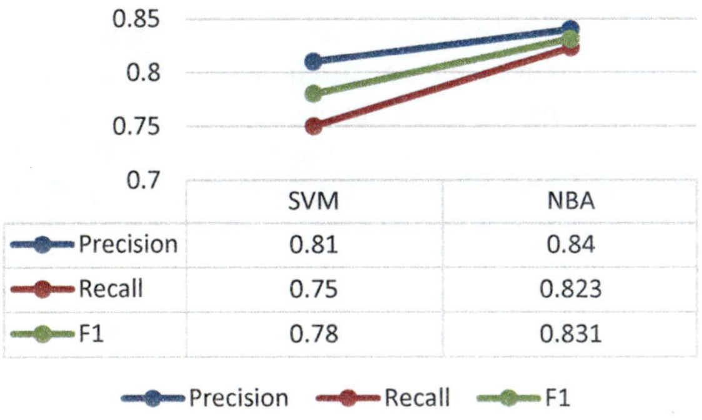

Fig. 3 Relation between three evaluation criteria and two classifiers

Recall, and F1 measure. The conundrum essentially converts to a $\|V\|$ reduction in order to improve the kernel, calculated decidedly as:

$$\Sigma aiyixi \tag{3}$$

where *ai* is numeric and *zi* is labelled with vector help, *Xi*. It concurs: if $zi = 1$ then $\sum vixi \geq 1$; if $zi = -1$ then $\sum vixi \geq -1$. If the data is inseparable linearly, to convert the data into a greater dimension, SVM allows the use of nonlinear simulation. Subsequently, it solves the conundrum by locating a linear kernel. Functions to achieve such conversions are known as kernel. The kernel attribute that our work promotes: the Gaussian radial base function (GRBF):

$$J(Yk, Yj) = \mathrm{e}^{|-b|Yk-Yj||2} \tag{4}$$

yk is a tuple test and β is a different parameter with a predefined value while *yj* is a support vector.

4 Discussion

This model proposes to build a predictive model for acquired sales data and predict sales of each product at a particular store. Customer Behavior can be analyzed and it can be found which products are in demand and which products are not doing well [28]. Through this, it will try to improve sales based on sentiments portrayed for each product. System should be robust with high precision and applications from various sources easy to add. Nonetheless, this analysis still has a few constraints. One downside is that the scheme for evaluating our emotions in this analysis concentrates on the nature of stimuli [23], the program may be inadequate for only unconscious emotional analysis. A definition is typically implicitly conveyed in neutral terms, which allows deciding its polarity complicated. A sentence like, for instance, "Item described is", positive evaluations often have neutral terms only. The future works on solving these issues in the context of these restrictions. Specific functionalities should be clearly derived and organized into functional vectors to increase the grouping of feedback. The next phase in the area of intrinsic sensation research is to determine that it exists in the continuum of a specific product. Future analysis will include the validation of our categorization framework with other data sets.

5 Conclusion

Sentiment Analysis is a precise science that facilitates the development of new ideas as well as promotions and the public. Data sets taken from Amazon.com have been used which contain product feedbacks [29]. Precision, Recall, and F1 have been used as evaluation measures [30]. Algorithms used are NBA and SVM. Through choosing a judgment limit that maximizes distance from the nearest data points in all categories, SVM varies from other conventional algorithms. SVM not only considers a limit for decision but finds the best limit for options. The best choice is to be as far away from the closest point of all groups as possible. The nearest points to the boundary of judgment, which maximizes the distance from the boundary for judgments, are considered to help vectors as shown. The decision boundary when using vector-assisted systems is regarded as a fixed margin or median margin when using vector-assisted devices. Through context-based sentiment analysis the model will analyze customer behavior for each product and then try to improve the sales based on the sentiments portrayed for each product whether they be positive or negative.

Appendix

See Figs. 2 and 3.

References

1. Bhadane C, Dalal H, Doshi H (2015) Sentiment analysis: measuring opinions. Int Conf Adv Comput Technol Appl 43:809–814
2. Unnamalai K (2012) Sentiment analysis of products using web. Int Conf Modell Optim Comput 38:2257–2262
3. Kumar S, Koolwal V, Mohbey KK (2019) Sentiment analysis of electronic product tweets using big data framework. Jordanian J Comput Inf Technol 5:43–59
4. Sengottuvelan P, Karthik AV (2017) A survey on web and rule based traffic sentiment analysis. Int J Comput Trends Technol 4:13–19
5. Salas-Zarate MDP, Medina-Moreira J, Lagos-Ortiz K, Luna-Aveiga H, Rodriguez-Garcia MH, Valencia-Garcia R (2017) Sentiment analysis on tweets about diabetes: an aspect-level approach. Comput Math Methods Med 3:1–10

6. Yang F, Changshun D, Huang L (2019) Ensemble sentiment analysis method based on R-CNN and C-RNN with fusion gate. Int J Comput Commun Control 14:272–285
7. Neviarouskaya A, Oni M (2013) Analyzing sentiment word relations with affect, judgment and appreciation. IEEE Trans Affect Comput 3:425–438
8. Aye YM, Aung SS (2018) Senti-lexicon and analysis for restaurant reviews of Myammar text. Int J Adv Eng Manage Sci 4:380–385
9. Fang X, Zhan J (2015) Sentiment analysis using product review data. J Big Data 3:1–14
10. Vanzo CD, Basili R (2014) A context-based model for sentiment analysis in Twitter. Int Conf Comput Logis Tech Pap 3:2345–2354
11. Vilares D, Alonso MA, Gomez-Rodriguez C (2017) Supervised sentiment analysis in multilingual environments. Inf Process Manage Int J 4:595–607
12. Xiaomei Z, Yang J, Zhang J (2018) Microblog sentiment analysis using social and topic context. Publ Libr Sci 3:1–24
13. Urologin S (2018) Sentiment analysis, visualization and classification of summarized news articles: a novel approach. Int J Adv Comput Sci Appl 9:616–625
14. Brody S, Elhadad N (2010) An unsupervised aspect-sentiment model for online reviews. Annu Conf North Am Chap ACL 3:804–812
15. Clavel C, Callejas Z (2015) Sentiment analysis: from opinion mining to human-agent interaction. IEEE Trans Affect Comput 6:1–20
16. Guixian X, Yueting M, Xiaoyu Q, Ziheng Y, Xu W (2019) Sentiment analysis of comment texts based on BiLSTM. IEEE Access 7:51522–51532
17. Bouazizi M, Ohtsuki T (2017) A pattern-based approach for multi-class sentiment analysis in Twitter. IEEE Access 5:20617–20637
18. Alaoui IE, Gahi Y, Messoussi R, Chaabi Y, Todoskoff A, Kobi A (2018) A novel adaptable approach for sentiment analysis on big social data. J Big Data 5:1–18
19. Naorem D, Anand D (2015) Semi-supervised aspect based sentiment analysis for movies using review filtering. Int Conf Intell Human Comput Inter 8:86–93
20. Vinodhini G, Chandrasekaran RM (2012) Sentiment analysis and opinion mining: a survey. Int J Adv Res Comput Sci Softw Eng 2:282–292
21. Soleymani M, Garcia D, Jou B, Schuller B, Chang S, Pantic M (2017) A survey of multimodal sentiment analysis. Elsevier Image Vis Comput J 4:1–12
22. Kiritchenko S, Zhu X, Mohammad SM (2014) Sentiment analysis of short informal texts. J Artif Intell Res 3:723–762
23. Zehra S, Wasi S, Jami I, Nazir A, Khan A, Waheed N (2017) Ontology-based sentiment analysis model for recommendation systems. Int Conf Knowl Eng Ontol Dev 3:155–159
24. Brar GS, Sharma A (2018) Sentiment analysis of movie review using supervised machine learning techniques. Int J Appl Eng Res 13:12788–12791
25. Saberi B, Saad S (2017) Sentiment analysis or opinion mining: a review. Int J Adv Sci Eng Inf Technol 7:1660–1666
26. Hutto CJ, Gilbert E (2014) VADER: a parsimonious rule-based model for sentiment analysis of social media text. Int AAAI Conf Weblogs Soc Media 3:216–225
27. Murali JA, Varghese CS (2015) A literature survey on web-based traffic sentiment analysis: methods and applications. Int J Sci Eng Res 6:926–930
28. Ceci F, Goncalves AL, Weber R (2016) A model for sentiment analysis based on ontology and cases. IEEE Latin Am Trans 14:4560–4566
29. Guixian X, Ziheng Y, Haishen Y, Fan L, Yueting M, Xu W (2019) Chinese text sentiment analysis based on extended sentiment dictionary. IEEE Access 7:43749–43762
30. Nan L, Desheng WD (2009) Using text mining and sentiment analysis for online forums hotspot detection and forecast. Decis Support Syst 4:354–368

A Study on Image Hashing Techniques for Implementing Deduplication

G. Sujatha and Jeberson Retna Raj

Abstract Image deduplication is an approach to avoid duplicate images to be stored in cloud, which is currently required to increase the efficient utilization of cloud storage. Most of the times, the images uploaded by different users of cloud storage service need not to be unique. There is a possibility of storing same images by different users. This may lead to poor usage of cloud storage. To find duplicates, various image hashing techniques are widely used. The purpose of this paper is to review various image hashing methods which are used to generate hash value of the images. Using the hash value, we can conclude the uniqueness of the image. This can be achieved by using the "collision resistance" property of hash value. To measure the feasibility, the general necessity for image deduplication in cloud is discussed in this work. Additionally, the performance of few selected image hashing methods is compared with respect to robustness and discrimination.

Keywords Image hashing · Robustness · Deduplication · Discrimination

1 Introduction

Because of the advent growth of information and communication, the need for storing information in the form of text, image, and video also grows tremendously. In this regard, images and videos occupy more space when compared to text data. Individuals or organizations cannot accommodate their information with their storage capacity. The cloud provides many services; among them, "storage as a service" is a boon to our clients.

The storage provided by the cloud is utilized by the clients to store their data. But the question is whether they are efficiently utilized or not? The answer is obviously "No" because cloud providers support huge number of clients. They cannot check

G. Sujatha (✉) · J. R. Raj
Sathyabama Institute of Science and Technology, Chennai, India
e-mail: sujathamurugan.srm@gmail.com

D. J. Hemanth et al. (eds.), *Artificial Intelligence Techniques for Advanced Computing Applications*, Lecture Notes in Networks and Systems 130,
https://doi.org/10.1007/978-981-15-5329-5_49

and allow only unique information to be stored in the cloud without any duplication always. As we know, the images need more memory than text, it is necessary to avoid duplicate images to be saved in the cloud. To achieve this, we can implement deduplication approach in the cloud storage. Hash function is a one way function which is used to map any sized input data to a fixed size value called hash for various purposes like retrieval of image [1], detection of tampered areas in the image [2], copy detection [3], image authentication [4], image indexing [5], image quality assessment [6], and image forensics [7]. We have many well-known algorithms such as MD5 and SHA-1. But these algorithms are not well suited for finding image hash because of their sensitivity to bit-level change.

For finding duplicate copies of images, many image hashing techniques are available. In this paper, the performance of various image hashing techniques is compared in various dimensions. The main requirements of image hashing are "robustness" and "discriminative capability." In real time, the digital representation of input image gets altered because of major content-preserving operations such as JPEG compression, image filtering, brightness or contrast adjustment, watermark embedding, and geometric transform. But these digital operations will not modify its visual content. In the sense, the processed image and its original copy are visually identical. So the hash value of the processed image is required to be the same or similar with that of its original copy. This property is called robustness, which means that visually similar images are required to have same or very similar hash values. The second requirement is discriminative capability or anti-collision capability, which requires the property that visually different images should be mapped to different image hashes.

2 Related Work

There are various techniques to generate image hash. One among is using discrete cosine transform coefficients. Using DCT, initially the image should be normalized, and then, the normalized image is divided into non-overlapping blocks. The feature matrix will be calculated from the extracted dominant DCT coefficients. Then, the feature matrix is compressed to generate the hash value. This technique seems robust against normal digital operations [8].

Another technique is using DCT and LLE. Using this technique, the source image is initially processed to a normalized image. From that image, color vector angles should be extracted. From DCT, they calculate a stable feature matrix. Then, apply the data reduction to that feature matrix with LLE and quantized the variance of LLE results. Then it is encrypted to construct the resultant hash. They collected more than 200 source images from the open datasets like USC-SIPI image database and copy days dataset. They modified the images to apply the robustness attacks. They applied the following operations, brightness adjustment and contrast. They calculated hashes for original and modified similar images and evaluate the maximum, minimum and mean value of Hamming distances of each pair. The mean value is much smaller than 10 except the operation of rotation, cropping and rescaling and the maximum mean value is 15. To validate the discrimination capability, they construct a large dataset from different sources like Ground truth database, Internet and captured images by digital cameras. They calculated the hashed for these images and found the hamming distances among them [9].

Many researchers are interested and have shown very good attention over image hashing and gave many algorithms on this area. They classified those available algorithms with respect to resistant to rotation into two categories. They are nothing but algorithms resistant to small and large angle rotation. For example, Venkatesan et al. [10] constructed image hash using coefficients of Discrete Wavelet Transform and it is robust to only 2° of rotation but good for contrast adjustment. Lin [11] proposed a robust hashing using invariant associations between DCT coefficients. Using these values they can be able to distinguish compressions and other attacks. The algorithms can be categorized as resistant to large angle rotation when they resilient to large degree rotation.

There are many techniques available for data analysis and one among the best technique is multidimensional scaling. It has been effectively used in various applications and that can also be efficiently used for image hashing. Zhenjun Tang generated hash value for image by initially preprocess the image to create normalized image. Then extract the stable feature from this normalized image. Feature matrix is created to resist the rotation and MDS is conducted to study the closed features from feature matrix. Then short hash will generated by compressing the closed features [12].

"Robust image hashing with embedding vector variance of LLE," in this paper they used the important property of LLE. They found that the vector variances of LLE are linearly changed during modification operations. The processing involves the following steps, initially by using bilinear interpolation, converting color space, extraction of block mean and Gaussian low-pass filtering, the input image is mapped to a normalized matrix. Then a secondary image is constructed from this normalized matrix. Finally, image hash is calculated by applying LLE to secondary image and the embedding vector variances of LLE. They took images from USC-SIPI image database. They generate visually similar images using Photoshop, MATLAB, and striMark. They tried with many digital modifications like brightness adjustment, contrast adjustment, gamma correction and more. They calculate image hashes of

Table 1 Comparison of image hashing techniques with respect to resistance to image rotation

S. No.	Author	Technic used	Resistant to image rotation (in angle)
1	Venkatesan [10]	DWT coefficients	2
2	Khelifi and Jiang [22]	Optimum watermark detection	5
3	Tang [23]	DCT and non-negative matrix factorization	1–2
4	Tang [9]	DCT and LLE	5
5	Swaminathan [24]	DFT	20

the test images and their identical versions and found that their hashing is robust to the above-mentioned operations. For testing discrimination property, they took 200 different color images from various sources. They compared similarity between each pair of hashes and found that minimum and maximum of correlation coefficients are -0.6891 and 0.6967 and only 0.05% of false positive exists [13].

Many of the known image hashing algorithms [14–20] were developed for handling gray images. But it is also required to handle color images. The significant feature which is widely used for processing color images is color vector angle (CVA). But there is no solution saying that how CVA is efficiently used to generate an image hash. Many algorithms used luminance component YCbCr to generate the hash but not concentrating much on the color information of the images. YCbCr is the combination of luminance component, blue-difference chroma component and red-difference chroma component. Practically for calculating the image hash, if the luminance component is only considered then it is very hard to capture the color information of the images. To overcome this, color vector angle is used to generate the robust hash. They generate the HCVA hashing with discrete cosine transform (DCT). The author employed 4 phase approach to generate the HCVA hash value. They are preprocessing phase, CVA extraction, histogram calculation, and compression phase. They conducted and presented few experimental results based on various digital operations like contrast adjustment, brightness and more; using that, the efficiency of HCVA hashing algorithm is evaluated [21].

3 Conclusion and Future Work

In this paper, we have represented literature review on several image hashing techniques that can be used for finding duplicate images to implement image deduplication. We have discussed about robustness and discrimination, the two important requirements to find the similar images and how they are used to find the similarity. We have also discussed the capability of the various algorithms to handle different versions of attacked images by applying different digital operations over the original images and also their deduction rate of finding the similar images. The review of this paper concludes the strength and weakness of various image hashing technique which will help us to decide the direction of our future research work towards the selection and improvisation of image hashing technique to achieve better results in image deduplication considering all types of attacks on the images. Future work will focus on the development of new image hashing technique to efficiently handle almost all attacks on original image and malicious change deduction.

Table 2 Survey on different image hashing technique for implementing deduplication

S. No.	Year	Technique used	Advantages	Robustness	Discrimination	Limitations
1	2014	DCT [8]	Work well against usual digital operations except rotations	98.5%	0.11% when the threshold is 10	Not supporting for rotation
2	2015	Locally linear embedding is used to form image hash [13]	Robust to content-preserving operations, common digital operations	Similarity among images with no rotations are identified correctly	99.91% (for 0.60 threshold)	Not supporting for rotation, cropping and rescaling
3	2016	DCT and LLE [9]	Resistant to normal digital operations except rotation	84.68% if there is no rotated image. When there is rotation then the deduction rate will be 74.52%	84.68% similar images are correctly identified for images with no rotation. When there is rotation then the deduction rate will be 74.52%	The correct deduction rate will get reduce when the similar images contain rotation
4	2017	Multidimensional scaling [12]	Robust to any angle rotation	94.50% similar images are correctly identified and 97.53% similar images are correctly identified when there is no rotation in the similar images	99.99% similar images are correctly identified (when the threshold is 0.96)	The correct deduction rate will get reduce when the similar images contain rotation
5	2018	Color vector angel [21]	Resist rotation with arbitrary angle and also resistant for combinational attacks	96.25% and even it is possible to get increased deduction rate of 98.5% when the threshold is increased	The threshold value should be selected properly. For example with Td = 12,000, 96.25% -correctly identified their similarity and 8.84% images are falsely judged as similar	Can be improved to achieve malicious change deduction

References

1. Wang K, Tang J, Wang N, Shao L (2016) Semantic boosting cross-modal hashing for efficient multimedia retrieval. Inf Sci 330:199–210
2. Yan C, Pun C, Yuan X (2016) Multi-scale image hashing using adaptive local feature extraction for robust tampering detection. Signal Process 121:1–16
3. Zou F, Chen Y, Song J, Zhou K, Yang Y, Sebe N (2015) Compact image fingerprint via multiple kernel hashing. IEEE Trans Multimed 17:1006–1018
4. Ahmed F, Siyal MY, Abbas VU (2010) A secure and robust hash-based scheme for image authentication. Signal Process 90:1456–1470
5. Winter C, Steinebach M, Yannikos Y (2014) Fast indexing strategies for robust image hashes. Digital Invest 11:S27–S35
6. Lv X, Wang ZJ (2009) Reduced-reference image quality assessment based on perceptual image hashing. In: Proceedings of IEEE international conferenceon image processing (ICIP2009), pp 4361–4364
7. Lu W, Wu M (2010) Multimedia forensic hash based on visual words. In: Proceedings of IEEE international conferenceon image processing, pp 989–992
8. Tang Z, Yang F, Huang L, Zhang X (2014) Robust image hashing with dominant DCT coefficients. Optik 125:5102–5107
9. Tang Z, Lao H, Zhang X, Liu K (2016) Robust image hashing via DCT and LLE. Comput Secur 62:133–148
10. Venkatesan R, Koon SM, Jakubowski MH, Moulin P (2000) Robust image hashing. In: Proceedings of the IEEE international conference on image processing (ICIP 2000), pp 664–666
11. Lin CY, Chang SF (2001) A robust image authentication system distinguishing JPEG compression from malicious manipulation. IEEE Trans Circ Syst Video Technol 11:153–168
12. Tang Z, Huang Z, Zhang X, Lao H (2017) Robust Image hashing with multidimensional scaling. Signal Process 137:240–250
13. Tang Z, Ruan L, Qin C, Zhang X, Yu C (2015) Robust image hashing with embedding vector variance of LLE. Dig Signal Process 43:17–27
14. Lefebvre F, Macq B, Legat JD (2002) RASH: Radon soft hash algorithm. In: Proceedings of European signal processing conference, Toulouse, France, 3–6, pp 299–302, Sept 2002
15. Monga V, Mihcak MK (2007) Robust and secure image hashing via non-negative matrix factorizations. IEEE Trans Inf Forensics Secur 2(3):376–390
16. Tang Z, Wang S, Zhang X, Wei W (2011) Structural feature-based image hashing and similarity metric for tampering detection. Fundam Inform 106(1):75–91
17. Sun R, Yan X, Ding Z (2011) Robust image hashing using locally linear embedding. In: Proceedings of the 2011 international conference on computer science and service system (CSSS), pp 715–718
18. Li Y, Lu Z, Zhu C, Niu X (2012) Robust image hashing based on random Gabor filtering and dithered lattice vector quantization. IEEE Trans Image Process 21(4):1963–1980
19. Vadlamudi LN, Vaddella RPV, Devara V (2016) Robust hash generation technique for content-based image authentication using histogram. Multimed Tools Appl 75(11):6585–6604
20. Davarzani R, Mozaffari S, Yaghmaie K (2016) Perceptual image hashing using center-symmetric local binary patterns. Multimed Tools Appl 75(8):4639–4667
21. Tang Z, Li X, Zhang X, Zhang S, Dai Y (2018) Image hashing with color vector angle. Neurocomputing 308:147–158
22. Khelifi and Jiang (2010) Perceptual image hashing based on virtual watermark detection – IEEE transactions on image processing
23. Tang (2011) Lexicographical framework for image hashing with implementation based on DCT and NMF – Multimedia tools and applications.
24. Swaminathan (2006) Robust and secure image hashing – IEEE transactions on information forensics and security